Primitive Forms and Related Subjects
—Kavli IPMU 2014

ADVANCED STUDIES
IN PURE MATHEMATICS 83

Chief Editor of the Series: Ken'ichi Ohshika (Gakushuin University)

Primitive Forms and Related Subjects — Kavli IPMU 2014

Edited by

Kentaro Hori (Kavli IPMU)
Changzheng Li (Sun Yat-sen University)
Si Li (Tsinghua University)
Kyoji Saito (Kavli IPMU)

Mathematical Society of Japan

This book was typeset by $\mathcal{A}_{\mathcal{M}}\mathcal{S}$-TEX and $\mathcal{A}_{\mathcal{M}}\mathcal{S}$-LATEX, the TEX macro systems of the American Mathematical Society, together with the style files aspm.sty *and* aspmfm.sty *for $\mathcal{A}_{\mathcal{M}}\mathcal{S}$-TEX written by Dr. Chiaki Tsukamoto and* aspmproc.sty *for $\mathcal{A}_{\mathcal{M}}\mathcal{S}$-LATEX written by Dr. Akihiro Munemasa and* aspm.cls *for $\mathcal{A}_{\mathcal{M}}\mathcal{S}$-LATEX provided by Livretech Co., Ltd.*

TEX is a trademark of the American Mathematical Society.

Edited by the Mathematical Society of Japan.

Published by the Mathematical Society of Japan.

Distributed by the Mathematical Society of Japan, the American Mathematical Society and the World Scientific Publishing Co., Ltd.
Distributed exclusively in North America by the American Mathematical Society.
Partially supported by Grant-in-Aid for Publication of Scientific Research Results, Japan Society for the Promotion of Science, Grant Number 17HP1002.

Advanced Studies in Pure Mathematics 83
ISBN 978-4-86497-085-3

PRINTED IN JAPAN
by Livretech Co., Ltd.

2010 Mathematics Subject Classification.
Primary 53D54.
Secondary 14N35.

Preface

The workshop "Primitive forms and related subjects" was held in Kavli IPMU, University of Tokyo, during February 10-14, 2014. There were over eighty participants, more than one half of which were from overseas such as Austria, China (including Hong Kong), Korea, France, Mexico, Russia and U.S.A.

The purpose of this workshop was to present the status of the rapidly developing subject of primitive forms. A primitive from is a top degree differential form on a family of (open) Calabi-Yau varieties with the defining property that it determines a filtration opposite to the semi-infinite Hodge structure and induces the flat (=Frobenius) structure on the deformation parameter space. Since late 1990s, primitive forms have drawn attention from both mathematicians and physicists. They are expected to describe objects in complex geometry that are mirror to invariants in symplectic geometry like Gromov-Witten invariants or Fan-Jarvis-Ruan-Witten invariants.

In this workshop, there were three basic lecture series as well as 16 conference talks by experts from both mathematicians and physicists with various viewpoints. There were fruitful discussions during the conference, with significant ideas exchanged and explored. The present volume is the proceedings of this workshop, and contains 13 contributions on the workshop topics. We hope that this volume will contribute to further developments on primitive forms and related subjects.

This workshop was financially supported Grant-in-Aid for Scientific Research (A) JP25247004, JSPS, and by Kavli Institute for the Physics and Mathematics of the Universe (Kavli IPMU), which is supported by World Premier International Research Center Initiative (WPI), MEXT, Japan.

August 2018
Kentaro Hori
Changzheng Li
Si Li
Kyoji Saito

All papers in this volume have been refereed and are in final form. No version of any of them will be submitted for publication elsewhere.

Primitive Forms and Related Subjects

Dates February 10 to – February 14, 2014
Venue Lecture Hall, Kavli IPMU main building, Kashiwa, Japan

Credit: Kavli IPMU

Program

Lecture series:

Lecture I: Frobenius manifold structure and Lagrangian Floer theory for toric manifolds.
Kenji Fukaya (SCGP), Yong-Geun Oh (IBS),
Hiroshi Ohta (Nagoya U), Kaoru Ono (KURIMS)

Lecture II: Introduction to FJRW theory and a mathematical approach to the Gauged Linear Sigma Model.
Huijun Fan (Peking U), Tyler Jarvis (Brigham Young U),
Yongbin Ruan (Michigan U)

Lecture III: LG-model via Kodaira-Spencer gauge theory.
Si Li (Boston U)

February 10 (Monday)

09:30-10:45 Lecture I-1: Kenji Fukaya (SCGP)
11:15-12:30 Lecture III-1: Si Li (Boston U)
13:45-15:00 Lecture II-1: Tyler Jarvis (Brigham Young U)
15:30-16:45 Serguei Barannikov (Jussieu)
On the noncommutative Batalin-Vilkovisky formalism and EA matrix integrals.
17:00-18:15 Yuuki Shiraishi (Osaka U.)
On Weyl group and Artin group associated to orbifold projective lines.
Nathan Priddis (U. Michigan)
A Landau-Ginzburg/Calabi-Yau correspondence for the mirror quintic.
Mohammad Reza Rahmati (CIMAT)
Hodge Theory of Isolated Hypersurface Singularities.

February 11 (Tuesday)

09:30-10:45 Lecture II-2: Tyler Jarvis (Brigham Young U)
11:15-12:30 Lecture I-2: Kenji Fukaya (SCGP)
13:45-15:00 Lecture III-2: Si Li (Boston U)
15:30-16:45 Atsushi Takahashi (Osaka U)
From Calabi-Yau dg categories to Frobenius manifolds via primitive forms.
17:00-18:15 Kentaro Hori (Kavli IPMU), Mauricio Romo (Kavli IPMU)
The parameter delta.

February 12 (Wednesday)

09:30-10:45	Huijun Fan (Peking U) Analytic construction of quantum invariant of singularity.
11:15-12:30	Boris Bychkov (HSE) On the number of coverings of the sphere ramified over given points. Michel van Garrel (KIAS) Integrality of relative BPS state counts of toric Del Pezzo surfaces. Satoshi Sugiyama (U. Tokyo) On the Fukaya-Seidel categories of surface Lefschetz fibrations. Alexey Bondal (Kavli IPMU), Ilya Zhdanovskiy (MITP) Critical points of a functional and orthogonal pairs of Cartan subalgebras.

February 13 (Thursday)

09:30-10:45	Lecture III-3: Si Li (Boston U)
11:15-12:30	Lecture II-3: Tyler Jarvis (Brigham Young U)
13:45-15:00	Lecture I-3: Kenji Fukaya (SCGP)
15:30-16:45	Youjin Zhang (Tsinghua U) On the genus two free energies for semisimple Frobenius manifolds.
17:00-18:15	Andrey Losev (HSE) K.Saito theory of primitive form, generalized harmonic theory and mirror symmetry.

February 14 (Friday)

09:30-10:45	Hiroshi Iritani (Kyoto U) Gamma Conjecture for Fano manifolds.
11:15-12:30	Yefeng Shen (Kavli IPMU) Mirror symmetry for exceptional unimodular singularities.
13:45-15:00	Todor Milanov (Kavli IPMU) The phase form in singularity theory.
15:30-16:45	Daniel Pomerleano (Kavli IPMU) Deformation theory of affine symplectic manifolds.

Organizers: Kentaro Hori (Kavli IPMU) Changzheng Li (Kavli IPMU) Si Li (Boston U) Kyoji Saito (Kavli IPMU)

Information of the affiliations shown here is current as of the date of the workshop.

CONTENTS

Advanced Studies in Pure Mathematics 83, 2019
Primitive Forms and Related Subjects — Kavli IPMU 2014
pp. 1–18

Symplectic geometry of unbiasedness and critical points of a potential

Alexey Bondal and Ilya Zhdanovskiy

Abstract.

The goal of these notes is to show that the classification problem of algebraically unbiased system of projectors has an interpretation in symplectic geometry. This leads us to a description of the moduli space of algebraically unbiased bases as critical points of a potential function, which is a Laurent polynomial in suitable coordinates. The Newton polytope of the Laurent polynomial is the classical Birkhoff polytope, the set of doubly stochastic matrices. Mirror symmetry interprets the polynomial as a Landau-Ginzburg potential for corresponding Fano variety and relates the symplectic geometry of the variety with systems of unbiased projectors.

§1. Introduction

The classification problem of systems of unbiased bases attracted a lot of attention in physics literature (cf. [BLDG], [Ru], [DEBZ], [FM]). In its algebraic version, this is about complete systems of orthogonal rank 1 projectors $\{p_i\}$ and $\{q_j\}$ in a finite dimensional vector space of dimension n satisfying the condition $\mathrm{Tr} p_i q_j = \frac{1}{n}$. An equivalent formulation of the problem is to classify pairs of Cartan subalgebras in the Lie algebra $sl(n)$ which are orthogonal with respect to Killing form. It is also important to classify bigger sets of systems of projectors which are pairwise algebraically unbiased.

In [BZh] we introduced algebras $B_\Lambda(\Gamma)$, where Γ is a simply laced graph with weights of edges Λ, and gave an interpretation of this problem

Received May 3, 2015.

Revised February 1, 2016.

2010 *Mathematics Subject Classification.* 53D12, 53D20, 53D37, 14J33, 14J45, 14M25, 81P45, 35Q56.

Key words and phrases. Mutually unbiased bases, Lagrangian submanifold, momentum map, symplectic reduction, mirror symmetry, Fano variety, toric varieties, Newton polyhedra, quantum information, Landau-Ginzburg potential.

in the context of representation theory of these algebras. For the case of two complete systems of orthogonal projectors the graph is the complete bipartite graph with n vertices in both rows, all weights on edges are $\frac{1}{n}$, and representations under consideration are of dimension n. More general weights appear when we put condition $\mathrm{Tr} p_i q_j = \lambda_{ij}$. Algebras happened to be homotopes of Poincare groupoids of graphs associated to discrete Laplace operators (see details in [BZh]). In that respect the theory is closely related to discrete harmonic analysis on graphs and to the gluing of t-structures in the style of [BBD].

In this paper, we develop a symplectic approach to the problem. The symplectic structure on the variety of projectors of rank 1, considered as a coadjoint orbit, is relevant to the problem. If we assume orthogonal projectors q_j to be fixed and vary the projectors p_i without requiring them to be orthogonal, then the equations on unbiased systems define a Lagrangian fibre of an integrable system on the product Y of coadjoint orbits corresponding to projectors. The locus of systems of orthogonal projectors defines another Lagrangian subvariety X in Y. The problem is then to find the locus of intersection of these two Lagrangian subvarieties. We construct a symplectic open embedding of the cotangent bundle to X into Y.

The Lagrangian fiber is then interpreted as a (multi-valued) section of the cotangent bundle to X given as a differential of a potential function on X. If the transition matrix from projectors $\{q_i\}$ to $\{p_i\}$ is g, then the exponential of the function is:

$$E = \frac{\det g}{\prod_{i=1}^{n} \prod_{j=1}^{n} (g_{ij})^{\frac{1}{n}}}.$$

It is a Laurent polynomial in a suitable system of coordinates. Critical points of the function are solutions to our problem.

We consider also some generalizations to the case of non-complete system of projectors $\{p_i\}$, $i = 1, \ldots, k$, and complete systems of projectors $\{q_j\}$, $j = 1, \ldots, n$. The corresponding graph Γ is the complete bipartite graph Γ_{kn} with k vertices in one row and n vertices in the other. Also we allow arbitrary constants λ_{ij} instead of $\frac{1}{n}$ in the above equation for traces. For our analysis, we introduce another algebra $A_{n,\bar{\Lambda}}$ together with a homomorphism $\psi : A_{n,\bar{\Lambda}} \to B_{\mathbf{\Lambda}}(\Gamma_{kn})$, which corresponds to summation of projectors p_i to a projector of rank k. We prove that the natural symplectic form that exists on the moduli of representations of $A_{n,\bar{\Lambda}}$ goes to zero when restricted to the moduli of $B_{\mathbf{\Lambda}}(\Gamma_{kn})$ representaions. We conjecture that the image of every component of the latter moduli is Lagrangian in the former moduli.

Critical points of Laurent polynomials appear in the study of mirror symmetry for toric varieties and their deformations. The cones of the fan of the toric variety are spanned by faces of the Newton polytope of the Laurent polynomial. In our case this polytope is the classical Birkhoff polytope. Its points are doubly stochastic matrices studied in probability theory. The corresponding toric variety is a Gorenstein Fano variety with terminal singularities. According to mirror symmetry, the symplectic geometry of this variety is captured by critical points of the potential function under an appropriate choice of symplectic form and desingularization of the variety.

This connection deserve further study.

Acknowledgements. We are grateful to Alexei Rosly for useful discussions. This work was partly done during authors visit to Kavli IPMU and was supported by World Premier International Research Center Initiative (WPI Initiative), MEXT, Japan. The reported study was partially supported by RFBR, research projects 15-51-50045, 16-01-00113, 18-01-00908. AB is partially supported by Laboratory of Mirror Symmetry NRU HSE, RF Government grant, ag. No. 14.641.31.0001. The study of IZ has been partially funded by the Russian Academic Excellence Project "5-100".

§2. Algebraically unbiased projectors and representation theory

Here we recall the notion of algebraically unbiased projectors and show how to interpret them in terms of representations of suitable algebras associated to graphs.

2.1. Algebraically unbiased projectors and orthogonal Cartan subalgebras

Let V be a n-dimensional space over field of characteristic zero.

Two minimal (i.e. rank 1) projectors p and q in V are said to be *algebraically unbiased* if

$$\operatorname{Tr}(pq) = \frac{1}{n}. \tag{1}$$

Equivalently, this reads as one of the two equivalent algebraic relations:

$$pqp = \frac{1}{n}p, \tag{2}$$

$$qpq = \frac{1}{n}q. \tag{3}$$

We will also consider *orthogonal* projectors. Orthogonality of p and q is algebraically expressed as

$$pq = qp = 0. \tag{4}$$

Two complete (i.e. of cardinality n) systems of minimal orthogonal projectors $(p_1, \ldots, p_n)$ and $(q_1, \ldots, q_n)$ are said to be *algebraically unbiased* if p_i and q_j are algebraically unbiased for all pairs (i, j).

Classification of pairs of algebraically unbiased complete systems of projectors is a notoriously difficult problem. Another difficult problem is to find the maximal sets of pairwise algebraically unbiased complete systems of projectors. Both problems have the following interpretation in the theory of Lie algebras.

Consider a simple Lie algebra L over an algebraically closed field of characteristic zero. Let K be the Killing form on L. The following definitions were introduced by J. G. Thompson in 1960, in course of his study of integer quadratic lattices.

Definition. Two Cartan subalgebras H_1 and H_2 in L are said to be *orthogonal* if $K(h_1, h_2) = 0$ for all $h_1 \in H_1, h_2 \in H_2$.

Definition. Decomposition of L into the direct sum of Cartan subalgebras $L = \oplus_{i=1}^{h+1} H_i$ is said to be *orthogonal* if H_i is orthogonal to H_j, for all $i \neq j$.

Intensive study of orthogonal decompositions has been undertaken since then (see the book [KT] and references therein). For Lie algebra $sl(n)$, A. I. Kostrikin et all arrived to the following conjecture, called *Winnie-the-Pooh Conjecture* (cf. *ibid.* where, in particular, the name of the conjecture is explained by a wordplay in the Milne's book in Russian translation).

Conjecture 1. *Lie algebra $sl(n)$ has an orthogonal decomposition if and only if $n = p^m$, for a prime number p.*

The non-existence of an orthogonal decomposition for $sl(6)$, when $n = 6$ is the first number which is not a prime power is still open. It is also important to find the maximal number of pairwise orthogonal Cartan subalgebras in $sl(n)$ for any given n as well as to classify them up to obvious symmetries.

Recall the interpretation of the problem in terms of algebraically unbiased systems of rank 1 projectors. This was discovered by the first author of the present paper about 25 years ago (cf. [KT]).

Let $sl(V)$ be the Lie algebra of traceless operators in V. Killing form is given by the trace of product of operators. A Cartan subalgebra H

in V defines a unique complete system of minimal orthogonal projectors in V. Indeed, H can be extended to the Cartan subalgebra H' in $gl(V)$ spanned by H and the identity operator E. Rank 1 projectors in H' are pairwise orthogonal and comprise the required system.

If p is a minimal projector in H', then trace of p is 1, hence, $p - \frac{1}{n}E$ is in H. If projectors p and q are associated to orthogonal Cartan subalgebras, then

$$\mathrm{Tr}\left(p - \frac{1}{n}E\right)\left(q - \frac{1}{n}E\right) = 0,$$

which is equivalent to p and q to be algebraically unbiased.

Therefore, an orthogonal pair of Cartan subalgebras is in one-to-one correspondence with a pair of algebraically unbiased complete systems of orthogonal projectors. Similarly, orthogonal decompositions of $sl(n)$ correspond to a set of $n+1$ of pairwise algebraically unbiased complete systems of orthogonal projectors. In particular, it is clear from this correspondence that the number of pairwise algebraically unbiased complete systems never exceeds $n+1$.

2.2. Graphs and algebra $B_{\mathbf{\Lambda}}(\Gamma)$

In the analysis of the problem, it makes sense to consider not only complete systems of orthogonal projectors, but also study mutual unbiasedness for other configurations of projectors. Thus, we come to the problem of studying the sets of projectors where every pair satisfies either conditions (2–3) or (4). This leads us to labelled graphs and representation theory of suitable algebras related to such graphs.

Assume we are given a simply laced graph Γ with a finite number of vertices and no loop (i.e. no edge with coinciding ends). Consider a finite dimensional vector space V. Assign a rank 1 projector in V to every vertex of the graph. If two vertices are related by an edge, then we require the corresponding two projectors to be algebraically unbiased. If there is no edge between the vertices, then we require the projectors to be orthogonal. The problem is to classify, for a given graph, all possible systems of projectors satisfying the required conditions modulo automorphisms of V.

It makes sense to consider the problem in a more general context. Let us assign to every edge (ij) of the graph a nonzero weight λ_{ij} of the base field. We put $\lambda_{ij} = \lambda_{ji}$. The set of all weights is denoted by $\mathbf{\Lambda}$.

Define algebra $B_{\mathbf{\Lambda}}(\Gamma)$ as a unital algebra over the base field with generators x_i, numbered by vertices i of Γ, which is subject to relations:

- $x_i^2 = x_i$, for every vertex i,
- $x_i x_j x_i = \lambda_{ij} x_i$, $x_j x_i x_j = \lambda_{ij} x_j$, if i and j are adjacent in Γ,
- $x_i x_j = x_j x_i = 0$, if there is no edge connecting (ij) in Γ.

Various properties of this algebra, as well as relation of it with Hecke algebras of graphs and Poincare groupoids of graphs, are discussed in [BZh].

We are interested in finite dimensional representations of $B_{\mathbf{\Lambda}}(\Gamma)$. If graph is connected, then all generators x_i are presented by projectors of the same rank. We will call this rank by the *rank of the representation.* The category of representation of $B_{\mathbf{\Lambda}}(\Gamma)$ can be understood (see [BZh]) in terms of gluing of t-structures [BBD].

Consider graph Γ the complete bipartite graph $\Gamma_{n,n}$ with n vertices in both rows. If we put all weights $\lambda_{ij} = \frac{1}{n}$, then the classification problem for rank 1 n-dimensional representations of $B_{\mathbf{\Lambda}}(\Gamma_{nn})$ is equivalent to the problem of classifying pairs of algebraically unbiased complete systems of orthogonal projectors. If Γ is the graph with $n+1$ rows and n vertices in every row, such that all vertices in different rows are connected by an edge and vertices in the same row have no connecting edge, and all weights are again $\lambda_{ij} = \frac{1}{n}$, then n-dimensional representations of algebra $B_{\mathbf{\Lambda}}(\Gamma)$ correspond to the maximal system of pairwise orthogonal projectors, i.e. orthogonal decomposition of the Lie algebra $sl(n)$.

2.3. Mutually unbiased bases and configurations of lines

Now assume that the vector space is over $\mathbb{C}$ and it is endowed with a hermitian metric. Let all projectors be Hermitian, i.e. $p = p^{\dagger}$. A version of the problem on classification of pairwise unbiased systems of projectors attracted a lot of attention from physicists. If orthonormal bases compatible with an algebraically unbiased pair of complete system of orthogonal Hermitian projectors is chosen, then we got what is called in physics literature mutually unbiased bases.

Let V be an n dimensional complex space with a fixed Hermitian metric $\langle\ ,\ \rangle$. Two orthonormal Hermitian bases $\{e_i\}$ and $\{f_j\}$ in V are said to be *mutually unbiased* if, for all (i, j),

$$|\langle e_i, f_j\rangle|^2 = \frac{1}{n}. \tag{5}$$

They have numerous applications in the quantum information theory and related subjects (cf. [BZh]).

The systems of orthogonal projectors related to mutually unbiased bases are algebraically unbiased systems of Hermitian projectors. The moduli spaces of algebraically unbiased systems of orthogonal projectors is a complexification of the moduli of mutually unbiased bases, in the sense that the latter moduli is an open subset in the stable locus of the anti-holomorphic involution on the former space, which is defined by action on the projectors by $p \mapsto p^{\dagger}$.

In the same spirit, the moduli of rank 1 representations of the algebra $B_{\mathbf{\Lambda}}(\Gamma)$ can be interpreted as a complexified moduli of solutions for the following problem. Assign a complex line in a Hermitian space to every vertex of the graph Γ. Put the condition that two lines are Hermitian orthogonal if there is no edge linking the corresponding vertices of the graph. Put condition on the angle α_{ij} between lines corresponding to vertices i and j:

$$\cos \alpha_{ij} = \lambda_{ij},$$

where λ_{ij} is the weight of the edge (ij) in the graph. Clearly, we should assume that λ_{ij} are real numbers in this case.

This problem for a relatively simple graph of one cycle of order 5 is the subject of intricate study in classical papers by J. Napier [N] and C. F. Gauss [G] (see also [Sch]).

§3. Algebraically unbiased projectors and symplectic geometry

Here we undertake the study of the moduli of representations for algebra $B_{\mathbf{\Lambda}}(\Gamma)$ for bipartite graphs Γ in the framework of symplectic geometry.

3.1. Bipartite graphs and an integrable system

Fix positive integers k and n with $k \leq n$. Let $\Gamma = \Gamma_{kn}$ be the complete bipartite graph with k vertices in one row and n vertices in the other. Here we discuss the moduli space, $\mathcal{M}_B$, of $B_{\mathbf{\Lambda}}(\Gamma_{kn})$ representations of dimension n and rank 1. By the moduli spaces of representations we will always mean the GIT quotient of the affine variety of representations in a space with given basis by the action of $\mathbf{GL}(V)$.

Denote by p_i projectors from the first row, and by q_j projectors from the second row. Our initial data include the matrix Λ of coefficients λ_{ij} the weights of the edge (ij), where i is the index of a vertex from the first row and j from the second one. If we are interested in n-dimensional representations, then clearly we should put conditions:

$$\sum_i \lambda_{ij} = 1 \tag{6}$$

for every j. If $k = n$, then we should put also conditions:

$$\sum_j \lambda_{ij} = 1 \tag{7}$$

for every i.

We shall assume from now on that the base field is $\mathbb{C}$. There is an integrable system related to the moduli space $\mathcal{M}_B$. Consider a symplectic variety Y, the product of k copies of the coadjoint orbit of projectors of rank 1, which we denote by $\mathcal{O}_{p_i}$:

$$Y = \mathcal{O}_{p_1} \times \cdots \times \mathcal{O}_{p_k}.$$

Despite of the notation, all the orbits $\mathcal{O}_{p_i}$ are naturally identified. Nevertheless, it appears to be convenient to ditinguish orbits by operators that run in them.

Condition $p_i q_j p_i = \lambda_{ij} p_i$ can be read as:

$$\mathrm{Tr} p_i q_j = \lambda_{ij}. \tag{8}$$

If we assume the projectors q_j to be fixed and every projector p_i to be varied in the i-th orbit $\mathcal{O}_{p_i}$, then functions $\mathrm{Tr} p_i q_j$ commute with respect to the Poisson structure on Y, in this sense, they define an integrable system. Fix a basis $\{e_i\}$ compatible with projectors q_i and denote by $T = (\mathbb{C}^*)^n$ the Cartan torus of diagonal matrices. The map into $\mathbb{C}^{kn}$ defined by $\mathrm{Tr} p_i q_j$ is actually the moment map for the action of T^k on Y by independent conjugations of projectors p_i. It takes values in the subspace $\mathbb{C}^{k(n-1)}$ in view of (6), for $k < n$, and in the subspace $\mathbb{C}^{(n-1)^2}$, for $k = n$, in view of (7). Thus, the Lagrangian fibers of the integrable system are algebraic tori, the orbits of the torus T^k.

Consider the product of Y with the coadjoint orbit $\mathcal{O}_{-P}$ of negated projectors $-P$ (where $P^2 = P$) of rank k. Let X be the pre-image of $\mathbf{0}$ under the moment map

$$\mu : Y \times \mathcal{O}_{-P} \to \mathfrak{g}^*$$

for the diagonal action of $G = \mathbf{GL}(n)$ on this product, where $\mathfrak{g} = \mathfrak{gl}(n)$. Note that $\mu(p_1, \dots, p_k, -P) = \sum p_i - P$.

We leave the proof of the following lemma for the reader.

Lemma 2. *The sum of k rank 1 projectors is a projector of rank k if and only if the rank 1 projectors are pairwise orthogonal.*

This implies that X is identified with the moduli space of systems of k orthogonal projectors in a space with fixed basis.

Lemma 3. *X is Lagrangian in $Y \times \mathcal{O}_{-P}$.*

Proof. Indeed, all systems of k orthogonal projectors are in one orbit of $\mathbf{GL}(n)$, hence they are in the fibre of the projection map $\pi : Y \times \mathcal{O}_{-P} \to Y \times \mathcal{O}_{-P}/\mathbf{GL}(n)$. Since the arrows μ and π comprise a Poisson pair of maps (i.e. pull-backs of algebras of functions along μ and

π are mutual annihilators), the symplectic form on Y restricts by zero to the intersection of μ-fiber with π-fiber. It implies that X is isotropic and it is easy to see that it is of half dimension of Y. Q.E.D.

Let $\mathrm{pr}_1 : Y \times \mathcal{O}_{-P} \to Y$ be the projection along the first coordinate. Denote by $F_{\lambda_{ij}} \subset Y$ the fiber of the above integrable system over the point (λ_{ij}).

Proposition 4. *Moduli space $\mathcal{M}_B$ is identified with $\mathrm{pr}_1^{-1} F_{\lambda_{ij}} \cap X$ modulo the action of one copy of T that conjugates all p_i's simultaneously.*

3.2. Algebra $A_{n,\bar{\Lambda}}$ and Lagrangian image of $\mathcal{M}_B$

We will also consider the algebra $A_{n,\bar{\Lambda}}$, where $\bar{\Lambda} = (\lambda_1, \dots, \lambda_n)$ is a sequence of elements in $\mathbb{C}^*$, with generators q_i, for $i = 1, \dots, n$, and P, with relations:

$$q_i^2 = q_i, \quad P^2 = P, \quad q_i q_j = 0, \text{ for } i \neq j, \ q_i P q_i = \lambda_i q_i.$$

An $A_{n,\bar{\Lambda}}$-representation is said to be *admissible* if it is of dimension n and q_i's are represented by projectors of rank 1 and P by a projector of rank k. Let $\mathcal{M}_A$ be the moduli of admissible $A_{n,\bar{\Lambda}}$-representations. One can see that it is an irreducible variety. If $\mathcal{M}_A$ is non-empty, then $\sum_{j=1}^n \lambda_j = k$.

The variety $\mathcal{M}_A$ has a natural symplectic form (on an open subvariety of smooth points). To see this, choose a basis in the representation vector space V which is compatible with q_i's. This choice is up to the action of the diagonal torus T. Projector P is an element of the coadjoint orbit $\mathcal{O}_P$ of $\mathbf{GL}(n)$. Conditions $q_i P q_i = \lambda_i q_i$ determine the fiber over $\bar{\Lambda}$ of the moment map $\mu_A : \mathcal{O}_P \to \mathfrak{t}^*$ for the action of T (= fix diagonal coefficients of the matrix P in the chosen basis), where $\mathfrak{t}$ is the Cartan subalgebra. This means that the moduli space $\mathcal{M}_A$ is the symplectic reduction of the coadjoint orbit. Hence it is endowed with a symplectic form.

Let $\lambda_j = \sum_{i=1}^k \lambda_{ij}$. Then there is a homomorphism $\psi : A_{n,\bar{\Lambda}} \to B_{\mathbf{\Lambda}}(\Gamma_{kn})$ that takes $q_i \mapsto q_i$ and $P \mapsto \sum_{i=1}^k p_i$, which is in fact monomorphic. It allows us to consider the pull-back morphism $\Psi : \mathcal{M}_B \to \mathcal{M}_A$.

Theorem 5. *The pull-back of the symplectic structure on $\mathcal{M}_A$ along Ψ to $\mathcal{M}_B$ is zero.*

Proof. According to Proposition 4, $\mathcal{M}_B$ is identified with the quotient of $\mathrm{pr}_1^{-1} F_{\lambda_{ij}} \cap X$ by the action of T. The map $\Psi : \mathcal{M}_B \to \mathcal{M}_A$ can be understood as the composite of the restriction to this intersection of the second projection $\mathrm{pr}_2 : Y \times \mathcal{O}_{-P} \to \mathcal{O}_{-P}$, which takes it to a fibre $\mu_A^{-1}(\bar{\Lambda})$ of the moment map μ_A, with the map on the fiber, the quotient by the action of T.

Since the symplectic form on $\mathcal{M}_A$ is obtained by symplectic reduction, it is pulled back to the fibre of the μ_A to a restriction of the standard symplectic form ω_{-P} on the orbit of $-P$ (up to sign). Since X is Lagrangian in $Y \times \mathcal{O}_{-P}$ the restriction of the form $\mathrm{pr}_2^*(\omega_{-P})$ to it coincides with the restriction of the form $\mathrm{pr}_1^*(\omega_Y)$, where ω_Y is the symplectic form on Y, the product of coadjoint orbits. The letter form is zero on $\mathrm{pr}_1^{-1} F_{\lambda_{ij}} \cap X$, because $F_{\lambda_{ij}}$ is Lagrangian in Y. Hence the result. Q.E.D.

Any admissible representation of $A_{n,\bar{\Lambda}}$ satisfies an extra relation $\sum_{i=1}^n q_i = 1$. Define a new algebra

$$A_{\bar{\Lambda}}(n) = A_{n,\bar{\Lambda}} / \left(\sum_{i=1}^n q_i - 1 \right).$$

Note that algebra $A_{\bar{\Lambda}}(n)$ is Morita equivalent to the deformed preprojective algebra for a star quiver (see [BZh1], [BZh2]). The star quiver Q is a quiver with one central vertex and n vertices at the boundary. The central vertex is connected with every vertex on the boundary by one out-bound arrow. Consider the double quiver $\bar{Q}$ with the same vertices as Q and add to each arrow a in Q an extra arrow a^* in the opposite direction. Let $k\bar{Q}$ be the path algebra of the double star quiver and $e_i \in k\bar{Q}$ the projector corresponding to the vertex i. Denote by $c \in k\bar{Q}$ the commutator:

$$c = \sum_{a \in \mathrm{arr} Q} [a, a^*].$$

Define deformed preprojective algebra as follows:

$$\Pi_{\bar{\Lambda}}(Q) = k\bar{Q} / \left(c + \sum_{i=1}^{n+1} \lambda_i e_i \right),$$

where we assign $\lambda_0 = -1$ to the central vertex v_0 of the quiver and the numbers λ_i that are involved in the definition of the algebras $A_{\bar{\Lambda}}(n)$ and $A_{n,\bar{\Lambda}}$ to the boundary vertices of Q.

In view of the Morita equivalence, moduli space $\mathcal{M}_A$ is identified with a moduli space of representations for the preprojective algebra $\Pi_{\bar{\Lambda}}(Q)$. The later one is naturally endowed with the symplectic form (cf. [CBEG]). Indeed, this moduli space can be constructed via simplectic reduction for the moment map related to the moduli space of representations of the double star-quiver. As the result of the symplectic reduction, it is endowed with the symplectic form, which coincides with the symplectic form on $\mathcal{M}_A$ discussed above. An interpretation of the symplectic form on the moduli of representations of the deformed preprojective algebra in terms of Noncommutative Differential Geometry see in [CBEG].

The theorem implies that the image of the map $\Psi : \mathcal{M}_B \to \mathcal{M}_A$ is isotropic with respect to the symplectic form.

Conjecture 6. *The image of Ψ is Lagrangian in $\mathcal{M}_A$ for every irreducible component of $\mathcal{M}_B$. If $k \leq \frac{n}{2}$, then the map is quasifinite on its image.*

It is not hard to check that the dimension of the irreducible components of $\mathcal{M}_B$ has the half of the dimension of $\mathcal{M}_A$ as its lower bound. The conjecture implies that this bound is saturated if $k \leq \frac{n}{2}$. An example when it is not so for $k > \frac{n}{2}$ is when $n = 6$. If $k = 4$, then $\mathcal{M}_B$ has two irreducible components, one of dimension 3 and the other one of dimension 4, while $\mathcal{M}_A$ is of dimension 6. If $k = 5$, or 6, there is a component $\mathcal{M}_B$ of dimension 4, this is the main result of [BZh1].

3.3. A pair of algebraically unbiased systems and a symplectic embedding of the cotangent bundle

The problem of classifications of two algebraically unbiased complete systems of projectors is about constructing the moduli of n-dimensional representations for algebra $B_{\mathbf{\Lambda}}(\Gamma_{nn})$ for the complete bipartite graph Γ_{nn} with two rows of the same length n.

For the case of the graph Γ_{nn}, we have a nicer interpretation of the moduli space as the intersection of two Lagrangian subvarieties in a symplectic variety. We will identify an open subset in the symplectic variety with the total space of the cotangent bundle to one of this Lagrangian subvariety. This leads the identification of the moduli space with critical points of a suitable potential function. The function is similar to Landau-Ginzburg potential for some Fano varieties.

In case $k = n$, the sum of n orthogonal projectors is the identity matrix. This means that the orbit $\mathcal{O}_{-P}$ consists of one point. Thus,

$$Y = \mathcal{O}_{p_1} \times \cdots \times \mathcal{O}_{p_n},$$

and X is a Lagrangian subvariety in Y, the pre-image of the identity matrix under moment map $\mu : Y \to \mathfrak{g}^*$.

Since the (formal) neighborhood of a Lagrangian subvariety L in a symplectic variety is symplectically isomorphic to the (formal) neighborhood of the Lagrangian variety embedded as the zero section into the cotangent bundle to L, it is tempting to identify a Zariski open subvariety in Y with the cotangent bundle to X. We construct an open symplectic embedding of the total space of the cotangent bundle to L into Y:

$$\Omega^1 X \to Y.$$

To understand this embedding, consider a projection from a Zariski open subset in Y to X. Fibers of this projection will correspond to fibers of the vector bundle $\Omega^1 X$ embedded in Y. A point in Y is a set of n rank 1 projectors. Let $U \subset Y$ be the open subset of those n-tuples of projectors whose images are linearly independent 1-dimensional vector spaces. To any such a set of projectors $\{r_i\}$, one can assign the unique complete system of *orthogonal* projectors $\{p_i\}$ with the same images:

$$\mathrm{Im} p_i = \mathrm{Im} r_i.$$

The fiber of the projection can be parameterized as follows. A projector with a given image, a line l, is uniquely defined by its kernel H, a hyperplane that does not contain l. Hence, one can parameterize projectors with the image $\mathrm{Im} p_i$ by points $h \in \mathbb{P}V^*$ which are in the complement to the hyperplane $\langle h, l\rangle = 0$. This is an affine space. The fibre over the point $\{p_i\}$ is the product of these affine spaces over all i. It has a distinguished point, $\{p_i\}$ itself, hence it has a structure of the vector space.

In calculations, we will use the fact that X is a homogenous G-space for $G = \mathbf{GL}(n)$. Fix a reference basis in V and consider the orthogonal projectors $\{q_i\}$ compatible with this basis. Let $g \in G$ be the transition matrix from the original basis to a basis compatible with a varying system of orthogonal projectors $\{p_i\}$:

$$p_i = g q_i g^{-1}.$$

Elements g in the diagonal torus $T \subset G$ preserve the basis $\{q_i\}$. This gives identification $X = G/T$.

Further, we identify the cotangent bundle to X as a symplectic reduction of the cotangent bundle to G with respect to the action of T. We identify elements of the tangent spaces to G (at every point g) with matrices in the basis compatible with $\{q_i\}$. Then elements of the contangent space can also be considered as matrices via Killing pairing $\mathrm{Tr} AB$. Thus, we consider elements of $\Omega^1 G$ to be pairs $(g, A) \in G \times \mathrm{Mat}_{n\times n}$. The symplectic form on the cotangent bundle reads:

$$\omega_X = \mathrm{Tr} dAdg \tag{9}$$

where $\mathrm{d}A$ and $\mathrm{d}g$ are understood as matrix valued 1-forms. The product is taken henceforth in the space of matrix valued skew-symmetric forms.

Action of the diagonal element $t \in T$ on the cotangent bundle is given on $(g, A) \in \Omega^1 G = G \times \mathrm{Mat}_{n\times n}$ by

$$(g, A) \mapsto (gt^{-1}, tA).$$

One can easily calculate the moment map $\mu_T : \Omega^1 G \to \mathfrak{t}^*$ with respect to the action of T:

$$\mu_T(g, A) = \mathrm{diag}(Ag),$$

where $\mathfrak{t}^*$ is identified with diagonal matrices and diag means the diagonal of the matrix. Via symplectic reduction, we have an identification $\Omega^1 X$ with $\mu^{-1}(0)/T$.

The map $\phi : \Omega^1 X \to Y$ is defined on an element $(g, A) \in \mu^{-1}(0)$ by the formula:

$$r_i = p_i(1 + gA) = gq_ig^{-1}(1 + gA). \tag{10}$$

Note that r_i is invariant with respect to the action of T. Hence the map is well-defined. Moreover, since the diagonal of Ag is zero, it follows that:

$$\begin{aligned} r_i^2 &= gq_ig^{-1}(1 + gA)gq_ig^{-1}(1 + gA) = gq_i(1 + Ag)q_ig^{-1}(1 + gA) \\ &= gq_ig^{-1}(1 + gA) = r_i, \end{aligned}$$

i.e. r_i's are projectors of rank 1.

Theorem 7. *Map $\phi : \Omega^1 X \to Y$ is an open symplectic embedding.*

Proof. The pull-back along ϕ of the symplectic form on Y reads as:

$$\mathrm{Tr} \sum_i r_i \mathrm{d}r_i \mathrm{d}r_i.$$

It is a tedious but straightforward calculation (which we skip) to see that if we substitute r_i's by formula (10), then the result coincides with the symplectic form (9). Q.E.D.

Remark. It would be interesting to have a more conceptual explanation for the coincidence of symplectic forms on $\Omega^1 X$.

§4. Critical points of the potential and mirror symmetry

Here we show that the solutions of our problem are critical points of a potential function, which can be chosen to be a Laurent polynomial. By mirror symmetry it corresponds to a toric variety, which we study here too.

4.1. The potential

Since fibers of the integrable system on Y are Lagrangian and the map ϕ identifies the cotangent bundle to X with the Zariski open set U in Y, then the pre-images of the fibers of the integrable system along ϕ are Lagrangian subvarieties in the cotangent bundle. It is well-known that locally around points where a Lagrangian subvariety projects biholomorphically on X, it is identified with the section of the cotangent bundle, the differential of a potential function. Such a function might be multi-valued, if we maximally holomorphically extend it to X. Let us calculate this potential function.

The equation of the fiber $F_{\lambda_{ij}}$ is $\mathrm{Tr}(r_i q_j) = \lambda_{ij}$, or, in view of (10):

$$\mathrm{Tr}(p_i(1+gA)q_j) = \mathrm{Tr}(gq_ig^{-1}(1+gA)q_j) = \lambda_{ij}. \tag{11}$$

Since $X = G/T$, a (local) holomorphic function F on X defines a (local) function on G:

$$F = F(g) = F(\{g_{ij}\}).$$

The chain rule $\mathrm{d}F = \sum_{ij} \frac{\partial F}{\partial g_{ij}} \mathrm{d}g_{ij}$ implies that

$$\mathrm{d}F = \mathrm{Tr}(A\mathrm{d}g)$$

where A is a matrix with coefficients

$$a_{ij} = \frac{\partial F}{\partial g_{ji}}.$$

Denote by $\{g_{ij}\}$ the coefficients of the matrix g and by $\{\hat{g}_{ij}\}$ those of the inverse matrix g^{-1}. If F is the potential of the fiber $F_{\lambda_{ij}}$ of the integrable system, then by plugging in the above formula for A in the equations (11), we obtain equations for F:

$$g_{ji}\hat{g}_{ij} + g_{ji}\frac{\partial F}{\partial g_{ji}} = \lambda_{ij}.$$

which imply:

$$\frac{\partial F}{\partial g_{ji}} = \frac{\lambda_{ij}}{g_{ji}} - \hat{g}_{ij}. \tag{12}$$

Note that

$$\hat{g}_{ij} = \frac{\partial(\mathrm{logdet} g)}{\partial g_{ji}}$$

This implies that up to a constant

$$F = \sum_{ij} \lambda_{ij} \log g_{ji} - \log\det g.$$

If we denote by Λ the matrix with coefficients λ_{ij}, then we can write the above equation in the matrix form:

$$F = \mathrm{Tr}(\Lambda \log g) - \log\det g,$$

which implies that the exponential of F has the form:

$$\mathrm{e}^F = \frac{\prod_{i=1}^n \prod_{j=1}^n (g_{ij})^{\lambda_{ji}}}{\det g}.$$

According to what was explained before, we have

Proposition 8. *The critical points of this function are solutions to the problem of classification of two complete systems of orthogonal projectors $\{p_i\}$ and $\{q_j\}$ satisfying condition* (8).

Note that (12) implies that critical points of F are solutions of the equations:

$$\frac{\lambda_{ij}}{g_{ji}} = \hat{g}_{ij}.$$

As $\hat{g}_{ij}$ are coefficients of the inverse matrix, the equation on coefficients of g are:

$$\sum_{j=1}^{n} \frac{\lambda_{jk} g_{ij}}{g_{kj}} = 0. \tag{13}$$

for all $i \neq k$.

4.2. A Laurent polynomial and Landau-Ginzburg model potentials

Denote by E the inverse of the exponential for the potential F:

$$E = \mathrm{e}^{-F} = \frac{\det g}{\prod_{i=1}^n \prod_{j=1}^n (g_{ij})^{\lambda_{ji}}}.$$

Note that critical points of this function are the same as those of F.

Function E is a multi-valued function on the n^2 dimensional torus $(\mathbb{C}^*)^{n^2}$ with coordinates g_{ij}. It is invariant with respect to the action of the diagonal torus T on both sides of the matrix g. Thus, we can consider E as a function on the torus $T \setminus (\mathbb{C}^*)^{n^2}/T$ of dimension $(n-1)^2$.

Now we will concentrate on the case when $\lambda_{ij} = \frac{1}{n}$, for all (ij). This corresponds to the case of algebraically mutually unbiased bases. Function E reads:

$$E = \frac{\det g}{\prod (g_{ij})^{\frac{1}{n}}}. \tag{14}$$

Consider the isogenic n-cover $\mathbb{T} \to T \setminus (\mathbb{C}^*)^{n^2}/T$ by another torus $\mathbb{T}$ of dimension $(n-1)^2$ defined by adding to the holomorphic functions the n-th root of $\prod g_{ij}$. Thus we have a new multiplicative coordinate on this torus:

$$z = \prod (g_{ij})^{\frac{1}{n}}.$$

The advantage of the new torus is that the lift of E to it (which we denote by the same letter E) is a Laurent polynomial in multiplicative coordinates on the torus. The right and left action of torus T lifts to the n-fold cover and E remains invariant with respect to it. Thus we can again diminish the dimension of the torus to $(n-1)^2$.

Example. Let $n = 2$. After killing the action of $T \times T$, the matrix g has the form:

$$g = \begin{pmatrix} 1 & 1 \\ 1 & x \end{pmatrix}. \tag{15}$$

Hence $z^2 = x$, and $E = z - \frac{1}{z}$. After changing the coordinate z for iz one get that $E = i(z + \frac{1}{z})$. In other words, E is (up to a multiplier) the Landau-Ginzburg potential corresponding to the projective line $\mathbb{P}^1$, the target space for the super-symmetric sigma-model.

The potential has finite number of critical points for $n = 2, 3, 5$. There is a curve of critical points for $n = 4$, and a 4-dimensional family for $n = 6$ (see [BZh1]). In view of Winnie-the-Pooh Conjecture, one can assume that the structure of critical points depend on the prime number decomposition for n. There is a counter-example to the conjecture that the number of critical points is always finite for prime n: there is a 1-parameter family of critical points for $n = 7$.

4.3. The toric variety of Birkhoff polytope

According to the prescription of mirror symmetry, we consider the toric variety X_E whose defining fan has cones spanned by faces of the Newton polytope of the function E on the torus $\mathbb{T}$ introduced above.

The Newton polytope is in fact the classically known Birkhoff polytope, the convex hull of permutation matrices in the space of real square matrices. We denote the *real* coordinates in the space of matrices by

λ_{ij} (this notation might look strange, but it has the reason that this coordinates are indeed dual to g_{ij}). Then Birkhoff polytope lies in the affine subspace L defined by equations (6) and (7), which reflects the fact that the torus $\mathbb{T}$ has dimension $(n-1)^2$. Let $\tilde{N}$ be the lattice generated by all integer matrices and the matrix Λ_o with all coefficients equal to $\frac{1}{n}$. Define an integer lattice N in L to be $L \cap \tilde{N}$. This is the lattice of characters of our torus $\mathbb{T}$ if we take the origin to be matrix Λ_o. The shift of the origin is justified by the denominator in (14).

According to Birkhoff-von Neumann theorem, Birkhoff polytope is the intersection of L with the set of matrices with non-negative coefficients. In other words, it is the locus of doubly stochastic matrices. It was intensively studied and has many applications. There are n^2 facets of the polytope, given by equation $\lambda_{ij} = 0$, for a particular choice of (ij).

Proposition 9. *Variety X_E is a Gorenstein toric Fano variety with terminal singularities.*

Proof. Function $l_{ij} = n\lambda_{ij} - 1$ is zero at the origin, integer-valued on $\tilde{N}$ and has value -1 on the facet $\lambda_{ij} = 0$. This means that it defines an integer vertex of the dual polytope. Hence, all vertices of the dual polytope are integer, i.e. Birkhoff polytope is reflexive. In terms of toric geometry, this means that our toric variety X_E is Fano and Gorenstein.

Clearly, the polytope lies in the unit hypercube $[0,1]^{n^2}$. The only points of the lattice $\tilde{N}$ in the unit cube are vertices of the cube and integer multiples of the matrix Λ_o with multiplier up to n. As Birkhoff polytope lies in L, this implies that the only integer points in Birkhoff polytope are Λ_o and the vertices of the polytope. This means that the toric variety has terminal singularities. Q.E.D.

The toric variety X_E is $\mathbb{P}^1$, for $n = 2$, it is $\mathbb{P}^2 \times \mathbb{P}^2$, for $n = 3$, and it is singular for higher n. The geometry of this toric variety as well as critical points of the potential deserve further study.

References

[BBD] A. Beilinson, J. Bernstein and P. Deligne, *Faisceaux pervers*, Astérisque, vol. 100, Soc. Math. France, 1983.

[BLDG] de Burgh M. D., Landford N. K., Doherty A. C., Gilchrist A., Choice of measurement sets in qubit tomography, Phys. Rev. A 78, 052122 (2008).

[BZh] A. Bondal, I. Zhdanovskiy, *Representation theory for system of projectors and discrete Laplace operators*, IPMU13-0001, IPMU, Kashiwa, Japan, 2013, 48 pp.

[BZh1] A. Bondal, I. Zhdanovskiy, *Orthogonal pairs for Lie algebra* $sl(6)$, IPMU14-0296, IPMU, Kashiwa, Japan, 2014, 89 pp.

[BZh2] A. Bondal, I. Zhdanovskiy, *Ortogonal pairs and mutually unbiased bases*, Representation theory, dynamical systems, combintorial methods, XXVI, Zapiski POMI, 437, POMI, St.-Peterburg, 2015, 35–61, arXiv: 1510.05317.

[BTSW] P. Oscar Boykin, Meera Sitharam, Pham Huu Tiep, Pawel Wocjan, *Mutually Unbiased Bases and Orthogonal Decompositions of Lie Algebras*, http://arxiv.org/abs/quant-ph/0506089.

[CBEG] W. Crawley-Boevey, P. Etingof, V. Ginzburg, *Noncommutative geometry and quiver algebras*, Adv. Math., v. 209, 2007, 1, 274–336.

[DEBZ] Durt T., Englert B.-G., Bengtsson I., Zyczkowski K., On mutually unbiased bases, Int. J. Quantum Inform., 08, 535 (2010)

[FM] Filippov S. N., Man'ko V. I., Mutually unbiased bases: tomography of spin states and the star-product scheme, Physica Scripta, vol. 2011, T143.

[G] C. F. Gauss, Pentagramma Mirificum, Werke, Bd. III, 481–490; Bd VIII, 106–111.

[Kr] Kraft, H., *Geometric methods in representation theory.* In: Representations of Algebras, Workshop Proceedings (Puebla, Mexico 1980), Lecture Notes in Math., 944, Springer, Berlin-New York, 1982.

[KT] A. I. Kostrikin, Pham Huu Tiep, *Orthogonal decompositions of Lie algebras and integral lattices.* Berlin: Walter de Gruyter, 1994.

[N] John Napier, *Mirifici Logarithmorum canonis descriptio*, Lugdini, 1619.

[Ru] Ruskai M. B., Some connections between frames, mutually unbiased bases, and POVM's in quantum information theory, Acta Applicandae Mathematicae 12/2009, 108(3), 709–719

[Sch] V. Schectman, *Pentagramma mirificum and elliptic functions*, Annales de la Faculte des Sciences de Toulouse. Serie VI. Mathematiques 01/2013; 2(2), arXiv:1106.3633.

Steklov Institute of Mathematics, Moscow,
and Kavli Institute for the Physics and Mathematics of the Universe (WPI),
The University of Tokyo, Kashiwa, Chiba 277-8583, Japan,
and National Research University Higher School of Economics,
Russian Federation,
and The Institute for Fundamental Science, Moscow,
and Laboratory AGHA, MIPT, 9 Institutskiy per.,
Dolgoprudniy, Moscow Region, Russia.

MIPT, 9 Institutskiy per., Dolgoprudniy, Moscow Region, Russia,
141701 and National Research University Higher School of Economics,
AG Laboratory, HSE, 6 Usacheva str., Moscow, Russia, 119048

Advanced Studies in Pure Mathematics 83, 2019
Primitive Forms and Related Subjects — Kavli IPMU 2014
pp. 19–53

A brief survey of FJRW theory

Amanda E. Francis, Tyler J. Jarvis and Nathan Priddis

Abstract.

In this paper we describe some of the constructions of FJRW theory. We also briefly describe its relation to Saito-Givental theory via Landau-Ginzburg mirror symmetry and its relation to Gromov-Witten theory via the Landau-Ginzburg/Calabi-Yau correspondence. We conclude with a discussion of some of the recent results in the field, including the gauged linear sigma model, which is expected to provide a geometric framework for unifying many of these ideas.

§1. Introduction

In this article we briefly describe what is now called *FJRW* (*Fan-Jarvis-Ruan-Witten*) *theory* and the part it plays in two deep relations, namely mirror symmetry and the Landau-Ginzburg/Calabi-Yau (LG/CY) correspondence.

FJRW theory is analogous, in many ways, to the better-known Gromov-Witten (GW) theory. It associates a cohomological field theory (and hence also Frobenius manifold) to each pair (W, G) of a nondegenerate quasihomogeneous polynomial W and an abelian group G of symmetries of W.

Some of the many similarities between FJRW theory and GW theory include the fact that the state space of each is the cohomology of a "target space" (see Section 3.2), both are equipped with stabilization and evaluation maps, and both have virtual cycles that satisfy certain properties (see Section 3.5) that facilitate computation and that, when pushed down to $\overline{\mathscr{M}}_{g,k}$, give a cohomological field theory satisfying the

Received March 3, 2015.
Revised October 12, 2016.
2010 *Mathematics Subject Classification.* Primary 14N35, 53D45; Secondary 32S05, 37K10, 37K20, 35Q53.
Key words and phrases. FJRW theory, GLSM, gauged linear sigma model, Gromov-Witten, quantum singularity, Saito-Givental, Landau-Ginzburg, Calabi-Yau.

WDVV equations, the string equation, the dilaton equation, and topological recursion relations. These similarities are not just superficial, as the two theories are highly related via the LG/CY correspondence.

To describe the LG/CY correspondence, we first remark that the pair (W, G) also defines a smooth hypersurface $X_W = \{W = 0\}$ in weighted projective space. And if we let $\langle J \rangle = G \cap \mathbb{C}^*$, where $\mathbb{C}^*$ is the 1-parameter group defining weighted projective space as $\mathbb{P}(c_1, \ldots, c_N) = [\mathbb{C}^N/\mathbb{C}^*]$, then there is an induced action of the group $\tilde{G} = G/\langle J \rangle$ on X_W. Under certain conditions the quotient orbifold $[X_W/\tilde{G}]$ is Calabi-Yau. The *Landau-Ginzburg/Calabi-Yau* (*LG/CY*) *Correspondence* (see [Mar90, VW89, Wit93b]) predicts that the analytic continuation of the orbifold GW potential of the orbifold $[X_W/\tilde{G}]$, after a suitable symplectic transformation, will match precisely with the FJRW potential of the pair (W, G) in all genera. We will discuss the progress that has been made on this conjecture (and some generalizations) in Section 5.

There are other theories which yield cohomological field theories, such as Saito-Givental theory. To any isolated singularity, such as defined by the polynomial W in affine space, we may construct the local algebra (or Milnor ring) of W, which is a Frobenius algebra. For each choice of primitive form, K. Saito constructed a Frobenius manifold deforming the Milnor ring (see [Sai81, Sai83a, Sai83b, ST08]). Since the Saito construction is semisimple, Givental's theory of "higher genus Frobenius manifolds" (see [Giv01, Giv04]) determines a potential function for all genera. This is known as Saito-Givental theory.

FJRW theory is closely related to Saito-Givental theory via mirror symmetry. In particular, FJRW provides a Landau-Ginzburg A model for the pair (W, G), and in one formulation of mirror symmetry, Saito-Givental provides the Landau-Ginzburg B model.

The *Landau-Ginzburg* (*LG*) *Mirror Symmetry Conjecture* (see [Kra10]) predicts that for a large class of polynomials W (called invertible) with a group G of admissible symmetries of W, there is a dual polynomial W^T and dual group G^T of symmetries of W^T such that FJRW Landau-Ginzburg A model for the pair (W, G) is isomorphic to the Landau-Ginzburg B model for the pair (W^T, G^T). In the case that G^T is trivial, the Landau-Ginzburg B model corresponds to the Saito-Givental theory. There is currently no uniform method for constructing a Frobenius manifold in case G^T is nontrivial (the *orbifolded* case). However, some work has been done in this direction in [BT14, Lee17].

There is a construction for a B model Frobenius algebra (see [IV90, Kau06, Kau02, Kau03, Kra10]), which should be thought of as the tangent space of a Frobenius manifold at the origin, so it has been conjectured that a sensible "orbifolded" Frobenius manifold (and hence cohomological field theory) can be constructed in all genera for the B model in this case as well. There are also some other formulations of Landau-Ginzburg mirror symmetry which we will discuss in Section 4, along with a brief survey of recent results.

The two mirror symmetry relations are themselves related: we can put mirror symmetry for both Calabi-Yau and Landau-Ginzburg theories together with the LG/CY correspondence to obtain the following schematic diagram.

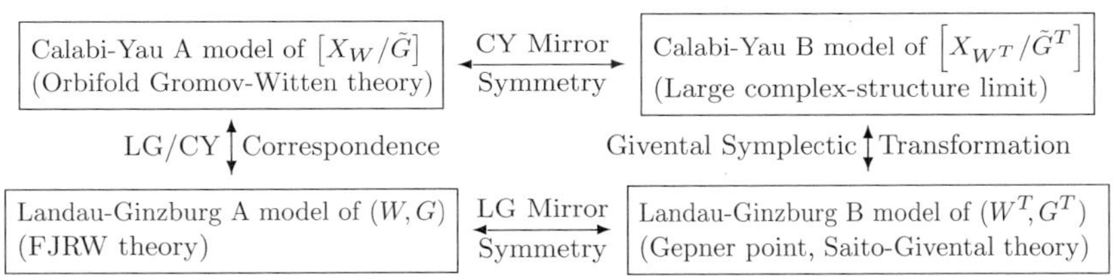

This diagram should not be thought of as fact or theorem, but rather as a way to organize the ideas behind mirror symmetry and the LG/CY correspondence into one single framework in order to better conceptualize the ideas presented here. In the following sections, we will give a more clear picture of what has been proven and what is still conjectural for each of the various parts of the diagram.

In Section 2, we give a brief summary of cohomological field theories, and in Section 3 we describe the construction of FJRW theory. In Sections 4 and 5 we describe mirror symmetry and the LG/CY correspondence in more detail. We conclude in Section 6 with a brief discussion of recent progress on a mathematical approach to the Gauged Linear Sigma Model, which generalizes FJRW theory and the LG/CY correspondence to complete intersections, toric varieties, and more general spaces.

1.1. Further application and remarks

We would like to conclude this section with one more example of an appearance of FJRW theory, which does not fit into the framework outlined above, namely the generation of tautological relations on the moduli space of curves.

The theory of r-spin curves corresponds to FJRW theory for the polynomial $W = x^r$ with the group $G = \mu_r$. Pandharipande-Pixton-Zvonkine (see [PPZ15]) used the grading on the cohomological field theory of 3-spin curves, along with the semisimplicity property and the reconstruction theorem of Teleman, to identify new cohomological relations on the stack of stable curves.

A similar argument applied to higher spin curves (i.e. $r > 3$) also produces relations (see [PPZ16]), although all of those are known to follow from the 3-spin relations. Janda (see [Jan14]) has shown (in the Chow ring) that those follow from similar relations arising from the Gromov-Witten cohomological field theory for projective space. In [PPZ16], Pandharipande-Pixton-Zvonkine also uses the $r = 4$ relations to bound the Betti numbers of the tautological ring $R^*(M_g)$.

One might hope to use this method of generating relations on the stack of stable curves for more general pairs (W, G) of polynomials and groups, but when the group G is not maximal, the corresponding theory is (generally) not semisimple, so the arguments of Pixton and his collaborators will not work for these pairs. Furthermore Janda (see [Jan15]) has proved that no more new relations can be obtained from any semisimple cohomological field theories.

Remarkably, however, J. Guéré has used a different method, not based on semisimplicity, to obtain additional relations from the FJRW theory of chain polynomials (see [Gué17]). It is not yet known whether these relations follow from previously known relations or not.

§2. Cohomological Field Theories

Given the data of a quasihomogeneous polynomial and a finite group of diagonal symmetries, FJRW theory produces a cohomological field theory (in the sense of Kontsevich and Manin as in [KM94]). We begin with a brief review of cohomological field theories. These are essentially collections of cohomology classes that behave well with respect to the gluing maps of pointed stable curves.

Definition 2.1. For any nonnegative integers g, k with $2g-2+k > 0$ let $\overline{\mathscr{M}}_{g,k}$ denote the stack of stable k-pointed curves of genus g. Given nonnegative integers g_1, g_2, k_1, k_2 with $2g_i - 1 + k_i > 0$ for $i \in \{1, 2\}$, we define a gluing map

$$\rho_{\text{tree}} : \overline{\mathscr{M}}_{g_1,k_1+1} \times \overline{\mathscr{M}}_{g_2,k_2+1} \to \overline{\mathscr{M}}_{g_1+g_2,k_1+k_2}$$

by attaching the two additional marked points to form a node, as in Figure 1. Similarly, for any nonnegative integers g, k with $2g + k > 0$ we define a gluing map

$$\rho_{\text{loop}} : \overline{\mathscr{M}}_{g,k+2} \to \overline{\mathscr{M}}_{g+1,k},$$

again by attaching the two additional marked points to form a node, as in Figure 2.

Finally, if $2g - 2 + k > 0$ we define a *forgetting tails* map

$$\tau : \overline{\mathscr{M}}_{g,k+1} \to \overline{\mathscr{M}}_{g,k}$$

by forgetting the last marked point and successively contracting any resulting unstable components.

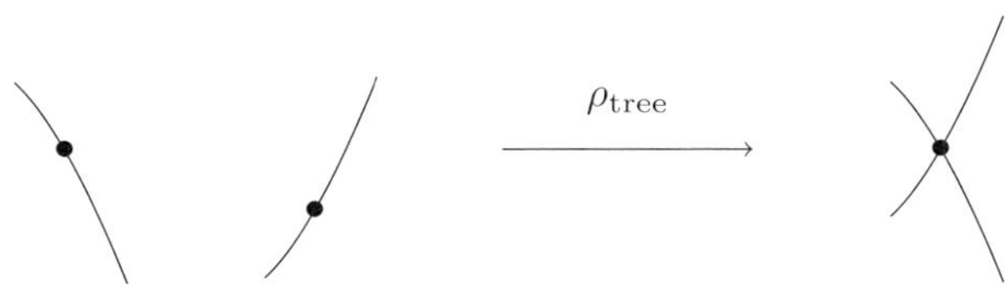

Fig. 1. The morphism ρ_{tree} glues a marked point on each of two stable curves to form a new, nodal curve.

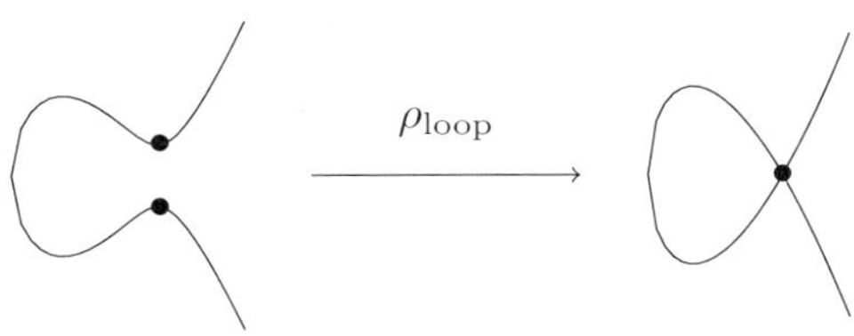

Fig. 2. The morphism ρ_{loop} glues a two marked points on one stable curve to form a new, nodal curve.

Definition 2.2. A *cohomological field theory with flat identity* (abbreviated CohFT) consists of the data of

(1) A vector space $\mathscr{H}$ (called the *state space*),

(2) A nondegenerate pairing:

$$\langle\ ,\ \rangle : \mathscr{H} \otimes \mathscr{H} \to \mathbb{C},$$

(3) An element $\mathbf{1} \in \mathscr{H}$ (called the *flat identity*),

(4) For each pair g, k, a k-linear form $\Lambda_{g,k} : \mathscr{H}^k \to H^*(\overline{\mathscr{M}}_{g,k})$, satisfying the following axioms:

   (a) Each form $\Lambda_{g,k}$ is equivariant under the action of the symmetric group S_k.

(b) For any $\alpha_1, \ldots, \alpha_{k_1+k_2} \in \mathscr{H}$ we have

$$\rho^*_{\text{tree}}\Lambda_{g_1+g_2,k_1+k_2}(\alpha_1, \ldots, \alpha_{k_1+k_2}) = \sum_{\sigma,\tau} \Lambda_{g_1,k_1+1}(\alpha_1, \ldots, \alpha_{k_1}, \sigma)\eta^{\sigma\tau}\Lambda_{g_2,k_2+1}(\tau, \alpha_{k_1+1}, \ldots, \alpha_{k_1+k_2}),$$

where σ, τ run over a basis of $\mathscr{H}$ and $\eta^{\sigma,\tau}$ is the appropriate entry in the inverse of the matrix expressing the pairing in terms of this basis.

(c) For any $\alpha_1, \ldots, \alpha_k \in \mathscr{H}$ we have

$$\rho^*_{\text{loop}}\Lambda_{g+1,k}(\alpha_1, \ldots, \alpha_k) = \sum_{\sigma,\tau} \Lambda_{g,k+2}(\alpha_1, \ldots, \alpha_k, \sigma, \tau)\eta^{\sigma\tau},$$

where again σ, τ run over a basis of $\mathscr{H}$ and $\eta^{\sigma,\tau}$ is the appropriate entry in the inverse of the matrix expressing the pairing in terms of this basis.

(d) For any $\alpha_1, \ldots, \alpha_k \in \mathscr{H}$ we have

$$\tau^*\Lambda_{g,k}(\alpha_1, \ldots, \alpha_k) = \Lambda_{g,k+1}(\alpha_1, \ldots, \alpha_k, \mathbf{1})$$

(e) $\int_{\overline{\mathscr{M}}_{0,3}} \Lambda_{0,3}(\alpha_1, \alpha_2, \mathbf{1}) = \langle \alpha_1, \alpha_2 \rangle$

As mentioned in the introduction, some of the most important examples of CohFTs are the Gromov-Witten theory of a variety X, the FJRW theory of a quasihomogeneous polynomial, and Saito-Givental theory for primitive forms.

§3. FJRW Theory

In this section we describe the construction of FJRW theory, including the state space, the moduli space of W-curves and the virtual class.

3.1. Inputs

The ingredients for FJRW theory are an admissible quasihomogeneous polynomial and an admissible group of diagonal symmetries. Let us now define what we mean by these.

Definition 3.1. A polynomial $W \in \mathbb{C}[x_1, \ldots, x_N]$, where $W = \sum_i b_i \prod_{j=1}^N x_j^{a_{i,j}}$ is called *quasihomogeneous* if there exist positive rational numbers q_j (called weights) for each variable x_j such that each monomial of W has weighted degree one. That is, for every i where $b_i \neq 0$,

$$\sum_j q_j a_{i,j} = 1.$$

A polynomial $W \in \mathbb{C}[x_1, \ldots, x_N]$ is called *nondegenerate* if it has an isolated singularity at the origin, otherwise it is called *degenerate*.

Any polynomial which is both quasihomogeneous and nondegenerate will be considered *admissible* if there are no monomials of W of the form $x_i x_j$ with $i \neq j$. This implies that that the weights $q_1, \ldots, q_N$ are uniquely determined and all weights lie in the interval $(0, 1/2] \cap \mathbb{Q}$.

We say that an admissible polynomial is *invertible* if it has the same number of variables and monomials.

Remark 3.2. It is often useful to use integer weights $c_1, \ldots, c_N, d$ for W with $\gcd(c_1, \ldots, c_N, d) = 1$ such that $q_i = c_i/d$.

Example 3.3. One may easily check that $x^3 + y^3$ is admissible and invertible, $x^3 + y^3 + xy^2$ is admissible but not invertible, $x^3 + y^3 + xy^4$ is not quasihomogeneous and $x^4 + x^2y^2$ is quasihomogeneous but degenerate.

Remark 3.4. Classical singularity theory studies singularities up to so-called *right equivalence*; that is, up to a smooth change of variables. The quantum singularity theory of FJRW theory appears to be more rigid, by comparison. The only changes of variables allowed in this theory are rescaling and reordering (or relabeling) variables of the same weight. Therefore, we often need classification results that are stronger than those of classical singularity theory. On the other hand, the quantum invariants of FJRW theory are known to be independent of the polynomial W and depend only on the weights and an *admissible group* G, as we will see in Theorem 3.18.

A fundamental tool for studying Landau-Ginzburg mirror symmetry is the following classification of invertible polynomials.

Proposition 3.5 ([KS92]). *A polynomial is (admissible and) invertible if, and only if, it can be written (up to relabeling, rearrangement, and rescaling of the variables) as a disjoint sum of invertible polynomials of one or more of the following three* atomic types:

$$\begin{aligned} \textit{Fermat type: } & x^a, \\ \textit{loop type: } & x_1^{a_1}x_2 + x_2^{a_2}x_3 + \cdots + x_N^{a_N}x_1, \\ \textit{chain type: } & x_1^{a_1}x_2 + x_2^{a_2}x_3 + \cdots + x_N^{a_N}, \end{aligned}$$

where $a \geq 2$ and $a_k \geq 2$ for all $k \in \{1, \ldots, N\}$.

In addition to an admissible polynomial, the other main ingredient of FJRW theory is the symmetry group. Indeed, the theory depends heavily on the choice of symmetry group. In this sense, it may be thought of as an orbifolded singularity or the orbifold Landau-Ginzburg theory of $[W/G]$.

Definition 3.6. Let W be an admissible polynomial. The *maximal diagonal symmetry group* $G_{\max}$ is the subgroup of $(\mathbb{C}^*)^N$ of elements that fix W:

$$G_{\max} = \{(\beta_1, \dots, \beta_N) \mid W(\beta_1 x_1, \dots, \beta_N x_N) = W(x_1, \dots, x_N)\}.$$

It is easy to see that every element $\gamma \in G_{\max}$ can be written in the form $\gamma = (e^{2\pi i\theta_1}, \dots, e^{2\pi i\theta_N})$ with each $\theta_i \in [0, 1)$. We call the rational numbers $\theta_1, \dots, \theta_N$ the *phases* of γ. The sum of the phases is the *age* of γ:

$$\operatorname{age}(\gamma) = \sum_{i=1}^{N} \theta_i.$$

Definition 3.7. If $q_1, q_2, \dots, q_N$ are the weights of W, then $J = (e^{2\pi i q_1}, \dots, e^{2\pi i q_N})$ is an element of $G_{\max}$, called the *grading element.* Any subgroup G of $G_{\max}$ which contains the element J is called *admissible.*

Definition 3.8. Given an admissible group G and an element $g \in G$, we let $\operatorname{Fix}(g)$ denote the subspace of $\mathbb{C}^N$ which is fixed by g, i.e.

$$\operatorname{Fix}(g) = \{(a_1, \dots, a_N) \mid g \cdot (a_1, \dots, a_N) = (a_1, \dots, a_N)\}.$$

We denote the fixed indices by $I_g = \{i \mid \theta_i^g = 0\}$, and we write $N_g = \#I_g = \dim(\operatorname{Fix}(g))$.

Finally, we write

$$W_g = W|_{\operatorname{Fix}(g)}$$

to denote the polynomial W restricted to $\operatorname{Fix}(g)$.

Notice that

$$\operatorname{age}(g) + \operatorname{age}(g^{-1}) = \sum_{i:\theta_i^g \neq 0} 1 = N - N_g.$$

Example 3.9. Consider $W = x^3 + y^3 + yz^2$. This is a sum of a Fermat polynomial and a (reverse) chain polynomial. In this case $q_x = q_y = q_z = \frac{1}{3}$, so $J = (e^{2\pi i/3}, e^{2\pi i/3}, e^{2\pi i/3})$ has order 3. However, $G_{\max} = \langle (e^{2\pi i/3}, 1, 1), (1, e^{2\pi i/3}, e^{2\pi i \cdot 5/6}) \rangle$, which is a group of order 18. It is easy to check that there are also admissible subgroups of order 6 and 9.

Each admissible symmetry group could potentially give a distinct FJRW theory, although there are cases where different groups do give the equivalent theories (see e.g. [BP18]).

3.2. State Space

In this section we define the state space of the theory for a given admissible polynomial $W \in \mathbb{C}[x_1, \dots, x_N]$ and an admissible group G of symmetries of W.

Definition 3.10. Because W is G-invariant, it defines a function $W : [\mathbb{C}^N/G] \to \mathbb{C}$, where $[\mathbb{C}^N/G]$ denotes the stack quotient of $\mathbb{C}^N$ by the action of G. We define

$$W^{\infty} = (\mathfrak{Re}(W))^{-1}(M, \infty) \text{ for } M >> 0.$$

Similarly, for $g \in G$, we have $W_g : [\mathrm{Fix}(g)/G] \to \mathbb{C}$, and we define

$$W_g^{\infty} = (\mathfrak{Re}(W_g))^{-1}(M, \infty).$$

Definition 3.11. Fix an admissible polynomial $W \in \mathbb{C}[x_1, \dots, x_N]$ with weights $q_1, \dots, q_N$ and a group of symmetries G, and let $q = \sum_{i=1}^N q_i$. The FJRW state space corresponding to W and G is the relative Chen-Ruan cohomology:

$$\begin{aligned} \mathscr{H}_{W,G} &= H_{CR}^{\bullet+2q}\left([\mathbb{C}^N/G], W^{\infty}\right) \\ &= \bigoplus_g H^{\bullet+2q-2\,\mathrm{age}(g)}\left([\mathrm{Fix}(g)/G], W_g^{\infty}\right) = \bigoplus_g \mathscr{H}_g, \end{aligned}$$

where for each $g \in G$, the subspace $\mathscr{H}_g$ is called the *g-sector* of the state space. Since G is abelian, we can also write $H^n([\mathrm{Fix}(g)/G], W^{\infty}, \mathbb{C}) = H^n(\mathrm{Fix}(g), W^{\infty}, \mathbb{C})^G$ in the definition above. For simplicity in what follows, we will assume all elements of the state space are even cohomology classes. The theory makes sense for odd classes as well, provided proper signs are inserted to account for supercommutativity.

Also in the definition above, the degree of each element is shifted by $2\,\mathrm{age}(g) - 2q$, so if $\varkappa \in H^m([\mathrm{Fix}(g)/G], W_g^{\infty}, \mathbb{C})$, then the degree of $\varkappa$ in $\mathscr{H}_{W,G}$ is $m - 2q + 2\,\mathrm{age}(g)$. This degree is often called the W-degree in the literature. Furthermore, it is a standard fact, that $H^n(\mathrm{Fix}(g), W^{\infty}, \mathbb{C}) = 0$ except when $n = N_g = \dim(\mathrm{Fix}(g))$, therefore every element of $\mathscr{H}_g$ has degree $N_g - 2q + 2\,\mathrm{age}(g)$.

Remark 3.12.

(1) If $\mathrm{Fix}(g) = \{\mathbf{0}\}$, then $W_g = 0$ and

$$\mathscr{H}_g = H^0([\mathbf{0}/G], \emptyset, \mathbb{C}) = \mathbb{C},$$

and the elements of $\mathscr{H}_g$ all have degree $2\,\mathrm{age}(g) - 2q$. We often write $\mathscr{H}_g = \mathbb{C}_g$ instead of $\mathbb{C}$, to indicate which sector of the state space we are describing. Such sectors are called *narrow* sectors.

(2) If $g = J$ (the grading element of W), then the sector $\mathscr{H}_J$ is narrow, and $\mathrm{age}(J) = q$, so

$$\mathscr{H}_J = H^0\left([\mathbf{0}/G], \mathbb{C}\right) = \mathbb{C}$$

is supported only in degree zero. We denote the element $1 \in \mathbb{C}_J$ by $\mathbf{1}$. This will be the flat identity of our theory (see Definition 2.2).

(3) Sectors with $\dim \mathrm{Fix}(g) > 0$ are called *broad*. It is difficult to deal with the broad sectors, in general. This is also the case in the generalization of FJRW theory, which we describe in Section 6. Despite the difficulty of dealing with broad sectors, some results are known about them, mostly due to Guéré, as we describe later in the paper.

We will often use the following alternative description of $\mathscr{H}_{W,G}$ in terms of germs of holomorphic N forms on $\mathbb{C}^N$ proven by Wall:

Theorem 3.13 ([Wal80a, Wal80b]). *Let Ω^N be the germs of holomorphic N-forms on $\mathbb{C}^N$ near zero. We have*

$$H^N(\mathbb{C}^N, W^\infty, \mathbb{C}) \cong \Omega^N/(dW \wedge \Omega^{N-1})$$

as G-modules.

Notice that

$$\frac{\Omega^N}{dW \wedge \Omega^{N-1}} \cong \frac{\mathbb{C}[x_1, \ldots, x_N]}{(\frac{\partial W}{\partial x_1}, \ldots, \frac{\partial W}{\partial x_N})} \cdot dx_1 \wedge \ldots \wedge dx_N,$$

so if $\mathscr{Q}_W = \mathbb{C}[x_1, \ldots, x_N]/(\frac{\partial W}{\partial x_1}, \ldots, \frac{\partial W}{\partial x_N})$ is the local algebra (a.k.a. Milnor ring) of W, and if $d\mathbf{x} = dx_1 \wedge \cdots \wedge dx_N$ is the obvious volume form, then

$$\mathscr{H}_1 = H^N(\mathbb{C}^N, W^\infty, \mathbb{C})^G \cong \left(\Omega^N/(dW \wedge \Omega^{N-1})\right)^G \cong (\mathscr{Q}_W d\mathbf{x})^G.$$

And, more generally, if the fixed indices of g are $\{i_1, \ldots, i_{N_g}\} = I_g$, we have $\mathscr{Q}_{W_g} = \mathbb{C}[x_{i_1}, \ldots, x_{i_{N_g}}]/(\frac{\partial W}{\partial x_{i_1}}, \ldots, \frac{\partial W}{\partial x_{i_{N_g}}})$ and $d\mathbf{x}_{\mathrm{Fix}(g)} = dx_{i_1} \wedge \cdots \wedge dx_{i_{N_g}}$, therefore

$$\begin{aligned}\mathscr{H}_{W,G} &= \bigoplus_g H^{N_g}([\mathrm{Fix}(g)/G], W_g^\infty, \mathbb{C}) \cong \bigoplus_g \left(\Omega^{N_g}/(dW_g \wedge \Omega^{N_g-1})\right)^G \\ &\cong \bigoplus_g (\mathscr{Q}_{W_g} d\mathbf{x}_{\mathrm{Fix}(g)})^G.\end{aligned}$$

Definition 3.14. Based on the previous discussion, we often write elements of the sector $\mathscr{H}_g$ as $\lceil \phi\,; g\rfloor$, where $g \in G$ and ϕ is an element of $\mathscr{Q}_{W_g} d\mathbf{x}_{\mathrm{Fix}(g)}$.

Remark 3.15. There is a natural $\mathbb{Q}$-grading on the local algebra $\mathscr{Q}_{W_g}$ defined by the weighted degree of monomials in $\mathscr{Q}_{W_g}$, and this induces a $\mathbb{Q}$-grading on $\mathscr{Q}_{W_g} d\mathbf{x}_{\mathrm{Fix}(g)}$ defined by giving the volume form $d\mathbf{x}_{\mathrm{Fix}(g)}$ degree $\sum_{j=1}^{N_g} q_{i_j}$. So if $\phi = m\, d\mathbf{x}_{\mathrm{Fix}(g)}$, then $\deg(\phi) = \deg(m) + \sum_{j=1}^{N_g} q_{i_j}$.

There is also a natural Hodge structure on the J-invariant part (and therefore also on the G-invariant part) of the relative cohomology. Tracking the Hodge structure through Wall's isomorphism connects the Hodge grading on relative cohomology with the grading on $(\mathscr{Q}_{W_g} d\mathbf{x}_{\mathrm{Fix}(g)})^{\langle J\rangle}$:

$$H^{N-p,p}(\mathbb{C}^{N_g}, W_g^{\infty}, \mathbb{C})^{\langle J\rangle} \cong \left\{\phi \in \mathscr{Q}_{W_g} d\mathbf{x}_{\mathrm{Fix}(g)} \mid \deg(\phi) = p\right\}^{\langle J\rangle}.$$

We use this to put a bigrading on the state space

$$\mathscr{H}^{a,b}_{W,G} = \bigoplus_g H^{a+q-\mathrm{age}\, g,\, b+q-\mathrm{age}\, g}(\mathbb{C}^{N_g}, W_g^{\infty}, \mathbb{C})^G$$

so if $\deg(\phi) = p$ in the usual grading on $\mathscr{Q}_{W_g} d\mathbf{x}_{\mathrm{Fix}(g)}$, then the bidegree of $\lceil \phi\,; g\rfloor$ is

$$(\deg_+(\lceil \phi\,; g\rfloor), \deg_-(\lceil \phi\,; g\rfloor)) = (N_g - p - q + \mathrm{age}(g), p - q + \mathrm{age}(g)),$$

and we have

$$\deg(\lceil \phi\,; g\rfloor) = \deg_+(\lceil \phi\,; g\rfloor) + \deg_-(\lceil \phi\,; g\rfloor).$$

Finally, we note that it is often useful to define the complex degree to be

$$\deg_{\mathbb{C}}(\lceil \phi\,; g\rfloor) = \frac{1}{2}\deg(\lceil \phi\,; g\rfloor).$$

Remark 3.16. The following are standard facts from classical singularity theory and classical geometry:

(1) $\mathscr{Q}_W$ is finite dimensional if and only if W has an isolated singularity at the origin.

(2) $\mathscr{Q}_{f_1} = \mathscr{Q}_{f_2}$ if and only if $f_1 = f_2 \circ h$ for some analytic isomorphism $h : \mathbb{C}^N \to \mathbb{C}^N$.

(3) $\dim \mathscr{Q}_W$ is determined solely by the weights $q_i = \frac{c_i}{d}$. In fact

$$\mu := \dim \mathscr{Q}_W = \prod \left(\frac{1}{q_i} - 1\right).$$

(4) The monomials of highest degree in $\mathcal{Q}_W$ have degree $\hat{c} = N - 2q = \sum_{i=1}^{N}(1-2q_i)$. This number is called the *central charge* of W. The subspace of $\mathcal{Q}_W$ of complex degree $\hat{c}$ is one-dimensional and is generated by $\mathrm{Hess}(W) = \det\left[\frac{\partial^2 W}{\partial x_i \partial x_j}\right]$.

(5) The hypersurface $X_W = \{W = 0\}$ in weighted projective space is Calabi-Yau, if and only if $\hat{c} = N - 2$. This condition on the central charge (and therefore also on the weights) is called the *Calabi-Yau condition*.

Example 3.17. Consider $W = x^3 + y^3 + yz^2$ as in Example 3.9 with the group $G = \langle\gamma\rangle$, where $\gamma = (e^{2\pi i/3}, e^{2\pi i/3}, e^{2\pi i \cdot 5/6})$. Note that in this case γ has order 6, and we see that $J = \gamma^4$.

We can find a basis for the state space $\mathscr{H}_{W,G} = \bigoplus_g \mathscr{H}_g^G$, in the following way.

For $g \in \gamma, \gamma^2, \gamma^4, \gamma^5$ we have $I_g = \emptyset$ and

$$\mathscr{H}_g = \mathrm{Span}\{\lceil 1\,;g\rfloor\} \cong \mathbb{C}.$$

These are the narrow sectors.

Since the usual grading of $1 \in \mathbb{C}$ is 0, and since $N_g = 0$ and $q = 1$, we have

$$(\deg_+ \lceil 1\,;g\rfloor, \deg_- \lceil 1\,;g\rfloor) = (0 - 0 - 1 + \mathrm{age}(g), 0 - 1 + \mathrm{age}(g))$$

From this, we easily compute

$$\begin{aligned}
(\deg_+ \lceil 1\,;\gamma\rfloor, \deg_- \lceil 1\,;\gamma\rfloor) &= (1/2, 1/2)\\
(\deg_+ \lceil 1\,;\gamma^2\rfloor, \deg_- \lceil 1\,;\gamma^2\rfloor) &= (1, 1)\\
(\deg_+ \lceil 1\,;\gamma^4\rfloor, \deg_- \lceil 1\,;\gamma^4\rfloor) &= (0, 0)\\
(\deg_+ \lceil 1\,;\gamma^5\rfloor, \deg_- \lceil 1\,;\gamma^5\rfloor) &= (1/2, 1/2)
\end{aligned}$$

For $g = \gamma^3 = (1, 1, -1)$ we are interested in those $x^i y^j dx \wedge dy$ in $\mathcal{Q}_g dx \wedge dy$ that are fixed by G, that is, where

$$\gamma \cdot x^i y^j dx \wedge dy = e^{2\pi i \cdot \frac{i+j+1+1}{3}} x^i y^j dx \wedge dy = x^i y^j dx \wedge dy.$$

So, $\mathscr{H}_g = \mathrm{Span}\{\lceil x\,;g\rfloor, \lceil y\,;g\rfloor\} \cong \mathbb{C}^2$. This is a broad sector.

Since the usual grading of both $x\,dx \wedge dy$ and $y\,dx \wedge dy$ in $\mathcal{Q}_{W_g} dx \wedge dy$ is 1, and since $N_g = 2$, $\mathrm{age}(\gamma^3) = 1/2$, and $q = 1$, we have

$$(\deg_+ \lceil x\,;\gamma^3\rfloor, \deg_- \lceil x\,;\gamma^3\rfloor) = (\deg_+ \lceil y\,;\gamma^3\rfloor, \deg_- \lceil y\,;\gamma^3\rfloor) = (1/2, 1/2).$$

Finally, for $g = \gamma^0 = (1, 1, 1)$, it is easy to see that none of the monomials $x^i y^j z^k dx \wedge dy \wedge dz$ in $\mathcal{Q}_g d\mathbf{x}$ are fixed by G, so $\mathscr{H}_g = \emptyset$, and

$$\mathscr{H}_{W,G} = \mathrm{Span}\left\{\lceil 1\,;\gamma\rfloor, \lceil 1\,;\gamma^2\rfloor, \lceil x\,;\gamma^3\rfloor, \lceil y\,;\gamma^3\rfloor, \lceil 1\,;\gamma^4\rfloor, \lceil 1\,;\gamma^5\rfloor\right\}.$$

Although our definition of the state space depends *a priori* on the polynomial W, as a bigraded vector space the state space really only depends on the weights $q_1, \ldots, q_N$ and the group G (thought of as a subgroup of $\mathrm{GL}(N)$).

Theorem 3.18. *As a bigraded vector space, the state space is determined only by the weights $q_1, \ldots, q_N$ and by the action of the group G on $\mathbb{C}^N$.*

Proof. If there is a continuous family of G-invariant, quasihomogeneous polynomials $W_t : \mathbb{C}^N \to \mathbb{C}$, all with the same weights $q_1, \ldots, q_N$, then for each t_1, t_2 we have $H^\bullet(\mathbb{C}^N, W_{t_1}^\infty, \mathbb{C})^G \cong H^\bullet(\mathbb{C}^N, W_{t_2}^\infty, \mathbb{C})^G$.

The result now follows from the fact that the space of G-invariant admissible polynomials with a given set of weights $q_1, \ldots, q_N$ is path connected. This can be seen from the fact that the locus of polynomials with singularities away from the origin is a subvariety of complex codimension at least 1 in the space of G-invariant quasihomogeneous polynomials. Q.E.D.

Remark 3.19. As an alternative (algebraic) proof to the previous theorem, one can consider the representation ring $\mathrm{Rep}(G)$ of G and the $\mathrm{Rep}(G)$-valued Poincaré polynomial $P(t)$ of the G-module $\Omega^N/(dW \wedge \Omega^{N-1})$. The Koszul resolution gives us

$$P(t) = \prod_{i=1}^{N} \frac{\rho_i - t^{d-q_i}}{1 - \rho_i t^{q_i}},$$

where each ρ_i is the representation of G on the span of the variable x_i. Since this expression for $P(t)$ is obviously independent of W, the result follows. The details are given in [Tay13].

3.3. Pairing

Next we define a nondegenerate pairing, as required for the construction of our CohFT. If $W^{-\infty}$ is defined to be

$$W^{-\infty} = (\mathfrak{Re}(W))^{-1}(-\infty, -M) \text{ for } M >> 0,$$

then there is a natural perfect pairing

$$\langle\ ,\rangle : H_N(\mathbb{C}^N, W^{-\infty}, \mathbb{Z}) \otimes H_N(\mathbb{C}^N, W^{\infty}, \mathbb{Z}) \to \mathbb{Z}$$

defined by intersecting the relative homology cycles. But we need a pairing on $H^N(\mathbb{C}^N, W^\infty, \mathbb{C})$.

Choose any ξ such that $\xi^d = -1$. Multiplication by the diagonal matrix $(\xi^{n_1}, \ldots, \xi^{n_N})$ defines a map $I : \mathbb{C}^N \longrightarrow \mathbb{C}^N$ sending $W^{\infty} \longrightarrow W^{-\infty}$. Hence, it induces an isomorphism

$$I_* : H_N(\mathbb{C}^N, W^{\infty}, \mathbb{C}) \longrightarrow H_N(\mathbb{C}^N, W^{-\infty}, \mathbb{C}).$$

We define a pairing on $H_N(\mathbb{C}^N, W^{\infty}, \mathbb{Z})$ by

$$\langle \Delta_i, \Delta_j \rangle = \langle \Delta_i, I_*(\Delta_j) \rangle.$$

This induces a pairing (denoted by $\langle\ ,\ \rangle$) on the dual space $H^N(\mathbb{C}^N, W^{\infty}, \mathbb{C})$. Changing the choice of ξ will change the isomorphism I by an element of the group $\langle J \rangle$, and $I^2 \in \langle J \rangle$. Therefore, the pairing is independent of the choice of I on the invariant subspace $H^N(\mathbb{C}^N, W^{\infty}, \mathbb{C})^{\langle J \rangle}$ or on $H^N(\mathbb{C}^N, W^{\infty}, \mathbb{C})^G$ for any admissible group G.

Carefully tracking this pairing through Wall's isomorphism shows that it is equivalent to the residue pairing $\langle\ ,\ \rangle_W$ on the Milnor ring, which can be computed by the following equation:

$$f_1 \cdot f_2 = \frac{\langle f_1, f_2 \rangle_W}{\mathrm{Hess}(W)} + \text{lower order terms}.$$

Now we define the pairing on the orbifolded state space $\mathscr{H}_{W,G} = \bigoplus_g \mathscr{H}_g^G$ by noting that there is a natural isomorphism $\beta_g : \mathscr{H}_g \cong \mathscr{H}_{g^{-1}}$, so we may define $\langle\ ,\ \rangle : \mathscr{H}_{W,G} \times \mathscr{H}_{W,G} \to \mathbb{C}$ by

$$\langle \lceil f_1 ; g_1 \rfloor , \lceil f_2 ; g_2 \rfloor \rangle = \begin{cases} \langle f_1, \beta_{g_1}^{-1} f_2 \rangle_{W_{g_1}} & g_1 = g_2^{-1} \\ 0 & g_1 \neq g_2^{-1}, \end{cases}$$

and extending linearly.

Note that the pairing has bigraded degree $(\hat{c}, \hat{c})$, so that for any p_1, p_2 we have:

$$\langle\ ,\ \rangle : \mathscr{H}_{W,G}^{\hat{c}-p_1, \hat{c}-p_2} \otimes \mathscr{H}_{W,G}^{p_1,p_2} \to \mathbb{C}.$$

3.4. The Moduli Space

A key part of FJRW theory is the stack $\mathscr{W}_{g,k}^{W,G}$ of pointed stable W-curves, which has a finite morphism to $\overline{\mathscr{M}}_{g,k}$. We define a virtual cycle on $\mathscr{W}_{g,k}^{W,G}$, and the CohFT classes $\Lambda_{g,k}^{W,G}$ will be defined as the pushforward to $\overline{\mathscr{M}}_{g,k}$ of the Poincaré dual of the virtual cycle.

The original definition of the moduli problem defining $\mathscr{W}_{g,k}^{W,G}$ was given in terms of a collection of line bundles satisfying certain conditions related to the *log-canonical bundle* $\omega_{\log,\mathscr{C}} = \omega_{\mathscr{C}}(\sum_{i=1}^N p_i)$, where p_i is the ith marked point on the curve. Roughly speaking $\mathscr{W}_{g,k}^{W,G_{\max}}$ is the stack of k-pointed, stable orbicurves* with one orbifold line bundle $\mathscr{L}_i$ for each variable x_i with $i \in \{1, \ldots, N\}$. These line bundles must satisfy the property that for each monomial W_j of W, we have

(1) $$W_j(\mathscr{L}_1, \ldots, \mathscr{L}_N) \cong \omega_{\log,\mathscr{C}}.$$

For more general admissible subgroups $G \subset G_{\max}$ the stack $\mathscr{W}_{g,k}^{W,G}$ consists of those pointed stable W-curves where the local group action at a marked point p_i on the fiber of $\mathscr{L}_1 \oplus \cdots \oplus \mathscr{L}_N$ factors through G. But this can be described more cleanly using a reformulation in terms of principal bundles due to Polishchuk-Vaintrob (see [PV16, §3.2]), as we now explain.

We will use the integer weights $c_1, \ldots, c_N, d$ for W with each $q_i = c_i/d$, as described in Remark 3.2. Let $G \subset \mathrm{Aut}(W)$ be an admissible group, and let Γ be the subgroup of $(\mathbb{C}^*)^N$ generated by G and $\mathbb{C}^* = \{(\lambda^{c_1}, \ldots, \lambda^{c_N}) \mid \lambda \in \mathbb{C}^*\}$, where this second factor $\mathbb{C}^*$ corresponds to the quasihomogeneity of W. It is easy to see that

(2) $$G \cap \mathbb{C}^* = \langle J \rangle.$$

We define a surjective homomorphism

$$\zeta : \Gamma \to \mathbb{C}^*$$

by sending G to 1 and $(\lambda^{c_1}, \ldots, \lambda^{c_N})$ to λ^d. Equation (2) shows that the map ζ is well-defined and that $\ker(\zeta) = G$. Let $\mathring{\omega}_{\log,\mathscr{C}}$ denote the principal $\mathbb{C}^*$-bundle associated to $\omega_{\log,\mathscr{C}}$.

Definition 3.20. A Γ-structure on an orbicurve $\mathscr{C}$ is

(1) A principal Γ-bundle $\mathscr{P}$ on $\mathscr{C}$ such that the corresponding map $\mathscr{C} \to B\Gamma$ to the classifying stack $B\Gamma = [pt/\Gamma]$ is representable.
(2) A choice of isomorphism $\varkappa : (\zeta)_*\mathscr{P} \cong \mathring{\omega}_{\log,\mathscr{C}}$. Here $(\zeta)_*\mathscr{P}$ denotes the principal $\mathbb{C}^*$ bundle on $\mathscr{C}$ induced from $\mathscr{P}$ by the map ζ.

An equivalent way to state (2) from the Definition 3.20 is to recognize that the homomorphism ζ induces a morphism of stacks $B\zeta : B\Gamma \to B\mathbb{C}^*$ and (2) is equivalent to the requirement that the composition $B\zeta \circ \mathscr{P} : \mathscr{C} \to B\mathbb{C}^*$ be equal to the morphism of stacks $\mathscr{C} \to B\mathbb{C}^*$ induced by the principal $\mathbb{C}^*$-bundle $\mathring{\omega}_{\log,\mathscr{C}}$.

*This is a very rough sketch of the ideas. For a more thorough discussion of orbicurves and orbifold line bundles, see [FJR13, §2.1]

Given a Γ-structure on $\mathscr{C}$, the projection $\pi_i : \Gamma \subseteq (\mathbb{C}^*)^N \to \mathbb{C}^*$ to the ith factor for each $i \in \{1, \dots, N\}$ defines a collection of line bundles

$$\mathscr{L}_1, \dots, \mathscr{L}_N.$$

It is easy to check that $\pi_i^d = \zeta^{c_i}$, and thus we have $\mathscr{L}_i^{\otimes d} = \omega_{\mathscr{C},\log}^{\otimes c_i}$.

Let $W = \sum_j \alpha_j W_j$, where each W_j is a monomial and $\alpha_j \in \mathbb{C}$ is a coefficient. The monomial W_j induces a homomorphism $(\mathbb{C}^*)^N \to \mathbb{C}^*$. Because W is G-invariant, we have $W_j|_G = 1$. Therefore, W_j defines a homomorphism $W_j : \Gamma/G \to \mathbb{C}^*$.

A straightforward check of how W_j acts on the subgroup $\mathbb{C}^* = \{(\lambda^{c_1}, \dots, \lambda^{c_N})\}$ shows that this homomorphism is an isomorphism. Hence we have $W_j(\pi_1, \dots, \pi_N) = \zeta$. This implies that for each monomial W_j Equation (1) holds.

Therefore, associated to each Γ-structure on a pointed stable orbicurve $\mathscr{C}$, there is a collection of line bundles $\mathscr{L}_1, \dots, \mathscr{L}_N$ on $\mathscr{C}$ and isomorphisms as in Equation (1).

Definition 3.21. Let $\mathscr{W}_{g,k}^{W,G}$ be the stack of tuples $(\mathscr{C}, p_1, \dots, p_k, \mathscr{P}, \varkappa)$, where $\mathscr{C}, p_1, \dots, p_k$ is a stable k-pointed orbicurve of genus g and $(\mathscr{P}, \varkappa)$ is a Γ-structure on $\mathscr{C}$.

Remark 3.22. As with the state space (see Theorem 3.18), it turns out the stack $\mathscr{W}_{g,k}^{W,G}$ depends only on g, k, G, and Γ and is completely independent of the polynomial W.

We have several natural morphisms connected with $\mathscr{W}_{g,k}^{W,G}$. The first of these is the forgetful morphism $st : \mathscr{W}_{g,k}^{W,G} \to \overline{\mathscr{M}}_{g,k}$ (analogous to the stabilization morphism of Gromov-Witten theory). The following result is fundamental.

Theorem 3.23 ([FJR13, Thm 2.2.6]). *If $J \in G \subseteq G_{\max}$, then for any nonnegative integers g and k with $2g - 2 + k > 0$, the morphism $st : \mathscr{W}_{g,k}^{W,G} \to \overline{\mathscr{M}}_{g,k}$ is proper and quasifinite (but not representable).*

Example 3.24. If $g = 0$ and $k = 1$, then for any admissible group G, for each choice of three-pointed genus-zero orbifold curve, there is at most one Γ-structure on that curve. Moreover, the automorphisms of the Γ-structure are exactly G. So for any particular choice of $\boldsymbol{\gamma} = (\gamma_1, \gamma_2, \gamma_3)$ the stack $\mathscr{W}_{0,3}^{W,G}(\boldsymbol{\gamma})$ is either empty (see the selection rule below in Equation (3)) or is isomorphic to BG. Since $\overline{\mathscr{M}}_{0,3}$ is a point, this illustrates that the stabilization morphism is not representable.

Remark 3.25. If the grading element J is not contained in G, it is not hard to show that the stack $\mathscr{W}_{g,k}^{W,G}$ is not proper for general g and k, hence the requirement that G be admissible.

We also have an analogue of the evaluation morphism. Let G_{p_i} be the local group of $\mathscr{C}$ at the marked point p_i. Since we are working over $\mathbb{C}$, it has a canonical generator. Let γ_{p_i} denote this canonical generator. The morphism $\mathscr{P} : \mathscr{C} \to B\Gamma$ implies that each G_{p_i} has a homomorphism to Γ. The fact that $\omega_{\mathscr{C},\log}$ has no orbifold structure at any marked point implies that G_{p_i} actually maps to $\ker(\zeta) = G \subset \Gamma$. And representability of the morphism $\mathscr{P} : \mathscr{C} \to B\Gamma$ implies that the map $G_{p_i} \to \ker(\zeta) = G$ is injective.

Thus for each $i \in \{1, \ldots, k\}$ we have an "evaluation" map ev_i mapping $\mathscr{W}^{W,G}_{g,k}$ to G, or rather to the inertia stack $I[\mathbb{C}^N/G] = [\mathbb{C}^N/G] \sqcup \coprod_{\gamma \neq 1 \in G} BG$. This map is given by sending $(\mathscr{C}, p_1, \ldots, p_k, \mathscr{P}, \varkappa)$ to $[0/G] \subset [\mathbb{C}^N/G]$ if $\gamma_{p_i} = 1$, and otherwise sending it to the copy of BG corresponding to the image of γ_{p_i} in G.

This gives us a decomposition of $\mathscr{W}^{W,G}_{g,k}$ into open and closed substacks

$$\mathscr{W}^{W,G}_{g,k} = \coprod_{\gamma_1, \ldots, \gamma_k} \mathscr{W}^{W,G}_{g,k}(\gamma_1, \ldots, \gamma_k),$$

where each $\mathscr{W}^{W,G}_{g,k}(\gamma_1, \ldots, \gamma_k)$ is the locus where the ith evaluation map lies in the copy of BG corresponding to the element $\gamma_i \in G$.

The decomposition into these substacks described above corresponds to the monodromies of the line bundles at the orbifold markings. In fact, from this description, we get a selection rule, namely, the substack $\mathscr{W}^{W,G}_{g,k}(\gamma_1, \ldots, \gamma_k)$ is empty, unless

$$\gamma_1 \cdots \gamma_k = J^{2g-2+n}. \tag{3}$$

Finally we have a "forgetting tails" morphism. If the canonical generator of G_{p_i} maps to $J \in G$, then it is not hard to show that we can "desingularize" the line bundles $(\zeta)_* \mathscr{P}$ and $\mathring{\omega}_{\log,\mathscr{C}}$ by forgetting the orbifold structure at p_i, and in a neighborhood of p_i the new line bundles are isomorphic to $\mathring{\omega}_{\mathscr{C}}$ (see [FJR13, §2.2.3] for details). Thus, if we start with a $k+1$-pointed curve marked with J at the ith point, then forgetting the marked point p_i means the homomorphism $\varkappa$ maps $\mathscr{P}$ to the log-canonical bundle of the corresponding k-pointed curve. This gives us a morphism

$$\mathscr{W}^{W,G}_{g,k+1}(\gamma_1, \ldots, J, \ldots, \gamma_k) \to \mathscr{W}^{W,G}_{g,k}(\gamma_1, \ldots, \gamma_k),$$

which we call *forgetting tails*. Note that we can only forget tails which are marked with the element J. Forgetting any other tail will fail to produce a Γ-structure on the resulting curve.

3.5. Virtual cycle

The virtual cycle $\left[\mathscr{W}_{g.k}^{W,G}(\gamma_1,\ldots,\gamma_k)\right]^{virt}$ of FJRW theory is a homology class on $\mathscr{W}_{g,k}^{W,G}(\gamma_1,\ldots,\gamma_k)$ after being capped with elements of the state space. This cycle satisfies certain key properties or "axioms," which we will describe below.

The original idea for the virtual class in the case where $W = x^r$ and $G = \mu_r$ (r-spin curves) is due to Witten (see [Wit93a]). An algebraic construction for r-spin curves was given by Polishchuk-Vaintrob (see [PV01]) and a K-theoretic construction given by Chiodo (see [Chi06a, Chi06b]). An analytic construction for the r-spin virtual class based on Witten's original outline was given by T. Mochizuki (see [Moc06]).

For an arbitrary admissible polynomial W and group G, an analytic construction of the virtual class was given in [FJR07] using the first-order elliptic PDE

$$\bar{\partial} s_i + \overline{\frac{\partial W}{\bar{\partial} s_i}} = 0, \tag{4}$$

called the *Witten Equation*, where each s_i is a C^∞-section of the ith line bundle $\mathscr{L}_i$ of the corresponding Γ-structure.

An algebraic definition of a virtual class was given by Polishchuk-Vaintrob in [PV16] using graded matrix factorizations. And more recently Chang-Li-Li (see [CLL15]) used the cosection localization technique of Chang-Kiem-Li (see [KL13, CL12]) to construct a virtual cycle on the stacks $\mathscr{W}_{g,k}^{W,G}(\gamma_1,\ldots,\gamma_k)$ when each γ_i is narrow (see Remark 3.12). They prove furthermore that their construction agrees with both the analytic construction of Fan-Jarvis-Ruan and the algebraic construction of Polishchuk-Vaintrob in the narrow case.

Polishchuk-Vaintrob showed that the algebraic and analytic virtual cycles agree for all simple (ADE) singularities (these are polynomials with central charge $\hat{c} < 1$). J. Guéré has shown that for most invertible polynomials with the maximal group $G_{\max}$ all these constructions of the virtual class coincide, up to a rescaling of the broad sectors (see [Gué16, Thm 3.25]). Unfortunately, it is not yet known whether the algebraic and analytic constructions agree with broad insertions present with arbitrary groups and arbitrary (noninvertible) polynomials.

In what follows, we will focus on the analytic virtual cycle of [FJR07].

As mentioned above, the virtual cycle $\left[\mathscr{W}_{g.k}^{W,G}(\gamma_1,\ldots,\gamma_k)\right]^{virt}$ of FJRW theory is a homology class on $\mathscr{W}_{g,k}^{W,G}(\gamma_1,\ldots,\gamma_k)$ after being capped with elements of the state space. Pushing the homology class down to $\overline{\mathscr{M}}_{g,k}$ gives the FJRW CohFT, as in the following definition. A more complete treatment can be found in the original articles [FJR07, FJR13].

Definition 3.26. For each $i \in \{1, \ldots, k\}$ let $\alpha_i \in \mathscr{H}_{\gamma_i}$. We define $\Lambda^{W,G}_{g,k}(\alpha_1, \ldots, \alpha_k) \in H^*(\overline{\mathscr{M}}_{g,k})$ as

$$(5)\quad \Lambda^{W,G}_{g,k}(\alpha_1, \ldots, \alpha_k) = \frac{|G|^g}{\deg st} PDst_* \left(\left[\mathscr{W}^{W.G}_{g,k}(\gamma_1, \ldots, \gamma_k) \right]^{virt} \cap \prod_{i=1}^{k} \alpha_i \right).$$

This definition is extended linearly to all choices of k-tuples $\alpha_1, \ldots, \alpha_k \in \mathscr{H}_{W,G}$.

Theorem 3.27 ([FJR13]). *The collection of cohomology classes given by $\{\Lambda^{W,G}_{g,k}\}$ as defined in Equation (5) gives a CohFT with flat identity element equal to $\mathbf{1} \in \mathscr{H}_J$, and satisfies the following additional properties:*

(1) ***Dimension:*** *If $\alpha_i \in \mathscr{H}_{\gamma_i}$ for each $i \in \{1, \ldots, k\}$, then the dimension of $\Lambda^{W,G}_{g,k}(\alpha_1, \ldots, \alpha_k)$ is given by*

$$dim \left(\Lambda^{W,G}_{g,k}(\alpha_1, \ldots, \alpha_k) \right) = (\hat{c} - 3)(1 - g) + k - \sum_{i=1}^{k} \deg_{\mathbb{C}} \alpha_i.$$

(2) ***Integral Degrees:*** *If $\alpha_i \in \mathscr{H}_{\gamma_i}$ for each $i \in \{1, \ldots, k\}$, and if the degree of the desingularization $|\mathscr{L}_i|$ of each line bundle $\mathscr{L}_i$ defined by the Γ-structure on $\mathscr{W}^{W,G}_{g,k}(\gamma_1, \ldots, \gamma_k)$ is not integral, then $\Lambda^{W,G}_{g,k}(\alpha_1, \ldots, \alpha_k) = 0$.*

(3) ***$G_{\max}$ Invariance:*** *Letting the group $G_{\max}$ act on $\mathscr{H}_{W,G}$ in the obvious way, the class $\Lambda^{W,G}_{g,k}(\alpha_1, \ldots, \alpha_k)$ is $G_{\max}$-invariant.*

(4) ***Disjoint Sums:*** *If $W_1 \in \mathbb{C}[x_1, \ldots, x_N]$ and $W_2 \in \mathbb{C}[y_1, \ldots, y_m]$ are each nondegenerate quasihomogeneous polynomials with G_1 admissible for W_1 and G_2 admissible for W_2, and if $\alpha_1, \ldots, \alpha_k \in \mathscr{H}_{W_1,G_1}$ and $\beta_1, \ldots, \beta_k \in \mathscr{H}_{W_2,G_2}$, then $\mathscr{H}_{W_1+W_2,G_1\times G_2} = \mathscr{H}_{W_1,G_1} \otimes \mathscr{H}_{W_2,G_2}$ and*

$$\begin{aligned} &\Lambda^{W_1+W_2,G_1\times G_2}(\alpha_1 \otimes \beta_1, \ldots, \alpha_k \otimes \beta_k) \\ &\quad = \Lambda^{W_1,G_1}_{g,k}(\alpha_1, \ldots, \alpha_k)\Lambda^{W_2,G_2}_{g,k}(\beta_1, \ldots, \beta_k). \end{aligned}$$

(5) ***Deformation Invariance:*** *Let W_t be a one-parameter family (one real parameter) of non-degenerate polynomials of a given weight $(q_1, \ldots, q_N)$, and let G be a symmetry group which is admissible for each W_t. The family W_t defines an isomorphism of the state spaces $\mathscr{H}_{W_t,G}$ to $\mathscr{H}_{W_{t_0},G}$ for any t_0. If $\alpha_1, \ldots, \alpha_k \in \mathscr{H}_{W_{t_0},G}$, then (using the corresponding family of isomorphisms) the class $\Lambda^{W_t,G}_{g,k}(\alpha_1, \ldots, \alpha_k)$ is independent of t.*

(6) ***Topological Euler Class:*** *Let* $\mathscr{L}_1, \ldots, \mathscr{L}_N$ *be the line bundles induced by the universal* Γ*-structure on the universal curve* $\pi : \mathscr{C} \to \mathscr{W}^{W,G}_{g,k}(\gamma_1, \ldots, \gamma_k)$.

If each γ_i *is narrow (i.e., the fixed locus of* γ_i *is* $0 \in \mathbb{C}^N$*) for each* $i \in \{1, \ldots, k\}$*, then we can define a particular map (the* Witten map*)*

$$\mathscr{D} : \pi_*(\mathscr{L}_1 \oplus \cdots \oplus \mathscr{L}_N) \to R^1\pi_*(\mathscr{L}_1 \oplus \cdots \oplus \mathscr{L}_N)$$

in the derived category of $\mathscr{W}^{W,G}_{g,k}(\gamma_1, \ldots, \gamma_k)$ *such that the virtual class* $\left[\mathscr{W}^{W,G}_{g,k}(\gamma_1, \ldots, \gamma_k)\right]^{virt}$ *is Poincaré dual to a topological Euler class for* $\mathscr{D}$*. This implies the following additional properties hold:*

(a) **Concavity:** *Suppose that each* γ_i *is narrow.*
If $\pi_*\left(\bigoplus_{i=1}^N \mathscr{L}_i\right) = 0$*, then the virtual cycle is given by capping the top Chern class of* $\left(R^1\pi_*\left(\bigoplus_{i=1}^N \mathscr{L}_i\right)\right)^*$ *with the usual fundamental cycle of the moduli space:*

$$\left[\mathscr{W}^{W,G}_{g,k}(\gamma_1, \ldots, \gamma_k)\right]^{vir} = c_{top}\left(\left(R^1\pi_* \bigoplus_{i=1}^N \mathscr{L}_i\right)^*\right) \cap \left[\mathscr{W}^{W,G}_{g,k}(\gamma_1, \ldots, \gamma_k)\right].$$

(b) **Index zero:** *Suppose that* $\dim(\mathscr{W}^{W,G}_{g,k}(\gamma_1, \ldots, \gamma_k)) = 0$ *and each* γ_i *is narrow.*
If the pushforwards $\pi_*\left(\bigoplus_{i=1}^N \mathscr{L}_i\right)$ *and* $R^1\pi_*\left(\bigoplus_{i=1}^N \mathscr{L}_i\right)$ *are both vector bundles of the same rank, then the virtual cycle is just the degree of the Witten map times the fundamental cycle.*

In addition to the selection rule in Equation (3), the dimension, integral degrees, and $G_{\max}$ invariance properties all give selection rules that force many of these classes to vanish. Another main tool for computing these classes is the Concavity Axiom, combined with reconstruction rules arising from the WDVV equation (see [Fra15] for many examples of how this works).

One can also integrate tautological classes against the virtual class to produce numbers, which are called FJRW invariants. In some cases, these can be computed even when the virtual class itself cannot. J. Guéré has also been able to use the Polishchuk-Vaintrob matrix factorization approach to compute all the invariants for any chain polynomial with group $G_{\max}$, even including invariants with broad insertions (see [Gué16]).

§4. Mirror Symmetry

We will now discuss three variations of mirror symmetry that involve FJRW theory, and we will also also attempt to make explicit what is known about the various arrows in Diagram 1. However, the body of knowledge is growing rapidly, and is in some ways not always conforming to the diagram.

4.1. BHHK Mirror Construction

First we describe a construction that provides a conjectural "mirror pair" to the pair (W, G). For each invertible polynomial W and each group $G \subseteq G_{\max,W}$ of diagonal symmetries of W, the Berglund-Hübsch-Henningson-Krawitz (BHHK) mirror construction produces a "transpose" polynomial W^T and a "transpose" group G^T, where W^T is again an invertible polynomial, and G^T is a diagonal group of symmetries of W^T.

The construction of the transpose is very simple (and justifies the use of the term "transpose"). First, since the polynomial W is invertible, we may rescale the variables so that W can be written as a sum of monomials all with coefficient 1.

$$W = \sum_{i=1}^{N} \prod_{j=1}^{N} x_j^{a_{ij}}.$$

Let the matrix $E = (a_{ij})$ have the exponents of the various monomials for its entries. It is not hard to show that this matrix is invertible when the polynomial W is invertible. The polynomial W^T is defined as the polynomial corresponding to the exponent matrix E^T:

$$W^T = \sum_{i=1}^{N} \prod_{j=1}^{N} x_j^{a_{ji}}.$$

It can easily be checked that this polynomial is also admissible and invertible.

To define the transpose group, note first that the group $G_{\max}$ is isomorphic to the additive group

$$\left\{\mathbf{g} = (g_1, \ldots, g_N)^T \in (\mathbb{Q}/\mathbb{Z})^N \mid E\mathbf{g} \in \mathbb{Z}^N\right\},$$

and we may write any subgroup G in additive notation. In this additive notation, the "transpose" group G^T is the group

$$G^T = \{\mathbf{h} \in \mid \mathbf{h}^T E\mathbf{g} \in \mathbb{Z}, \forall \mathbf{g} \in G\}.$$

In this definition, we think of $h \in G_{\max,W^T}$ written additively as an element of $(\mathbb{Q}/\mathbb{Z})^N$, as above.

The following properties of G^T are easy to verify:

(1) $(G^T)^T = G$.
(2) $(G_{\max,W})^T = \{0\}$.
(3) $G \subseteq G_{\max,W}$ contains J if and only if $G^T \subset G_{\max,W^T} \cap \mathrm{SL}(N)$.
(4) For any subgroups $G_1 \subseteq G_2 \subseteq G_{\max,W}$ we have $G_2^T \subseteq G_1^T$ and

$$G_2/G_1 \cong G_1^T/G_2^T.$$

A guiding conjecture for Landau-Ginzburg mirror symmetry is the following (see [Kra10]).

Conjecture 4.1 (Landau-Ginzburg Mirror Conjecture). *If W is an invertible polynomial and G is an admissible group of symmetries of W, then there should be a B-model associated to W^T and G^T such that the FJRW theory (LG A-model) of (W, G) is isomorphic to the B model of (W^T, G^T).*

As mentioned previously, an important part of this conjecture is to define an appropriate B-model. Different choices of B-model, or surrogates for the B-model, give different forms of mirror symmetry.

4.2. *I*- and *J*-functions

In this first formulation of mirror symmetry, we replace the Landau-Ginzburg B model in Diagram 1 with a Calabi-Yau B model; Chiodo-Ruan referred to this as LG to CY mirror symmetry. In fact, this formulation fits into a *global* mirror symmetry framework, namely, there is a global B model, and mirror symmetry happens at two special points of the global model—the *large structure limit*, mirror to the Calabi-Yau A-model (GW theory) and the *Gepner point*, mirror to the LG A-model (FJRW theory); see [CR11a] for more details. This formulation of mirror symmetry is very closely connected with the LG/CY correspondence (see also Section 5).

This version of mirror symmetry was inspired by the work of Greene-Plesser (see [GR90]), Candelas-de la Ossa-Green-Parks (see [CdlOGP92]), Givental (see [Giv96]), and Lian-Liu-Yau (see [LLY99]), where mirror symmetry was first demonstrated and then proved for the quintic threefold. This should be considered as the Calabi-Yau side. The idea is as follows. On the A side, Givental defined his J-function, which is a generating function for the genus-zero GW invariants of the quintic (corresponding to a quintic polynomial W with a group of symmetries $G = \mathbb{Z}/5$). On the B side there is a family of Calabi-Yau manifolds (called the mirror quintic), corresponding to a one–parameter deformation of the BHHK mirror (W^T, G^T).

The corresponding B model is given by the period integrals for some particularly chosen three-forms over paths around a special point in the parameter space called the *large structure limit*. These period integrals, when evaluated, yield Givental's I-function. This I-function satisfies the Picard-Fuchs differential equations coming from the variation of Hodge structure on this family in a neighborhood of the special point.

The original formulation of mirror symmetry for the quintic threefold can, in this case, be formulated as a change of variables (the mirror map) so that the I-function on the mirror family is identified with the J-function on the quintic. This formulation can be generalized to other hypersurfaces in weighted projective space.

FJRW theory comes into play at the *Gepner point* of the global B-model. Following the ideas just outlined, on the A side we can consider a J-function as the generating function of the genus zero FJRW invariants for a pair (W, G). On the B side, we consider a family of Calabi-Yau manifolds determined by the pair (W, G). The I-function is constructed as period integrals of certain three-forms on this family of Calabi-Yau manifolds. This is again a full set of solutions to the Picard-Fuchs equation in a neighborhood of the Gepner point. The mirror conjecture in this setting is that there is an explicit change of variables (a "mirror map") which identifies the I-function with the J-function.

Chiodo-Ruan first proved this form of mirror symmetry for the pair consisting of the Fermat quintic polynomial and the group generated by the exponential grading operator J (see [CR10]). This example exactly corresponds to the mirror symmetry for the quintic threefold described above. Chiodo-Iritani-Ruan then extended this result to any Fermat polynomial satisfying the CY condition with symmetry group generated by the exponential grading operator J (see [CIR14]).

Several others have also identified the FJRW I- and J-functions for various cases of polynomials and groups as a step on the way to proving various cases of the LG/CY correspondence. We will list these in the next section when we discuss the LG/CY correspondence. In fact, this is currently the most common approach to verifying specific cases of the LG/CY correspondence.

Most of the cases mentioned in the next section consider only polynomials satisfying the condition that each weight c_i divides the degree d. This implies that the polynomials can be taken as Fermat polynomials. Because the CohFT then satisfies condition 6a of Theorem 3.27 in this case, we refer to this case as the *concave* case. When the Calabi–Yau condition is satisfied, this corresponds with the so–called *Gorenstein condition*—that is that each weight c_i divides the sum of the weights. In this case, the ambient weighted projective stack on the CY side is Gorenstein (as a Deligne-Mumford stack).

Apart from the concave case, J. Guéré has used this same method to prove that this form of mirror symmetry (again in genus zero) holds for all pairs consisting of a chain-type polynomial and symmetry group $G_{\max}$ (see [Gué16, Thm 0.5]). The concavity condition is not satisfied in general for such polynomials.

Finally, we note that this form of mirror symmetry can be extended beyond the Calabi-Yau case, as was done by P. Acosta in [Aco14], where he proved that the J-function was determined uniquely by the I-function and its derivatives.

4.3. Mirror Symmetry with Saito-Givental

In this second formulation of mirror symmetry, which is often called *LG to LG* mirror symmetry (see e.g. [CR11a]), the B-model is given by the singularity theory of $[W^T/G^T]$.

We first consider a reduction of the mirror symmetry conjecture to the special case of Frobenius algebras. This Frobenius algebra mirror conjecture is that the Frobenius algebra corresponding to the FJRW A-model in genus 0 with 3 marked points should be isomorphic to a B model Frobenius algebra constructed from the (orbifolded) singularity defined by $[W^T/G^T]$. In the "unorbifolded" case, that is, when the group G^T is trivial, the B-model Frobenius algebra is just the local algebra (a.k.a. Milnor ring) $\mathscr{Q}_{W^T}$ of the singularity defined by W^T. In the orbifolded case, the B-model Frobenius algebra was constructed by Kaufmann and Krawitz (see [Kra10, Kau06, Kau02, Kau03]), following ideas of Intriligator-Vafa (see [IV90]).

Krawitz (see [Kra10]) first proved the mirror conjecture for Frobenius algebras when the group G^T is trivial. In the orbifolded case (when G^T is nontrivial) the Landau-Ginzburg Mirror Symmetry conjecture is proved in [FJJS12] at the level of Frobenius algebras for a large class of invertible polynomials and groups. But for certain chain-type polynomials and certain groups of symmetries this conjecture remains open.

Moving from genus zero with three marked points to the general genus zero FJRW theory produces a Frobenius manifold that agrees with the FJRW Frobenius algebra at the origin. There is also a mirror conjecture for Frobenius manifolds. For any isolated singularity f and a choice of primitive form, K. Saito (see [Sai81, Sai83a, Sai83b, ST08]) has constructed a semisimple Frobenius manifold from the local algebra $\mathscr{Q}_f$ that agrees with $\mathscr{Q}_f$ at the origin (this is for "unorbifolded" singularities, that is, without any additional group action). And for general genus, since the Saito Frobenius manifold is generically semisimple, one can (with some work) use Givental's theory of higher genus Frobenius manifolds (see [Giv01, Giv04]) to construct a potential function that behaves as if it came from a CohFT. The full (unorbifolded) LG to LG mirror conjecture is that the potential function of the FJRW CohFT associated to $W, G_{\max}$ should agree with the Saito-Givental potential function for W^T for a suitable choice of primitive form.

This version of the Landau-Ginzburg Mirror Symmetry conjecture was proved for the A_n singularity (corresponding to the polynomial $W = x^{n+1}$ with group $G_{\max} = \langle J \rangle$) in genus 0 in [Wit93a] (see also [JKV01]), in genus 1 and 2 by Y. P. Lee (see [Lee06]), and in higher genus by Faber-Shadrin-Zvonkine (see [FSZ10]). The conjecture was proved for all remaining simple singularities in [FJR13]. For all the simple singularities except D_n with n even, the only admissible group on the A side is the maximal group $G_{\max}$, corresponding to a trivial group on the B side. Moreover, for simple singularities all primitive forms differ only by scalar multiplication. So for these singularities, there is no orbifolding on the B side and there is a clear choice of Saito-Givental construction to play the role of the B model (see [FFJ$^+$16]).

However, in the case of D_n with n even, the group $\langle J \rangle$ has index 2 inside of $G_{\max}$, and so the B side should be orbifolded by $\mathbb{Z}/2$. In this special case, it turns out that the orbifolded Frobenius algebra on the B side is isomorphic to the unorbifolded Frobenius algebra for D_n again, so one might expect that in this case the entire orbifolded B model should be isomorphic to the unorbifolded B model for D_n, and indeed in [FJR13, FFJ$^+$16] it is proved that the FJRW theory of D_n with the group $\langle J \rangle$ (and n even) is isomorphic to the unorbifolded Saito-Givental B model for D_n.

For more general polynomials the situation is complicated by the fact that one must identify the correct primitive form to define the Saito Frobenius manifold on the B side. In the unorbifolded case, this was done and the Mirror Conjecture proved for simple elliptic singularities by Krawitz, Milanov and Shen in [KS11, MS16] (see also [ST11]). The Mirror Conjecture was proved for Arnold's 14 exceptional unimodular singularities by Li-Li-Saito-Shen (see [LLSS17, Sai14]). Finally, He-Li-Shen-Webb (see [HLSW15]) have identified the appropriate primitive form for the Saito theory in general and have proved that Landau-Ginzburg Mirror Symmetry holds for arbitrary invertible polynomials with $G = G_{\max}$ (and therefore G^T trivial). Since Saito's Frobenius manifolds are generically semisimple, one can then use Telemann's theorem [Tel12] to show that Givental's higher genus construction also agrees with the FJRW theory.

For the orbifolded case, before we can even hope to have a mirror symmetry result at the level of Frobenius manifolds, we will need a general construction of an orbifolded version of Saito's Frobenius manifold. In a few special cases candidate Frobenius manifolds have been constructed (see [Lee17]), but there is still no general construction of an appropriate orbifolded B model Frobenius manifold. In (see [BT14]), the authors identified axiomatically the properties that should define the B model for polynomials defining simple elliptic singularities with nontrivial symmetry groups, and then proved that there is exactly one Frobenius manifold with these properties.

Note also that, unlike the unorbifolded Saito Frobenius manifolds, we have no reason to expect that the orbifolded Frobenius manifolds would be generically semisimple, which means that proving mirror symmetry in genus zero would not automatically give the result in higher genera. This phenomenon is mirrored on the A side, where the FJRW theory is generally not expected to be generically semisimple if $G \neq G_{\max}$.

4.4. Integrable Hierarchies

The original Landau-Ginzburg mirror symmetry conjecture was the so-called "Generalized Witten Conjecture" (see [Wit93a]) that suggested the potential function of r-spin theory (FJRW theory for the polynomial x^r) should satisfy the r-KdV (a.k.a. Gelfand-Dikii) hierarchy. This was proved by Witten in genus 0 (see [Wit93a] and also [JKV01]), by Lee (see [Lee06]) for $g = 1, 2$, and by Faber-Shadrin-Zvonkine (see [FSZ10]) for higher genera. It can be thought of as an alternative form of mirror symmetry for the A_r singularity.

In fact, there are several ways to associate an integrable hierarchy to each of the simple singularities (see [DS84, KW89, Giv03, GM05]) but these all turn out to be equivalent (e.g. [HM93, FGM10, LWZ11]). Witten conjectured that the potential function of FJRW theory for any simple singularity should be a tau-function of the corresponding integrable hierarchy. This turned out to be almost, but not quite correct. In [FJR13, FFJ$^+$16] it is shown that the potential function of the FJRW theory for each of the simple singularities satisfies the corresponding integrable hierarchy of the transposed polynomial, rather than the original polynomial. This almost matched the original Witten conjecture because many of the simple singularities are self dual in the BHHK mirror construction.

The obvious generalization of the conjecture here is that for each pair (W, G) there should be an associated integrable hierarchy and the potential function of FJRW theory for (W, G) should be a tau-function for that hierarchy. Dubrovin and Zhang (see [DZ01]) have identified a way to construct an integrable hierarchy associated to a generically semisimple CohFT, and furthermore, they proved that the potential function of the semisimple CohFT is a tau-function of the integrable hierarchy. Thus, in the unorbifolded case, this integrable hierarchies mirror conjecture follows from the LG to LG mirror theorem of He-Li-Shen-Webb (see [HLSW15]).

In the orbifolded case, in [BPS12, BR16] the authors have constructed an integrable hierarchy associated to any CohFT, without the semisimplicity restriction, and Buryak conjectures that in the semisimple case it is equivalent to the Dubrovin-Zhang integrable hierarchy after a change of variables called a Miura transform. This conjecture is proved in some cases, including the A_{r+1} singularity (r-spin) for $r \leq 5$ [BG16].

As a final remark, in [BDGR16] there is a conjecture that the potential function of any CohFT is a tau-function of the DR hierarchy after a Miura transformation. For a general pair (W, G) one could construct the FJRW CohFT and corresponding integrable hierarchy, but without the BHHK mirror transpose operation. This is not entirely in the spirit of mirror symmetry, unless the Miura transformation can be thought of as some sort of mirror map.

§5. Landau-Ginzburg/Calabi-Yau Correspondence

As mentioned in the introduction, a quasihomogeneous polynomial W with an isolated singularity at the origin defines a smooth hypersurface $X_W = \{W = 0\}$ in weighted projective space. And if G is an admissible group of automorphisms of W, there is an induced action of the group $\tilde{G} = G/\langle J \rangle$ on X_W.

The hypersurface X_W is Calabi-Yau precisely when the sum $\sum_{i=1}^{N} q_i$ of the charges is equal to 1. This is the *Calabi-Yau condition* from Remark 3.16. The quotient orbifold $[X_W/\tilde{G}]$ is also Calabi-Yau when $G \subseteq \mathrm{SL}(N)$.

The *Landau-Ginzburg/Calabi-Yau (LG/CY) Correspondence* predicts that when $[X_W/\tilde{G}]$ is Calabi-Yau, then the FJRW theory of (W, G) should agree with the Gromov-Witten theory of $[X_W/\tilde{G}]$ in all genera (after analytic continuation and a symplectic transformation).

This conjecture is interesting for its own sake, but it is also important because FJRW invariants are generally believed to be easier to compute than Gromov-Witten invariants (see [Fra15, Gué16] for examples). So the LG/CY correspondence could provide new tools for computing Gromov-Witten invariants.

The first result on the LG/CY correspondence is the isomorphism of state spaces of [CR11b]. More precisely, they proved that the Gromov-Witten state space $H^*_{CR}([X_W/\tilde{G}])$ and the FJRW state space $\mathscr{H}_{W,G}$ match as bigraded vector spaces for all W satisfying the Calabi-Yau condition and for all admissible G (not only those in $\mathrm{SL}(N)$). They further showed that the narrow sectors of the FJRW state space correspond to the ambient cohomology of $H^*_{CR}([X_W/\tilde{G}])$.

In genus zero, the LG/CY correspondence has been verified for the Fermat Quintic with $\tilde{G} = \{1\}$ in [CR10]; for the narrow/ambient part of a general hypersurface in Gorenstein weighted projective space with $\tilde{G} = \{1\}$ in [CIR14]; for the mirror quintic (with group $\tilde{G} = (\mathbb{Z}/5)^3$) in [PS16, LS14]; and for arbitrary Fermat polynomials with arbitrary admissible groups in $\mathrm{SL}(N)$ in [LPS16]. This last paper [LPS16] is also interesting because it connects the LG/CY correspondence to the better-understood Crepant Transformation Conjecture. Each of these results matches the genus zero GW invariants with the genus zero FJRW invariants.

In each of the papers mentioned so far, the method of proof is closely tied to the formulation of mirror symmetry outlined in Section 4.2. Namely, one considers on the one hand Givental's I-function for GW theory (which we have seen is identified via the mirror map with the generating function of GW invariants). This function can be analytically continued from the large complex limit to the Gepner point. Then there is a symplectic transformation that matches the analytic continuation with the I-function for FJRW theory (which we have seen is identified via the mirror map with the generating function of FJRW invariants). Thus the generating function of GW theory is identified with the generating function for FJRW theory in genus zero.

The LG/CY correspondence has been verified in all genera for elliptic curves with $G = G_{\max}$ (see [KS11, MR11, MS16]). These results match certain "primary" invariants (both for GW and FJRW theory) with the corresponding invariants from Saito-Givental theory in genus zero. The analytic continuation from the previous results is replaced by a change of primitive form that matches the theories at the various special points. Since these theories are all generically semisimple, the genus zero correspondence can then be extended to all genera.

In fact, the conjecture can be extended beyond the Calabi-Yau case. In [Aco14], P. Acosta has recently shown how to generalize the correspondence to Fano and general-type hypersurfaces by replacing analytic continuation with asymptotic expansions. He has proved an LG/Fano and an LG/General-Type correspondence in genus zero for arbitrary nondegenerate polynomials for the group $\langle J \rangle$.

§6. Gauged Linear Sigma Model

As discussed in the previous section, the LG/CY correspondence relates FJRW theory for certain pairs (W, G) to the orbifold GW theory of hypersurfaces in weighted projective space. Several years ago, E. Clader generalized FJRW theory with a theory known as the *hybrid model* (see [Cla17]). She has shown that the LG/CY correspondence also gives a relation between the hybrid model and the GW theory for certain Calabi-Yau complete intersections in projective space. It is natural to ask whether there are further generalizations of the LG/CY correspondence to more general geometric objects.

The answer is "yes," and physically the generalization is given by Witten's *Gauged Linear Sigma Model (GLSM)*, which he developed in the early 1990s (see [Wit92]). His physical arguments, when translated into a mathematical setting, should extend the LG/CY correspondence to complete intersections, toric varieties, and more general quotients by nonabelian groups, and allow us to prove such correspondences mathematically.

From the point of view of partial differential equations, the gauged linear sigma model generalizes the Witten Equation (4) to the *Gauged Witten Equation*

$$\bar{\partial}_A u_i + \overline{\frac{\partial W}{\partial u_i}} = 0,$$
$$*F_A = \mu,$$

where A is a connection of certain principal bundle, and μ is the moment map of a certain Hamiltonian group action. Both the Gromov-Witten theory of a Calabi-Yau complete intersection and the corresponding "dual" Landau-Ginzburg theory are generally expressible as gauged linear sigma models. Furthermore, the LG/CY correspondence can be interpreted as a variation of the moment map μ in the GLSM.

In the special case where $W = 0$, the Gauged Witten Equation is called the *symplectic vortex equation*, and it has been widely studied, both from an algebraic and a symplectic point of view. An important development in this direction is the theory of stable quotients [MOP11] and stable quasimaps [CFKM14, CFK10, Kim12, CCFK15].

Recently Fan-Jarvis-Ruan in [FJR18] have described a mathematical approach to the GLSM. Their construction is essentially a union of FJRW theory (in the Polishchuk-Vaintrob formulation) with stable quasimaps. The relation between the GLSM and FJRW theory can be viewed as a generalization from a finite abelian gauge group G (FJRW theory) to an arbitrary reductive group (GLSM).

Just as in FJRW theory, the GLSM has both narrow and broad sectors (see Remark 3.12) in its state space, and the theory for narrow sectors admits a purely algebraic construction of the virtual cycle, using the cosection localization techniques of [CLL15, KL13, CL12] mentioned in Section 3.5. As special cases, one recovers Clader's hybrid model and Chang-Li's stable maps with p-fields [CL12]. In the broad case, cosection localization does not apply, but there is an analytic construction of the virtual cycle [FJR16] (see also [TX18]) that is expected to agree with the algebraic construction on the narrow sectors.

References

[Aco14] P. Acosta. Asymptotic Expansion and the LG/(Fano, General Type) Correspondence. *ArXiv e-prints*, November 2014, 1411.4162.

[BDGR16] A. Buryak, B. Dubrovin, J. Guéré and P. Rossi. Tau-structure for the Double Ramification Hierarchies. Communications in Mathematical Physics, 363(1) 191–260, 2018.

[BG16] A. Buryak and J. Guéré. Towards a description of the double ramification hierarchy for Witten's r-spin class. *Journal de Mathématiques Pures et Appliquées*, 2016.

[BP18] A. Basalaev and N. Priddis. Givental-type reconstruction at a nonsemisimple point. *Michigan Math. J.*, 67(2):333–369, 2018.

[BPS12] A. Buryak, H. Posthuma and S. Shadrin. A polynomial bracket for the Dubrovin-Zhang hierarchies. *J. Differential Geom.*, 92(1):153–185, 2012.

[BR16] A. Buryak and P. Rossi. Double ramification cycles and quantum integrable systems. *Lett. Math. Phys.*, 106(3):289–317, 2016.

[BT14] A. Basalaev and A. Takahashi. On rational Frobenius manifolds of rank three with symmetries. *Journal of Geometry and Physics*, 84:73–86, October 2014, 1401.3505.

[CCFK15] D. Cheong, I. Ciocan-Fontanine and B. Kim. Orbifold quasimap theory. *Math. Ann.*, 363(3-4):777–816, 2015.

[CdlOGP92] P. Candelas, X. C. de la Ossa, P. S. Green and L. Parkes. A pair of Calabi–Yau manifolds as an exactly soluble superconformal theory. In *Essays on mirror manifolds*, pages 31–95. Int. Press, Hong Kong, 1992.

[CFK10] I. Ciocan-Fontanine and B. Kim. Moduli stacks of stable toric quasimaps. *Adv. Math.*, 225(6):3022–3051, 2010.

[CFKM14] I. Ciocan-Fontanine, B. Kim and D. Maulik. Stable quasimaps to GIT quotients. *J. Geom. Phys.*, 75:17–47, 2014.

[Chi06a] A. Chiodo. A construction of Witten's top Chern class in K-theory. In *Gromov-Witten theory of spin curves and orbifolds*, volume

403 of *Contemp. Math.*, pages 21–29. Amer. Math. Soc., Providence, RI, 2006.

[Chi06b] A. Chiodo. The Witten top Chern class via K-theory. *J. Algebraic Geom.*, 15(4):681–707, 2006.

[CIR14] A. Chiodo, H. Iritani and Y. Ruan. Landau-Ginzburg/Calabi-Yau correspondence, global mirror symmetry and Orlov equivalence. *Publ. Math. Inst. Hautes Études Sci.*, 119:127–216, 2014.

[CL12] H.-L. Chang and J. Li. Gromov-Witten invariants of stable maps with fields. *Int. Math. Res. Not. IMRN*, (18):4163–4217, 2012.

[Cla17] E. Clader. Landau-Ginzburg/Calabi-Yau correspondence for the complete intersections X_{3,3} and X_{2,2,2,2}. *Adv. Math.*, 307:1–52, 2017.

[CLL15] H.-L. Chang, J. Li and W.-P. Li. Witten's top Chern class via cosection localization. *Invent. Math.*, 200(3):1015–1063, 2015.

[CR10] A. Chiodo and Y. Ruan. Landau-Ginzburg/Calabi-Yau correspondence for quintic three-folds via symplectic transformations. *Invent. Math.*, 182(1):117–165, 2010.

[CR11a] A. Chiodo and Y. Ruan. A global mirror symmetry framework for the Landau-Ginzburg/Calabi-Yau correspondence. *Ann. Inst. Fourier (Grenoble)*, 61(7):2803–2864, 2011.

[CR11b] A. Chiodo and Y. Ruan. LG/CY correspondence: the state space isomorphism. *Adv. Math.*, 227(6):2157–2188, 2011.

[DS84] V. G. Drinfel′d and V. V. Sokolov. Lie algebras and equations of Korteweg-de Vries type. In *Current problems in mathematics, Vol. 24*, Itogi Nauki i Tekhniki, pages 81–180. Akad. Nauk SSSR, Vsesoyuz. Inst. Nauchn. i Tekhn. Inform., Moscow, 1984.

[DZ01] B. Dubrovin and Y. Zhang. Normal forms of hierarchies of integrable PDEs, Frobenius manifolds and Gromov-Witten invariants. *ArXiv Mathematics e-prints*, August 2001, math/0108160.

[FFJ+16] H. Fan, A. Francis, T. Jarvis, E. Merrell and Y. Ruan. Witten's D_4 integrable hierarchies conjecture. *Chin. Ann. Math. Ser. B*, 37(2):175–192, 2016.

[FGM10] E. Frenkel, A. Givental and T. Milanov. Soliton equations, vertex operators, and simple singularities. *Funct. Anal. Other Math.*, 3(1):47–63, 2010.

[FJJS12] A. Francis, T. Jarvis, D. Johnson and R. Suggs. Landau-Ginzburg mirror symmetry for orbifolded Frobenius algebras. In *String-Math 2011*, volume 85 of *Proc. Sympos. Pure Math.*, pages 333–353. Amer. Math. Soc., Providence, RI, 2012.

[FJR07] H. Fan, T. J. Jarvis and Y. Ruan. The Witten equation and its virtual fundamental cycle. *ArXiv e-prints*, December 2007, 0712.4025.

[FJR13] H. Fan, T. Jarvis, and Y. Ruan. The Witten equation, mirror symmetry, and quantum singularity theory. *Ann. of Math. (2)*, 178(1):1–106, 2013.

[FJR16] H. Fan, T. Jarvis, and Y. Ruan. The analytic theory of gauged Witten equation and its virtual fundamental cycle. *In preparation*, 2016.

[FJR18] H. Fan, T. Jarvis and Y. Ruan. A Mathematical Theory of the Gauged Linear Sigma Model. *Geom. Topol.*, 22(1):235–303, 2018.

[Fra15] A. Francis. Computational techniques in FJRW theory with applications to Landau-Ginzburg mirror symmetry. *Adv. Theor. Math. Phys.*, 19(6):1339–1383, 2015.

[FSZ10] C. Faber, S. Shadrin, and D. Zvonkine. Tautological relations and the r-spin Witten conjecture. *Ann. Sci. Éc. Norm. Supér.* (4), 43(4):621–658, 2010.

[Giv96] A. B. Givental. Equivariant Gromov–Witten invariants. *Internat. Math. Res. Notices*, (13):613–663, 1996.

[Giv01] A. B. Givental. Gromov-Witten invariants and quantization of quadratic Hamiltonians. *Mosc. Math. J.*, 1(4):551–568, 645, 2001. Dedicated to the memory of I. G. Petrovskii on the occasion of his 100th anniversary.

[Giv03] A. Givental. A_{n-1} singularities and nKdV hierarchies. *Mosc. Math. J.*, 3(2):475–505, 743, 2003. Dedicated to Vladimir I. Arnold on the occasion of his 65th birthday.

[Giv04] A. B. Givental. Symplectic geometry of Frobenius structures. In *Frobenius manifolds*, Aspects Math., E36, pages 91–112. Friedr. Vieweg, Wiesbaden, 2004.

[GM05] A. B. Givental and T. E. Milanov. Simple singularities and integrable hierarchies. In *The breadth of symplectic and Poisson geometry*, volume 232 of *Progr. Math.*, pages 173–201. Birkhäuser Boston, Boston, MA, 2005.

[GR90] B. Greene and P. Ronen. Duality in Calabi–Yau moduli space. *Nuclear Physics B.*, 338(1):15–37, 1990.

[Gué16] J. Guéré. A Landau–Ginzburg mirror theorem without concavity. *Duke Math. J.*, 165(13):2461–2527, 2016.

[Gué17] J. Guéré. Hodge integrals in FJRW theory. *Michigan Math. J.*, 66(4):831–854, 2017.

[HLSW15] W. He, S. Li, Y. Shen and R. Webb. Landau-Ginzburg Mirror Symmetry Conjecture. *ArXiv e-prints*, March 2015, 1503.01757.

[HM93] T. Hollowood and J. L. Miramontes. Tau-functions and generalized integrable hierarchies. *Comm. Math. Phys.*, 157(1):99–117, 1993.

[IV90] K. Intriligator and C. Vafa. Landau-ginzburg orbifolds. *Nuclear Physics B*, 339(1):95–120, 1990.

[Jan14] F. Janda. Comparing tautological relations from the equivariant Gromov-Witten theory of projective spaces and spin structures. *ArXiv e-prints*, July 2014, 1407.4778.

[Jan15] F. Janda. Relations in the Tautological Ring and Frobenius Manifolds near the Discriminant. *ArXiv e-prints*, May 2015, 1505.03419.

[JKV01] T. J. Jarvis, T. Kimura and A. Vaintrob. Moduli spaces of higher spin curves and integrable hierarchies. *Compositio Mathematica*, 126:157, 2001.

[Kau02] R. M. Kaufmann. Orbifold frobenius algebras, cobordisms and monodromies. In D. DeTurck, editor, *Contemporary Mathematics*, pages 135–161, Providence, R.I, 2002. American Mathematical Society.

[Kau03] R. M. Kaufmann. Orbifolding Frobenius algebras. *Internat. J. Math.*, 14(6):573–617, 2003.

[Kau06] R. M. Kaufmann. Singularities with Symmetries, orbifold Frobenius algebras and Mirror Symmetry. *Gromov-Witten theory of spin curves and orbifolds*, volume 403 of *Contemp. Math.*, pages 67–116. Amer. Math. Soc., Providence, RI, 2006.

[Kim12] B. Kim. Stable quasimaps. *Commun. Korean Math. Soc.*, 27(3):571–581, 2012.

[KL13] Y.-H. Kiem and J. Li. Localizing virtual cycles by cosections. *J. Amer. Math. Soc.*, 26(4):1025–1050, 2013.

[KM94] M. Kontsevich and Y. Manin. Gromov-Witten classes, quantum cohomology, and enumerative geometry. *Comm. Math. Phys.*, 164(3):525–562, 1994.

[Kra10] M. Krawitz. *FJRW Rings and Landau-Ginzburg Mirror Symmetry*. PhD thesis, The University of Michigan, 2010.

[KS92] M. Kreuzer and H. Skarke. On the classification of quasihomogeneous functions. *Communications in Mathematical Physics*, 150:137–147, November 1992, arXiv:hep-th/9202039.

[KS11] M. Krawitz and Y. Shen. Landau-Ginzburg/Calabi-Yau Correspondence of all Genera for Elliptic Orbifold $\mathbb{P}^1$. *ArXiv e-prints*, June 2011, 1106.6270.

[KW89] V. G. Kac and M. Wakimoto. Exceptional hierarchies of soliton equations. In *Theta functions—Bowdoin 1987, Part 1 (Brunswick, ME, 1987)*, volume 49 of *Proc. Sympos. Pure Math.*, pages 191–237. Amer. Math. Soc., Providence, RI, 1989.

[Lee06] Y.-P. Lee. Witten's conjecture and the Virasoro conjecture for genus up to two. In *Gromov-Witten theory of spin curves and orbifolds*, volume 403 of *Contemp. Math.*, pages 31–42. Amer. Math. Soc., Providence, RI, 2006.

[Lee17] B. Lee. G-Frobenius Manifolds. *J. Algebra*, 479:35–77, 2017.

[LLSS17] C. Li, S. Li, K. Saito and Y. Shen. Mirror symmetry for exceptional unimodular singularities. *J. Eur. Math. Soc. (JEMS)*, 19(4):1189–1229, 2017.

[LLY99] B. H. Lian, K. Liu and S.-T. Yau. Mirror principle. I [MR1621573 (99e:14062)]. In *Surveys in differential geometry: differential geometry inspired by string theory, volume 5* of *Surv. Differ. Geom.*, pages 405–454. Int. Press, Boston, MA, 1999.

[LPS16] Y.-P. Lee, N. Priddis and M. Shoemaker. A proof of the Landau-Ginzburg/Calabi-Yau correspondence via the crepant transformation conjecture. *Ann. Sci. Éc. Norm. Supér.* (4), 49(6):1403–1443, 2016.

[LS14] Y.-P. Lee and M. Shoemaker. A mirror theorem for the mirror quintic. *Geom. Topol.*, 18(3):1437–1483, 2014.

[LWZ11] S.-Q. Liu, C.-Z. Wu and Y. Zhang. On the Drinfeld-Sokolov hierarchies of D type. *Int. Math. Res. Not. IMRN*, (8):1952–1996, 2011.

[Mar90] E. J. Martinec. Criticality, catastrophes, and compactifications. In *Physics and mathematics of strings*, pages 389–433. World Sci. Publ., Teaneck, NJ, 1990.

[Moc06] T. Mochizuki. The virtual class of the moduli stack of stable r-spin curves. *Comm. Math. Phys.*, 264(1):1–40, 2006.

[MOP11] A. Marian, D. Oprea and R. Pandharipande. The moduli space of stable quotients. *Geom. Topol.*, 15(3):1651–1706, 2011.

[MR11] T. Milanov and Y. Ruan. Gromov-Witten theory of elliptic orbifold $\mathbb{P}^1$ and quasi-modular forms. *ArXiv e-prints*, June 2011, 1106.2321.

[MS16] T. Milanov and Y. Shen. Global mirror symmetry for invertible simple elliptic singularities. *Ann. Inst. Fourier (Grenoble)*, 66(1):271–330, 2016.

[PPZ15] R. Pandharipande, A. Pixton and D. Zvonkine. Relations on $\overline{\mathscr{M}}_{g,n}$ via 3-spin structures. *J. Amer. Math. Soc.*, 28(1):279–309, 2015.

[PPZ16] R. Pandharipande, A. Pixton and D. Zvonkine. Tautological relations via r-spin structures. *ArXiv e-prints*, July 2016, 1607.00978.

[PS16] N. Priddis and M. Shoemaker. A Landau-Ginzburg/Calabi-Yau correspondence for the mirror quintic. *Ann. Inst. Fourier (Grenoble)*, 66(3):1045–1091, 2016.

[PV01] A. Polishchuk and A. Vaintrob. Algebraic construction of Witten's top Chern class. In *Advances in algebraic geometry motivated by physics (Lowell, MA, 2000)*, volume 276 of *Contemp. Math.*, pages 229–249. Amer. Math. Soc., Providence, RI, 2001.

[PV16] A. Polishchuk and A. Vaintrob. Matrix factorizations and cohomological field theories. *J. Reine Angew. Math.*, 714:1–122, 2016.

[Sai81] K. Saito. Primitive forms for a universal unfolding of a function with an isolated critical point. *J. Fac. Sci. Univ. Tokyo Sect. IA Math.*, 28(3):775–792 (1982), 1981.

[Sai83a] K. Saito. The higher residue pairings $K_F^{(k)}$ for a family of hypersurface singular points. In *Singularities, Part 2 (Arcata, Calif., 1981)*, volume 40 of *Proc. Sympos. Pure Math.*, pages 441–463. Amer. Math. Soc., Providence, RI, 1983.

[Sai83b] K. Saito. Period mapping associated to a primitive form. *Publ. Res. Inst. Math. Sci.*, 19(3):1231–1264, 1983.

[Sai14] K. Saito. From primitive form to mirror symmetry. *ArXiv e-prints*, August 2014, 1408.4208.

[ST08] K. Saito and A. Takahashi. From primitive forms to Frobenius manifolds. In *From Hodge theory to integrability and TQFT tt*-geometry, volume 78* of *Proc. Sympos. Pure Math.*, pages 31–48. Amer. Math. Soc., Providence, RI, 2008.

[ST11] I. Satake and A. Takahashi. Gromov-Witten invariants for mirror orbifolds of simple elliptic singularities. *Ann. Inst. Fourier (Grenoble)*, 61(7):2885–2907, 2011.

[Tay13] J. Tay. *Poincaré Polynomial of FJRW Rings and the Group-weights Conjecture*. Brigham Young University. Department of Mathematics, 2013.

[Tel12] C. Teleman. The structure of 2D semi-simple field theories. *Invent. Math.*, 188(3):525–588, 2012.

[TX18] G. Tian and G. Xu. Analysis of gauged Witten equation. *J. Reine Angew. Math.*, 740:187–274, 2018.

[VW89] C. Vafa and N. Warner. Catastrophes and the classification of conformal theories. *Physics Letters B*, 218:51–58, February 1989.

[Wal80a] C. T. C. Wall. A note on symmetry of singularities. *Bulletin of the London Mathematical Society*, 12(3):169–175, 1980, http://blms.oxfordjournals.org/content/12/3/169.full.pdf+html.

[Wal80b] C. T. C. Wall. A second note on symmetry of singularities. *Bulletin of the London Mathematical Society*, 12(5):347–354, 1980, http://blms.oxfordjournals.org/content/12/5/347.full.pdf+html.

[Wit92] E. Witten. The N-matrix model and gauged WZW models. *Nuclear Physics B*, 371:191–245, March 1992.

[Wit93a] E. Witten. Algebraic geometry associated with matrix models of two dimensional gravity. In *Topological methods in modern mathematics (Stony Brook, NY, 1991)*, pages 235–269. Publish or Perish, Houston, TX, 1993.

[Wit93b] E. Witten. Phases of N=2 theories in two dimensions. *Nucl. Phys.*, B403:159–222, 1993, hep-th/9301042.

Mathematical Reviews, American Mathematical Society, Ann Arbor, MI 48103, USA
E-mail address: aef@ams.org

Department of Mathematics, Brigham Young University, Provo, UT 84602, USA
E-mail address: jarvis@mathematics.byu.edu

Department of Mathematics, Brigham Young University, Provo, UT 84602, USA
E-mail address: priddis@mathematics.byu.edu

Advanced Studies in Pure Mathematics 83, 2019
Primitive Forms and Related Subjects — Kavli IPMU 2014
pp. 55–115

Gamma conjecture via mirror symmetry

Sergey Galkin and Hiroshi Iritani

Abstract.

The asymptotic behaviour of solutions to the quantum differential equation of a Fano manifold F defines a characteristic class A_F of F, called the principal asymptotic class. Gamma conjecture [29] of Vasily Golyshev and the present authors claims that the principal asymptotic class A_F equals the Gamma class $\widehat{\Gamma}_F$ associated to Euler's Γ-function. We illustrate in the case of toric varieties, toric complete intersections and Grassmannians how this conjecture follows from mirror symmetry. We also prove that Gamma conjecture is compatible with taking hyperplane sections, and give a heuristic argument how the mirror oscillatory integral and the Gamma class for the projective space arise from the polynomial loop space.

Contents

Received June 14, 2015.

Revised June 23, 2017.

2010 *Mathematics Subject Classification.* Primary 53D37; Secondary 14N35, 14J45, 14J33, 11G42.

Key words and phrases. Fano varieties, quantum cohomology, mirror symmetry, Dubrovin's conjecture, Gamma class, Apery constant, derived category of coherent sheaves, exceptional collection, Landau–Ginzburg model.

§1. Introduction

1.1. Gamma conjecture

Gamma conjecture [29] is a conjecture which relates quantum cohomology of a Fano manifold with its topology. The small quantum cohomology of F defines a flat connection (quantum connection) over $\mathbb{C}^\times$ and its solution is given by a (multi-valued) cohomology-valued function $J_F(t)$ called the *J-function*. Under a certain condition (Property $\mathcal{O}$), the limit of the J-function:

$$A_F := \lim_{t\to+\infty} \frac{J_F(t)}{\langle[\mathrm{pt}], J_F(t)\rangle} \in H^\bullet(F)$$

exists and defines the *principal asymptotic class* A_F of F. *Gamma conjecture* I says that A_F equals the Gamma class $\widehat{\Gamma}_F = \widehat{\Gamma}(TF)$ of the tangent bundle of F (see §3.2):

$$A_F = \widehat{\Gamma}_F.$$

Suppose that the quantum cohomology of F is semisimple. In this case, we can define *higher* asymptotic classes $A_{F,i}$, $1 \leqslant i \leqslant N = \dim H^\bullet(F)$ from exponential asymptotics of flat sections of the quantum connection. *Gamma conjecture* II says that there exists a full exceptional collection $E_1, E_2, \ldots, E_N$ of $D^b_{\mathrm{coh}}(F)$ such that we have

$$A_{F,i} = \widehat{\Gamma}_F \cdot \mathrm{Ch}(E_i) \qquad i = 1, \ldots, N.$$

Here we write $\mathrm{Ch}(E) := (2\pi\sqrt{-1})^{\frac{\deg}{2}} \mathrm{ch}(E) = \sum_{p=0}^{\dim F} (2\pi\sqrt{-1})^p \mathrm{ch}_p(E)$ for the $(2\pi\sqrt{-1})$-modified version of the Chern character. The principal asymptotic class A_F corresponds to the exceptional object $E = \mathcal{O}_F$.

1.2. Riemann–Roch

One may view Gamma conjecture as a *square root* of the index theorem. Recall the Hirzebruch–Riemann–Roch formula:

$$\chi(E_1, E_2) = \int_F \mathrm{ch}(E_1^\vee) \cdot \mathrm{ch}(E_2) \cdot \mathrm{td}_F$$

for vector bundles E_1, E_2 on F, where $\chi(E_1, E_2) = \sum_{i=0}^{\dim F} (-1)^i \dim \mathrm{Ext}^i(E_1, E_2)$ is the Euler pairing and $\mathrm{td}_F = \mathrm{td}(TF)$ is the Todd class of F. The famous identity

$$\frac{x}{1-e^{-x}} = e^{x/2}\Gamma\left(1 - \frac{x}{2\pi\sqrt{-1}}\right)\Gamma\left(1 + \frac{x}{2\pi\sqrt{-1}}\right)$$

gives the factorization $(2\pi\sqrt{-1})^{\frac{\deg}{2}}\,\mathrm{td}_F = e^{\pi\sqrt{-1}c_1(F)}.\widehat{\Gamma}_F\cdot\widehat{\Gamma}_F^*$ of the Todd class[1], which in turn factorizes the Hirzebruch–Riemann–Roch formula as:

$$\chi(E_1, E_2) = \left[\mathrm{Ch}(E_1)\cdot\widehat{\Gamma}_F, \mathrm{Ch}(E_2)\cdot\widehat{\Gamma}_F\right). \tag{1}$$

Here we set $\widehat{\Gamma}_F^* = (-1)^{\frac{\deg}{2}}\widehat{\Gamma}_F = \sum_{p=0}^{\dim F}(-1)^p\gamma_p$ (if we write $\widehat{\Gamma}_F = \sum_{p=0}^{\dim F}\gamma_p$ with $\gamma_p \in H^{2p}(F)$) and $[\cdot,\cdot)$ is a non-symmetric pairing on $H^\bullet(F)$ given by

$$[\alpha, \beta) = \frac{1}{(2\pi)^{\dim F}}\int_F (e^{\pi\sqrt{-1}c_1(X)}e^{\pi\sqrt{-1}\mu}\alpha)\cup\beta \tag{2}$$

with $\mu \in \mathrm{End}(H^\bullet(F))$ defined by $\mu(\phi) = (p - \frac{\dim F}{2})\phi$ for $\phi \in H^{2p}(F)$. Via the factorization (1), Gamma conjecture II implies part of Dubrovin's conjecture [21]: the Euler matrix $\chi(E_i, E_j)$ of the exceptional collection equals the Stokes matrix $S_{ij} = [A_{F,i}, A_{F,j})$ of the quantum differential equation.

1.3. Mirror symmetry

In the B-side of mirror symmetry, solutions to Picard–Fuchs equations are often given by hypergeometric series whose coefficients are ratios of Γ-functions. Recall that a mirror of a quintic threefold $Q \subset \mathbb{P}^4$ is given by the pencil of hypersurfaces $Y_t = \{f(x) = t^{-1}\}$ in the torus $(\mathbb{C}^\times)^4$ which can be compactified to smooth Calabi–Yau threefolds $\overline{Y}_t$ [13], where f is the Laurent polynomial given by

$$f(x) = x_1 + x_2 + x_3 + x_4 + \frac{1}{x_1x_2x_3x_4}.$$

For the holomorphic volume form $\Omega_t = d\log x_1 \wedge \cdots \wedge d\log x_4/df$ on $\overline{Y}_t$ and a real 3-cycle $C \subset \overline{Y}_t$, the period $\int_C \Omega_t$ satisfies the Picard–Fuchs equations

$$\left(\theta^4 - 5^5t^5(\theta+1)(\theta+2)(\theta+3)(\theta+4)\right)\int_C \Omega_t = 0$$

with $\theta = t\frac{\partial}{\partial t}$. The Frobenius method yields the following solution to this differential equation:

$$\Phi(t) = \sum_{n=0}^{\infty}\frac{\Gamma(1+5n+5\epsilon)}{\Gamma(1+n+\epsilon)^5}t^{5n+5\epsilon}$$

---

[1] We define $(2\pi\sqrt{-1})^{\frac{\deg}{2}}\,\mathrm{td}_F := \sum_{p=0}^{\dim F}(2\pi\sqrt{-1})^p t_p$ if $\mathrm{td}_F = \sum_{p=0}^{\dim F} t_p$ with $t_p \in H^{2p}(F)$.

where ϵ is an infinitesimal parameter satisfying $\epsilon^4 = 0$. Regarding ϵ as a hyperplane class on the quintic Q, we may identify the leading term of the series $\Phi(t)$ with the inverse Gamma class of Q:

$$\frac{\Gamma(1+5\epsilon)}{\Gamma(1+\epsilon)^5} = \frac{1}{\widehat{\Gamma}_Q}.$$

This is how the Gamma class originally arose in the context of mirror symmetry [48, 60]. Hosono [47] conjectured (more generally for a complete intersection Calabi–Yau) that the period $\int_C \Omega_t$ of an integral 3-cycle $C \subset \overline{Y}_t$ should be written in the form

$$\int_Q \Phi(t) \cdot \mathrm{Ch}(V) \cdot \mathrm{Td}_Q \tag{3}$$

for a vector bundle $V \to Q$ which is "mirror" to C, where $\mathrm{Td}_Q = (2\pi\sqrt{-1})^{\frac{\deg}{2}} \mathrm{td}_Q$. In physics terminology, the period $\int_C \Omega_t$ (or the quantity (3)) is called the *central charge* of the D-branes C (resp. V). Hosono's conjecture has been answered affirmatively in [52] by showing that the natural integral structure $H^3(\overline{Y}_t, \mathbb{Z})$ agrees with the $\widehat{\Gamma}$-integral structure in quantum cohomology of Q, see also [46, 11, 50, 39].

A main purpose of this article is to explain a relationship between Gamma conjecture and mirror symmetry. In fact, Hosono's conjecture for a quintic Q is closely related to the truth of Gamma conjecture for the ambient Fano manifold $\mathbb{P}^4$. A mirror of $\mathbb{P}^4$ is given by the Landau–Ginzburg model $f\colon (\mathbb{C}^\times)^4 \to \mathbb{C}$ and the quantum differential equation for $\mathbb{P}^4$ has a solution given by the oscillatory integral:

$$\int_\Gamma \exp(tf(x)) \frac{dx_1}{x_1} \wedge \cdots \wedge \frac{dx_4}{x_4}.$$

When the cycle Γ is a Lefschetz thimble of $f(x)$ and C is the associated vanishing cycle, this oscillatory integral can be written as a Laplace transform of the period $\int_C \Omega_t$. Correspondingly, by quantum Lefschetz principle [33, 56, 59, 15], the quantum differential equation for Q arises from a Laplace transform of the quantum differential equation for $\mathbb{P}^4$ [22, 53]. Gamma conjecture relates a Lefschetz thimble of f with an exceptional object E in $D^b_{\mathrm{coh}}(\mathbb{P}^4)$ via the exponential asymptotics of the corresponding oscillatory integral; then the vanishing cycle C associated with Γ corresponds to the spherical object $V = i^*E$ on Q under Hosono's conjecture, see [52, Theorem 6.9]. See the comparison table below for quantum differential equations (QDE) of a Fano manifold F and its anti-canonical section Q.

	Fano		Calabi–Yau
space X	F		$Q \in \lvert -K_F \rvert$
mirror	$f\colon (\mathbb{C}^\times)^n \to \mathbb{C}$		$\overline{Y}_t = \overline{f^{-1}(1/t)}$
singularities of QDE	irregular		regular
solutions (central charges)	oscillatory integral of f	$\overset{\text{Laplace}}{\leftrightarrow}$	period integral of $\overline{Y}_t$
cycles on the mirror	Lefschetz thimble	$\overset{\text{fiber}}{\rightarrow}$	vanishing cycle
objects of $D^b_{\mathrm{coh}}(X)$	exceptional object E_i	$\overset{i^*}{\rightarrow}$	spherical object V_i
monodromy data	Stokes $S_{ij} = \chi(E_i, E_j)$		reflection $S_{ij} - (-1)^n S_{ji}$

Table 1. We expect a mirror correspondence between objects of $D^b_{\mathrm{coh}}(X)$ and integration cycles on the mirror; when a vanishing cycle C arises as a fiber of a Lefschetz thimble Γ, the spherical object V on Q mirror to C should be the pull-back of the exceptional object E on F mirror to Γ. A Lefschetz thimble of f gives a solution to the quantum differential equation of F which has a specific exponential asymptotics as $t \to \infty$.

1.4. Plan of the paper

In §2, we review definitions and basic facts on quantum cohomology and quantum connection. In §3 and §4, we discuss Gamma conjecture I and II respectively. This part is mostly a review of our previous paper [29] with Vasily Golyshev, but we provide a new sufficient condition which implies the Gamma conjecture I' under the Gamma conjecture I. In §5, we give a heuristic argument which gives mirror oscillatory integral and the Gamma class in terms of polynomial loop spaces. In §6 and §7, we discuss Gamma conjecture for toric varieties and toric complete intersections using Batyrev–Borisov/Givental/Hori–Vafa mirrors. In §8, we discuss compatibility of Gamma conjecture I with taking hyperplane sections (quantum Lefschetz). In §9, we discuss Gamma conjecture for Grassmannians $\mathrm{Gr}(r,n)$ using the Hori–Vafa mirror which is the rth alternate product of the mirrors of $\mathbb{P}^{n-1}$. The Gamma conjectures I, II for Grassmannian were proved in [29], but here we present a different approach based on the Hori–Vafa mirror, focusing on the Gamma conjecture I'. In Appendix A, we discuss eigenvalues of the quantum multiplication by $c_1(F)$ on odd cohomology (and on $H^{p,q}(F)$ with $p \neq q$).

Acknowledgments. We thank Vasily Golyshev for insightful discussions during the collaboration [29]. We also thank an anonymous referee for very helpful comments. H.I. thanks Kentaro Hori, Mauricio Romo and Kazushi Ueda for the discussion on the papers [8, 19, 44]. This project was supported by JSPS and Russian Foundation for Basic Research under the Japan–Russia Research Cooperative Program

"Categorical and Analytic Invariants in Algebraic Geometry". S.G. was partially supported by the Russian Academic Excellence Project '5-100', Dynasty Foundation, Simons IUM fellowship, and by RFBR, research project No.15-51-50045яфа. H.I. was supported by JSPS Kakenhi Grant number 25400069, 23224002, 26610008, 24224001, 16K05127, 16H06337, 16H06335.

§2. Quantum cohomology and quantum connection

Let $H^\bullet(F) = H^\bullet(F,\mathbb{C})$ denote the even degree part of the Betti cohomology group of F (see Appendix A). The (small) quantum product $\star_0$ on $H^\bullet(F)$ is defined by the formula:

$$(\alpha \star_0 \beta, \gamma)_F = \sum_{d\in H_2(F,\mathbb{Z})} \langle \alpha, \beta, \gamma\rangle_{0,3,d}$$

where $(\cdot,\cdot)_F$ is the Poincaré pairing on F and $\langle\cdots\rangle_{0,3,d}$ is the genus-zero three point Gromov–Witten invariants, which roughly speaking counts the number of rational curves in F passing through the cycles Poincaré dual to α, β and γ (see e.g. [62]). By the dimension axiom, these Gromov–Witten invariants are non-zero only if $2c_1(F)\cdot d + 2\dim F = \deg\alpha + \deg\beta + \deg\gamma$; since F is Fano, there are finitely many such curve classes d and the above sum is finite. The product $\star_0$ is associative and commutative, and $(H^\bullet(F),\star_0)$ becomes a finite dimensional algebra. More generally we can define the big quantum product $\star_\tau$ for $\tau\in H^\bullet(F)$:

$$(\alpha \star_\tau \beta, \gamma) = \sum_{d\in H_2(F,\mathbb{Z})} \sum_{n=0}^{\infty} \frac{1}{n!} \langle \alpha, \beta, \gamma, \tau, \dots, \tau\rangle_{0,3+n,d}\,.$$

This defines a formal[2] deformation of the small quantum cohomology as a commutative ring. In this paper, we will restrict our attention to the small quantum product $\star_0$.

The quantum connection is a meromorphic flat connection on the trivial $H^\bullet(F)$-bundle over $\mathbb{P}^1$. It is given by:

$$\nabla_{z\partial_z} = z\frac{\partial}{\partial z} - \frac{1}{z}(c_1(F)\star_0) + \mu \tag{4}$$

where z is an inhomogeneous co-ordinate on $\mathbb{P}^1$ and $\mu\in\operatorname{End}(H^\bullet(F))$ is defined by $\mu(\phi) = (p - \frac{\dim F}{2})\phi$ for $\phi\in H^{2p}(F)$. This is regular singular

[2]The convergence of the big quantum product is not known in general.

(logarithmic) at $z = \infty$ and irregular singular at $z = 0$. We have a canonical fundamental solution around $z = \infty$ as follows:

Proposition 2.1 ([20, 29]). *There exists a unique* $\operatorname{End}(H^\bullet(F))$*-valued power series* $S(z) = \mathrm{id} + S_1 z^{-1} + S_2 z^{-2} + \cdots$ *which converges over the whole* z^{-1}*-plane such that*

$$\nabla(S(z)z^{-\mu}z^{c_1(F)}\phi) = 0 \qquad \forall \phi \in H^\bullet(F)$$

$$T(z) = z^{\mu}S(z)z^{-\mu} \text{ is regular at } z = \infty \text{ and } T(\infty) = \mathrm{id}.$$

Here we set $z^{c_1(F)} = \exp(c_1(F)\log z)$ *and* $z^{-\mu} = \exp(-\mu \log z)$.

The fundamental solution $S(z)z^{-\mu}z^{c_1(F)}$ identifies the space of flat sections with the cohomology group $H^\bullet(F)$. In other words, a basis of the cohomology group $H^\bullet(F)$ yields a basis of flat sections via the map $S(z)z^{-\mu}z^{c_1(F)}$ — this amounts to solving the quantum differential equation by the *Frobenius method* around $z = \infty$.

Remark 2.2. The quantum connection can be extended to a meromorphic flat connection on the trivial $H^\bullet(F)$-bundle over $H^\bullet(F) \times \mathbb{P}^1$ via the big quantum product:

$$\begin{aligned} \nabla_\alpha &= \partial_\alpha + \frac{1}{z}(\alpha \star_\tau) \qquad \alpha \in H^\bullet(F), \\ \nabla_{z\partial_z} &= z\partial_z - \frac{1}{z}(E\star_\tau) + \mu \end{aligned} \tag{5}$$

where $E = c_1(F) + \sum_{i=1}^N (1 - \frac{\deg \phi_i}{2})\tau^i \phi_i$ with $\{\phi_i\}_{i=1}^N$ a homogeneous basis of $H^\bullet(F)$ and $\tau = \sum_{i=1}^N \tau^i \phi_i$.

Remark 2.3 ([29]). Via the gauge transformation z^μ and the change $z = t^{-1}$ of variables, we have

$$z^{\mu}\nabla_{z\partial_z} z^{-\mu} = z\frac{\partial}{\partial z} - (c_1(F)\star_{-c_1(F)\log z}) = -\left(t\frac{\partial}{\partial t} + c_1(F)\star_{c_1(F)\log t}\right).$$

This gives the quantum connection along the "anticanonical line" $\mathbb{C}c_1(F)$.

§3. Gamma conjecture I (and I')

3.1. Property $\mathcal{O}$

We start with Property $\mathcal{O}$ (or Conjecture $\mathcal{O}$) for a Fano manifold. This is a supplementary condition we need to formulate Gamma conjecture I.

Definition 3.1 (Property $\mathcal{O}$ [29]). Let F be a Fano manifold and define a non-negative real number T as

$$T := \max\{|u| : u \in \mathbb{C} \text{ is an eigenvalue of } (c_1(F)\star_0)\} \in \overline{\mathbb{Q}}. \tag{6}$$

We say that F satisfies *Property* $\mathcal{O}$ if the following conditions are satisfied:

(1) T is an eigenvalue of $(c_1(F)\star_0)$ of multiplicity one;
(2) if u is an eigenvalue of $(c_1(F)\star_0)$ with $|u| = T$, there exists a complex number ζ such that $\zeta^r = 1$ and that $u = \zeta T$, where r is the Fano index of F, i.e. the maximal integer $r > 0$ such that $c_1(F)/r$ is an integral class.

Remark 3.2. '$\mathcal{O}$' means the structure sheaf of F. Under mirror symmetry, it is conjectured that the set of eigenvalues of $(c_1(F)\star_0)$ agrees with the set of critical values of the mirror Landau–Ginzburg potential f (see, e.g. [31], [4, Theorem 6.1]); for example, this holds for Fano toric manifolds. Under Property $\mathcal{O}$, number T should be a critical value of f and the Lefschetz thimble corresponding to T should be mirror to the structure sheaf $\mathcal{O}$. Conjecture $\mathcal{O}$ [29] says that every Fano manifold satisfies Property $\mathcal{O}$. Some Fano *orbifolds* with non-trivial $\pi_1^{\mathrm{orb}}(F)$ do not satisfy Property $\mathcal{O}$ [28, 29]. We also note that Part (1) of Property $\mathcal{O}$ implies that the small quantum cohomology ring $(H^\bullet(F), \star_0)$ has a field as a direct summand; this weaker conjecture is open as well.

Remark 3.3. Perron–Frobenius theorem says that an irreducible square matrix with non-negative entries has a positive eigenvalue with the biggest norm whose multiplicity is one. It is likely that $c_1(F)\star_0$ is represented by a non-negative matrix if we have a basis of $H^\bullet(F)$ consisting of 'positive' (algebraic) cycles. This remark is due to Kaoru Ono.

Remark 3.4. Property $\mathcal{O}$ for homogeneous spaces G/P was recently proved by Cheong–Li [14].

3.2. Gamma class

The Gamma class [60, 61, 50] is a characteristic class defined for an almost complex manifold F. Let $\delta_1, \dots, \delta_n$ be the Chern roots of the tangent bundle TF of F such that $c(TF) = (1+\delta_1)(1+\delta_2)\cdots(1+\delta_n)$. The Gamma class $\widehat{\Gamma}_F$ is defined to be

$$\widehat{\Gamma}_F = \prod_{i=1}^{n} \Gamma(1+\delta_i) \in H^\bullet(F, \mathbb{R})$$

where $\Gamma(z) = \int_0^\infty e^{-t} t^{z-1} dt$ is Euler's Γ-function. Since $\Gamma(z)$ is holomorphic at $z = 1$, via the Taylor expansion, the right-hand side makes sense as a symmetric power series in $\delta_1, \dots, \delta_n$, and therefore as a (real) cohomology class of F. This is a transcendental class and is given by

$$\widehat{\Gamma}_F = \exp\left(-\gamma c_1(F) + \sum_{k=2}^{\infty} (-1)^k (k-1)! \zeta(k) \operatorname{ch}_k(TF) \right)$$

where $\zeta(z)$ is the Riemann zeta function. As explained in the Introduction (see §1.2), the formula $(2\pi\sqrt{-1})^{\frac{\deg}{2}} \operatorname{td}_F = e^{\pi\sqrt{-1}c_1(F)} \widehat{\Gamma}_F \widehat{\Gamma}_F^*$ shows that the Gamma class $\widehat{\Gamma}_F$ can be regarded as a square root of the Todd class td_F. There is also an interpretation of the Gamma class in terms of the free loop space $\mathcal{L}F$ of F due to Lu [61] (see also [51, 29]): $\widehat{\Gamma}_F$ arises from the ζ-function regularization of the S^1-equivariant Euler class $e_{S^1}(\mathcal{N}_+)$, where $\mathcal{N}_+$ is the positive normal bundle of the locus F of constant loops in $\mathcal{L}F$.

3.3. Principal asymptotic class and Gamma conjecture I

Consider the space of flat sections for the quantum connection ∇ (4) over the positive real line $\mathbb{R}_{>0}$. We introduce the subspace $\mathcal{A}$ of flat sections having the smallest asymptotics $\sim e^{-T/z}$ as $z \to +0$.

$$\mathcal{A} := \left\{ s \colon \mathbb{R}_{>0} \to H^\bullet(F) : \begin{array}{l} \nabla s(z) = 0, \\ \|e^{T/z} s(z)\| = O(z^{-m}) \text{ as } z \to +0 \ (\exists m) \end{array} \right\}$$

where $T > 0$ is the number in Definition 3.1.

Proposition 3.5 ([29, Proposition 3.3.1]). *Suppose that a Fano manifold F satisfies Property $\mathcal{O}$. We have $\dim_{\mathbb{C}} \mathcal{A} = 1$. Moreover, for every element $s(z) \in \mathcal{A}$, the limit $\lim_{z\to+0} e^{T/z} s(z)$ exists and lies in the T-eigenspace $E(T)$ of $(c_1(F)\star_0)$.*

The principal asymptotic class of F is defined to be the class corresponding to a generator of the one-dimensional space $\mathcal{A}$.

Definition 3.6 ([29]). Suppose that a Fano manifold F satisfies Property $\mathcal{O}$. A cohomology class $A_F \in H^\bullet(F)$ satisfying

$$\mathcal{A} = \mathbb{C}\left[S(z) z^{-\mu} z^{c_1(F)} A_F \right]$$

is called the *principal asymptotic class*. Here $S(z)z^{-\mu}z^{c_1(F)}$ is the fundamental solution in Proposition 2.1. The principal asymptotic class is determined up to multiplication by a non-zero complex number; when $\langle [\mathrm{pt}], A_F \rangle \neq 0$, we can normalize A_F so that $\langle [\mathrm{pt}], A_F \rangle = 1$.

Since the space $\mathcal{A}$ is identified via the asymptotics near the irregular singular point $z = 0$ and the fundamental solution $S(z)z^{-\mu}z^{c_1(F)}$ is normalized at the regular singular point $z = \infty$, the definition of the class A_F involves *analytic continuation* along the positive real line $\mathbb{R}_{>0}$ on the z-plane. Note that $S(z)z^{-\mu}z^{c_1(F)}$ has a standard determination for $z \in \mathbb{R}_{>0}$ given by $z^{-\mu}z^{c_1(F)} = \exp(-\mu \log z)\exp(c_1(F)\log z)$ and $\log z \in \mathbb{R}$.

Conjecture 3.7 (Gamma Conjecture I [29]). *Let F be a Fano manifold satisfying Property $\mathcal{O}$. The principal asymptotic class A_F of F is given by the Gamma class $\widehat{\Gamma}_F$ of F.*

There is another description of the principal asymptotic class A_F in terms of solutions to the quantum differential equation. We introduce Givental's J-function [33] by the formula:

$$
\begin{aligned}
J_F(t) &= z^{\frac{\dim F}{2}} \left(S(z)z^{-\mu}z^{c_1(F)}\right)^{-1} 1 \qquad \text{with } t = z^{-1} \\
&= e^{c_1(F)\log t}\left(1 + \sum_{i=1}^{N} \sum_{d \in H_2(F,\mathbb{Z}), d \neq 0} \left\langle \frac{\phi_i}{1-\psi} \right\rangle_{0,1,d} \phi^i t^{c_1(F)\cdot d}\right)
\end{aligned}
\tag{7}
$$

where $S(z)z^{-\mu}z^{c_1(F)}$ is the fundamental solution in Proposition 2.1 and ψ is the first Chern class of the universal cotangent line bundle over the moduli space of stable maps. This is a cohomology-valued (and multi-valued) function. Since $S(z)z^{-\mu}z^{c_1(F)}$ is a fundamental solution of the quantum connection and by Remark 2.3, we have

$$P(t, \nabla_{c_1(F)})1 = 0 \iff P(t, t\tfrac{\partial}{\partial t})J_F(t) = 0$$

for any differential operator $P(t, t\frac{\partial}{\partial t}) \in \mathbb{C}\langle t, t\frac{\partial}{\partial t}\rangle$, where $\nabla_{c_1(F)} = t\frac{\partial}{\partial t} + c_1(F)\star_{c_1(F)\log t}$ is the quantum connection along the anticanonical line. In other words, $J_F(t)$ satisfies all the differential relations satisfied by the identity class 1 with respect to the connection $\nabla_{c_1(F)}$; in this sense $J_F(t)$ is a solution of the quantum connection. Differential operators $P(t, t\frac{\partial}{\partial t})$ annihilating $J_F(t)$ are called *quantum differential operators*. The principal asymptotic class A_F can be computed by the $t \to +\infty$ asymptotics of the J-function:

Proposition 3.8 ([29]). *Suppose that a Fano manifold F satisfies Property $\mathcal{O}$ and let A_F be the principal asymptotic class. Then we have an asymptotic expansion of the form:*

$$J_F(t) = Ct^{-\frac{\dim F}{2}}e^{Tt}(A_F + \alpha_1 t^{-1} + \alpha_2 t^{-2} + \cdots)$$

as $t \to +\infty$ on the positive real line, where $C \neq 0$ is a non-zero constant and $\alpha_i \in H^\bullet(F)$.

Proof. It follows from [29, Proposition 3.6.2] that $\lim_{t\to\infty} t^{\frac{\dim F}{2}} e^{-Tt} J_F(t)$ exists and is proportional to A_F. The fact that the remainder admits an asymptotic expansion of the form $\alpha_1 t^{-1} + \alpha_2 t^{-2} + \cdots$ follows from the proof there, in particular from [29, Proposition 3.2.1]. Q.E.D.

This proposition says that $\mathbb{C}J_F(t)$ converges to $\mathbb{C}A_F$ in the projective space $\mathbb{P}(H^\bullet(F))$ as $t \to +\infty$. Since $\langle [\mathrm{pt}], \widehat{\Gamma}_F\rangle = 1$, we obtain the following corollary.

Corollary 3.9 ([29, Corollary 3.6.9]). *Let F be a Fano manifold satisfying Property $\mathcal{O}$. Gamma conjecture* I *holds for F if and only if we have*

$$\widehat{\Gamma}_F = \lim_{t\to\infty} \frac{J_F(t)}{\langle [\mathrm{pt}], J_F(t)\rangle}.$$

We can replace the continuous limit in the above corollary with a discrete limit of ratios of the Taylor coefficients. Expand the J-function as:

$$J_F(t) = e^{c_1 \log t} \sum_{n=0}^{\infty} J_n t^n. \tag{8}$$

Note that $J_n = 0$ if n is not divisible by the Fano index r of F. We have the following:

Proposition 3.10 ([29, Theorem 3.7.1]). *Suppose that a Fano manifold F satisfies Property $\mathcal{O}$ and Gamma conjecture* I. *Let r be the Fano index of F. Then we have*

$$\liminf_{n\to\infty} \left| \frac{\langle \alpha, J_{rn}\rangle}{\langle [\mathrm{pt}], J_{rn}\rangle} - \langle \alpha, \widehat{\Gamma}_F\rangle \right| = 0$$

for every $\alpha \in H_\bullet(F)$ with $\alpha \cap c_1(F) = 0$.

Define the (unregularized and regularized) *quantum period* of F [16] to be

$$\begin{aligned} G_F(t) &= \langle [\mathrm{pt}], J_F(t)\rangle = \sum_{n=0}^{\infty} G_n t^n \\ \widehat{G}_F(\kappa) &= \sum_{n=0}^{\infty} n! G_n \kappa^n = \frac{1}{\kappa}\int_0^\infty G_F(t) e^{-t/\kappa} dt \end{aligned} \tag{9}$$

where $G_n = \langle [\mathrm{pt}], J_n \rangle$. It is shown in [29, Lemma 3.7.6] that if F satisfies Property $\mathcal{O}$ and if $\langle [\mathrm{pt}], A_F \rangle \neq 0$, the convergence radius of $\widehat{G}_F(\kappa)$ equals $1/T$. In particular

$$\limsup_{n\to\infty} \sqrt[rn]{(rn)!|G_{rn}|} = T. \tag{10}$$

Suppose that this limit sup (10) can be replaced with the limit, i.e. $\lim_{n\to\infty} \sqrt[rn]{(rn)!|G_{rn}|} = T$. Then the argument in the proof of [29, Theorem 3.7.1] shows, under the same assumption as in Proposition 3.10, that

$$\lim_{n\to\infty} \frac{\langle \alpha, J_{rn} \rangle}{\langle [\mathrm{pt}], J_{rn} \rangle} = \langle \alpha, \widehat{\Gamma}_F \rangle \tag{11}$$

for a class $\alpha \in H_\bullet(F)$ such that $\alpha \cap c_1(F) = 0$. Therefore we can consider the following variant of Gamma conjecture I:

Conjecture 3.11 (Gamma conjecture I'). *For a Fano manifold F and a class $\alpha \in H_\bullet(F)$ with $\alpha \cap c_1(F) = 0$, the limit formula* (11) *holds.*

Remark 3.12. By the above discussion, Gamma conjecture I' holds if Gamma conjecture I holds and one has $\lim_{n\to\infty} \sqrt[rn]{(rn)!|G_{rn}|} = T$.

The discrete limit on the left-hand side of (11) is called the *Apéry constant* (or Apéry limit) and was studied by Almkvist–van-Straten–Zudilin [1] in the context of Calabi–Yau differential equations and by Golyshev [37] and Galkin [26] for Fano manifolds. Under Gamma conjecture I', these limits are expressed in terms of the zeta values $\zeta(2), \zeta(3), \zeta(4), \dots$. For some Fano manifolds, they are precisely the limits which Apéry used to prove the irrationality of $\zeta(2)$ and $\zeta(3)$: an Apéry limit for the Grassmannian $G(2,5)$ gives a fast approximation of $\zeta(2)$ and an Apéry limit for the orthogonal Grassmannian $OG(5,10)$ gives a fast approximation of $\zeta(3)$ [26, 37]. Most of the Apéry limits of Fano manifolds are not fast enough to prove irrationality (see [26]). It would be very interesting to find a Fano manifold which gives a fast approximation of $\zeta(5)$, for example.

We give a sufficient condition that ensures that the limit sup in (10) can be replaced with the limit. When a Laurent polynomial $f(x) \in \mathbb{C}[x_1^\pm, \dots, x_m^\pm]$ is mirror to F, we expect that the quantum period $G_F(t)$ (9) for F should be given by the *constant term series* of f:

$$G_F(t) = \frac{1}{(2\pi\sqrt{-1})^m} \int_{(S^1)^m} e^{tf(x)} \frac{dx_1 \cdots dx_m}{x_1 \cdots x_m} = \sum_{n=0}^{\infty} \frac{1}{n!} \mathrm{Const}(f^n) t^n$$

where $\operatorname{Const}(f^n)$ denotes the constant term of the Laurent polynomial $f(x)^n$. When this holds, $f(x)$ is said to be a *weak Landau–Ginzburg model* of F [69].

Lemma 3.13. *Let F be a Fano manifold of index r. Suppose that F admits a weak Landau–Ginzburg model $f(x) \in \mathbb{C}[x_1^{\pm}, \dots, x_m^{\pm}]$ whose coefficients are non-negative real numbers. Suppose also that* $\operatorname{Const}(f^{rn}) \neq 0$ *for all but finitely many $n \in \mathbb{Z}_{\geqslant 0}$. Then the coefficients G_n of the quantum period* (9) *are non-negative and the limit*

$$\lim_{n\to\infty} \sqrt[rn]{(rn)!|G_{rn}|} = \lim_{n\to\infty} \sqrt[rn]{\operatorname{Const}(f^{rn})}$$

exists.

Proof. Some of the techniques here are borrowed from [27]. We set $\alpha_n = \log(\sqrt[rn]{\operatorname{Const}(f^{rn})})$. It suffices to show that $\lim_{n\to\infty} \alpha_n$ exists. By assumption, there exists $n_0 \in \mathbb{Z}_{\geqslant 0}$ such that α_n is well-defined for all $n \geqslant n_0$. Since $\operatorname{Const}(f^{r(n+m)}) \geqslant \operatorname{Const}(f^{rn}) \operatorname{Const}(f^{rm})$, we have

$$\alpha_{n+m} \geqslant \frac{n}{n+m}\alpha_n + \frac{m}{n+m}\alpha_m.$$

Set $\alpha := \limsup_{n\to\infty} \alpha_n$. For any $\epsilon > 0$, there exists $n_1 \geqslant 1$ such that $\alpha_{n_1} \geqslant \alpha - \epsilon$. Then we have, for all $k \geqslant 1$ and $0 \leqslant i < n_1$,

$$\alpha_{kn_1+n_0+i} \geqslant \frac{n_0+i}{kn_1+n_0+i}\alpha_{n_0+i} + \frac{kn_1}{kn_1+n_0+i}\alpha_{n_1}.$$

The right-hand side converges to α_{n_1} as $k \to \infty$. This implies that there exists $n_2 \geqslant n_0$ such that for all $n \geqslant n_2$, we have

$$\alpha_n \geqslant \alpha - 2\epsilon.$$

Since $\epsilon > 0$ was arbitrary, this implies $\liminf_{n\to\infty} \alpha_n \geqslant \alpha$ and the conclusion follows. Q.E.D.

Remark 3.14. When discussing the continuous limit, we only need to assume Part (1) of Property $\mathcal{O}$ (Definition 3.1). More precisely, Propositions 3.5, 3.8 and Corollary 3.9 hold for Fano manifolds satisfying only Part (1) of Property $\mathcal{O}$, and Gamma conjecture I (Conjecture 3.7) makes sense for such Fano manifolds. On the other hand, we need Part (2) of Property $\mathcal{O}$ in the proof of Proposition 3.10.

Remark 3.15. Golyshev–Zagier [40] proved Gamma conjecture I for Fano threefolds of Picard rank one (there are 17 families of such). More precisely, they showed that the limit formula in Corollary 3.9 holds for those varieties. In this case, it is straightforward to check Property $\mathcal{O}$ using Golyshev's multiplication table [35], [36, §5.6] for $(c_1(F)\star_0)$. For rank one Fano threefolds, the characteristic polynomial $\det(1-tc_1(F)\star_0)$ equals the symbol of the regularized quantum differential equation. Hence one can easily check Property $\mathcal{O}$ also from the list [17].

§4. Gamma conjecture II

The small quantum cohomology $(H^\bullet(F), \star_0)$ of a Fano manifold F is said to be semisimple if it is isomorphic to the direct sum of $\mathbb{C}$ as a ring. This is equivalent to the condition that $(H^\bullet(F), \star_0)$ has no nilpotent elements. In this section, we give a refinement of Gamma conjecture I for a Fano manifold with semisimple quantum cohomology.

4.1. Formal fundamental solution

Suppose that $(H^\bullet(F), \star_0)$ is semisimple. Then we have an idempotent basis $\psi_1, \dots, \psi_N$ of $H^\bullet(F)$ such that $\psi_i \star_0 \psi_j = \delta_{ij}\psi_i$. Define the normalized idempotent basis $\Psi_1, \dots, \Psi_N$ to be $\Psi_i = \psi_i/\sqrt{(\psi_i, \psi_i)}$, where (ψ_i, ψ_i) denotes the Poincaré pairing of ψ_i with itself. Note that $(\Psi_1, \dots, \Psi_N)$ is unique up to sign and ordering. We set

$$\Psi = \begin{pmatrix} | & | & & | \\ \Psi_1 & \Psi_2 & \cdots & \Psi_N \\ | & | & & | \end{pmatrix}.$$

This is a matrix with column vectors Ψ_i. We may regard it as a linear map $\mathbb{C}^N \to H^\bullet(F)$. Let $u_1, \dots, u_N$ be the eigenvalues of $(c_1(F)\star_0)$ such that $c_1(F) \star_0 \Psi_i = u_i \Psi_i$. Let U be the diagonal matrix with entries $u_1, \dots, u_N$:

$$U = \begin{pmatrix} u_1 & & & \\ & u_2 & & \\ & & \ddots & \\ & & & u_N \end{pmatrix}.$$

The following proposition is well-known in the context of Frobenius manifolds.

Proposition 4.1 ([20, Lectures 4,5], [74, Theorem 8.15]). *Suppose that the small quantum cohomology $(H^\bullet(F), \star_0)$ is semisimple. The quantum connection (4) near the irregular singular point $z = 0$ admits a formal fundamental matrix solution of the form:*

$$\Psi R(z) e^{-U/z}$$

where $R(z) = \mathrm{id} + R_1 z + R_2 z^2 + \cdots \in \mathrm{End}(\mathbb{C}^N)[\![z]\!]$ *is a matrix-valued formal power series. The formal solution* $\Psi R(z) e^{-U/z}$ *is unique up to multiplication by a signed permutation matrix from the right (which corresponds to the ambiguity of* $\Psi_1, \ldots, \Psi_N$*).*

Remark 4.2. Since $R(z)$ is a formal power series in positive powers of z, the product $R(z)e^{-U/z}$ does not make sense as a power series. The meaning of the proposition is that the formal gauge transformation by $\Psi R(z)$ turns $\nabla_{z\partial_z}$ into $z\partial_z - U/z$.

Remark 4.3. We do not need to assume that $u_1, \ldots, u_N$ are mutually distinct.

4.2. Lift to an analytic solution

By choosing an angular sector, the above formal solution can be lifted to an actual analytic solution. This is an instance of the *Hukuhara–Turrittin theorem* for irregular connections (see e.g. [77, Theorem 19.1], [72, II, 5.d]). We say that a phase $\phi \in \mathbb{R}$ (or $e^{\sqrt{-1}\phi} \in S^1$) is *admissible* for a multiset $\{u_1, u_2, \ldots, u_N\} \subset \mathbb{C}$ if $\mathrm{Im}(u_i e^{-\sqrt{-1}\phi}) \neq \mathrm{Im}(u_j e^{-\sqrt{-1}\phi})$ for every pair (u_i, u_j) with $u_i \neq u_j$, i.e. $e^{\sqrt{-1}\phi}$ is not parallel to any non-zero difference $u_i - u_j$.

Proposition 4.4 ([77, Theorem 12.2], [5, Theorem A], [20, Lectures 4,5], [12, §8], [29, Proposition 2.5.1])**.** *Let* $\phi \in \mathbb{R}$ *be an admissible phase for the spectrum* $\{u_1, u_2, \ldots, u_N\}$ *of* $(c_1(F)\star_0)$*. There exist* $\epsilon > 0$ *and an analytic fundamental solution* $Y_\phi(z) = (y_1^\phi(z), \ldots, y_N^\phi(z))$ *for the quantum connection* (4) *on the angular sector* $|\arg(z) - \phi| < \frac{\pi}{2} + \epsilon$ *around* $z = 0$ *such that one has the asymptotic expansion*

$$Y_\phi(z) e^{U/z} \sim \Psi R(z) \tag{12}$$

as $z \to 0$ *in the sector* $|\arg(z) - \phi| < \frac{\pi}{2} + \epsilon$*, where* $\Psi R(z) e^{-U/z}$ *is the formal fundamental solution in Proposition* 4.1*. Such an analytic solution* $Y_\phi(z)$ *is unique when we fix the sign and the ordering of* $\Psi_1, \ldots, \Psi_N$*.*

Remark 4.5. Notice that each flat section $y_i^\phi(z)$ has the exponential asymptotics $\sim e^{-u_i/z}\Psi_i$ as $z \to 0$ in the sector $|\arg(z) - \phi| < \frac{\pi}{2} + \epsilon$.

Remark 4.6. The precise meaning of the asymptotic expansion (12) is as follows. For any $0 < \epsilon' < \epsilon$ and $n \in \mathbb{Z}_{\geqslant 0}$, there exists a constant $C = C(\epsilon', n)$ such that

$$\left\| Y_\phi(z) e^{-U/z} - \sum_{k=0}^{n} \Psi R_k z^k \right\| \leqslant C|z|^{n+1}$$

for all z with $|\arg z - \phi| \leqslant \frac{\pi}{2} + \epsilon'$ and $|z| \leqslant 1$, where we write $R(z) = \sum_{k=0}^{\infty} R_k z^k$.

4.3. Asymptotic basis and Gamma conjecture II

Let $Y_\phi(z) = (y_1^\phi(z), \dots, y_N^\phi(z))$ be the analytic fundamental solution in Proposition 4.4. We regard $Y_\phi(z)$ as a function defined on the universal cover of $\mathbb{C}^\times$; initially it is defined on the angular sector $|\arg(z) - \phi| < \pi + \epsilon$ with $|z| \ll 1$, but can be analytically continued to the whole universal cover since it is a solution to a linear differential equation. For an admissible phase ϕ for $\{u_1, \dots, u_N\}$, we define the higher asymptotic classes $A_{F,i}^\phi \in H^\bullet(F)$, $i = 1, \dots, N$ by

$$y_i^\phi(z)\Big|_{\substack{\text{parallel translate}\\ \text{to } \arg(z)=0}} = \frac{1}{(2\pi)^{\dim F/2}} S(z) z^{-\mu} z^{c_1(F)} A_{F,i}^\phi$$

where $S(z)z^{-\mu}z^{c_1(F)}$ is the fundamental solution for the quantum connection in Proposition 2.1. We call $\{A_{F,1}^\phi, A_{F,2}^\phi, \dots, A_{F,N}^\phi\}$ the *asymptotic basis* at the phase ϕ. The asymptotic basis is the same as what Dubrovin [20] called the *central connection matrix.*

Remark 4.7. Suppose that F satisfies Property $\mathcal{O}$. When we take an admissible phase ϕ from the interval $(-\frac{\pi}{2}, \frac{\pi}{2})$, the class $A_{F,i}^\phi$ corresponding to the eigenvalue $u_i = T$ is proportional to the principal asymptotic class A_F.

Remark 4.8. The asymptotic basis at a phase ϕ is unique up to sign and ordering. The sign and the ordering depend on those of the normalized idempotents $\Psi_1, \dots, \Psi_N$. Each asymptotic class $A_{F,i}^\phi$ is marked by the eigenvalue u_i of $(c_1(F)\star_0)$. With respect to the pairing $[\cdot,\cdot)$ in (2), these data $\{(A_{F,i}^\phi, u_i)\}_{i=1}^N$ form a *marked reflection system* [29], see also Remark 4.13.

Conjecture 4.9 (Gamma conjecture II [29]). *Suppose that a Fano manifold F has a semisimple small quantum cohomology and that $D^b_{\mathrm{coh}}(F)$ has a full exceptional collection. Let ϕ be an admissible phase for the spectrum $\{u_1, \dots, u_N\}$ of $(c_1(F)\star_0)$. We number the eigenvalues $u_1, \dots, u_N$ so that $\mathrm{Im}(e^{-\sqrt{-1}\phi}u_1) \geqslant \mathrm{Im}(e^{-\sqrt{-1}\phi}u_2) \geqslant \cdots \geqslant \mathrm{Im}(e^{-\sqrt{-1}\phi}u_N)$. There exists a full exceptional collection $E_1^\phi, \dots, E_N^\phi$ such that $A_{F,i}^\phi = \widehat{\Gamma}_F \cdot \mathrm{Ch}(E_i^\phi)$.*

Recall that $\mathrm{Ch}(E) = \sum_{p=0}^{\dim F} (2\pi\sqrt{-1})^p \operatorname{ch}_p(E)$ is the modified Chern character. We stated Gamma conjecture II under a slightly restrictive assumption: as in the original formulation [29, Conjecture 4.6.1], we can state Gamma conjecture II under the assumption that the big quantum cohomology is analytic and semisimple at some $\tau \in H^\bullet(F)$. We restrict to the small quantum cohomology just for exposition. Note that there are Fano manifolds (symplectic isotropic Grassmannians $IG(2,2n)$) such that the small quantum cohomology is not semisimple, but that the big quantum cohomology is generically semisimple [30, 67, 65, 18].

Remark 4.10. Part (3) of Dubrovin's conjecture [21, Conjecture 4.2.2] says that the columns of the central connection matrix are given by $C'(\mathrm{Ch}(E_i))$ for some linear operator $C' \in \mathrm{End}(H^\bullet(F))$ commuting with $c_1(F)\cup$. Gamma conjecture II says that $C' = \widehat{\Gamma}_F \cup$. Recently Dubrovin [23] also proposed the same conjecture as Gamma conjecture II.

4.4. Stokes matrix

Let $Y_\phi(z)$ and $Y_\phi^-(z)$ be the analytic fundamental solutions from Proposition 4.4 associated respectively to admissible directions $e^{\sqrt{-1}\phi}$ and $-e^{\sqrt{-1}\phi}$. The domains of definitions of Y_ϕ and Y_ϕ^- are shown in Figure 1. Let $\Pi_\pm$ be the angular regions as in Figure 1 which are components of the intersection of the domains of Y_ϕ and Y_ϕ^-. The *Stokes matrices* are the constant matrices S^ϕ and S_-^ϕ satisfying

$$Y_\phi(z) = Y_\phi^-(z) S^\phi \qquad \text{for } z \in \Pi_+;$$
$$Y_\phi(z) = Y_\phi^-(z) S_-^\phi \qquad \text{for } z \in \Pi_-.$$

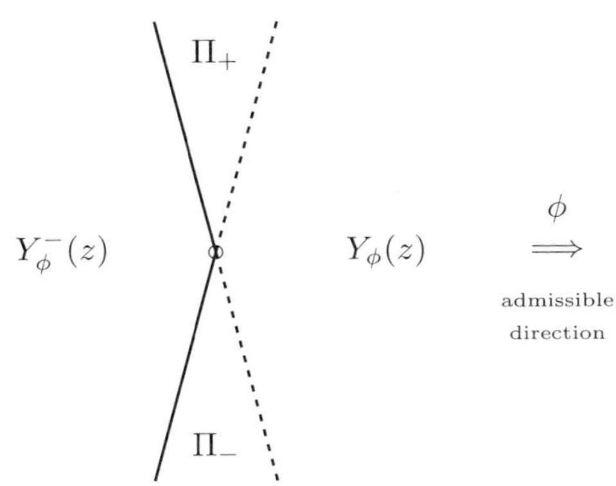

Fig. 1. $Y_\phi^-(z)$ is defined on the left side of the dotted line and $Y_\phi(z)$ is defined on the right side of the solid line.

Proposition 4.11 ([20, Theorem 4.3], [29, Proposition 2.6.4]). *Let $Y_\phi(z) = (y_1^\phi(z), \dots, y_N^\phi(z))$ be the fundamental solution from Proposition 4.4. The Stokes matrices at phase ϕ are given by $S_{ij}^\phi = S_{-,ji}^\phi = (y_i^\phi(e^{-\pi\sqrt{-1}}z), y_j^\phi(z))$, where $y_i^\phi(e^{-\pi\sqrt{-1}}z)$ denotes the analytic continuation of $y_i(z)$ along the path $[0,\pi] \ni \theta \mapsto e^{-\sqrt{-1}\theta}z$. The flat sections $y_1^\phi, \dots, y_N^\phi$ are semi-orthogonal in the following sense:*

$$S_{ij}^\phi = \begin{cases} 0 & \text{if } (i \neq j \text{ and } u_i = u_j) \text{ or } \operatorname{Im}(e^{-\sqrt{-1}\phi}u_i) < \operatorname{Im}(e^{-\sqrt{-1}\phi}u_j); \\ 1 & \text{if } i = j. \end{cases}$$

Using the non-symmetric pairing $[\cdot,\cdot)$ given in (2), we have

$$\begin{aligned} S_{ij}^\phi &= \left(y_i^\phi(e^{-\sqrt{-1}\pi}z), y_j^\phi(z)\right) \\ &= \frac{1}{(2\pi)^{\dim F}}\left(S(-z)z^{-\mu}e^{\pi\sqrt{-1}\mu}z^{c_1(F)}e^{-\pi\sqrt{-1}c_1(F)}A_{F,i}^\phi, S(z)z^{-\mu}z^{c_1(F)}A_{F,j}^\phi\right) \\ &= [A_{F,i}^\phi, A_{F,j}^\phi) \end{aligned}$$

where we used the fact that $(S(-z)\alpha, S(z)\beta) = (\alpha,\beta)$ and $(z^{-\mu}\alpha, z^{-\mu}\beta) = (\alpha,\beta)$ (see [20]). Therefore, the factorization (1) of the Hirzebruch–Riemann–Roch formula implies the following corollary.

Corollary 4.12. *Suppose that a Fano manifold F satisfies Gamma conjecture* II. *Then there exists a full exceptional collection $E_1, \dots, E_N$ of $D^b_{\mathrm{coh}}(F)$ such that $\chi(E_i, E_j)$ equals the Stokes matrix S_{ij}. (This conclusion is part* (2) *of Dubrovin's conjecture* [21, Conjecture 4.2.2]).

Remark 4.13. The asymptotic basis can be defined similarly for the big quantum product $\star_\tau$ with $\tau \in H^\bullet(F)$ as far as $\star_\tau$ is semisimple. We have an asymptotic basis $A_{F,i}^{\phi,\tau}$ depending on both ϕ and τ, and Gamma conjecture II makes sense at general $\tau \in H^\bullet(F)$. The truth of the Gamma conjecture II is, however, independent of the choice of (τ,ϕ). This is because the asymptotic basis changes (discontinuously) by mutation as (τ,ϕ) varies, and we can consider the corresponding mutation for exceptional collections. The right mutation of asymptotic bases takes the form

$$\begin{aligned} &(A_1, A_2, \dots, \overset{i}{A_i}, \overset{i+1}{A_{i+1}}, \dots A_N) \\ &\qquad\mapsto (A_1, A_2, \dots, \overset{i}{A_{i+1}}, \overset{i+1}{A_i - [A_i, A_{i+1})A_{i+1}}, \cdots, A_N) \end{aligned}$$

if we order the asymptotic basis so that $\operatorname{Im}(e^{-\sqrt{-1}\phi}u_1) \geqslant \operatorname{Im}(e^{-\sqrt{-1}\phi}u_2) \geqslant \cdots \geqslant \operatorname{Im}(e^{-\sqrt{-1}\phi}u_N)$. The right mutation happens when the eigenvalue u_{i+1} crosses the ray $u_i + \mathbb{R}_{\geqslant 0} e^{\sqrt{-1}\phi}$. (The left mutation is the inverse of the right mutation). The braid group action on asymptotic bases and Stokes matrices has been studied by Dubrovin, see [20, 29] for more details.

Remark 4.14. In general, the quantum connection of a smooth projective variety (or a projective orbifold) X is underlain by the integral local system consisting of flat sections $(2\pi)^{-\frac{\dim F}{2}} S(z) z^{-\mu} z^{c_1(F)} \widehat{\Gamma}_X \operatorname{Ch}(E)$ with $E \in K^0(X)$. This is called the $\widehat{\Gamma}$-*integral structure* [50, 55]. In this language, Gamma conjecture implies that this integral structure should be compatible with the Stokes structure at the irregular singular point $z = 0$ of the quantum connection. Katzarkov–Kontsevich–Pantev [55] imposed the compatibility of rational structure with Stokes structure as part of conditions for nc-Hodge structure. Our Gamma conjecture can be viewed as an adaptation of their compatibility condition to the quantum cohomology of Fano manifolds. Note that Gamma conjecture II implies the integrality of the Stokes matrix S_{ij}.

Remark 4.15. Recently, Sanda–Shamoto [73] proposed a generalization of Gamma Conjecture II to the non-semisimple case, and called it Dubrovin type conjecture. (Among the authors of [29], such a conjecture was locally called "Gamma Conjecture III", although we did not have a precise formulation.)

§5. Mirror heuristics

In this section we give a heuristic argument which gives the mirror oscillatory integral and the Gamma class from the polynomial loop space (quasi map space) for $\mathbb{P}^{N-1}$. This is motivated by Givental's equivariant Floer theory heuristics [31, 32]. The argument in this section can be applied more generally to toric varieties to yield their mirrors and Gamma classes.

5.1. Polynomial loop space

Givental [31, 32] conjectured that the quantum D-module (i.e. quantum connection) should be identified with the S^1-equivariant Floer theory for the free loop space (see also [76, 49, 2]). Following Givental, we consider an algebraic version of the loop space instead of the actual free loop space. The (Laurent) *polynomial loop space* of $\mathbb{P}^{N-1}$ is defined to be

$$L_{\mathrm{poly}}\mathbb{P}^{N-1} = \left(\mathbb{C}[\zeta, \zeta^{-1}]^N \setminus \{0\}\right) / \mathbb{C}^\times$$

where $\mathbb{C}^\times$ acts on $\mathbb{C}[\zeta,\zeta^{-1}]^N$ by scalar multiplication. The polynomial loop space $L_{\mathrm{poly}}\mathbb{P}^{N-1}$ can be also described as a symplectic reduction of $\mathbb{C}[\zeta,\zeta^{-1}]^N \cong \mathbb{C}^{\oplus\infty}$ by the diagonal S^1-action. For a point $(a_1(\zeta),\dots,a_N(\zeta)) \in \mathbb{C}[\zeta,\zeta^{-1}]^N$, we write $a_i(\zeta) = \sum_{n\in\mathbb{Z}} a_{i,n}\zeta^n$ and regard $(a_{i,n} : 1 \leqslant i \leqslant N, n \in \mathbb{Z})$ as a co-ordinate system on $\mathbb{C}[\zeta,\zeta^{-1}]^N$. With respect to the standard Kähler form

$$\omega = \frac{\sqrt{-1}}{2}\sum_{i=1}^{N}\sum_{n\in\mathbb{Z}} da_{i,n}\wedge d\overline{a_{i,n}} \tag{13}$$

on $\mathbb{C}[\zeta,\zeta^{-1}]^N$, the diagonal S^1-action admits a moment map[3] $\mu\colon \mathbb{C}[\zeta,\zeta^{-1}]^N \to \mathrm{Lie}(S^1)^* \cong \mathbb{R}$ given by

$$\mu(a_1(\zeta),\dots,a_N(\zeta)) = \pi\sum_{i=1}^{N}\sum_{n\in\mathbb{Z}} |a_{i,n}|^2.$$

This satisfies $\iota_X\omega + d\mu = 0$ for the vector field $X = 2\pi\sqrt{-1}\sum_{i=1}^N\sum_{n\in\mathbb{Z}} (a_{i,n}\frac{\partial}{\partial a_{i,n}} - \overline{a_{i,n}}\frac{\partial}{\partial \overline{a_{i,n}}})$ generating the S^1-action. Then we have

$$L_{\mathrm{poly}}\mathbb{P}^{N-1} \cong \mu^{-1}(u)/S^1$$

for every $u \in \mathbb{R}_{>0}$. Via the symplectic reduction, $L_{\mathrm{poly}}\mathbb{P}^{N-1}$ is equipped with the reduced symplectic form ω_u such that the pull-back of ω_u to $\mu^{-1}(u)$ equals the restriction $\omega|_{\mu^{-1}(u)}$. The class ω_u represents the cohomology class $uc_1(\mathcal{O}(1))$ on $L_{\mathrm{poly}}\mathbb{P}^{N-1}$.

The loop rotation defines the S^1-action on $L_{\mathrm{poly}}\mathbb{P}^{N-1}$ given by $[a_1(\zeta),\dots,a_N(\zeta)] \mapsto [a_1(\lambda\zeta),\dots,a_N(\lambda\zeta)]$ with $\lambda \in S^1$. With respect to the reduced symplectic form ω_u on $L_{\mathrm{poly}}\mathbb{P}^{N-1}$, this S^1-action admits a moment map H_u given by

$$H_u([a_1(\zeta),\dots,a_N(\zeta)]) = \pi\sum_{i=1}^{N}\sum_{n\in\mathbb{Z}} n|a_{i,n}|^2$$

for $(a_1(\zeta),\dots,a_N(\zeta)) \in \mu^{-1}(u)$. The function H_u is an analogue of the *action functional* on the free loop space.

[3]We identify $\mathrm{Lie}(S^1)^*$ with $\mathbb{R}$ so that the radian angular form $d\theta$ on $S^1 = \{e^{\sqrt{-1}\theta} : \theta \in \mathbb{R}\}$ corresponds to 2π. With this choice, the reduced symplectic form ω_u on $\mu^{-1}(u)/S^1$ represents an integral cohomology class precisely when $u \in \mathbb{Z}$.

Remark 5.1. More precisely, the polynomial loop space should be regarded as an analogue of the *universal cover* of the free loop space $\mathcal{L}\mathbb{P}^{N-1}$. In fact, we have an analogue of the deck transformation on $L_{\mathrm{poly}}\mathbb{P}^{N-1}$ given by $[a_1(\zeta), \dots, a_N(\zeta)] \mapsto [\zeta a_1(\zeta), \dots, \zeta a_N(\zeta)]$; this corresponds to a generator of $\pi_1(\mathcal{L}\mathbb{P}^{N-1}) \cong \pi_2(\mathbb{P}^{N-1}) = \mathbb{Z}$. Recall that the action functional is defined on the universal cover of $\mathcal{L}\mathbb{P}^{N-1}$.

5.2. Solution as a path integral

Recall that symplectic Floer theory is an infinite-dimensional analogue of the Morse theory with respect to the action functional on the loop space. We consider the Morse theory on $L_{\mathrm{poly}}\mathbb{P}^{N-1}$ with respect to the Bott–Morse function H_u. Note that the critical set of H_u is a disjoint union of infinite copies of $\mathbb{P}^{N-1}$ given by $(\mathbb{P}^{N-1})_n = \{[a_1\zeta^n, \dots, a_N\zeta^n]\} \subset L_{\mathrm{poly}}\mathbb{P}^{N-1}$ for each $n \in \mathbb{Z}$. The Floer fundamental cycle Δ is defined to be the closure of the stable manifold associated to the critical component $(\mathbb{P}^{N-1})_0$. This is given by

$$\Delta = \Big\{[a_1(\zeta), \dots, a_N(\zeta)] \in L_{\mathrm{poly}}\mathbb{P}^{N-1} : a_i(\zeta) \in \mathbb{C}[\zeta]\Big\}.$$

Under the isomorphism between the Floer homology and the quantum cohomology [68], the Floer fundamental cycle corresponds to the identity class $1 \in H^\bullet(\mathbb{P}^{N-1})$. Consider the following equivariant 2-form on $L_{\mathrm{poly}}\mathbb{P}^{N-1}$ (in the Cartan model)

$$\Omega_u = \omega_u - zH_u$$

where $z \in H^2_{S^1}(\mathrm{pt}, \mathbb{Z})$ is a positive generator. Note that Ω is equivariantly closed since H_u is a Hamiltonian for the S^1-action. Givental [31, 32] proposed that the infinite-dimensional integral (which could be viewed as a Feynman path integral)

$$\int_\Delta e^{\Omega_u/z} \tag{14}$$

should give a solution to the quantum differential equation for $\mathbb{P}^{N-1}$. This may be viewed as the image of the Floer fundamental cycle Δ under the homomorphism $C \mapsto \int_C e^{\Omega_u/z}$. The integral does not have a rigorous definition in mathematics, but we can heuristically compute this quantity in two different ways — one is by a direct computation and the other is by localization. The former method yields a mirror oscillatory integral and the latter (due to Givental [31, 32]) yields the J-function of $\mathbb{P}^{N-1}$. In Givental's original calculation, however, an infinite (constant) factor corresponding to the Gamma class has been ignored.

From this calculation we obtain a (mirror) integral representation of the $\widehat{\Gamma}_{\mathbb{P}^{N-1}}$-component of the J-function.

5.3. Direct calculation

We compute the infinite dimensional integral (14) directly. We regard z as a positive real parameter. Since $\Delta = (\mu^{-1}(u) \cap \mathbb{C}[\zeta]^N)/S^1$, we have

$$(14) = \int_{(\mu^{-1}(u)\cap\mathbb{C}[\zeta]^N)/S^1} e^{-H_u} e^{\omega_u/z} = \int_{\mu^{-1}(u)\cap\mathbb{C}[\zeta]^N} e^{-H_u} \frac{d\theta}{2\pi} \wedge e^{\omega/z}$$

where $d\theta$ is the angular form (connection form) on the principal S^1-bundle $\mu^{-1}(u) \to \mu^{-1}(u)/S^1$ given by $d\theta = \frac{\pi}{2\mu\sqrt{-1}} \sum_{i=1}^N \sum_{k\in\mathbb{Z}} (\overline{a_{i,n}} da_{i,n} - a_{i,n} d\overline{a_{i,n}})$ (satisfying $d\theta(X) = 2\pi$) and ω is the Kähler form (13). Changing co-ordinates $a_{i,n} \to \sqrt{z} a_{i,n}$, we find that this equals

$$\int_{\mu^{-1}(u/z)\cap\mathbb{C}[\zeta]^N} e^{-zH_u} \frac{d\theta}{2\pi} \wedge e^{\omega}.$$

If $\mathbb{C}[\zeta]^N$ were a finite dimensional vector space, the top-degree component of the differential form $d\mu \wedge \frac{d\theta}{2\pi} \wedge e^{\omega}$ would equal the Liouville volume form on $\mathbb{C}[\zeta]^N$ associated to ω. Therefore we may write this as

$$\int_{\mathbb{C}[\zeta]^N} \delta(\mu - u/z) e^{-zH_u} d\,\mathrm{vol}$$

where $d\,\mathrm{vol} = \bigwedge_{i=1}^N \bigwedge_{n=0}^\infty \left(\frac{\sqrt{-1}}{2} da_{i,n} \wedge d\overline{a_{i,n}}\right)$ and $\delta(x)$ is the Dirac delta-function. Using $\delta(x) = \frac{1}{2\pi}\int_{-\infty}^\infty e^{\sqrt{-1}x\xi} d\xi$, we can compute this as:

$$\begin{aligned}
&\frac{1}{2\pi}\int_{\mathbb{C}[\zeta]^N} d\,\mathrm{vol} \int_{-\infty}^{\infty} d\xi e^{\sqrt{-1}\xi(\mu - u/z)} e^{-zH_u} \\
(15) \qquad &= \frac{1}{2\pi}\int_{-\infty}^{\infty} d\xi e^{-\sqrt{-1}\xi u/z} \int_{\mathbb{C}[\zeta]^N} d\,\mathrm{vol} \prod_{i=1}^N \prod_{n=0}^\infty e^{-\pi|a_{i,n}|^2(nz - \sqrt{-1}\xi)} \\
&= \frac{1}{2\pi}\int_{-\infty}^{\infty} d\xi e^{-\sqrt{-1}\xi u/z} \prod_{i=1}^N \prod_{n=0}^\infty \frac{1}{nz - \sqrt{-1}\xi}
\end{aligned}$$

By the ζ-function regularization, we can regularize the infinite product to get:

$$\frac{1}{\prod_{n=0}^\infty (nz - \sqrt{-1}\xi)} \sim (2\pi z)^{-1/2} z^{-\sqrt{-1}\xi/z} \Gamma(-\sqrt{-1}\xi/z).$$

Therefore (15) should equal, after the change $\xi \to z\xi$ of co-ordinates,

$$\frac{z}{2\pi(2\pi z)^{N/2}} \int_{-\infty}^{\infty} d\xi e^{-\sqrt{-1}(u+N\log z)\xi}\Gamma(-\sqrt{-1}\xi)^N. \tag{16}$$

This integral makes sense if we perturb the integration contour so that it avoids the singularity at $\xi = 0$. We will consider the perturbed contour from $-\infty+\sqrt{-1}\epsilon$ to $\infty+\sqrt{-1}\epsilon$ with $\epsilon > 0$. Then the integral (16) is just a (finite-dimensional) Fourier transform and the following discussion can be made completely rigorous. Using the integral representation $\Gamma(z) = \int_0^\infty e^{-x}z^{x-1}dx$ of the Γ-function, we find that (16) equals

(17)

$$\begin{aligned}
&\frac{z}{2\pi(2\pi z)^{N/2}} \\
&\times \int_{-\infty}^{\infty} d\xi \int_{[0,\infty)^N} \frac{dx_1\cdots dx_N}{x_1\cdots x_N} e^{-\sqrt{-1}\xi(u+N\log z+\sum_{i=1}^N \log x_i)} e^{-(x_1+\cdots+x_N)} \\
&= \frac{z}{(2\pi z)^{N/2}} \\
&\quad \times \int_{[0,\infty)^N} \frac{dx_1\cdots dx_N}{x_1\cdots x_N} \delta\left(u+N\log z+\textstyle\sum_{i=1}^N \log x_i\right) e^{-(x_1+\cdots+x_N)} \\
&= \frac{z}{(2\pi z)^{N/2}} \int_{[0,\infty)^{N-1}} \frac{dx_1}{x_1}\cdots\frac{dx_{N-1}}{x_{N-1}} e^{-\left(x_1+\cdots+x_{N-1}+\frac{e^{-u}}{x_1\cdots x_{N-1}}\right)/z}
\end{aligned}$$

where in the last line we considered the co-ordinate change $x_i \to x_i/z$. The Landau–Ginzburg mirror of $\mathbb{P}^{N-1}$ is given by the Laurent polynomial function $f(x_1,\ldots,x_{N-1}) = x_1+\cdots+x_{N-1}+\frac{e^{-u}}{x_1\cdots x_{N-1}}$ and this is the associated oscillatory integral [31, 45].

5.4. Calculation by localization

Next we calculate the quantity (14) using the localization formula of equivariant cohomology (or Duistermaat–Heckman formula) [24, 9, 3]. The S^1-fixed set in Δ is the disjoint union of $(\mathbb{P}^{N-1})_n$ with $n \geqslant 0$ and we have $[\Omega_u]|_{(\mathbb{P}^{N-1})_n} = uh - znu$, where $h := c_1(\mathcal{O}(1)) \in H^2(\mathbb{P}^{N-1})$ is the hyperplane class. Therefore

$$\int_\Delta e^{\Omega_u/z} = \sum_{n=0}^{\infty} \int_{(\mathbb{P}^{N-1})_n} \frac{e^{uh/z-nu}}{e_{S^1}(\mathcal{N}_n)}$$

where $\mathcal{N}_n$ is the (infinite-rank) normal bundle of $(\mathbb{P}^{N-1})_n$ in Δ and

$$e_{S^1}(\mathcal{N}_n) = \prod_{k\geqslant -n, k\neq 0} (h+kz)^N = \prod_{k=1}^{\infty}(h+kz)^N \prod_{k=1}^{n}(h-kz)^N.$$

The infinite factor $\prod_{k=1}^{\infty}(h+kz)^N$ was discarded in the original calculation of Givental [32]. Using again the ζ-function regularization, we find that this factor yields the Gamma class:

$$\frac{1}{\prod_{k=1}^{\infty}(h+kz)} \sim \left(\frac{z}{2\pi}\right)^{1/2} z^{h/z}\Gamma(1+h/z).$$

Thus we should have

$$\begin{aligned}\int_{\Delta} e^{\Omega_u/z} &= \sum_{n=0}^{\infty}\int_{\mathbb{P}^{N-1}}\left(\frac{z}{2\pi}\right)^{N/2} z^{Nh/z}\Gamma(1+h/z)^N \frac{e^{uh/z-nu}}{\prod_{k=1}^{n}(h-kz)^N} \\ &= \frac{z}{(2\pi z)^{N/2}}\int_{\mathbb{P}^{N-1}}\Gamma(1+h)^N\sum_{n=0}^{\infty}\frac{(e^u z^N)^{h-n}}{\prod_{k=1}^{n}(h-k)^N}.\end{aligned}$$

Recall that the J-function (7) of $\mathbb{P}^{N-1}$ is given by [33]

$$J_{\mathbb{P}^{N-1}}(t) = \sum_{n=0}^{\infty}\frac{t^{N(h+n)}}{\prod_{k=1}^{n}(h+k)^N}. \tag{18}$$

Thus, using $\widehat{\Gamma}_{\mathbb{P}^{N-1}} = \Gamma(1+h)^N$, we obtain

$$\int_{\Delta} e^{\Omega_u/z} = \frac{z}{(2\pi z)^{N/2}}(2\pi\sqrt{-1})^{N-1}\left[J_{\mathbb{P}^{N-1}}(e^{\pi\sqrt{-1}}t), \widehat{\Gamma}_{\mathbb{P}^{N-1}}\right) \tag{19}$$

under the identification $t = e^{-u/N}z^{-1}$, where $[\cdot,\cdot)$ is the non-symmetric pairing defined in (2).

Remark 5.2. The quantity (19) coincides, up to a factor, with the *quantum cohomology central charge* of $\mathcal{O}_{\mathbb{P}^{N-1}}$ [50, 29]. For a vector bundle E on a Fano manifold F, the quantum cohomology central charge $Z(E)$ is defined to be:

$$\begin{aligned}Z(E) &= (2\pi\sqrt{-1})^{\dim F}\left[J_F(e^{\pi\sqrt{-1}}t), \widehat{\Gamma}_F \operatorname{Ch}(E)\right) \\ &= z^{\frac{\dim F}{2}}\left(1, S(z)z^{-\mu}z^{c_1(F)}\widehat{\Gamma}_F \operatorname{Ch}(E)\right).\end{aligned}$$

where $t = z^{-1}$.

5.5. Comparison

We computed the infinite dimensional integral (14) in two ways. Comparing (17) and (19), we should have the equality

$$\begin{aligned}&\int_{[0,\infty)^{N-1}} e^{-\left(x_1+\cdots+x_{N-1}+\frac{e^{-u}}{x_1\cdots x_{N-1}}\right)/z}\frac{dx_1}{x_1}\cdots\frac{dx_{N-1}}{x_{N-1}} \\ &\qquad = (2\pi\sqrt{-1})^{N-1}\left[J_{\mathbb{P}^{N-1}}(e^{\pi\sqrt{-1}}t), \widehat{\Gamma}_{\mathbb{P}^{N-1}}\right)\end{aligned} \tag{20}$$

with $t = e^{-u/N} z^{-1}$. This oscillatory integral representation yields the asymptotic expansion (as $t \to +\infty$)

$$Z(\mathcal{O}_{\mathbb{P}^{N-1}}) = (2\pi\sqrt{-1})^{N-1} \left[J_{\mathbb{P}^{N-1}}(e^{\pi\sqrt{-1}}t), \widehat{\Gamma}_{\mathbb{P}^{N-1}} \right) \sim \text{const} \times t^{-\frac{N-1}{2}} e^{-Nt}$$

which can be used to prove the Gamma conjecture for $\mathbb{P}^{N-1}$. See §6 and [29, §3.8].

Remark 5.3. We have a rigorous independent proof of the equality (20) (see [50, 55]). Recall that we can write the left-hand side as the Fourier transform (16) of $\Gamma(-\sqrt{-1}\xi)^N$; by closing the integration contour in the lower half ξ-plane and writing the integral as the sum of residues at $\xi = 0, -\sqrt{-1}, -2\sqrt{-1}, -3\sqrt{-1}, \dots$, we arrive at the expression on the right-hand side.

Remark 5.4. A similar regularization of an infinite dimensional integral appears in the computation of (sphere or hemisphere) partition functions of (2,2) supersymmetric gauge theories, see Benini–Cremonesi [8], Doroud–Gomis–Le-Floch–Lee [19] and Hori–Romo [44]. It appears that the computations in §5.3 and §5.4 correspond, in the terminology of [8, 19], to the localization on the *Coulomb branch* and on the *Higgs branch* respectively.

§6. Toric Manifold

In this section, we discuss Gamma conjecture for Fano toric manifolds. We prove Gamma conjecture I by assuming a certain condition for the mirror Laurent polynomial f which is analogous to Property $\mathcal{O}$.

Let X be an n-dimensional Fano toric manifold. A mirror of X is given by the Laurent polynomial [31, 45, 34]:

$$f(x) = x^{b_1} + x^{b_2} + \cdots + x^{b_m}$$

where $x = (x_1, \dots, x_n) \in (\mathbb{C}^\times)^n$, $b_1, \dots, b_m \in \mathbb{Z}^n$ are primitive generators of the 1-dimensional cones of the fan of X and $x^{b_i} = x_1^{b_{i1}} \cdots x_n^{b_{in}}$ for $b_i = (b_{i1}, \dots, b_{in})$. By mirror symmetry [34, 50], the small quantum cohomology ring $(H^\bullet(X), \star_0)$ is isomorphic to the Jacobian ring $\mathbb{C}[x_1^\pm, \dots, x_n^\pm]/(\partial_{\log x_1} f, \dots, \partial_{\log x_n} f)$ and the first Chern class $c_1(X)$ corresponds to the class of f in the Jacobian ring under this isomorphism. Therefore, the set of eigenvalues of $(c_1(X)\star_0)$ coincides with the set of critical values of f; moreover their multiplicities also coincide. The restriction of f to the real locus $(\mathbb{R}_{>0})^n$ is strictly convex since the logarithmic Hessian

$$\frac{\partial^2 f}{\partial \log x_i \partial \log x_j}(x) = \sum_{k=1}^{m} b_{kj} b_{ki} x^{b_k}$$

is positive definite for any $x \in (\mathbb{R}_{>0})^n$. One can also show that $f|_{(\mathbb{R}_{>0})^n}$ is proper and bounded from below since the convex hull of $b_1, \dots, b_m$ contains the origin in its interior (cf. Remark 7.4). Therefore $f|_{(\mathbb{R}_{>0})^n}$ admits a global minimum at a unique point $x_{\rm con} \in (\mathbb{R}_{>0})^n$. We call $x_{\rm con}$ the *conifold point* [28, 29]. Consider the following condition:

Condition 6.1 (analogue of Property $\mathcal{O}$ for toric manifolds)**.** Let X be a Fano toric manifold and f be its mirror Laurent polynomial. Let $T_{\rm con} = f(x_{\rm con})$ be the value of f at the conifold point $x_{\rm con}$. One has

(a) every critical value u of f satisfies $|u| \leqslant T_{\rm con}$;

(b) the conifold point is the unique critical point of f contained in $f^{-1}(T_{\rm con})$.

From the relationship between the quantum cohomology of X and the Jacobian ring of f as above, we know that Condition 6.1 implies[4] Part (1) of Property $\mathcal{O}$ (Definition 3.1). Recall that Part (1) was enough to make sense of Gamma Conjecture I (see Remark 3.14).

Remark 6.2. One can define the conifold point for every Laurent polynomial such that the Newton polytope contains the origin in its interior and that all the coefficients are positive, but Condition 6.1 does not always hold. For instance, the one-dimensional Laurent polynomial $f(x) = x^{-1} + x + tx^2$ with a sufficiently small $t > 0$ does not satisfy Part (a) of the condition. Also, if the lattice generated by $b_1, \dots, b_m$ is not equal to $\mathbb{Z}^n$, then $f(x) = x^{b_1} + x^{b_2} + \cdots + x^{b_m}$ has non-trivial diagonal symmetry $\{\zeta \in (\mathbb{C}^\times)^n : f(x) = f(\zeta x)\} \neq \{1\}$ and Part (b) of the condition fails.

Theorem 6.3. *Suppose that a Fano toric manifold X satisfies Condition* 6.1. *Then X satisfies Gamma Conjecture* I.

Proof. It follows from the argument in [50, §4.3.1] that

$$z^{n/2}\left(\phi, S(z) z^{-\mu} z^{c_1(X)} \widehat{\Gamma}_X\right)_X = \int_{(\mathbb{R}_{>0})^n} e^{-f(x)/z} \varphi(x,z) \frac{dx_1 \cdots dx_n}{x_1 \cdots x_n}$$

[4]Note that Condition 6.1 means slightly more than what Part (1) of Property $\mathcal{O}$ (Definition 3.1) says for Fano toric manifolds; Condition 6.1 additionally claims that $T = T_{\rm con}$.

for $z > 0$, where $\phi \in H^\bullet(X)$ and $\varphi(x,z) \in \mathbb{C}[x_1^\pm, \dots, x_n^\pm, z]$ is such that the class $[\varphi(x,z)e^{-f(x)/z}dx_1 \cdots dx_n/(x_1 \cdots x_n)]$ corresponds to ϕ under the mirror isomorphism in [50, Proposition 4.8] and $n = \dim X$. When $\varphi = 1$, one has $\phi = 1$ and this gives the integral representation

$$Z(\mathcal{O}_X) = \int_{(\mathbb{R}_{>0})^n} e^{-f(x)/z} \frac{dx_1 \cdots dx_n}{x_1 \cdots x_n}$$

of the quantum cohomology central charge $Z(\mathcal{O}_X)$ from Remark 5.2. We already saw this in (20) for $X = \mathbb{P}^{N-1}$. This integral representation implies that the flat section $S(z)z^{-\mu}z^{c_1(X)}\widehat{\Gamma}_X$ has the following asymptotics as $z \to +0$, i.e.

$$\left\| e^{T_{\mathrm{con}}/z} S(z) z^{-\mu} z^{c_1(X)} \widehat{\Gamma}_X \right\| = O(1)$$

where recall that T_{con} is the global minimum of f on $(\mathbb{R}_{>0})^n$. Condition 6.1 ensures that $T = T_{\mathrm{con}}$ and hence the conclusion follows. Q.E.D.

A toric manifold has a generically semisimple quantum cohomology. We note that the following weaker version of Gamma Conjecture II (Conjecture 4.9) can be shown for a toric manifold.

Theorem 6.4 ([50]). *Let X be a Fano toric manifold. We choose a semisimple point $\tau \in H^\bullet(X)$ and an admissible phase ϕ. There exists a $\mathbb{Z}$-basis $\{[E_1], \dots, [E_N]\}$ of the K-group such that the matrix $(\chi(E_i, E_j))_{1 \leqslant i,j \leqslant N}$ is uni-uppertriangular and that the asymptotic basis (see §4.3) of X at τ with respect to ϕ is given by $\widehat{\Gamma}_X \operatorname{Ch}(E_i)$, $i = 1, \dots, N$.*

Proof. Recall from Remark 4.13 that the asymptotic basis makes sense also for the big quantum product $\star_\tau$ and that it changes by mutation as τ varies. By the theory of mutation, it suffices to prove the theorem at a single semisimple point τ. If τ is in the image of the mirror map, we can take $[E_i]$ to be the mirror images of the Lefschetz thimbles (with phase ϕ) under the isomorphism between the integral structures of the A-model and of the B-model, given in [50, Theorem 4.11]. Q.E.D.

Remark 6.5 ([50]). Results similar to Theorems 6.3, 6.4 hold for a weak-Fano toric orbifold. In the weak-Fano case, however, we need to take into consideration the effect of the mirror map.

Remark 6.6. Kontsevich's homological mirror symmetry suggests that the basis $[E_1], \dots, [E_N]$ in the K-group should be lifted to an exceptional collection in $D^b_{\mathrm{coh}}(X)$ because the corresponding Lefschetz thimbles form an exceptional collection in the Fukaya–Seidel category of the mirror.

§7. Toric complete intersections

In this section we discuss Gamma conjecture I for a Fano complete intersection Y in a toric manifold X. Again the problem comes down to the truth of a mirror analogue of Property $\mathcal{O}$.

We begin with the remark that the number T can be evaluated on the ambient part. Let

$$H^\bullet_{\rm amb}(Y) := \operatorname{Im}(i^*\colon H^\bullet(X) \to H^\bullet(Y))$$

denote the *ambient part* of the cohomology group, where $i\colon Y \to X$ is the inclusion. It is shown in [66, Proposition 4], [52, Corollary 2.5] that the ambient part $H^\bullet_{\rm amb}(Y)$ is closed under quantum multiplication $\star_\tau$ when $\tau \in H^\bullet_{\rm amb}(Y)$. Note that Givental's mirror theorem [34] determines this ambient quantum cohomology $(H^\bullet_{\rm amb}(Y), \star_\tau)$.

Proposition 7.1. *The spectrum of $(c_1(Y)\star_0)$ on $H^\bullet(Y)$ is the same as the spectrum of $(c_1(Y)\star_0)$ on $H^\bullet_{\rm amb}(Y)$ as a set (when we ignore the multiplicities). In particular the number T for Y (see (6)) can be evaluated on the ambient part.*

Proof. It follows from the fact that $c_1(Y)$ belongs to $H^\bullet_{\rm amb}(Y)$ and the fact that $H^\bullet(Y)$ is a module over the algebra $(H^\bullet_{\rm amb}(Y), \star_0)$. Q.E.D.

We assume that Y is a complete intersection in X obtained from the following *nef partition* [6, 34]. Let $D_1, \dots, D_m$ denote prime toric divisors of X. Each prime toric divisor D_i corresponds to a primitive generator $b_i \in \mathbb{Z}^n$ of a 1-dimensional cone of the fan of X. We assume that there exists a partition $\{1, \dots, m\} = I_0 \sqcup I_1 \sqcup \cdots \sqcup I_l$ such that $\sum_{i\in I_k} D_i$ is nef for $k = 1, \dots, l$ and $\sum_{i\in I_0} D_i$ is ample. Let $\mathcal{L}_k := \mathcal{O}(\sum_{i\in I_k} D_i)$ be the corresponding nef line bundle. We assume that Y is the zero-locus of a transverse section of $\bigoplus_{k=1}^l \mathcal{L}_k$ over X. Then $c_1(Y) = c_1(X) - \sum_{k=1}^l c_1(\mathcal{L}_k) = \sum_{i\in I_0}[D_i]$ is ample and Y is a Fano manifold. Define the Laurent polynomial functions $f_0, f_1, \dots, f_l\colon (\mathbb{C}^\times)^n \to \mathbb{C}$ as follows:

$$f_k(x) = -c_0\delta_{0,k} + \sum_{i\in I_k} x^{b_i}$$

where c_0 is the constant arising from Givental's mirror map [34]:

$$c_0 := \sum_{\substack{d\in \mathrm{Eff}(X): c_1(Y)\cdot d = 1\\ [D_i]\cdot d \geqslant 0\ (\forall i)}} \frac{\prod_{k=1}^l (c_1(\mathcal{L}_k)\cdot d)!}{\prod_{i=1}^m ([D_i]\cdot d)!}.$$

A mirror of Y [45, 34] is the affine variety $Z := \{x \in (\mathbb{C}^\times)^n : f_1(x) = \cdots = f_l(x) = 1\}$ equipped with a function $f_0 \colon Z \to \mathbb{C}$ and a holomorphic volume form

$$\omega_Z := \frac{\frac{dx_1}{x_1} \wedge \cdots \wedge \frac{dx_n}{x_n}}{df_1 \wedge \cdots \wedge df_l}.$$

Consider the following condition, which implies Part (1) of Property $\mathcal{O}$ for Y.

Condition 7.2 (analogue of Property $\mathcal{O}$ for Y)**.** Let Y, Z and $f_0 \colon Z \to \mathbb{C}$ be as above. Let $Z_{\mathrm{real}} := \{(x_1, \ldots, x_n) \in Z : x_i \in \mathbb{R}_{>0}\ (\forall i)\}$ be the positive real locus of Z. We have:

(a) Z is a smooth complete intersection of dimension $n - l$;
(b) $f_0|_{Z_{\mathrm{real}}} \colon Z_{\mathrm{real}} \to \mathbb{R}$ attains global minimum at a unique critical point $x_{\mathrm{con}} \in Z_{\mathrm{real}}$; we call it *the conifold point*;
(c) the conifold point x_{con} is a non-degenerate critical point of f_0;
(d) the number T (6) for Y coincides with $T_{\mathrm{con}} := f_0(x_{\mathrm{con}})$;
(e) T_{con} is an eigenvalue of $(c_1(Y)\star_0)$ with multiplicity one.

Variant. We can consider the weaker version of Condition 7.2 where part (e) is replaced with

(e′) T_{con} is an eigenvalue of $(c_1(Y)\star_0)|_{H_{\mathrm{amb}}(Y)}$ with multiplicity one.

When (a), (b), (c), (d), (e′) hold, we say that Y satisfies Condition 7.2 *on the ambient part.*

Example 7.3. Let Y be a degree d hypersurface in $\mathbb{P}^n$ with $d < n + 1$. The mirror is given by the function

$$f_0(x) = \frac{1}{x_1 x_2 \cdots x_n} + x_1 + \cdots + x_{n-d} - \delta_{d,n} d!$$

on the affine variety $Z = \{x \in (\mathbb{C}^\times)^n : f_1(x) = x_{n-d+1} + x_{n-d+2} + \cdots + x_n = 1\}$. Following Przyjalkowski [70], introduce homogeneous co-ordinates $[y_{n-d+1} : \cdots : y_n]$ of $\mathbb{P}^{n-d-1}$ and consider the change of variables:

$$x_i = \frac{y_i}{y_{n-d+1} + \cdots + y_n} \quad \text{for} \quad n - d + 1 \leqslant i \leqslant n.$$

Then we have

$$f_0(x) = \frac{(y_{n-d+1} + \cdots + y_n)^d}{\prod_{i=1}^{n-d} x_i \cdot \prod_{j=n-d+1}^{n} y_j} + x_1 + \cdots + x_{n-d} - \delta_{n,d} d!$$

Setting $y_n = 1$ we obtain a Laurent polynomial expression (with positive coefficients) for f_0 in the variables $x_1, \ldots, x_{n-d}, y_{n-d+1}, \ldots, y_{n-1}$. Note that the change of variables maps Z_{real} isomorphically onto the real locus $\{x_i > 0, y_j > 0 : 1 \leqslant i \leqslant n-d, n-d+1 \leqslant j \leqslant n-1\}$. As in the case of mirrors of toric manifolds (see §6), the Laurent polynomial expression of f_0 shows the existence of a conifold point in Condition 7.2. (Similarly, the mirrors of complete intersections in weighted projective spaces have Laurent polynomial expressions [70].)

We have $T_{\mathrm{con}} = (n+1-d)d^{d/(n+1-d)} - \delta_{n,d}d!$. It is easy to check that T_{con} coincides with T for Y and gives a simple eigenvalue of $(c_1(Y)\star_0)|_{H^\bullet_{\mathrm{amb}}(Y)}$ restricted to the ambient part (using Givental's mirror theorem [34]). Therefore Y satisfies Condition 7.2 on the ambient part.

Furthermore, if the Fano index $n+1-d$ is greater than one, we have $c_1(Y)\star_0 \theta = 0$ for every primitive class $\theta \in H^{n-1}(Y)$. This follows from the fact that $(c_1(Y)\star_0)$ preserves the primitive part in $H^{n-1}(Y)$ and for degree reasons. Also if $\dim Y = n-1$ is odd, Y has no even degree primitive classes (see Appendix A for the treatment of odd classes). Therefore we conclude that Y satisfies Condition 7.2 except possibly for even-dimensional index-one hypersurfaces.

Remark 7.4 ([52]). Using an inequality of the form $\beta_1 u_1 + \cdots + \beta_m u_m \geqslant u_1^{\beta_1} \cdots u_m^{\beta_m}$ for $u_i > 0$, $\beta_i > 0$, $\sum_i \beta_i = 1$, we find that $f_0|_{Z_{\mathrm{real}}}$ is bounded (from below) by a convex function:

$$f_0(x) + l = f_0(x) + f_1(x) + \cdots + f_l(x) \geqslant -c_0 + \epsilon \sum_{i=1}^{n} (x_i^{1/k} + x_i^{-1/k})$$

for all $x \in Z_{\mathrm{real}}$, where $\epsilon > 0$ and $k > 0$ are constants. In particular $f_0|_{Z_{\mathrm{real}}} : Z_{\mathrm{real}} \to \mathbb{R}$ is proper and attains global minimum at some point. It also follows that the oscillatory integral (21) below converges.

Theorem 7.5. *Let Y be a Fano complete intersection in a toric manifold constructed from a nef partition as above. If Y satisfies Condition* 7.2, *Y satisfies Gamma Conjecture* I. *If Y satisfies Condition* 7.2 *on the ambient part, the ambient quantum cohomology of Y satisfies Gamma conjecture* I.

Proof. In [52, Theorem 5.7], it is proved that the quantum cohomology central charge (see Remark 5.2) of $\mathcal{O}_Y$ has an integral representation:

$$Z(\mathcal{O}_Y) = z^{\frac{\dim Y}{2}} \left(1, S(z) z^{-\mu} z^{c_1(Y)} \widehat{\Gamma}_Y\right)_X = \int_{Z_{\mathrm{real}}} e^{-f_0/z} \omega_Z. \tag{21}$$

We note that the constant term c_0 in f_0 comes from the value of the mirror map at $\tau = 0$. In [52], a slightly more general statement was proved: we have mirrors $(Z_\tau, f_{0,\tau}, \omega_{Z,\tau})$ parameterized by $\tau \in H^2_{\mathrm{amb}}(Y)$ and

$$z^{\frac{\dim Y}{2}} \left(1, S(\tau,z)z^{-\mu}z^{c_1(Y)}\widehat{\Gamma}_Y\right)_Y = \int_{Z_{\tau,\mathrm{real}}} e^{f_{0,\tau}/z}\omega_{Z,\tau}$$

where $S(\tau,z)z^{-\mu}z^{c_1(Y)}$ is the fundamental solution for the big quantum connection (5) which restricts to the one in Proposition 2.1 at $\tau = 0$. By differentiating this in the τ-direction, we obtain

$$z^{\frac{\dim Y}{2}} \left(z\nabla_{\alpha_1}\cdots z\nabla_{\alpha_k}1, S(\tau,z)z^{-\mu}z^{c_1(Y)}\widehat{\Gamma}_Y\right)_Y = z\partial_{\alpha_1}\cdots z\partial_{\alpha_k}\int_{Z_{\tau,\mathrm{real}}} e^{-f_{0,\tau}/z}\omega_{Z,\tau}$$

for $\alpha_1,\dots,\alpha_k \in H^2(Y)$. Since $H^\bullet_{\mathrm{amb}}(Y)$ is generated by $H^2_{\mathrm{amb}}(Y)$, we obtain an integral representation of the form:

$$z^{\frac{\dim Y}{2}} \left(\phi, S(z)z^{-\mu}z^{c_1(Y)}\widehat{\Gamma}_Y\right)_Y = \int_{Z_{\mathrm{real}}} \varphi(x,z)e^{-f_0/z}\omega_Z$$

for every $\phi \in H^\bullet_{\mathrm{amb}}(Y)$ (for some function $\varphi(x,z)$). Since the flat section $S(z)z^{-\mu}z^{c_1(Y)}\widehat{\Gamma}_Y$ takes values in the ambient part $H^\bullet_{\mathrm{amb}}(Y)$, we obtain integral representations for all components of $S(z)z^{-\mu}z^{c_1(Y)}\widehat{\Gamma}_Y$. These integral representations and Condition 7.2 show that $\|e^{T/z}S(z)z^{-\mu}z^{c_1(Y)}\widehat{\Gamma}_Y\|$ grows at most polynomially as $z \to +0$.

Q.E.D.

§8. Quantum Lefschetz

In this section we show that Gamma conjecture I is compatible with the quantum Lefschetz principle.

Let X be a Fano manifold of index $r \geqslant 2$. We write $-K_X = rh$ for an ample class h. Let $Y \subset X$ be a degree-a Fano hypersurface in the linear system $|ah|$ with $0 < a < r$. Assuming the truth of Gamma Conjecture I for X, we study Gamma Conjecture I for Y.

We write the J-function of X (7) as

$$J_X(t) = e^{rh\log t}\sum_{n=0}^{\infty} J_{rn}t^{rn}$$

with $J_d \in H^\bullet(X)$. By the quantum Lefschetz theorem [59, 15], the J-function of Y is

$$J_Y(t) = e^{(r-a)h\log t - c_0 t}\sum_{n=0}^{\infty}(ah+1)\cdots(ah+an)(i^*J_{rn})t^{(r-a)n}, \tag{22}$$

where $i\colon Y \to X$ is the natural inclusion and c_0 is the constant:

$$c_0 = \begin{cases} a!\,\langle [\mathrm{pt}], J_r\rangle = a!\sum_{h\cdot d=1}\langle [\mathrm{pt}]\psi^{r-2}\rangle_{0,1,d}^X & \text{if } r-a=1; \\ 0 & \text{if } r-a>1. \end{cases} \tag{23}$$

The quantum Lefschetz principle can be rephrased in terms of the Laplace transformation:

Lemma 8.1. *We have*

$$J_Y(t = u^{a/(r-a)}) = \frac{e^{-c_0 t}}{\Gamma(1+ah)u}\int_0^\infty i^* J_X(q^{a/r})e^{-q/u}dq.$$

Remark 8.2. The Laplace transformation in the above lemma converges for sufficiently small $u > 0$ because of the exponential asymptotics as $t \to +\infty$ in Proposition 3.8 and the growth estimate $\|J_X(t)\| \leqslant C|\log t|^{\dim X}$ as $t \to +0$.

Suppose that X satisfies Gamma Conjecture I. Recall from Proposition 3.8 that we have the following limit formula for J_X:

$$\widehat{\Gamma}_X \propto \lim_{t\to+\infty} t^{\frac{\dim X}{2}} e^{-T_X t} J_X(t). \tag{24}$$

Here T_X is the number T in (6) for X. Using the stationary phase approximation in Lemma 8.1, we obtain the following result.

Theorem 8.3. *Suppose that the J-function of X satisfies the limit formula* (24) *and let Y be a Fano hypersurface in the linear system $|-(a/r)K_X|$, where r is the Fano index of X and $0 < a < r$. Then the J-function of Y satisfies the limit formula*:

$$\widehat{\Gamma}_Y \propto \lim_{t\to+\infty} t^{\frac{\dim Y}{2}} e^{-(T_0 - c_0)t} J_Y(t)$$

where c_0 is given in (23) *and the positive number $T_0 > 0$ is determined by the relation*:

$$\left(\frac{T_0}{r-a}\right)^{r-a} = a^a\left(\frac{T_X}{r}\right)^r.$$

In particular, if Y satisfies Property $\mathcal{O}$ (*Definition* 3.1), *then Y satisfies Gamma conjecture* I *by Corollary* 3.9.

Proof. We write $n = \dim X$ and $T = T_X$ for simplicity. We set $\widetilde{J}(t) := t^{\frac{n}{2}} e^{-Tt} i^* J_X(t)$. The limit formula (24) gives $\lim_{t\to+\infty}\widetilde{J}(t) =$

$C_1 i^*\widehat{\Gamma}_X$ for some $C_1 \neq 0$. By Lemma 8.1, we have

$$\begin{aligned} t^{\frac{n-1}{2}} e^{-(T_0-c_0)t} J_Y(t = u^{a/(r-a)}) \\ &= \frac{t^{\frac{n-1}{2}} e^{-T_0 t}}{\Gamma(1+ah)u} \int_0^\infty q^{-an/(2r)} e^{Tq^{a/r}-q/u} \widetilde{J}(q^{a/r}) dq \\ &= \frac{\sqrt{t}}{\Gamma(1+ah)} \int_0^\infty q^{-an/(2r)} e^{-(q-Tq^{a/r}+T_0)t} \widetilde{J}(tq^{a/r}) dq \end{aligned}$$

where in the second line we performed the change of variables $q \to u^{r/(r-a)}q$ and used $t = u^{a/(r-a)}$. Consider the function $\theta(q) = q - Tq^{a/r} + T_0$ on $[0,\infty)$. This function has a unique critical point at $q_0 := (\frac{a}{r}T)^{\frac{r}{r-a}}$ and attains a global minimum at $q = q_0$; we have $\theta(q_0) = 0$ by the definition of T_0. By the stationary phase approximation, the $t \to +\infty$ asymptotics of this integral is determined by the behaviour of the integrand near $q = q_0$. To establish the asymptotics rigorously, we divide the interval $[0,\infty)$ of integration into $[0, q_0/2]$ and $[q_0/2, \infty)$. We first estimate the integral over $[0, q_0/2]$.

$$\begin{aligned} &\left\| \sqrt{t} \int_0^{q_0/2} q^{-an/(2r)} e^{-\theta(q)t} \widetilde{J}(tq^{a/r}) dq \right\| \\ &\leqslant e^{-t\theta(q_0/2)} \sqrt{t} \int_0^{q_0/2} q^{-an/(2r)} \|\widetilde{J}(tq^{a/r})\| dq \qquad (25) \\ &\leqslant e^{-t\theta(q_0/2)} t^{\frac{n+1}{2}-\frac{r}{a}} \frac{r}{a} \int_0^{t(q_0/2)^{a/r}} y^{-n/2} \|\widetilde{J}(y)\| y^{(r/a)-1} dy. \end{aligned}$$

where we set $y = tq^{a/r}$. Note that $y^{-n/2}\|\widetilde{J}(y)\|y^{(r/a)-1}$ is integrable near $y = 0$ (by the definition of $\widetilde{J}$) and is of polynomial growth as $y \to +\infty$. Therefore the integral

$$\int_0^{t(q_0/2)^{a/r}} y^{-n/2} \|\widetilde{J}(y)\| y^{(r/a)-1} dy.$$

is of polynomial growth as $t \to +\infty$. Since $\theta(q_0/2) > 0$, the integral (25) over $[0, q_0/2]$ goes to zero (exponentially) as $t \to +\infty$. Next we consider the integral over $[q_0/2, \infty)$. By the change of variables $q = q_0 e^{x/\sqrt{t}}$, the integral over $[q_2/2, \infty)$ can be written as

$$\frac{q_0^{1-\frac{an}{2r}}}{\Gamma(1+ah)} \int_{-\sqrt{t}\log 2}^\infty e^{(1-\frac{an}{2r})x/\sqrt{t}} e^{-t\theta(q_0 e^{x/\sqrt{t}})} \widetilde{J}\left(tq_0^{a/r} e^{\frac{a}{r}x/\sqrt{t}}\right) dx. \qquad (26)$$

Since we have an expansion of the form $\theta(q_0 e^y) = a_2 y^2 + \sum_{n=3}^{\infty} a_n y^n$, we have for a fixed $x \in \mathbb{R}$,

$$e^{(1-\frac{an}{2r})x/\sqrt{t}} \to 1, \quad e^{-t\theta(q_0 e^{x/\sqrt{t}})} \to e^{-a_2 x^2}, \quad \widetilde{J}\left(tq_0^{a/r} e^{\frac{a}{r}x/\sqrt{t}}\right) \to C_1 i^* \widehat{\Gamma}_X$$

as $t \to +\infty$. On the other hand, since $\theta(q_0 e^y)$ grows exponentially as $y \to +\infty$, we have an estimate of the form

$$\theta(q_0 e^y) \geqslant C_2 y^2$$

on $y \in [-\log 2, \infty)$ for some $C_2 > 0$. Therefore, when $t \geqslant 1$, the integrand of (26) can be estimated as

$$\left\| e^{(1-\frac{an}{2r})x/\sqrt{t}} e^{-t\theta(q_0 e^{x/\sqrt{t}})} \widetilde{J}\left(tq_0^{a/r} e^{\frac{a}{r}x/\sqrt{t}}\right)\right\| \leqslant C_3 e^{|1-\frac{an}{2r}||x|} e^{-C_2 x^2}$$

for all $x \in [-\sqrt{t}\log 2, \infty)$ for some $C_3 > 0$. Thus the integrand is uniformly bounded by an integrable function and we can apply Lebesgue's dominated convergence theorem to (26) to see that the limit of (26) as $t \to +\infty$ is proportional to $i^*\widehat{\Gamma}_X/\Gamma(1+ah)$. The conclusion follows by noting that

$$\widehat{\Gamma}_Y = \frac{i^*\widehat{\Gamma}_X}{\Gamma(1+ah)}.$$

Q.E.D.

It is natural to ask the following questions:

Problem 8.4. Check that $T_0 - c_0$ is the number T (6) for the hypersurface Y. Also prove that Y satisfies Property $\mathcal{O}$ (assuming that X satisfies Property $\mathcal{O}$).

Remark 8.5 ([26]). It is easy to see that Apéry limits (11) (or Gamma conjecture I') are compatible with quantum Lefschetz. Suppose that X is a Fano manifold of index r satisfying Gamma conjecture I' and let $Y \in |-(a/r)K_X|$ be a Fano hypersurface of index $r - a > 1$. Then for any $\alpha \in H_\bullet(Y)$ with $c_1(Y) \cap \alpha = 0$, we have the limit formula:

$$\lim_{n\to\infty} \frac{\langle i_*\alpha, J_{X,rn}\rangle}{\langle [\mathrm{pt}], J_{X,rn}\rangle} = \langle i_*\alpha, \widehat{\Gamma}_X\rangle = \langle \alpha, \widehat{\Gamma}_Y\rangle$$

where $J_{X,n}$ is the Taylor coefficients of the J-function of X (as in (8)) and we used $h \cap i_*\alpha = 0$ in the second equality. Since the index $r - a$ is greater than one, the quantum Lefschetz (22) gives:

$$\left\langle \alpha, J_{Y,(r-a)n}\right\rangle = (an)!\,\langle i_*\alpha, J_{X,rn}\rangle, \quad \left\langle [\mathrm{pt}], J_{Y,(r-a)n}\right\rangle = (an)!\,\langle [\mathrm{pt}], J_{X,rn}\rangle$$

and thus Y also satisfies the Gamma conjecture I'.

§9. Grassmannians

In this section, we prove Gamma conjecture I' (Conjecture 3.11) for Grassmannians $\mathrm{Gr}(r,n)$ using the Hori–Vafa mirror [45] and abelian/non-abelian correspondence [10]. The discussion in this section extends the proof of Dubrovin conjecture for Grassmannians by Ueda [75]. In the course of the proof, we obtain a formula for quantum cohomology central charges of $\mathrm{Gr}(r,n)$ in terms of mirror oscillatory integrals. We note that Gamma conjecture I' for $\mathrm{Gr}(2,n)$ was proved by Golyshev [37]. Gamma conjectures I and II were proved for general $\mathrm{Gr}(r,n)$ in [29] by a different (but closely related) method based on the quantum Satake principle [38].

9.1. Abelian quotient and non-abelian quotient

Let $\mathbb{G} = \mathrm{Gr}(r,n)$ denote the Grassmann variety of r-dimensional subspaces in $\mathbb{C}^n$ and let $\mathbb{P} = \mathbb{P}^{n-1} \times \cdots \times \mathbb{P}^{n-1}$ (r times) denote the product of r copies of the projective space $\mathbb{P}^{n-1}$. We relate these two spaces in the framework of Martin [64]: $\mathbb{G}$ and $\mathbb{P}$ arise as non-abelian and abelian quotients of the same vector space $\mathrm{Hom}(\mathbb{C}^r, \mathbb{C}^n)$:

$$\mathbb{G} = \mathrm{Hom}(\mathbb{C}^r, \mathbb{C}^n)/\!/GL(r), \quad \mathbb{P} = \mathrm{Hom}(\mathbb{C}^r, \mathbb{C}^n)/\!/(\mathbb{C}^\times)^r.$$

We have a rational map

$$\mathbb{P} \dashrightarrow \mathbb{G}$$

sending a collection of lines $l_1, \dots, l_r$ to the subspace spanned by $l_1, \dots, l_r$. We can also relate $\mathbb{P}$ and $\mathbb{G}$ by the following diagram [64]:

$$\begin{array}{ccc} \mathbb{F} & \xrightarrow{p} & \mathbb{G} \\ {\scriptstyle\iota}\downarrow & & \\ \mathbb{P} & & \end{array} \tag{27}$$

where $\mathbb{F} := \mathrm{Fl}(1,2,\dots,r,n)$ is the partial flag variety, $p\colon \mathbb{F} \to \mathbb{G}$ is the natural projection and ι is a real-analytic embedding which sends a flag $0 \subset V_1 \subset V_2 \subset \cdots \subset V_r \subset \mathbb{C}^n$ to the lines $L_1, \dots, L_r$ such that L_i is the orthogonal complement of V_{i-1} in V_i. Here we need to choose a Hermitian metric on $\mathbb{C}^n$ for the definition of ι. The diagram (27) naturally comes from the following description of $\mathbb{G}$ and $\mathbb{P}$ as symplectic reductions of $\mathrm{Hom}(\mathbb{C}^r, \mathbb{C}^n)$:

$$\mathbb{G} \cong \mu_G^{-1}(\eta)/G, \quad \mathbb{F} \cong \mu_G^{-1}(\eta)/T, \quad \mathbb{P} \cong \mu_T^{-1}(\eta_0)/T. \tag{28}$$

Here μ_G and μ_T are the moment maps of $G = U(r)$ and $T = (S^1)^r$-actions on $\mathrm{Hom}(\mathbb{C}^r, \mathbb{C}^n)$ respectively and $\eta \in \mathrm{Lie}(G)^\star$ and $\eta_0 \in \mathrm{Lie}(T)^\star$

are non-zero central elements (i.e. scalar multiples of the identify matrix) such that $\pi(\eta) = \eta_0$ (where $\pi\colon \operatorname{Lie}(G)^\star \to \operatorname{Lie}(T)^\star$ is the natural projection):

$$\mu_G(A) = A^\dagger A, \ \mu_T(A) = \left((A^\dagger A)_{i,i}\right)_{i=1}^r, \quad \begin{array}{ccc} \operatorname{Hom}(\mathbb{C}^r, \mathbb{C}^n) & \xrightarrow{\mu_G} & \operatorname{Lie}(G)^\star \\ & \searrow^{\mu_T} & \downarrow \pi \\ & & \operatorname{Lie}(T)^\star \end{array}$$

The map ι in (27) is then induced from the inclusion $\mu_G^{-1}(\eta) \subset \mu_T^{-1}(\eta_0)$.

We describe the cohomology groups of $\mathbb{G}$ and $\mathbb{P}$. Let $\mathcal{V}_1 \subset \mathcal{V}_2 \subset \cdots \subset \mathcal{V}_r$ denote the tautological bundles (with $\operatorname{rank} \mathcal{V}_i = i$) over $\mathbb{F}$ and set

$$\mathcal{L}_i := (\mathcal{V}_i/\mathcal{V}_{i-1})^\vee.$$

The line bundle $\mathcal{L}_i$ is the pull-back of the line bundle $\mathcal{O}(1)$ over the ith factor $\cong \mathbb{P}^{n-1}$ of $\mathbb{P}$ under the non-algebraic map ι. We denote the corresponding line bundle over $\mathbb{P}$ also by $\mathcal{L}_i$, i.e. $\iota^\star \mathcal{L}_i = \mathcal{L}_i$. We set

$$x_i := c_1(\mathcal{L}_i) \in H^\bullet(\mathbb{P}) \text{ or } H^\bullet(\mathbb{F}).$$

A symmetric polynomial in $x_1, \dots, x_r$ can be written in terms of Chern classes of $\mathcal{V}_r$ and thus makes sense as a cohomology class on $\mathbb{G}$. The cohomology rings of $\mathbb{P}$, $\mathbb{G}$ and $\mathbb{F}$ are described in terms of x_i as follows:

$$\begin{aligned} H^\bullet(\mathbb{P}) &\cong \mathbb{C}[x_1, \dots, x_r] \big/ \langle x_1^n, \dots, x_r^n \rangle, \\ H^\bullet(\mathbb{G}) &\cong \mathbb{C}[x_1, \dots, x_r]^{\mathfrak{S}_r} \big/ \langle h_{n-r+1}, \dots, h_n \rangle, \\ H^\bullet(\mathbb{F}) &\cong \mathbb{C}[x_1, \dots, x_r] \big/ \langle h_{n-r+1}, \dots, h_n \rangle. \end{aligned}$$

Here $h_i = c_i(\mathbb{C}^n/\mathcal{V}_r)$ is the ith complete symmetric function of $x_1, \dots, x_r$. An additive basis of $H^\bullet(\mathbb{G})$ is given by the Schubert classes:

$$\sigma_\mu := \frac{\det\left(x_i^{\mu_j + r - j}\right)_{1 \leqslant i,j \leqslant r}}{\det\left(x_i^{r-j}\right)_{1 \leqslant i,j \leqslant r}} \tag{29}$$

where μ ranges over partitions such that $n - r \geqslant \mu_1 \geqslant \mu_2 \geqslant \cdots \geqslant \mu_r \geqslant 0$. This is the Poincaré dual of the Schubert cycle

$$\Omega_\mu = \{V \in \operatorname{Gr}(r,n) : \dim(V \cap \mathbb{C}^{n-r+i-\mu_i}) \geqslant i, \ 1 \leqslant i \leqslant r\}$$

and $\deg \sigma_\mu = 2|\mu| = 2\sum_{i=1}^r \mu_i$ (degree as a cohomology class), see [41, §6, Chapter 1].

Let $J_{\mathbb{P}^{n-1}}(t)$ be the J-function (see (7), (18)) of $\mathbb{P}^{n-1}$. We define the multivariable J-function of $\mathbb{P}$ by

$$J_{\mathbb{P}}(t_1, \dots, t_r) = J_{\mathbb{P}^{n-1}}(t_1) \otimes J_{\mathbb{P}^{n-1}}(t_2) \otimes \cdots \otimes J_{\mathbb{P}^{n-1}}(t_r).$$

This takes values in $H^\bullet(\mathbb{P}) \cong H^\bullet(\mathbb{P}^{n-1})^{\otimes r}$. Bertram–Ciocan-Fontanine–Kim [10] proved the following *abelian/non-abelian correspondence* between the J-functions of $\mathbb{G}$ and $\mathbb{P}$.

Theorem 9.1 ([10]). *The J-function $J_{\mathbb{G}}(t)$ of $\mathbb{G}$ is given by*

$$p^\star J_{\mathbb{G}}(t) = e^{-\sigma_1 \pi\sqrt{-1}(r-1)} \left[\frac{\left(\prod_{1\leqslant i<j\leqslant r}(\theta_i - \theta_j)\right) \iota^\star J_{\mathbb{P}}(t_1, \dots, t_r)}{n^{\binom{n}{r}}\Delta} \right]_{t_1=\cdots=t_r=\xi t}$$

where $\Delta := \prod_{i<j}(x_i - x_j)$, $\theta_i := t_i \frac{\partial}{\partial t_i}$, $\xi := e^{\pi\sqrt{-1}(r-1)/n}$ *and* $\sigma_1 := x_1 + \cdots + x_r$.

9.2. Preliminary lemmas

We discuss elementary topological properties of the maps in (27). Martin [64] has shown similar results for general abelian/non-abelian quotients.

Lemma 9.2. *Let $\mathcal{N}_\iota \to \mathbb{F}$ denote the normal bundle of ι and let $\mathcal{T}_p = \mathrm{Ker}(dp) \subset T\mathbb{F}$ denote the relative tangent bundle of p. Then the complex structure I on $\mathbb{P}$ induces an isomorphism:*

$$I\colon \mathcal{T}_p \cong \mathcal{N}_\iota.$$

In particular, we have an isomorphism

$$\iota^\star T\mathbb{P} \cong p^\star T\mathbb{G} \oplus \left(\mathcal{T}_p \otimes_{\mathbb{R}} \mathbb{C}, \sqrt{-1}\right)$$

of topological complex vector bundles, where the complex structure $\sqrt{-1}$ on $\mathcal{T}_p \otimes_{\mathbb{R}} \mathbb{C}$ is induced from that on the $\mathbb{C}$-factor.

Proof. Recall that $\mathbb{P}$, $\mathbb{G}$, $\mathbb{F}$ are described as symplectic reductions (28). Note that $\operatorname{rank} \mathcal{N}_\iota = \operatorname{rank} \mathcal{T}_p = \dim G - \dim T$. Let $\underline{X}$ denote the fundamental vector field associated with $X \in \mathrm{Lie}(G)$. For $x \in \mu_G^{-1}(\eta)$, it follows from the property of the moment map that $I\underline{X}_x$ is perpendicular to $T_x\mu_G^{-1}(\eta)$ for $X \in \mathrm{Lie}(G)$. Thus we have an orthogonal decomposition:

$$T_x \operatorname{Hom}(\mathbb{C}^r, \mathbb{C}^n) = T_x\mu_G^{-1}(\eta) \oplus IT_x(G \cdot x).$$

Similarly we have an orthogonal decomposition

$$T_x \operatorname{Hom}(\mathbb{C}^r, \mathbb{C}^n) = T_x \mu_T^{-1}(\eta_0) \oplus I T_x (T \cdot x).$$

Note that the tangent space $T_{[x]}\mathbb{P}$ at $[x] \in \mathbb{P} = \mu_T^{-1}(\mu_0)/T$ is identified with the orthogonal complement of $T_x(T \cdot x)$ in $T_x \mu_T^{-1}(\eta_0)$, which is preserved by the complex structure I. Under this identification, the orthogonal complement of $T_x(T \cdot x)$ in $T_x(G \cdot x)$ is identified with the fiber $(\mathcal{T}_p)_{[x]}$; by the above decompositions, the complex structure I sends this to the orthogonal complement of $T_x \mu_G^{-1}(\eta)$ in $T_x \mu_T^{-1}(\eta_0)$. This shows $I((\mathcal{T}_p)_{[x]}) = (\mathcal{N}_\iota)_{[x]}$ and the conclusion follows. Q.E.D.

Remark 9.3. Note that as complex vector bundles

$$\left(\mathcal{T}_p \otimes_{\mathbb{R}} \mathbb{C}, \sqrt{-1}\,\right) \cong \mathcal{T}_p \oplus \overline{\mathcal{T}}_p$$

where $\overline{\mathcal{T}}_p$ denotes the conjugate of the complex vector bundle $\mathcal{T}_p$.

Because $\mathbb{F}, \mathbb{P}$ are compact oriented manifolds, the push-forward map $\iota_* \colon H^\bullet(\mathbb{F}) \to H^\bullet(\mathbb{P})$ is well-defined (even though ι is not algebraic). Here we need to be a little careful about the orientation of the normal bundle of ι.

Lemma 9.4. *The push-forward map $\iota_\star \colon H^\bullet(\mathbb{F}) \to H^\bullet(\mathbb{P})$ is given by*

$$\iota_\star(f(x)) = f(x) \cup \overline{\Delta}$$

for any polynomial $f(x) = f(x_1, \dots, x_r)$. Here $\overline{\Delta} = \prod_{i<j}(x_j - x_i) = (-1)^{\binom{r}{2}} \Delta$.

Proof. The normal bundle $\mathcal{N}_\iota$ of ι is not a complex vector bundle, but can be written as the formal difference of complex vector bundles:

$$\mathcal{N}_\iota = \iota^\star T\mathbb{P} \ominus T\mathbb{F}$$

and thus defines an element of the K-group $K^0(\mathbb{P})$ of topological complex vector bundles. Using Lemma 9.2, Remark 9.3 and a topological isomorphism $T\mathbb{F} \cong \mathcal{T}_p \oplus p^\star T\mathbb{G}$, we have:

$$\mathcal{N}_\iota = \mathcal{T}_p \otimes \mathbb{C} \ominus \mathcal{T}_p = \overline{\mathcal{T}}_p.$$

On the other hand, we have a topological isomorphism:

$$\mathcal{T}_p \cong \bigoplus_{i<j} \operatorname{Hom}(\mathcal{L}_i^\vee, \mathcal{L}_j^\vee).$$

Thus $\overline{\mathcal{T}}_p \to \mathbb{F}$ is isomorphic to the restriction of the following complex vector bundle over $\mathbb{P}$:

$$\bigoplus_{i<j} \overline{\mathcal{L}_i} \otimes \mathcal{L}_j \longrightarrow \mathbb{P}$$

and $\mathbb{F}$ is the zero locus of the section of this bundle defined by the given Hermitian metric on $\mathbb{C}^n$. Therefore, $\iota_\star(f(x)) = f(x) \cup \overline{\Delta}$. Q.E.D.

Lemma 9.5. *For a polynomial $f(x) = f(x_1, \ldots, x_r)$, we have*

$$p_\star(f(x)) = \frac{1}{\Delta}\left(\sum_{\sigma \in \mathfrak{S}_r} \operatorname{sgn}(\sigma) f(x_{\sigma(1)}, \ldots, x_{\sigma(r)})\right).$$

Note that the right-hand side is a symmetric polynomial in $x_1, \ldots, x_r$.

Proof. The $\mathfrak{S}_r$-action on $\mathbb{P}$ induces a non-algebraic $\mathfrak{S}_r$ action on $\mathbb{F}$ which preserves the fibration $p\colon \mathbb{F} \to \mathbb{G}$. The action of $\sigma \in \mathfrak{S}_r$ on $\mathbb{F}$ changes the orientation of each fiber of p by $\operatorname{sgn}(\sigma)$. Therefore $p_\star(f(x)) = \operatorname{sgn}(\sigma) p_\star(f(x_{\sigma(1)}, \ldots, x_{\sigma(r)}))$. Thus we have

$$\begin{aligned} p_\star(f(x)) &= \frac{1}{r!} p_\star\left(\sum_{\sigma \in \mathfrak{S}_r} \operatorname{sgn}(\sigma) f(x_{\sigma(1)}, \ldots, x_{\sigma(r)})\right) \\ &= \frac{1}{r!} \frac{\sum_{\sigma \in \mathfrak{S}_r} \operatorname{sgn}(\sigma) f(x_{\sigma(1)}, \ldots, x_{\sigma(r)})}{\Delta} p_\star(\Delta). \end{aligned}$$

The conclusion follows from $p_\star(\Delta) = p_\star(\operatorname{Euler}(\mathcal{T}_p)) = r!$. Q.E.D.

9.3. Comparison of cohomology and K-groups

Using the diagram (27), we identify the cohomology (or the K-group) of $\mathbb{G}$ with the anti-symmetric part of the cohomology (resp. the K-group) of $\mathbb{P}$. This identification was proved by Martin [64] for cohomology groups of general abelian/non-abelian quotients and has been generalized to K-groups by Harada–Landweber [42]. We compare the cohomological and K-theoretic identifications via the map $\mathcal{E} \mapsto \widehat{\Gamma}_F \operatorname{Ch}(\mathcal{E})$.

We identify the rth wedge product $\wedge^r H^\bullet(\mathbb{P}^{n-1})$ with the anti-symmetric part of $H^\bullet(\mathbb{P})$ (with respect to the $\mathfrak{S}_r$-action) via the map:

$$\begin{aligned} &\wedge^r H^\bullet(\mathbb{P}^{n-1}) \ni \alpha_1 \wedge \alpha_2 \wedge \cdots \wedge \alpha_r \\ &\qquad \mapsto \sum_{\sigma \in \mathfrak{S}_r} \operatorname{sgn}(\sigma) \alpha_{\sigma(1)} \otimes \alpha_{\sigma(2)} \otimes \cdots \otimes \alpha_{\sigma(r)} \in H^\bullet(\mathbb{P}). \end{aligned}$$

Similarly we identify $\wedge^r K^0(\mathbb{P})$ with the anti-symmetric part of $K^0(\mathbb{P})$, where $K^0(\cdot)$ denotes the topological K-group. For a non-increasing sequence $\mu_1 \geqslant \mu_2 \geqslant \cdots \geqslant \mu_r$ of integers, we set

$$\mathcal{E}_\mu := p_\star\left(\mathcal{L}_1^{\mu_1} \otimes \mathcal{L}_2^{\mu_2} \otimes \cdots \otimes \mathcal{L}_r^{\mu_r}\right).$$

By the Borel–Weil theory, $\mathcal{E}_\mu$ is the vector bundle on $\mathbb{G}$ associated to the irreducible $GL(r)$-representation of highest weight μ. Shifting μ_i by some number $\mu_i \mapsto \mu_i + \ell$ simultaneously corresponds to twisting the $GL(r)$-representation by $\det^{\otimes \ell}$, where $\det\colon GL(r) \to \mathbb{C}^\times$ corresponds to $\mathcal{O}(1)$ on $\mathbb{G}$. These vector bundles span the K-group of $\mathbb{G}$. We define a line bundle $\mathcal{R}$ on $\mathbb{F}$ by

$$\mathcal{R} = \mathcal{L}_1^{-(r-1)} \otimes \mathcal{L}_2^{-(r-2)} \otimes \cdots \otimes \mathcal{L}_{r-1}^{-1}.$$

This restricts to a half-canonical bundle[5] on each fiber of p. One can easily check that:

$$\det(\mathcal{T}_p^\vee) = p^\star(\mathcal{O}(r-1)) \otimes \mathcal{R}^{\otimes 2} \tag{30}$$

where $p^\star\mathcal{O}(1) = \mathcal{L}_1\mathcal{L}_2\cdots\mathcal{L}_r$. We have the following result:

Proposition 9.6. *Let $\mu_1 \geqslant \mu_2 \geqslant \cdots \geqslant \mu_r$ be a non-increasing sequence and let $k_1 > k_2 > \cdots > k_r$ be the strictly decreasing sequence given by $k_i = \mu_i + r - i$. Then we have:*

(1) *The map $(r!)^{-1}p_\star\iota^\star\colon \wedge^r H^\bullet(\mathbb{P}) \to H^\bullet(\mathbb{G})$ sends the class $x_1^{k_1} \wedge x_2^{k_2} \wedge \cdots \wedge x_r^{k_r}$ to the Schubert class σ_μ (29) and is an isomorphism.*

(2) *The map $(r!)^{-1}p_\star(\mathcal{R}\otimes\iota^\star(\cdot))\colon \wedge^r K^0(\mathbb{P}^{n-1}) \to K^0(\mathbb{G})$ sends the class $\mathcal{L}_1^{k_1}\wedge\mathcal{L}_2^{k_2}\wedge\cdots\wedge\mathcal{L}_r^{k_r}$ to the class $\mathcal{E}_\mu$ and is an isomorphism of vector spaces preserving the natural $\mathbb{Z}$-lattices.*

(3) *We have the commutative diagram:*

$$\begin{array}{ccc} \wedge^r K^0(\mathbb{P}^{n-1}) & \xrightarrow{\ \frac{1}{r!}p_\star(\mathcal{R}\otimes\iota^\star(\cdot))\ } & K^0(\mathbb{G}) \\ \Big\downarrow{\scriptstyle \widehat{\Gamma}_{\mathbb{P}}\,\mathrm{Ch}(\cdot)} & & \Big\downarrow{\scriptstyle \widehat{\Gamma}_{\mathbb{G}}\,\mathrm{Ch}(\cdot)} \\ \wedge^r H^\bullet(\mathbb{P}^{n-1}) & \xrightarrow{\ \frac{\epsilon_r}{r!}e^{-\pi\sqrt{-1}(r-1)\sigma_1}p_\star\iota^\star\ } & H^\bullet(\mathbb{G}) \end{array} \tag{31}$$

where $\epsilon_r = (2\pi\sqrt{-1})^{-\binom{r}{2}}$ and $\mathrm{Ch}(E) = \sum_{p\geqslant 0}(2\pi\sqrt{-1})^p\,\mathrm{ch}_p(E)$ is the modified Chern character.

Proof. (1): This follows from Lemma 9.5 and the definition of the Schubert class σ_μ (29).

[5] corresponding to the half-sum ρ of positive roots.

(2): Using the Grothendieck–Riemann–Roch theorem, we have

$$\begin{aligned}
\operatorname{ch}\Big(p_\star(\mathcal{R}\otimes\iota^\star(\mathcal{L}_1^{k_1}\otimes\cdots\otimes\mathcal{L}_r^{k_r}))\Big) &= p_\star(\operatorname{ch}(\mathcal{R}\otimes\mathcal{L}_1^{k_1}\otimes\cdots\otimes\mathcal{L}_r^{k_r})\cdot\operatorname{td}(\mathcal{T}_p))\\
&= p_\star\left(e^{\mu_1x_1+\cdots+\mu_rx_r}\prod_{i<j}\frac{x_i-x_j}{1-e^{x_i-x_j}}\right)\\
&= p_\star\left(e^{k_1x_1+\cdots+k_rx_r}\prod_{i<j}\frac{x_i-x_j}{e^{x_j}-e^{x_i}}\right).
\end{aligned}$$

By Lemma 9.5, it follows that this is anti-symmetric in $k_1,\dots,k_r$. Therefore

$$\begin{aligned}
\frac{1}{r!}\operatorname{ch}\Big(p_\star(\mathcal{R}\otimes\iota^\star(\mathcal{L}_1^{k_1}\wedge\cdots\wedge\mathcal{L}_r^{k_r}))\Big) &= \operatorname{ch}\Big(p_\star(\mathcal{R}\otimes\iota^\star(\mathcal{L}_1^{k_1}\otimes\cdots\otimes\mathcal{L}_r^{k_r}))\Big)\\
&= \operatorname{ch}\big(p_\star(\mathcal{L}_1^{\mu_1}\otimes\cdots\otimes\mathcal{L}_1^{\mu_r})\big) = \operatorname{ch}(\mathcal{E}_\mu).
\end{aligned}$$

This shows that $(r!)^{-1}p_\star(\mathcal{R}\otimes\iota^\star(\mathcal{L}_1^{k_1}\wedge\cdots\wedge\mathcal{L}_r^{k_r}))=\mathcal{E}_\mu$.

(3): It suffices to prove the commutativity of the following two diagrams:

$$\begin{array}{ccc}
\wedge^rK^0(\mathbb{P}^{n-1}) & \xrightarrow{\ \iota^\star\ } & K^0(\mathbb{F})\\
\Big\downarrow{\scriptstyle\widehat{\Gamma}_{\mathbb{P}}\operatorname{Ch}(\cdot)} & & \Big\downarrow{\scriptstyle(\iota^\star\widehat{\Gamma}_{\mathbb{P}})\operatorname{Ch}(\cdot)}\\
\wedge^rH^0(\mathbb{P}^{r-1}) & \xrightarrow{\ \iota^\star\ } & H^0(\mathbb{G})
\end{array}
\qquad
\begin{array}{ccc}
K^0(\mathbb{F}) & \xrightarrow{\ p_\star(\mathcal{R}\otimes(\cdot))\ } & K^0(\mathbb{G})\\
\Big\downarrow{\scriptstyle(\iota^\star\widehat{\Gamma}_{\mathbb{P}})\operatorname{Ch}(\cdot)} & & \Big\downarrow{\scriptstyle\widehat{\Gamma}_{\mathbb{G}}\operatorname{Ch}(\cdot)}\\
H^\bullet(\mathbb{F}) & \xrightarrow{\ \epsilon_re^{-\pi\sqrt{-1}(r-1)\sigma_1}p_\star\ } & H^\bullet(\mathbb{G})
\end{array}$$

The commutativity of the left diagram is obvious. The commutativity of the right diagram follows from Grothendieck–Riemann–Roch. For $\alpha\in K^0(\mathbb{F})$, we have

$$\operatorname{Ch}(p_\star(\mathcal{R}\otimes\alpha)) = \epsilon_rp_\star(\operatorname{Ch}(\mathcal{R}\otimes\alpha)\operatorname{Td}(\mathcal{T}_p)).$$

From $\iota^\star T\mathbb{P}\cong p^\star T\mathbb{G}\oplus\mathcal{T}_p\oplus\overline{\mathcal{T}_p}$ (Lemma 9.2 and Remark 9.3), we have

$$\iota^\star\widehat{\Gamma}_{\mathbb{P}} = (p^\star\widehat{\Gamma}_{\mathbb{P}})\widehat{\Gamma}(\mathcal{T}_p)\widehat{\Gamma}(\mathcal{T}_p^\vee) = (p^\star\widehat{\Gamma}_{\mathbb{P}})\operatorname{Td}(\mathcal{T}_p)e^{-\pi\sqrt{-1}c_1(\mathcal{T}_p)}$$

where we used the fact that the Gamma class is a square root of the Todd class (see §1.2). Therefore

$$\begin{aligned}
\widehat{\Gamma}_{\mathbb{G}}\operatorname{Ch}(p_\star(\mathcal{R}\otimes\alpha)) &= \epsilon_rp_\star\Big((p^\star\widehat{\Gamma}_{\mathbb{G}})\operatorname{Ch}(\mathcal{R}\otimes\alpha)\operatorname{Td}(\mathcal{T}_p)\Big)\\
&= \epsilon_rp_\star\Big((\iota^\star\widehat{\Gamma}_{\mathbb{P}})\operatorname{Ch}(\alpha)\operatorname{Ch}(\mathcal{R})e^{-\pi\sqrt{-1}c_1(\mathcal{T}_p)}\Big).
\end{aligned}$$

The relationship (30) gives $\operatorname{Ch}(\mathcal{R})e^{-\pi\sqrt{-1}c_1(\mathcal{T}_p)}=e^{-\pi\sqrt{-1}(r-1)\sigma_1}$ and the commutativity follows. Q.E.D.

We define the Euler pairing on $\wedge^r K^0(\mathbb{P}^{n-1})$ as

$$\chi(\alpha_1 \wedge \alpha_2 \wedge \cdots \wedge \alpha_r, \beta_1 \wedge \beta_2 \wedge \cdots \wedge \beta_r) := \det(\chi_{\mathbb{P}^{n-1}}(\alpha_i, \beta_j))_{1 \leqslant i,j \leqslant r}.$$

This is $1/r!$ of the Euler pairing induced from $K^0(\mathbb{P})$. Recall the non-symmetric pairing $[\cdot,\cdot)$ defined in (2). Similarly we define the pairing $[\cdot,\cdot)$ on $\wedge^r H^\bullet(\mathbb{P}^{n-1})$ by

$$(32) \qquad [\alpha_1 \wedge \alpha_2 \wedge \cdots \wedge \alpha_r, \beta_1 \wedge \beta_2 \wedge \cdots \wedge \beta_r) := \det([\alpha_i, \beta_j))_{1 \leqslant i,j \leqslant r}.$$

This is again $1/r!$ of the pairing $[\cdot,\cdot)$ on $H^\bullet(\mathbb{P})$.

Proposition 9.7. *The horizontal maps in the diagram* (31) *preserves the pairings. More precisely, the map*

$$(r!)^{-1} p_\star(\mathcal{R} \otimes \iota^\star(\cdot)) \colon \wedge^r K^0(\mathbb{P}^{n-1}) \to K^0(\mathbb{G})$$

preserves the Euler pairing χ and the map

$$(r!)^{-1} \epsilon_r e^{-\pi\sqrt{-1}(r-1)\sigma_1} p_\star \iota^\star \colon \wedge^r H^\bullet(\mathbb{P}^{n-1}) \to H^\bullet(\mathbb{G})$$

preserves the pairing $[\cdot,\cdot)$.

Proof. Since the vertical maps in (31) intertwines the pairings χ and $[\cdot,\cdot)$ (see (1)), it suffices to show that the map $(r!)^{-1}\epsilon_r e^{-\pi\sqrt{-1}(r-1)\sigma_1} p_\star\iota^\star$ on cohomology preserves the pairing $[\cdot,\cdot)$. For $\alpha, \beta \in \wedge^r H^\bullet(\mathbb{P}^{n-1})$, we have

$$\begin{aligned}
&\Big[(r!)^{-1}\epsilon_r e^{-\pi\sqrt{-1}(r-1)\sigma_1} p_\star\iota^\star(\alpha), (r!)^{-1}\epsilon_r e^{-\pi\sqrt{-1}(r-1)\sigma_1} p_\star\iota^\star(\beta)\Big) \\
&\quad = \left(\frac{\epsilon_r}{r!}\right)^2 \frac{1}{(2\pi\sqrt{-1})^{\dim \mathbb{G}}} \int_{\mathbb{G}} e^{\pi\sqrt{-1}c_1(\mathbb{G})}(e^{\pi\sqrt{-1}\deg/2} p_\star\iota^\star\alpha) \cup p_\star\iota^\star(\beta) \\
&\quad = \frac{1}{(r!)^2} \frac{(-1)^{\binom{r}{2}}}{(2\pi\sqrt{-1})^{\dim \mathbb{P}}} \int_{\mathbb{G}} (p_\star\iota^\star e^{\pi\sqrt{-1}c_1(\mathbb{P})} e^{\pi\sqrt{-1}\deg/2}\alpha) \cup p_\star\iota^\star(\beta) \\
&\quad = \frac{1}{(r!)^2} \frac{(-1)^{\binom{r}{2}}}{(2\pi\sqrt{-1})^{\dim \mathbb{P}}} \int_{\mathbb{P}} (e^{\pi\sqrt{-1}c_1(\mathbb{P})} e^{\pi\sqrt{-1}\deg/2}\alpha) \cup \iota_\star p^\star p_\star\iota^\star(\beta).
\end{aligned}$$

By the formulas in Lemma 9.4 and Lemma 9.5, it follows easily that $\iota_\star p^\star p_\star \iota^\star\beta = r!(-1)^{\binom{r}{2}}\beta$ for the antisymmetric element β. Therefore this gives $1/r!$ of the pairing $[\cdot,\cdot)$ on $\mathbb{P}$ and the conclusion follows. Q.E.D.

Corollary 9.8 ([75]). *For $\mu_1 \geqslant \mu_2 \geqslant \cdots \geqslant \mu_r$ and $\nu_1 \geqslant \nu_2 \geqslant \cdots \geqslant \nu_r$*

$$\chi(\mathcal{E}_\mu, \mathcal{E}_\nu) = \det\left(\chi(\mathcal{O}_{\mathbb{P}^{n-1}}(l_i), \mathcal{O}_{\mathbb{P}^{n-1}}(k_i))_{1 \leqslant i,j \leqslant r}\right)$$

where $l_i = \mu_i + r - i$ and $k_i = \nu_i + r - i$.

Remark 9.9. The map $(r!)^{-1}p_\star\iota^\star\colon \wedge^r H^\bullet(\mathbb{P}^{n-1}) \to H^\bullet(\mathbb{G})$ sending $x_1^{\mu_1+r-1}\wedge x_2^{\mu_2+r-2}\wedge\cdots\wedge x_r^{\mu_r}$ to σ_μ is called the *Satake identification* in [29].

9.4. Quantum cohomology central charge

Here we restate the abelian/non-abelian correspondence of the J-functions in terms of quantum cohomology central charges. Recall from Remark 5.2 that the central charge $Z^{\mathbb{G}}(E)$ of a vector bundle $E\to\mathbb{G}$ is given by

$$Z^{\mathbb{G}}(E)(t) = (2\pi\sqrt{-1})^{\dim\mathbb{G}}\left[J_{\mathbb{G}}(e^{\pi\sqrt{-1}}t), \widehat{\Gamma}_{\mathbb{G}}\operatorname{Ch}(E)\right), \tag{33}$$

where $[\cdot,\cdot)$ is the pairing defined in (2). This is a function on the universal cover of the punctured t-plane $\mathbb{C}^\times$. A branch of this function is determined when we specify $\arg t\in\mathbb{R}$. When $t\in\mathbb{R}_{>0}$, we regard $\arg t = 0$ unless otherwise specified. For $\alpha\in K^0(\mathbb{P})$, we define

$$Z^{\mathbb{P}}(\alpha) := (2\pi\sqrt{-1})^{\dim\mathbb{P}}\left[J_{\mathbb{P}}(e^{\pi\sqrt{-1}}t_1,\dots,e^{\pi\sqrt{-1}}t_r), \widehat{\Gamma}_{\mathbb{P}}\operatorname{Ch}(\alpha)\right)$$

where $[\cdot,\cdot)$ is the pairing on $H^\bullet(\mathbb{P})$. When $\alpha = \alpha_1\wedge\alpha_2\wedge\cdots\wedge\alpha_r\in\wedge^r K^0(\mathbb{P}^{n-1})$, we have

$$Z^{\mathbb{P}}(\alpha) = \det\left(Z^{\mathbb{P}^{n-1}}(\alpha_i)(t_j)\right)_{1\leqslant i,j\leqslant r}$$

where $Z^{\mathbb{P}^{n-1}}$ is the quantum cohomology central charge for $\mathbb{P}^{n-1}$.

Proposition 9.10. *Let $Z^{\mathbb{G}}$ and $Z^{\mathbb{P}}$ denote the quantum cohomology central charges of $\mathbb{G}$ and $\mathbb{P}$ respectively. Then we have the equality:*

$$Z^{\mathbb{G}}(p_\star(\mathcal{R}\otimes\iota^\star\alpha)) = \frac{1}{(2\pi\sqrt{-1}n)^{\binom{r}{2}}}\left[\prod_{i<j}(\theta_i-\theta_j)\cdot Z^{\mathbb{P}}(\alpha)\right]_{t_1=\cdots=t_r=\xi t},$$

where $\alpha\in\wedge^r K^0(\mathbb{P}^{n-1})$ and $\xi := e^{\pi\sqrt{-1}(r-1)/n}$.

Proof. By Lemma 9.5, the abelian/non-abelian correspondence (Theorem 9.1) can be written in the form:

$$J_{\mathbb{G}}(t) = \frac{1}{r!}e^{-\pi\sqrt{-1}(r-1)\sigma_1}p_\star\iota^\star\widetilde{J}(t)$$

with

$$\widetilde{J}(t) = \left[n^{-\binom{r}{2}}\prod_{i<j}(\theta_i-\theta_j)J_{\mathbb{P}}(t_1,\dots,t_r)\right]_{t_1=\cdots=t_r=\xi t}.$$

By Proposition 9.6 (3), we have

$$\widehat{\Gamma}_{\mathbb{G}} \operatorname{Ch}(p_\star(\mathcal{R} \otimes \iota^\star \alpha)) = \epsilon_r e^{-\pi\sqrt{-1}(r-1)\sigma_1} p_\star \iota^\star \left(\widehat{\Gamma}_{\mathbb{P}} \operatorname{Ch}(\alpha)\right).$$

By the above equations and the fact that $\frac{\epsilon_r}{r!} e^{-\pi\sqrt{-1}(r-1)\sigma_1} p_\star \iota^\star$ preserves the pairing $[\cdot,\cdot)$ (Proposition 9.7), we obtain

$$Z^{\mathbb{G}}(p_\star(\mathcal{R} \otimes \iota^\star \alpha)) = (2\pi\sqrt{-1})^{\dim \mathbb{G} + \binom{r}{2}} r! \left[\widetilde{J}(t), \widehat{\Gamma}_{\mathbb{P}} \operatorname{Ch}(\alpha)\right)$$

where the pairing $[\cdot,\cdot)$ on the right-hand side denotes the pairing (32) on $\wedge^r H^\bullet(\mathbb{P}^{n-1})$ (which is $1/r!$ of the pairing $[\cdot,\cdot)$ on $H^\bullet(\mathbb{P})$). Using $\dim G + \binom{r}{2} = \dim \mathbb{P} - \binom{r}{2}$, we obtain the formula in the proposition. Q.E.D.

9.5. Integral representation of quantum cohomology central charges

In this section we give integral representations of quantum cohomology central charges in terms of the Hori–Vafa mirrors. Let $f\colon (\mathbb{C}^\times)^{n-1} \to \mathbb{C}$ be the mirror Laurent polynomial of $\mathbb{P}^{n-1}$ (see §5):

$$f(y) = y_1 + \cdots + y_{n-1} + \frac{1}{y_1 y_2 \cdots y_{n-1}}.$$

The critical values of f are given as

$$v_k := n e^{-2\pi\sqrt{-1}k/n}, \quad k \in \mathbb{Z}/n\mathbb{Z}.$$

Let $\phi \in \mathbb{R}$ be an admissible phase for $\{v_0, v_1, \dots, v_{n-1}\}$ (see §4.2). Let $\Gamma_k(\phi) \subset (\mathbb{C}^\times)^{n-1}$ denote the Lefschetz thimble for f associated with the critical value v_k and the vanishing path $v_k + \mathbb{R}_{\geqslant 0} e^{\sqrt{-1}\phi}$ (see Figure 2). The Lefschetz thimble $\Gamma_k(\phi)$ is homeomorphic to $\mathbb{R}^{n-1}$ and fibers over the straight half-line $f(\Gamma_k) = v_k + \mathbb{R}_{\geqslant 0} e^{\sqrt{-1}\phi}$; the fiber is a vanishing cycle homeomorphic to S^{n-2}. We write $\Gamma_k^\vee(\phi)$ for the "opposite" Lefschetz thimble associated to the critical value v_k and the path $v_k - \mathbb{R}_{\geqslant 0} e^{\sqrt{-1}\phi}$. We choose orientation of Lefschetz thimbles such that $\sharp(\Gamma_k(\phi) \cap \Gamma_k^\vee(\phi)) = (-1)^{(n-1)(n-2)/2}$.

Proposition 9.11. *Let $\phi \in \mathbb{R}$ be an admissible phase for $\{v_0, v_1, \dots, v_{n-1}\}$. There exist K-classes $\mathcal{F}_k(\phi), \mathcal{G}_k(\phi) \in K^0(\mathbb{P}^{n-1})$ for $k \in \mathbb{Z}/n\mathbb{Z}$ such that*

$$\begin{aligned} Z^{\mathbb{P}^{n-1}}(\mathcal{F}_k(\phi))(t) &= \int_{\Gamma_k(\phi)} e^{-tf(y)} \frac{dy}{y}, \\ Z^{\mathbb{P}^{n-1}}(\mathcal{G}_k(\phi))(e^{-\pi\sqrt{-1}} t) &= \int_{\Gamma_k^\vee(\phi)} e^{tf(y)} \frac{dy}{y}, \end{aligned} \tag{34}$$

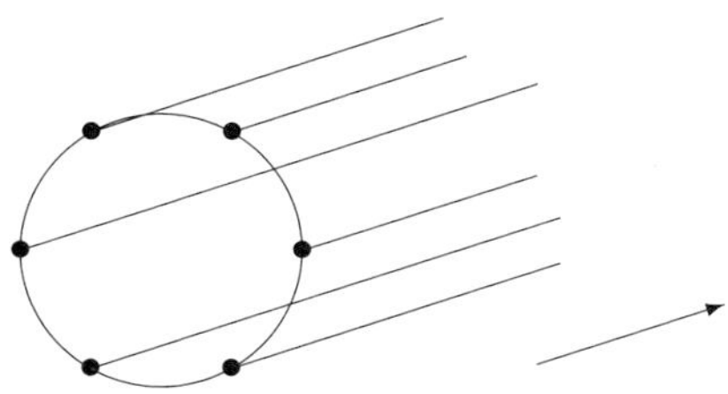

Fig. 2. Vanishing paths in the admissible direction ϕ

hold when $|\arg t + \phi| < \frac{\pi}{2}$, *where* $\frac{dy}{y} = \bigwedge_{i=1}^{n-1} \frac{dy_i}{y_i}$. *Moreover, we have:*

(1) $\chi(\mathcal{F}_k(\phi), \mathcal{G}_l(\phi)) = \delta_{kl}$;

(2) *when* $|k\frac{2\pi}{n} + \phi| < \frac{\pi}{2} + \frac{\pi}{n}$, *we have* $\mathcal{F}_k(\phi) = \mathcal{O}_{\mathbb{P}^{n-1}}(k)$.

Proof. This follows from the result in [50, 55]. (See §6 and [29, §5] for a closely related discussion). By [50, Theorems 4.11, 4.14], we have an isomorphism

$$K^0(\mathbb{P}^{n-1}) \cong H_{n-1}\left((\mathbb{C}^\times)^{n-1}, \{y : \operatorname{Re}(tf(y)) \geqslant M\}\right), \quad \alpha \mapsto \Gamma(\alpha, \arg t)$$

depending on $\arg t \in \mathbb{R}$, such that $\Gamma(\alpha, \arg t)$ is Gauss–Manin flat with respect to the variation of $\arg t$ and that

$$Z^{\mathbb{P}^{n-1}}(\alpha)(t) = \int_{\Gamma(\alpha, \arg t)} e^{-tf(y)} \frac{dy}{y}.$$

Here M is a sufficiently big positive number (it is sufficient that $M \geqslant 2n|t|$). Moreover, this isomorphism intertwines the Euler pairing with the intersection pairing:

$$\chi(\alpha, \beta) = (-1)^{\frac{(n-1)(n-2)}{2}} \sharp\left(\Gamma(\alpha, \arg t + \pi) \cap \Gamma(\beta, \arg t)\right).$$

Therefore equation (34) and part (1) of the proposition hold for $\mathcal{F}_k, \mathcal{G}_k$ such that

$$\Gamma(\mathcal{F}_k(\phi), -\phi) = \Gamma_k(\phi), \quad \Gamma(\mathcal{G}_k(\phi), -\phi - \pi) = \Gamma_k^\vee(\phi).$$

(We only need to check (34) at $\arg t = -\phi$; it then follows by analytic continuation for other $t \in (-\phi - \frac{\pi}{2}, -\phi + \frac{\pi}{2})$.) Recall from (20) that we have

$$Z^{\mathbb{P}^{n-1}}(\mathcal{O}_{\mathbb{P}^{n-1}})(t) = \int_{\Gamma_0(0)} e^{-tf(y)} \frac{dy}{y}$$

when $\arg t = 0$. Therefore, when $\arg t = 2\pi k/n$, we have by setting $t' = e^{-2\pi\sqrt{-1}k/n} t$,

$$\begin{aligned} Z^{\mathbb{P}^{n-1}}(\mathcal{O}_{\mathbb{P}^{n-1}}(k))(t) &= Z^{\mathbb{P}^{n-1}}(\mathcal{O}_{\mathbb{P}^{n-1}})(t') \\ &= \int_{\Gamma_0(0)} e^{-t' f(y)} \frac{dy}{y} = \int_{\Gamma_k(-2\pi k/n)} e^{-tf(y)} \frac{dy}{y} \end{aligned}.$$

where in the first line we used (18) and the definition of the central charge, and in the second line we performed the change $y_i \to e^{2\pi\sqrt{-1}k/n} y_i$ of variables and used the fact that $e^{-2\pi\sqrt{-1}k/n}\Gamma_0(0) = \Gamma_k(-2\pi k/n)$. From this it follows that

$$Z^{\mathbb{P}^{n-1}}(\mathcal{O}_{\mathbb{P}^{n-1}}(k))(t) = \int_{\Gamma_k(\phi)} e^{-tf(y)} \frac{dy}{y}$$

when $|\phi + k\frac{2\pi}{n}| < \frac{\pi}{2} + \frac{\pi}{n}$ and $\arg t = -\phi$. This implies $\mathcal{F}_k(\phi) = \mathcal{O}_{\mathbb{P}^{n-1}}(k)$ for such ϕ and k and part (2) follows. Q.E.D.

Remark 9.12. The sign $(-1)^{(n-1)(n-2)/2}$ in the intersection pairing was missing in [50], and this has been corrected in [52, footnote 16].

Remark 9.13. Part (2) of the proposition determines roughly half of $\mathcal{F}_k(\phi)$'s. The whole collection $\{\mathcal{F}_k(\phi) : k \in \mathbb{Z}/n\mathbb{Z}\}$ is obtained from the exceptional collection $\{\mathcal{O}(k) : -\phi - \pi < k\frac{2\pi}{n} < -\phi + \pi\}$ by a sequence of mutations (see [29, §5]), and thus is an exceptional collection. Because $\mathcal{F}_k(\phi)$ is mirror to the Lefschetz thimble $\Gamma_k(\phi)$, the set $\{\widehat{\Gamma}_{\mathbb{P}^{n-1}} \operatorname{Ch}(\mathcal{F}_k(\phi)) : k \in \mathbb{Z}/n\mathbb{Z}\}$ gives the asymptotic basis of $\mathbb{P}^{n-1}$ at phase ϕ (see the proof of Theorem 6.4) and the Gamma conjecture II (§4.3) holds for $\mathbb{P}^{n-1}$.

The Hori–Vafa mirror $\tilde{g}$ of $\mathbb{P} = (\mathbb{P}^{n-1})^r$ is given by:

$$\begin{aligned} \tilde{g}(y) &:= f(\vec{y}_1) + f(\vec{y}_2) + \cdots + f(\vec{y}_r) \\ &= \sum_{i=1}^{r} \left(y_{i,1} + y_{i,2} + \cdots + y_{i,n-1} + \frac{1}{y_{i,1} y_{i,2} \cdots y_{i,n-1}} \right) \end{aligned}$$

where $y = (\vec{y}_1, \ldots, \vec{y}_r) \in (\mathbb{C}^\times)^{r(n-1)}$ and $\vec{y}_i = (y_{i,1}, \ldots, y_{i,n-1}) \in (\mathbb{C}^\times)^{n-1}$. Critical values of $\tilde{g}$ are given by

$$\tilde{v}_K := v_{k_1} + \cdots + v_{k_r} \quad \text{for } K = (k_1, \ldots, k_r) \in \mathbb{Z}^r.$$

The product $\Gamma_{k_1}(\phi) \times \cdots \times \Gamma_{k_r}(\phi)$ of Lefschetz thimbles for f gives a Lefschetz thimble for $\tilde{g}$ associated to the critical value $\tilde{v}_K$ and the path $\tilde{v}_K + \mathbb{R}_{\geqslant 0} e^{\sqrt{-1}\phi}$.

The Hori–Vafa mirror of $\mathbb{G}$ is obtained from the mirror of $\mathbb{P}$ by shifting the phase by $(r-1)\pi/n$ and restricting to "anti-symmetrized" Lefschetz thimbles (see [45, 75, 57]). We set

$$g(y) := \xi \tilde{g}(y)$$

with $\xi = e^{\pi\sqrt{-1}(r-1)/n}$. Then critical values of $g(y)$ are

$$v_K := \xi \tilde{v}_K \quad \text{for } K = (k_1, \dots, k_r) \in \mathbb{Z}^r.$$

For a tuple $K = (k_1, \dots, k_r) \in \mathbb{Z}^r$ and $\phi \in \mathbb{R}$, we define the anti-symmetrized Lefschetz thimble for $g(y)$ as:

$$\Gamma_K(\phi) := \Gamma_{k_1}(\phi') \wedge \cdots \wedge \Gamma_{k_r}(\phi') = \sum_{\sigma \in \mathfrak{S}_r} \operatorname{sgn}(\sigma) \Gamma_{k_{\sigma(1)}}(\phi') \times \cdots \times \Gamma_{k_{\sigma(r)}}(\phi')$$
$$\Gamma_K^\vee(\phi) := \Gamma_{k_1}^\vee(\phi') \wedge \cdots \wedge \Gamma_{k_r}^\vee(\phi') = \sum_{\sigma \in \mathfrak{S}_r} \operatorname{sgn}(\sigma) \Gamma_{k_{\sigma(1)}}^\vee(\phi') \times \cdots \times \Gamma_{k_{\sigma(r)}}^\vee(\phi')$$

with $\phi' = \phi - \frac{(r-1)\pi}{n}$. They are elements of the relative homology groups

$$H_{r(n-1)}\left((\mathbb{C}^\times)^{r(n-1)}, \{y : \pm \operatorname{Re}(e^{-\sqrt{-1}\phi} g(y)) \geqslant M\}\right)$$

with M sufficiently large. We also define K-classes $\mathcal{F}_K(\phi), \mathcal{G}_K(\phi) \in K^0(\mathbb{G})$ as:

$$\begin{aligned} \mathcal{F}_K(\phi) &:= \frac{1}{r!} p_\star\Big(\mathcal{R} \otimes \iota^\star(\mathcal{F}_{k_1}(\phi') \wedge \cdots \wedge \mathcal{F}_{k_r}(\phi'))\Big) \\ \mathcal{G}_K(\phi) &:= \frac{1}{r!} p_\star\Big(\mathcal{R} \otimes \iota^\star(\mathcal{G}_{k_1}(\phi') \wedge \cdots \wedge \mathcal{G}_{k_r}(\phi'))\Big) \end{aligned} \tag{35}$$

with $\phi' = \phi - \frac{(r-1)\pi}{n}$, where $\mathcal{F}_k(\phi)$ and $\mathcal{G}_k(\phi)$ are as in Proposition 9.11. Note that $\{\mathcal{F}_K(\phi)\}_K$ or $\{\mathcal{G}_K(\phi)\}_K$ gives an integral basis of $K^0(\mathbb{G})$ by Proposition 9.6.

Remark 9.14. By the change of variables $y_{i,j} = \xi y'_{i,j}$, we can also write:

$$g(y) = \sum_{i=1}^r \left(y'_{i,1} + y'_{i,2} + \cdots + y'_{i,n-1} + \frac{(-1)^{r-1}}{y'_{i,1} y'_{i,2} \cdots y'_{i,n-1}} \right)$$

Theorem 9.15. *Let $\phi \in \mathbb{R}$ be such that $\phi' = \phi - \frac{(r-1)\pi}{n}$ is admissible for $\{v_0, v_1, \dots, v_{n-1}\}$. For a mutually distinct r-tuple $K = (k_1, \dots, k_r) \in$*

$(\mathbb{Z}/n\mathbb{Z})^r$, *we have*

$$\begin{aligned}&Z^{\mathbb{G}}(\mathcal{F}_K(\phi))(t)\\&\quad=\frac{1}{r!}\left(\frac{-\xi t}{2\pi\sqrt{-1}n}\right)^{\binom{r}{2}}\int_{\Gamma_K(\phi)}\frac{dy}{y}e^{-tg(y)}\cdot\prod_{i<j}(f(\vec{y}_i)-f(\vec{y}_j))\\&Z^{\mathbb{G}}(\mathcal{G}_K(\phi))(e^{-\pi\sqrt{-1}}t)\\&\quad=\frac{1}{r!}\left(\frac{\xi t}{2\pi\sqrt{-1}n}\right)^{\binom{r}{2}}\int_{\Gamma_K^\vee(\phi)}\frac{dy}{y}e^{tg(y)}\cdot\prod_{i<j}(f(\vec{y}_i)-f(\vec{y}_j))\end{aligned}$$

for $|\arg t+\phi|<\frac{\pi}{2}$, *where* $\xi:=e^{\pi\sqrt{-1}(r-1)/n}$ *and* $\frac{dy}{y}:=\bigwedge_{i=1}^r\bigwedge_{j=1}^{n-1}\frac{dy_{i,j}}{y_{i,j}}$. *Moreover, we have*

$$\chi(\mathcal{F}_K(\phi),\mathcal{G}_L(\phi))=\delta_{K,L}$$

when the tuples K, L *are ordered with respect to a fixed choice of a total order of* $\mathbb{Z}/n\mathbb{Z}$.

Proof. Combining Proposition 9.10 and Proposition 9.11, we have that

$$\begin{aligned}&Z^{\mathbb{G}}(\mathcal{F}_K(\phi))(t)\\&=\frac{1}{r!(2\pi\sqrt{-1}n)^{\binom{r}{2}}}\left[\prod_{i<j}(\theta_i-\theta_j)\cdot Z^{\mathbb{P}}(\mathcal{F}_{k_1}(\phi')\wedge\cdots\wedge\mathcal{F}_{k_r}(\phi'))\right]_{t_1=\cdots=t_r=\xi t}\\&=\frac{1}{r!(2\pi\sqrt{-1}n)^{\binom{r}{2}}}\left[\prod_{i<j}(\theta_i-\theta_j)\cdot\int_{\Gamma_K(\phi)}e^{-(t_1f(\vec{y}_1)+\cdots+t_rf(\vec{y}_r))}\frac{dy}{y}\right]_{t_1=\cdots=t_r=\xi t}\\&=\frac{1}{r!}\left(\frac{-\xi t}{2\pi\sqrt{-1}n}\right)^{\binom{r}{2}}\int_{\Gamma_K(\phi)}\frac{dy}{y}e^{-tg(y)}\prod_{i<j}(f(\vec{y}_i)-f(\vec{y}_j)),\end{aligned}$$

where in the last line, we used Lemma 9.16 below. The formula for $Z^{\mathbb{G}}(\mathcal{G}_K(\phi))(e^{-\pi\sqrt{-1}}t)$ follows similarly. The orthogonality $\chi(\mathcal{F}_K(\phi),\mathcal{G}_L(\phi))=\delta_{K,L}$ follows easily from Propositions 9.7 and 9.11. Q.E.D.

Lemma 9.16. *We have*

$$\left(\prod_{i<j}(\theta_i-\theta_j)\right)e^{\sum_{i=1}^r\alpha_it_i}=e^{\sum_{i=1}^r\alpha_it_i}\prod_{i<j}(\alpha_it_i-\alpha_jt_j)$$

where $\theta_i=t_i\frac{\partial}{\partial t_i}$.

Proof. The left-hand side can be written in the form $\varphi(\alpha_1 t_1, \dots, \alpha_r t_r) e^{\sum_{i=1}^r \alpha_i t_i}$ for some polynomial φ and the highest order term of φ is $\prod_{i<j}(\alpha_i t_i - \alpha_j t_j)$. On the other hand, φ is anti-symmetric in the arguments, and thus should be divisible by $\prod_{i<j}(\alpha_i t_i - \alpha_j t_j)$. This implies the lemma. Q.E.D.

Remark 9.17. The collection $\{\mathcal{F}_K(\phi)\}_K$, where K ranges over distinct r elements of $\mathbb{Z}/n\mathbb{Z}$, yields the asymptotic basis $\{\widehat{\Gamma}_{\mathbb{G}} \operatorname{Ch}(\mathcal{F}_K(\phi))\}_K$ of $\mathbb{G}$ at phase ϕ. This follows either by studying mirror oscillatory integrals in more details or by combining [29, Proposition 6.5.1], Remark 9.13 and Proposition 9.6. It follows from the deformation argument in [29, §6] that this is mutation equivalent to Kapranov's exceptional collection $\{\mathcal{E}_\mu : n - r \geqslant \mu_1 \geqslant \cdots \geqslant \mu_r \geqslant 0\}$ [54].

9.6. Quantum periods via mirrors of Eguchi–Hori–Xiong, Rietsch and Marsh–Rietsch

In this section, using mirrors of Eguchi–Hori–Xiong [25], Rietsch [71] and Marsh–Rietsch [63], we show that the limit sup in (10) can be replaced with the limit for Grassmannians.

Eguchi–Hori–Xiong's mirror [25] for Grassmannians is a Laurent polynomial in $r(n-r) = \dim \mathbb{G}$ variables. Let $X_{i,j}$ with $1 \leqslant i \leqslant r$, $1 \leqslant j \leqslant n-r$ be $r(n-r)$ independent variables. Define the Laurent polynomial W in these variables by

$$W = \sum_{i=1}^{r} \sum_{j=1}^{n-r-1} \frac{X_{i,j+1}}{X_{i,j}} + \sum_{j=1}^{n-r} \sum_{i=1}^{r-1} \frac{X_{i+1,j}}{X_{i,j}} + \frac{1}{X_{1,n-r}} + \frac{1}{X_{r,1}}.$$

Batyrev–Ciocan-Fontanine–Kim–van-Straten [7] showed that the Newton polytope of W equals the fan polytope of a toric degeneration of $\mathbb{G} = \operatorname{Gr}(r,n)$ and conjectured that W is a weak Landau–Ginzburg model of $\mathbb{G}$. Recently, Marsh–Rietsch [63] proved this conjecture by constructing a compactification of the Eguchi–Hori–Xiong mirror and showing an isomorphism between the quantum connection and the Gauss–Manin connection. Indeed, the mirror of Marsh–Rietsch is a reinterpretation of the earlier construction by Rietsch [71].

Theorem 9.18 ([63]). *The quantum period* (9) $G_{\mathbb{G}}(t) = \langle [\mathrm{pt}], J_{\mathbb{G}}(t) \rangle$ *of the Grassmannian* $\mathbb{G}$ *is given by the constant term series of the Eguchi–Hori–Xiong mirror* W, *i.e.* $G_{\mathbb{G}}(t) = \sum_{i=0}^{\infty} \frac{1}{i!} \operatorname{Const}(W^i) t^i$.

It is easy to check that the constant term series of W is of the form $\sum_{k=0}^{\infty} a_k t^{kn}$ with $a_k \neq 0$ for all $k \in \mathbb{Z}_{\geqslant 0}$ (see [7, §5.2] for the explicit form). Therefore, by applying Lemma 3.13, we obtain the following. (It seems difficult to deduce this from Hori–Vafa mirrors).

Proposition 9.19. *Let $G_{\mathbb{G}}(t) = \langle [\mathrm{pt}], J_{\mathbb{G}}(t)\rangle = \sum_{k=0}^{\infty} G_{kn} t^{kn}$ be the quantum period of $\mathbb{G}$. Then $G_{kn} > 0$ for all $k \in \mathbb{Z}_{\geqslant 0}$ and $\lim_{k\to\infty} \sqrt[kn]{(kn)! G_{kn}}$ exists.*

9.7. Apéry constants

In this section we prove Gamma conjecture I' (Conjecture 3.11) for $\mathbb{G} = \mathrm{Gr}(r, n)$.

Theorem 9.20. *The Grassmannian* $\mathrm{Gr}(r, n)$ *satisfies Gamma conjecture* I'.

The rest of the paper is devoted to the proof of Theorem 9.20. Since $\mathrm{Gr}(r, n) \cong \mathrm{Gr}(n-r, n)$, we may assume that $r \leqslant n/2$. We fix a sufficiently small phase $\phi > 0$ such that $\phi' = \phi - \frac{(r-1)\pi}{n}$ is admissible for $\{v_0, v_1, \ldots, v_{n-1}\}$. Let Λ denote the index set of mutually distinct r-tuples K of elements of $\mathbb{Z}/n\mathbb{Z}$:

$$\Lambda = \{K = (k_1, \ldots, k_r) \in (\mathbb{Z}/n\mathbb{Z})^r : n-1 \geqslant k_1 > k_2 > \cdots > k_r \geqslant 0\}.$$

For $K \in \Lambda$, we write $\mathcal{F}_K = \mathcal{F}_K(\phi)$ and $\mathcal{G}_K = \mathcal{G}_K(\phi)$ for the elements in (35).

Lemma 9.21. *Set $T := \max\{|v_K| : K \in \Lambda\}$.*

(1) *We have $T = n\frac{\sin(\pi r/n)}{\sin(\pi/n)}$.*
(2) *If $T = |v_K|$ for $K \in \Lambda$, then K is given by a consecutive r-tuple of elements in $\mathbb{Z}/n\mathbb{Z}$ and $v_K = Te^{-2\pi\sqrt{-1}k/n}$ for some $k \in \mathbb{Z}$.*
(3) *We have $T = v_{K_0}$ for $K_0 := (r-1, r-2, \ldots, 1, 0)$. Moreover, $\mathcal{F}_{K_0} = \mathcal{O}_{\mathbb{G}}$.*

Proof. Parts (1) and (2) follow easily from the definition of v_K. To see Part (3), note that Proposition 9.11 (2) gives $\mathcal{F}_k(\phi') = \mathcal{O}_{\mathbb{P}^{n-1}}(k)$ for $0 \leqslant k \leqslant r-1$ and $\phi' = \phi - \frac{(r-1)\pi}{n}$ (since $|\phi|$ is small and $r \leqslant n/2$). Then Proposition 9.6 (2) implies the conclusion. Q.E.D.

Let $\alpha \in H_{2|\alpha|}(\mathbb{G})$ be a homology class such that $c_1(\mathbb{G}) \cap \alpha = 0$. We want to show that the limit formula (see (11))

$$\lim_{n\to\infty} \frac{\langle \alpha, J_{kn}\rangle}{\langle [\mathrm{pt}], J_{kn}\rangle} = \langle \alpha, \widehat{\Gamma}_{\mathbb{G}}\rangle$$

holds where we set $J_{\mathbb{G}}(t) = e^{c_1(\mathbb{G})\log t} \sum_{k=0}^{\infty} J_{kn} t^{kn}$. We start with noting that it suffices to show this formula when $\langle \alpha, \widehat{\Gamma}_{\mathbb{G}}\rangle = 0$. Indeed, we obtain the general case by applying the formula to $\alpha' = \alpha - \langle \alpha, \widehat{\Gamma}_{\mathbb{G}}\rangle[\mathrm{pt}]$ (which satisfies $\langle \alpha', \widehat{\Gamma}_{\mathbb{G}}\rangle = 0$).

Let $\widehat{\alpha} \in K^0(\mathbb{G}) \otimes \mathbb{C}$ be a complexified K-class such that

$$\mathrm{PD}(\alpha) = \widehat{\Gamma}_{\mathbb{G}} \operatorname{Ch}(\widehat{\alpha}).$$

Then we have, using the definition of $Z^{\mathbb{G}}$ (see (33) and (2)),

$$\begin{aligned}\langle \alpha, J_{\mathbb{G}}(t)\rangle &= \int_{\mathbb{G}} J_{\mathbb{G}}(t) \cup \mathrm{PD}(\alpha) \\ &= (\sqrt{-1})^{\dim \mathbb{G} - 2|\alpha|} \int_{\mathbb{G}} J_{\mathbb{G}}(t) \cup e^{-\pi\sqrt{-1}\mu} e^{\pi\sqrt{-1}c_1(\mathbb{G})} \mathrm{PD}(\alpha) \\ &= (-1)^{|\alpha|}(2\pi\sqrt{-1})^{\dim \mathbb{G}} \left[J_{\mathbb{G}}(t), \widehat{\Gamma}_{\mathbb{G}} \operatorname{Ch}(\widehat{\alpha}) \right) \\ &= (-1)^{|\alpha|} Z^{\mathbb{G}}(\widehat{\alpha})(e^{-\pi\sqrt{-1}}t).\end{aligned}$$

Using the dual bases $\{\mathcal{F}_K\}_{K\in\Lambda}$ and $\{\mathcal{G}_K\}_{K\in\Lambda}$ of $K^0(\mathbb{G})$, we can expand

$$\widehat{\alpha} = \sum_{K\in\Lambda} \chi(\mathcal{F}_K, \widehat{\alpha})\mathcal{G}_K.$$

Thus by Theorem 9.15, we obtain the integral representation for $\langle \alpha, J_{\mathbb{G}}(t)\rangle$:

$$\begin{aligned}&\langle \alpha, J_{\mathbb{G}}(t)\rangle \\ &= (-1)^{|\alpha|} c_{n,r} \sum_{K\in\Lambda} \chi(\mathcal{F}_K, \widehat{\alpha}) \cdot t^{\binom{r}{2}} \int_{\Gamma_K^\vee(\phi)} \frac{dy}{y} e^{tg(y)} \prod_{i<j} (f(\vec{y}_i) - f(\vec{y}_j))\end{aligned} \tag{36}$$

for $|\arg t + \phi| < \pi/2$, where we set $c_{n,r} = \frac{1}{r!}(\xi/(2\pi\sqrt{-1}n))^{\binom{r}{2}}$. Let $C_K(\lambda)$ denote the "anti-symmetrized" vanishing cycle in the fiber $g^{-1}(\lambda)$:

$$C_K(\lambda) = \Gamma_K^\vee(\phi) \cap g^{-1}(\lambda) = \sum_{\sigma\in\mathfrak{S}_r} \operatorname{sgn}(\sigma) C_{k_{\sigma(1)}}(\lambda) \times \cdots \times C_{k_{\sigma(r)}}(\lambda)$$

where $\lambda \in v_K - \mathbb{R}_{\geqslant 0} e^{\sqrt{-1}\phi}$ and $C_{k_i}(\lambda) = \Gamma_{k_i}(\phi') \cap f^{-1}(\xi^{-1}\lambda)$ is the vanishing cycle for f. We define the period integral $P_K(\lambda)$ as:

$$P_K(\lambda) := \int_{C_K(\lambda)\subset g^{-1}(\lambda)} \prod_{i<j} (f(\vec{y}_i) - f(\vec{y}_j)) \left. \frac{\bigwedge_{i=1}^r \bigwedge_{j=1}^{n-1} \frac{dy_{i,j}}{y_{i,j}}}{dg} \right|_{g^{-1}(\lambda)}.$$

Then we may rewrite (36) as the Laplace transform of the period:

$$\langle \alpha, J_{\mathbb{G}}(t)\rangle = (-1)^{|\alpha|} c_{n,r} \sum_{K\in\Lambda} \chi(\mathcal{F}_K, \widehat{\alpha}) \int_{v_K - e^{\sqrt{-1}\phi}\mathbb{R}_{\geqslant 0}} d\lambda \cdot t^{\binom{r}{2}} e^{t\lambda} P_K(\lambda). \tag{37}$$

Applying the Laplace transformation

$$\varphi(t) \mapsto \int_0^\infty \varphi(t) e^{-ut} dt$$

to both sides of (37), we obtain

$$\begin{aligned} &\sum_{k=0}^{\infty} (kn)! \langle \alpha, J_{kn} \rangle u^{-kn-1} \\ &\qquad = (-1)^{|\alpha|} c'_{n,r} \sum_{K \in \Lambda} \chi(\mathcal{F}_K, \widehat{\alpha}) \int_{v_K - e^{\sqrt{-1}\phi}\mathbb{R}_{\geqslant 0}} d\lambda \frac{P_K(\lambda)}{(u-\lambda)^{\binom{r}{2}+1}} \end{aligned} \tag{38}$$

when $\mathrm{Re}(u - v_K) > 0$ for all $K \in \Lambda$, where $c'_{n,r} = \binom{r}{2}! c_{n,r}$. Note that the right-hand side can be analytically continued to a holomorphic function outside the branch cut:

$$\bigcup_{K \in \Lambda} v_K - e^{\sqrt{-1}\phi}\mathbb{R}_{\geqslant 0}.$$

See Figure 3. Moreover, this can be analytically continued to the universal cover of $\mathbb{C} \setminus \{v_K : K \in \Lambda\}$. Since the left-hand side is regular at

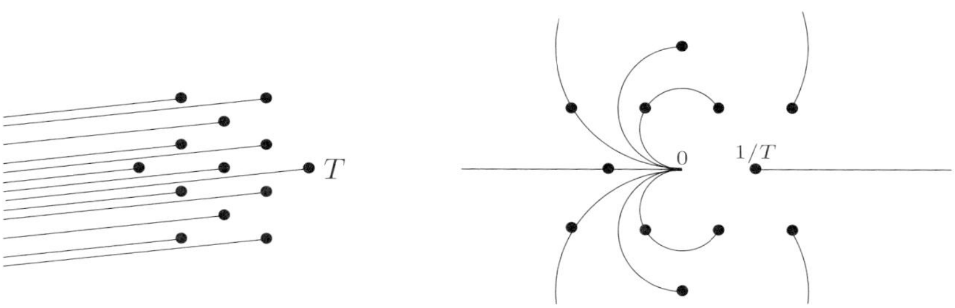

Fig. 3. Branch cut in the u-plane (left) and the u^{-1}-plane (right) ($n = 6$, $r = 2$). In the right picture, we set $\phi = 0$ for simplicity.

$u = \infty$, it follows from Lemma 9.21 that the convergence radius of the left-hand side of (38) (as a power series in u^{-1}) is bigger than or equal to

$$\frac{1}{T} = \frac{1}{n} \frac{\sin(\pi/n)}{\sin(\pi r/n)}.$$

By Cauchy's integral formula, it follows that the jump of the function (38) across the cut $v_K - e^{\sqrt{-1}\phi}\mathbb{R}_{\geqslant 0}$ is proportional to $\chi(\mathcal{F}_K, \widehat{\alpha}) \cdot (\partial_u)^{\binom{r}{2}} P_K(u)$. From this it follows that:

(a) if $\alpha = [\mathrm{pt}]$, we have $\chi(\mathcal{F}_{K_0}, \widehat{\alpha}) = (2\pi\sqrt{-1})^{-\dim \mathbb{G}}\chi(\mathcal{O}_{\mathbb{G}}, \mathcal{O}_{\mathrm{pt}}) \neq 0$ and thus the convergence radius of $\sum_{k=0}^{\infty}(kn)!\,\langle [\mathrm{pt}], J_{kn}\rangle\, x^{kn}$ is exactly $1/T$;

(b) if $\langle \alpha, \widehat{\Gamma}_{\mathbb{G}}\rangle = 0$, then we have $\chi(\mathcal{F}_{K_0}, \widehat{\alpha}) = \chi(\mathcal{O}_{\mathbb{G}}, \widehat{\alpha}) = [\widehat{\Gamma}_{\mathbb{G}}, \alpha) = 0$ (by (1)); this implies that (38) is holomorphic at $u = v_{K_0} = T$; since the left-hand side of (38) is a power series in u^{-n} (multiplied by u^{-1}), it is holomorphic at any other v_K with $|v_K| = T$ and the series $\sum_{k=0}^{\infty}(kn)!\,\langle \alpha, J_{kn}\rangle\, x^{kn}$ has a convergence radius $R(\alpha) > 1/T$.

Here $K_0 = (r-1, r-2, \dots, 1, 0)$ and recall from Lemma 9.21 that $\mathcal{F}_{K_0} = \mathcal{O}_{\mathbb{G}}$. From part (a) and Proposition 9.19, it follows that

$$\lim_{k\to\infty} \sqrt[kn]{(kn)!\,\langle [\mathrm{pt}], J_{kn}\rangle} = T.$$

From part (b), it follows that, when $\langle \alpha, \widehat{\Gamma}_{\mathbb{G}}\rangle = 0$,

$$\lim_{k\to\infty} \frac{\langle \alpha, J_{kn}\rangle}{\langle [\mathrm{pt}], J_{kn}\rangle} = \lim_{k\to\infty} \frac{(kn)!\,\langle \alpha, J_{kn}\rangle}{(kn)!\,\langle [\mathrm{pt}], J_{kn}\rangle} = 0.$$

This completes the proof of Theorem 9.20.

§Appendix A. Eigenvalues of quantum multiplication by $c_1(F)$ on odd cohomology

In the main body of the text, we restrict to the even part of cohomology group and ignore the odd part. The restriction to the even part was partly for the sake of simplicity: in general, the big quantum product defines a super-commutative algebra structure on the full cohomology group, and the big quantum connection (5) should be treated in the formalism of supermanifolds (see [62]). In this appendix, answering a question of an anonymous referee, we discuss Property $\mathcal{O}$ and Gamma conjectures on the *full* cohomology group, and show that they are in fact equivalent to Property $\mathcal{O}$ and Gamma conjectures on the even part (discussed in the main body of the text). Moreover, we argue that it is sufficient to consider Property $\mathcal{O}$ and Gamma conjectures on the subspace $\bigoplus_{p=0}^{\dim F} H^{p,p}(F)$ of diagonal Hodge type. In this section, we write $H^{\mathrm{even}}(F)/H^{\mathrm{odd}}(F)$ for the even/odd part of cohomology group, and $H^{\mathrm{full}}(F) = H^{\mathrm{even}}(F) \oplus H^{\mathrm{odd}}(F)$ for the full cohomology group. (We wrote $H^{\bullet}(F) = H^{\mathrm{even}}(F)$ in the main body of the text.)

Remark A.1. After we wrote this appendix, we noticed that Sanda–Shamoto [73, Remark 6.5, Lemma 6.6] had already discussed the same issue on odd cohomology. Our new observation is the discussion concerning $\bigoplus_{p=0}^{\dim F} H^{p,p}(F)$.

For Gamma Conjecture II, the restriction to the even part was superfluous. Recall from Hertling–Manin–Teleman [43] that if the quantum cohomology of F is semisimple, F is necessarily of Hodge–Tate type (i.e. $H^{p,q}(F) = 0$ unless $p = q$), and in particular has no odd cohomology classes.

Let us study Property $\mathcal{O}$ and Gamma Conjecture I including the odd part. For a subspace $V \subset H^{\mathrm{full}}(F)$ preserved by $(c_1(F)\star_0)$, we define

$$T_V := \max\{|u| : u \text{ is an eigenvalue of } (c_1(F)\star_0)\colon V \to V\} \in \overline{\mathbb{Q}}.$$

The number T (6) in the main body of the text equals $T_{H^{\mathrm{even}}(F)}$. By replacing T with T_V in Definition 3.1 and regarding $(c_1(F)\star_0)$ as an operator on V, we can similarly define "Property $\mathcal{O}$ *on the subspace* V". Note that the original Property $\mathcal{O}$ is the "Property $\mathcal{O}$ on $H^{\mathrm{even}}(F)$" in this terminology. We define

$$H^{(j)} = \bigoplus_{p-q=j} H^{p,q}(F).$$

By the motivic axiom [58], $H^{(j)}$ is preserved by $(c_1(F)\star_0)$ and $H^{(j)} \star_0 H^{(k)} \subset H^{(j+k)}$. We consider Property $\mathcal{O}$ on $H^{(0)} = \bigoplus_{p=0}^{\dim F} H^{p,p}(F)$, $H^{\mathrm{even}}(F) = \bigoplus_{j\in 2\mathbb{Z}} H^{(j)}$ and $H^{\mathrm{full}}(F)$.

Theorem A.2. *For every Fano manifold, we have* $T_{H^{(0)}} = T_{H^{\mathrm{full}}(F)} = T_{H^{\mathrm{even}}(F)} = T$. *Moreover, the following are equivalent*:

(1) *F has Property $\mathcal{O}$ on $H^{(0)}$.*
(2) *F has Property $\mathcal{O}$ on $H^{\mathrm{even}}(F)$.*
(3) *F has Property $\mathcal{O}$ on $H^{\mathrm{full}}(F)$.*

This theorem follows from the following two lemmas.

Lemma A.3. *For all j, the spectrum of $(c_1(F)\star_0)$ on $H^{(j)}$ is a subset of the spectrum of $(c_1(F)\star_0)$ on $H^{(0)}$.*

Proof. The reason is the same as in Proposition 7.1; it suffices to note that $H^{(j)}$ is a module over the algebra $(H^{(0)}, \star_0)$ and that $c_1(F) \in H^{(0)}$. Q.E.D.

Lemma A.4. *Let λ be an eigenvalue of $(c_1(F)\star_0)$ on $H^{(0)}$ of multiplicity one. Then λ is not an eigenvalue of $(c_1(F)\star_0)$ on $H^{(j)}$ with $j \neq 0$.*

Proof. The proof here is analogous to the argument of Hertling–Manin–Teleman [43]. Let $\psi_0 \in H^{(0)}$ be a non-zero eigenvector of $(c_1(F)\star_0)$ of eigenvalue λ. By assumption, such ψ_0 is unique up to constant. Since we have $H^{(0)} = \mathbb{C}\psi_0 \oplus \mathrm{Im}((c_1(F)\star_0) - \lambda)$, we can write $1 = c\psi_0 + ((c_1(F)\star_0) - \lambda)\gamma$ for some $\gamma \in H^{(0)}$. Applying $(\psi_0\star_0)$ and noting that $\psi_0 \star (c_1(F) \star_0 \gamma - \lambda\gamma) = \lambda\psi_0 \star_0 \gamma - \lambda\psi_0 \star_0 \gamma = 0$, we obtain

$$\psi_0 = c\psi_0 \star_0 \psi_0.$$

Thus $c \neq 0$. By replacing ψ_0 with $c\psi_0$, we may assume that $c = 1$, i.e. $\psi_0 \star_0 \psi_0 = \psi_0$.

Suppose on the contrary that there exists a non-zero $\theta \in H^{(j)}$ with $j \neq 0$ such that $c_1(F) \star_0 \theta = \lambda\theta$. By Poincaré–Serre duality, we can find $\sigma \in H^{(-j)}$ such that $(\theta, \sigma) \neq 0$. We now consider $\beta := \theta \star_0 \sigma \in H^{(0)}$. Since $(\beta, 1) = (\theta, \sigma) \neq 0$, we have $\beta \neq 0$. We also have $c_1(F) \star_0 \beta = c_1(F) \star_0 \theta \star_0 \sigma = T\theta \star_0 \sigma = T\beta$. Therefore β is a non-zero eigenvector of $(c_1(F)\star_0)$. Our assumption implies that $\beta = e\psi_0$ for some $e \neq 0$. On the other hand, $\beta = \theta \star_0 \psi_0$ is nilpotent in quantum cohomology because θ is nilpotent (recall that $H^{(j)} \star_0 H^{(k)} \subset H^{(j+k)}$). This contradicts the fact that ψ_0 is a non-zero idempotent. Q.E.D.

Remark A.5. Property $\mathcal{O}$ on the full cohomology group is useful when discussing Property $\mathcal{O}$ for the product. From the quantum Künneth isomorphism and Theorem A.2, one can easily deduce that if Fano manifolds F_1, F_2 satisfy Property $\mathcal{O}$ (say, on H^{even}), then $F_1 \times F_2$ also satisfies Property $\mathcal{O}$. Note that $H^{\mathrm{odd}}(F_1) \otimes H^{\mathrm{odd}}(F_2)$ contributes to $H^{\mathrm{even}}(F_1 \times F_2)$ under the Künneth isomorphism, and we need Property $\mathcal{O}$ on the full cohomology group.

Finally we discuss Gamma Conjecture I on the full cohomology group. The small quantum connection (4) on the full cohomology group is still a flat connection in the usual sense, but it splits into the direct sum of flat connections on $H^{\mathrm{even}}(F)$ and $H^{\mathrm{odd}}(F)$; moreover it decomposes into flat connections on $H^{(j)}$, $j \in \mathbb{Z}$. If a Fano manifold F satisfies Property $\mathcal{O}$, the spectrum of $(c_1(F)\star_0)$ on $\bigoplus_{j\neq 0} H^{(j)}$ is contained in $\{u : \Re(u) < T\}$ by the above lemmas. The general property of irregular connections of Poincaré rank one (see [29, Propositions 3.2.1, 3.3.1]) shows that the subspace $\mathcal{A}$ of flat sections and the principal asymptotic class A_F appearing in §3.3 remain the same if we consider the small quantum connection on $H^{(0)}$ or on $H^{\mathrm{full}}(F)$. This implies that A_F lies in $H^{(0)} = \bigoplus_{p=0}^{\dim F} H^{p,p}(F)$, and that *Gamma Conjecture I on the subspaces $H^{(0)}$, $H^{\mathrm{even}}(F)$, $H^{\mathrm{full}}(F)$ are all equivalent to each other,* provided that F satisfies Property $\mathcal{O}$.

Remark A.6. The discussion in this appendix suggests that the even part of quantum cohomology knows something about the odd part. Gamma conjecture also suggests a similar relation. If F satisfies Gamma conjecture I, the even part of quantum cohomology knows the Gamma class $\widehat{\Gamma}_F$; the Gamma class then gives the Chern classes $\mathrm{ch}_k(TF)$ (if the Euler constant γ and $\zeta(k), k \geqslant 2$ are linearly independent over $\mathbb{Q}$) and in particular, the Euler number and $\dim H^{\mathrm{odd}}(F)$.

References

[1] Gert Almkvist, Duco van Straten and Wadim Zudilin, *Apéry limits of differential equations of order* 4 *and* 5, Yui, Noriko (ed.) et al., Modular forms and string duality. Proceedings of a workshop, Banff, Canada, June 3–8, 2006. Providence, RI: American Mathematical Society (AMS); Toronto: The Fields Institute for Research in Mathematical Sciences. Fields Institute Communications 54, 105–123 (2008)., 2008.

[2] Sergey Arkhipov and Mikhail Kapranov, *Toric arc schemes and quantum cohomology of toric varieties*, Math. Ann. 335 (2006), no. 4, 953–964.

[3] Michael Francis Atiyah and Raoul Bott, *The moment map and equivariant cohomology*, Topology 23 (1984), no. 1, 1–28.

[4] Denis Auroux, *Mirror symmetry and T-duality in the complement of an anticanonical divisor*, J. Gökova Geom. Topol. GGT 1 (2007), 51–91.

[5] Welner Balser, Wolfgang B. Jurkat and Donald A. Lutz, *Birkhoff invariants and Stokes' multipliers for meromorphic linear differential equations*, J. Math. Anal. Appl. 71 (1979), no. 1, 48–94.

[6] Victor V. Batyrev and Lev A. Borisov, *On Calabi–Yau complete intersections in toric varieties*, Higher-dimensional complex varieties (Trento, 1994), pp. 39–65, de Gruyter, Berlin, 1996, arXiv:alg-geom/9412017.

[7] Victor Batyrev, Ionut Ciocan-Fontanine, Bumsig Kim and Duco van Straten, *Conifold transitions and mirror symmetry for Calabi–Yau complete intersections in Grassmannians*, Nuclear Phys. B 514 (1998), no. 3, 640–666, arXiv:alg-geom/9710022.

[8] Francesco Benini and Stefano Cremonesi, *Partition functions of $\mathcal{N} = (2,2)$ gauge theories on S^2 and vortices*, Comm. Math. Phys. 334 (2015), no. 3, 1483–1527.

[9] Nicole Berline and Michele Vergne, *Classes caractéristiques équivariantes. Formule de localisation en cohomologie équivariante*, C. R. Acad. Sci. Paris Ser. I Math. 295 (1982), no. 9, 539–541.

[10] Aaron Bertram, Ionut Ciocan-Fontanine and Bumsig Kim, *Two proofs of a conjecture of Hori and Vafa*, Duke Math. J. 126, No. 1, 101–136 (2005), arXiv:math.AG/0304403.

[11] Lev Borisov and Richard Paul Horja, *Mellin–Barnes integrals as Fourier–Mukai transforms*, Adv. Math. 207 (2006), no. 2, 876–927, arXiv:math/0510486.

[12] Tom Bridgeland and Valerio Toledano-Laredo, *Stokes factors and multilogarithms*, J. Reine Angew. Math. 682 (2013), 89–128, arXiv:1006.4623v6.

[13] Philip Candelas, Xenia C. de la Ossa, Paul S. Green and Linda Parkes, *An exactly soluble superconformal theory from a mirror pair of Calabi–Yau manifolds*, Phys. Lett. B 258 (1991), no. 1–2, pp. 118–126.

[14] Daewoong Cheong, Changzheng Li, *On the Conjecture $\mathcal{O}$ of GGI for G/P*, Adv. Math. 306 (2017), 704–721. arXiv:1412.2475.

[15] Tom Coates and Alexander B. Givental, *Quantum Riemann–Roch, Lefschetz and Serre*, Ann. of Math. (2) 165 (2007), no. 1, pp. 15–53, arXiv:math/0110142.

[16] Tom Coates, Alessio Corti, Sergey Galkin, Vasily Golyshev, Al Kasprzyk, *Mirror symmetry and Fano manifolds*, European Congress of Mathematics (Kraków, 2-7 July, 2012), November 2013 (824 pages), pp. 285–300, ISBN 978-3-03719-120-0, DOI:10.4171/120-1/16, arXiv:1212.1722.

[17] Tom Coates, Alessio Corti, Sergey Galkin, Vasily Golyshev, Al Kasprzyk, *Fano Varieties and Extremal Laurent Polynomials*, A collaborative research blog, 3D Minkowski Period Sequences, `http://coates.ma.ic.ac.uk/3DMinkowskiPeriodSequences.html`.

[18] John Alexander Cruz Morales, Anton Mellit, Nicolas Perrin and Maxim Smirnov, *On quantum cohomology of Grassmannians of isotropic lines, unfoldings of An-singularities, and Lefschetz exceptional collections*, with appendix by Alexander Kuznetsov, arXiv:1705.01819.

[19] Nima Doroud, Jaume Gomis, Bruno Le Floch and Sungjay Lee, *Exact results in $D = 2$ supersymmetric gauge theories*, J. High Energy Phys. 2013, no. 5, 093, front matter+69 pp.

[20] Boris Dubrovin, *Painlevé transcendents and two-dimensional topological field theory*, In *The Painlevé property*, CRM Ser. Math. Phys., 287–412. Springer, New York, 1999, arXiv:math/9803107.

[21] Boris Dubrovin, *Geometry and analytic theory of Frobenius manifolds*, In *Proceedings of the International Congress of Mathematicians, Vol.* II (*Berlin,* 1998), 315–326, arXiv:math/9807034.

[22] Boris Dubrovin, *On almost duality for Frobenius manifolds*, Geometry, topology, and mathematical physics, 75–132, Amer. Math. Soc. Transl. Ser. 2, 212, Amer. Math. Soc., Providence, RI, 2004.

[23] Boris Dubrovin, *Quantum cohomology and isomonodromic deformation*, Lecture at "Recent Progress in the Theory of Painlevé Equations: Algebraic, asymptotic and topological aspects", Strasbourg, November 2013.

[24] J. J. Duistermaat and G. J. Heckman, *On the variation in the cohomology of the symplectic form of the reduced phase space*, Invent. Math. 69 (2): 259–268; Addendum, Invent. Math. 72, 153–158.

[25] Toru Eguchi, Kentaro Hori and Chuan-Sheng Xiong, *Gravitational quantum cohomology*, Internat. J. Modern Phys. A 12 (1997), no. 9, 1743–1782, arXiv:hep-th/9605225.

[26] Sergey Galkin, *Apery constants of homogeneous varieties*, preprint SFB45 (2008), arXiv:1604.04652.

[27] Sergey Galkin, *Split notes (on non-commutative mirror symmetry)*, based on a lecture at Homological Mirror Symmetry and Category Theory workshop in Split, July 2011, IPMU 12-0112, `http://db.ipmu.jp/ipmu/sysimg/ipmu/884.pdf`.

[28] Sergey Galkin, *The conifold point*, arXiv:1404.7388.

[29] Sergey Galkin, Vasily Golyshev and Hiroshi Iritani, *Gamma classes and quantum cohomology of Fano manifolds: Gamma conjectures*, Duke Math. J. 165 (2016), no. 11, 2005–2077, arXiv:1404.6407.

[30] Sergey Galkin, Anton Mellit, Maxim Smirnov, *Dubrovin's conjecture for* $IG(2,6)$. Int. Math. Res. Not. IMRN 2015, no. 18, 8847–8859.

[31] Alexander B. Givental, *Homological geometry and mirror symmetry*, In: Proceedings of the ICM, Zürich, 1994, Birkhäuser, Basel, 1995, vol 1, pp. 472–480.

[32] Alexander B. Givental, *Homological geometry* I. *Projective hypersurfaces*, Selecta Math. (N.S.), vol 1 (1995), no. 2, pp. 325–345.

[33] Alexander B. Givental, *Equivariant Gromov–Witten invariants*, Internat. Math. Res. Notices 1996, no. 13, 613–663.

[34] Alexander B. Givental, *A mirror theorem for toric complete intersections*, Topological field theory, primitive forms and related topics, (Kyoto, 1996), pp. 141–175, Progr. Math., 160, Birkhäuser. Boston, Boston, MA, 1998, arXiv:alg-geom/9701016.

[35] Vasily Golyshev, *Modularity of the D*3 *equations and the Iskovskikh classification*, Dokl. Akad. Nauk 396 (2004), no. 6, 733–739.

[36] Vasily Golyshev, *Classification problems and mirror duality*, Surveys in geometry and number theory: reports on contemporary Russian mathematics, London Math. Soc. Lecture Note Ser., vol. 338, Cambridge Univ. Press, Cambridge, 2007, 88–121.

[37] Vasily Golyshev, *Deresonating a Tate period*, arXiv:0908.1458.

[38] Vasily Golyshev and Laurent Manivel, *Quantum cohomology and the Satake isomorphism*, arXiv:1106.3120.

[39] Vasily Golyshev and Anton Mellit, *Gamma structures and Gauss's contiguity*, J. Geom. Phys. **78** (2014), 12–18, arXiv:0902.2003.

[40] Vasily Golyshev and Don Zagier, *Proof of the gamma conjecture for Fano* 3*-folds of Picard rank* 1, Izv. Ross. Akad. Nauk Ser. Mat., 80 (2016), no. 1, 27–54.

[41] Phillip Griffiths and Joseph Harris, *Principles of Algebraic Geometry*, Pure and Applied Mathematics. Wiley-Interscience [John Wiley & Sons], New York, 1978.

[42] Megumi Harada and Gregory D. Landweber, *The* K*-theory of abelian versus nonabelian symplectic quotients*, arXiv:math/0601294.

[43] Claus Hertling, Yuri I. Manin, Constantine Teleman, *An update on semisimple quantum cohomology and* F*-manifolds*, Tr. Mat. Inst. Steklova 264 (2009), Mnogomernaya Algebraicheskaya Geometriya, 69–76; translation in Proc. Steklov Inst. Math. 264 (2009), no. 1, 62–69.

[44] Kentaro Hori and Mauricio Romo, *Exact results in two-dimensional* $(2, 2)$ *supersymmetric gauge theory with boundary*, arXiv:1308.2438.

[45] Kentaro Hori and Cumrun Vafa, *Mirror symmetry*, arXiv:hep-th/0002222.

[46] Richard Paul Horja, *Hypergeometric functions and mirror symmetry in toric varieties*, arXiv:math/9912109.

[47] Shinobu Hosono, *Central charges, symplectic forms, and hypergeometric series in local mirror symmetry*, Mirror symmetry. V, pp. 405–439, AMS/IP Stud. Adv. Math., 38, Amer. Math. Soc., Providence, RI, 2006, arXiv:hep-th/0404043.

[48] Shinobu Hosono, Albrecht Klemm, Stefan Theisen and Shing-Tung Yau, *Mirror symmetry, mirror map and applications to complete intersection Calabi–Yau spaces*, Nuclear Phys. B 433 (1995), no. 3, 501–552, arXiv:hep-th/9406055.

[49] Hiroshi Iritani, *Quantum D-modules and equivariant Floer theory for free loop spaces.* Math. Z. 252, no. 3 (2006), pp. 577–622, arXiv:math.DG/0410487.

[50] Hiroshi Iritani, *An integral structure in quantum cohomology and mirror symmetry for toric orbifolds*, Adv. Math. **222** (2009), no. 3, 1016–1079, arXiv:0903.1463.

[51] Hiroshi Iritani, *tt^*-geometry in quantum cohomology*, arXiv:0906.1307.

[52] Hiroshi Iritani, *Quantum cohomology and periods*, Ann. Inst. Fourier 61, No. 7, 2909–2958 (2011), arXiv:1101.4512.

[53] Hiroshi Iritani, Etienne Mann, Thierry Mignon, *Quantum Serre theorem as a duality between quantum D-modules*, Int. Math. Res. Not. IMRN 2016, no. 9, 2828–2888. arXiv:1412.4523.

[54] Mikhail Kapranov, *The derived category of coherent sheaves on Grassmann varieties*, Funktsional. Anal. i Prilozhen. 17 (1983), no. 2, 78–79.

[55] Ludmil Katzarkov, Maxim Kontsevich and Tony Pantev, *Hodge theoretic aspects of mirror symmetry*, From Hodge theory to integrability and TQFT tt^*-geometry, 87–174, Proc. Sympos. Pure Math., 78, Amer. Math. Soc., Providence, RI, 2008. arXiv:0806.0107.

[56] Bumsig Kim, *Quantum hyperplane section theorem for homogeneous spaces*, Acta Math. 183 (1999), no. 1, 71–99.

[57] Bumsig Kim and Claude Sabbah, *Quantum cohomology of the Grassmannian and alternate Thom–Sebastiani*, Compos. Math. 144 (2008), no. 1, 221–246, arXiv:math/0611475.

[58] Maxim Kontsevich and Yuri I. Manin, *Gromov–Witten classes, quantum cohomology, and enumerative geometry*, Comm. Math. Phys. Volume 164, Number 3 (1994), 525–562.

[59] Yuan-Pin Lee, *Quantum Lefschetz hyperplane theorem*, Invent. Math. 145 (2001), no. 1, 121–149.

[60] Anatoly S. Libgober, *Chern classes and the periods of mirrors*, Math. Res. Lett., 6 (1999), 141–149, arXiv:math/9803119.

[61] Rongmin Lu, *The $\widehat{\Gamma}$-genus and a regularization of an S^1-equivariant Euler class*, J. Phys. A 41 (2008), no. 42, 425204 (13pp), arXiv:0804.2714.

[62] Yuri I. Manin, *Frobenius manifolds, quantum cohomology, and moduli spaces*, American Mathematical Society Colloquium Publications, 47. American Mathematical Society, Providence, RI, 1999.

[63] Robert J. Marsh and Konstanze Rietsch, *The B-model connection and mirror symmetry for Grassmannians*, arXiv:1307.1085.

[64] Shaun Martin, *Symplectic quotients by a nonabelian group and by its maximal torus*, arXiv:math/0001002.

[65] Anton Mellit, Nicolas Perrin and Maxim Smirnov, *Update on quantum cohomology of $IG(2, 2n)$*, arXiv:1510.07903.

[66] Rahul Pandharipande, *Rational curves on hypersurfaces (after A. Givental)*, Séminaire Bourbaki. Vol. 1997/98, Astérisque, 252, 1998, Exp. No. 848, 5, 307–340, arXiv:math/9806133.

[67] Nicolas Perrin, *Semisimple quantum cohomology of some Fano varieties*, arXiv:1405.5914.

[68] S. Piunikhin, D. Salamon and M. Schwarz, *Symplectic Floer–Donaldson Theory and Quantum Cohomology*, Contact and symplectic geometry (Cambridge, 1994), 171–200, Publ. Newton Inst., 8, Cambridge Univ. Press, Cambridge, 1996.

[69] Victor Przyjalkowski, *Weak Landau–Ginzburg models for smooth Fano threefolds*, Izv. Ross. Akad. Nauk Ser. Mat. 77 (2013), no. 4, 135–160; translation in Izv. Math. 77 (2013), no. 4, 772–794. arXiv:0902.4668.

[70] Victor Przyjalkowski, *Hori–Vafa mirror models for complete intersections in weighted projective spaces and weak Landau–Ginzburg models*, Cent. Eur. J. Math. 9 (2011), no. 5, 972–977, arXiv:1003.5200.

[71] Konstanze Rietsch, *A mirror symmetric construction for $qH^*_T(G/P)_{(q)}$*, Advances in Mathematics 217 (2008), 2401–2442.

[72] Claude Sabbah, *Isomonodromic deformations and Frobenius manifolds. An Introduction*, Translated from the 2002 French edition, Universitext, Springer-Verlag London, Ltd., London; EDP Sciences, Les Ulis, 2007.

[73] Fumihiko Sanda and Yota Shamoto, *An analogue of Dubrovin's conjecture*, arXiv:1705.05989.

[74] Constantin Teleman, *The structure of 2D semisimple field theories*, Invent. Math. 188, No. 3, 525–588 (2012), arXiv:0712.0160.

[75] Kazushi Ueda, *Stokes matrices for the quantum cohomologies of Grassmannians*, Int. Math. Res. Not. 2005, No. 34, 2075–2086 (2005), arXiv:math/0503355.

[76] Yiannis Vlassopoulos, *Quantum Cohomology and Morse Theory on the Loop Space of Toric Varieties*, arXiv:math.AG/0203083.

[77] Wolfgang Wasow, *Asymptotic expansions for ordinary differential equations*, Pure and Applied Mathematics, Vol. XIV Interscience Publishers John Wiley & Sons, Inc., New York-London-Sydney 1965.

National Research University Higher School of Economics
Faculty of Mathematics
119048, Usacheva 6, Moscow, Russian Federation
E-mail address: Sergey.Galkin@phystech.edu

Department of Mathematics
Graduate School of Science
Kyoto University
Kitashirakawa-Oiwake-cho, Sakyo-ku, Kyoto
606-8502, Japan
E-mail address: iritani@math.kyoto-u.ac.jp

Advanced Studies in Pure Mathematics 83, 2019
Primitive Forms and Related Subjects — Kavli IPMU 2014
pp. 117–126

On a conjecture by N. Takahashi on log mirror symmetry for the projective plane

Michel van Garrel

Abstract.

In this survey, the relationship between local and relative BPS state counts is explored in light of a conjecture on log mirror symmetry by N. Takahashi. This is based on the paper [14] by the author and the integrality result of [14] is reported.

§1. Introduction

Log mirror symmetry for the pair of projective plane and elliptic curve on it was explored by N. Takahashi in [27]. The A-model invariants considered are counts of rational curves in the plane meeting the elliptic curve in a prescribed point of prescribed order. A relationship of these invariants with local BPS state counts is conjectured, the latter being the A-model invariants of local $\mathbb{P}^2$. In the author's paper [14], a virtual version of Takahashi's curve counts, derived from relative Gromov-Witten invariants, are considered. Namely, the relative BPS state counts introduced by Gross-Pandharipande-Siebert in [17]. The main result of [14] is the integrality of these invariants for toric del Pezzo surfaces. The key result towards proving that integrality claim is a linear relationship between the local and relative BPS state counts. Incidentally, this provides a virtual reformulation of Takahashi's conjecture, which we describe in the present survey. We start in Section 2 by defining local BPS state counts, putting extra care into justifying the definition. We then introduce Takahashi's conjecture in Section 3. In [27], the conjectural B-model to log mirror symmetry is described. Namely, the mirror family and the associated periods are defined. We

Received December 31, 2014.
Revised June 7, 2015.
2010 *Mathematics Subject Classification.* 05A15, 14J33, 14J45, 14N35.
Key words and phrases. Gromov-Witten theory, BPS state counts, Mirror Symmetry.

do not review them as our correspondence of BPS numbers concerns the A-model. Finally, in Section 4 we define relative BPS state counts and explain how they satisfy an analogue to Takahashi's conjecture.

§2. Preliminaries

Let S be a smooth del Pezzo surface. In Definition 2.2 below, local BPS state counts of S are introduced. These are the A-model invariants of local mirror symmetry, which was developed in [8]. See also [20] for a description in terms of Yukawa couplings.

Denote moreover by D a smooth effective anticanonical divisor on S, by K_S the total space of the canonical bundle $\mathcal{O}_S(-D)$, and let $\beta \in \mathrm{H}_2(S,\mathbb{Z})$. Denote by $\overline{M}_{0,0}(S,\beta)$, resp. by $\overline{M}_{0,1}(S,\beta)$, the moduli stack of genus 0 stable maps

$$f : C \to S$$

with no, resp. one, marked point and such that $f_*([C]) = \beta$. Denote by

$$\pi : \overline{M}_{0,1}(S,\beta) \to \overline{M}_{0,0}(S,\beta)$$

the forgetful morphism and by

$$ev : \overline{M}_{0,1}(S,\beta) \to S$$

the evaluation map. This determines the bundle $R^1\pi_* ev^* K_S$ whose fiber over a stable map $f : C \to S$ is $\mathrm{H}^1(C, f^*K_S)$. Consider moreover the virtual fundamental class $[\overline{M}_{0,0}(S,\beta)]^{vir}$, which intersection-theoretically is the correct class to be looking at, and denote by e the Euler class.

The idea of obtaining *local* Gromov-Witten invariants of S comes from assuming that S was embedded in a compact Calabi-Yau threefold X. That is, X is a 3-dimensional Kähler manifold with trivial canonical bundle. If $S \subset X$, then the local Gromov-Witten invariants of S should be the contribution of stable maps mapping to S to the Gromov-Witten invariants of X. Through excess intersection calculations, the authors of [8] arrive at the following definition.

Definition 2.1. *The* genus 0 degree β local Gromov-Witten invariant of S *is*

$$I_{K_S}(\beta) := \int_{[\overline{M}_{0,0}(S,\beta)]^{vir}} \mathrm{e}\left(R^1\pi_* ev^* K_S\right) \in \mathbb{Q}.$$

The degree map on a proper Deligne-Mumford stack is $\mathbb{Q}$-valued and thus in general, the invariants $I_{K_S}(\beta)$ are $\mathbb{Q}$-valued. This makes a direct enumerative interpretation impossible. Local BPS state counts aim at remedying this by extracting integer-valued invariants from the rational number $I_{K_S}(\beta)$. These new invariants would under some idealised (usually not satisfied) conditions be counts of rational curves.

As above, assume that S is embedded in a compact Calabi-Yau threefold X. Then $\mathrm{N}_{S/X} = K_S$, so that K_S is the local geometry of X near S. Moreover, the $I_{K_S}(\beta)$ can be regarded as counts of maps into K_S since genus 0 stable maps $C \to K_S$ factor through S. Indeed, for simplicity assume that $C = \mathbb{P}^1$. Then giving a map $\phi : \mathbb{P}^1 \to K_S$ is the same as a map $f : \mathbb{P}^1 \to S$ and a section of $f^*(\mathcal{O}(-D))$. Since S is del Pezzo however, the latter is negative on $\mathbb{P}^1$ and thus has no sections. It follows that the image of $\mathbb{P}^1$ in K_S is contained in the 0-section. Therefore, the moduli space of stable genus 0 maps into K_S is identical to the moduli space of stable genus 0 maps into S. However, their deformation theories differ yielding different virtual fundamental classes. For one, the virtual fundamental class of maps into K_S is in degree 0, whereas it is in degree non-zero for S.

Via $S \subseteq X$, we model stable maps to S from the situation for X, which we describe now. Let $\phi : \mathbb{P}^1 \to X$ be a holomorphic immersion. That is, we are looking at a rational curve $\phi(\mathbb{P}^1)$ in X (with possibly nodes), and ϕ is the normalization map. Then the normal sheaf of ϕ is

$$\mathrm{N}_\phi = \frac{\phi^* \mathrm{T}_X}{\mathrm{T}_{\mathbb{P}^1}} \cong \mathcal{O}_{\mathbb{P}^1}(a) \oplus \mathcal{O}_{\mathbb{P}^1}(b),$$

where by the adjunction formula $a + b = -2$. Moreover, $a = b = -1$ if and only if N_ϕ has no sections if and only if $\phi(\mathbb{P}^1)$ can not be infinitesimally deformed. Hence, $\phi(\mathbb{P}^1)$ is infinitesimally rigid if and only if $a = b = -1$.

We come to the Aspinwall-Morrison formula, which was proved in [22], see also [28]: Assume that $\phi(\mathbb{P}^1)$ is infinitesimally rigid of degree β, so that $\mathrm{N}_\phi \cong \mathcal{O}_{\mathbb{P}^1}(-1) \oplus \mathcal{O}_{\mathbb{P}^1}(-1)$. Consider maps $\mathbb{P}^1 \to X$ with image $\phi(\mathbb{P}^1)$. Such a map factors through the normalization of $\phi(\mathbb{P}^1)$, i.e. we are looking at multiple covers of $\mathbb{P}^1$. For the purpose of calculating Gromov-Witten invariants, we are considering multiple covers of $\mathbb{P}^1$ into N_ϕ. The Aspinwall-Morrison formula calculates the contributions of such maps to the Gromov-Witten invariants of X. It states that the degree k covers

$$\mathbb{P}^1 \xrightarrow{k:1} \mathbb{P}^1,$$

and their degenerations, (after composing with ϕ) contribute $\frac{1}{k^3}$ to the Gromov-Witten invariant $I_X(k\beta)$.

We come to the three idealised assumptions that underlie the definition of BPS numbers:

- Each rational curve in X is infinitesimally rigid.
- Any two irreducible rational curves do not intersect.
- For each curve class β, the number of infinitesimally rigid rational curves of class β is finite.

Supposing that these three idealised assumptions are satisfied, denote by n'_β the number of rational curves in X of degree β. Let γ be a curve class. Because of our assumptions, the only way to obtain a stable map of degree γ is as follows: Decompose $\gamma = k \cdot \beta$, for k a positive integer and β a curve class. Then consider the maps consisting of a degree k cover of $\mathbb{P}^1$ followed by a holomorphic immersion to a rational curve of degree β in X. By the Aspinwall-Morrison formula, all possible such decompositions contribute to $I_X(\gamma)$ as follows:

$$I_X(\gamma) = \sum_{\gamma = k \cdot \beta} \frac{1}{k^3}\, n'_\beta.$$

In particular, this stipulates that the rational numbers of Gromov-Witten invariants come from multiple covers.

The preceding heuristic argument justifies the definition below of local BPS state counts. Note that in absence of the prior mentioned ideal conditions, the meaning of the n_β is no longer clear. Moreover, the previous discussion assumes a compact Calabi-Yau threefold, whereas we are dealing with the open Calabi-Yau threefold K_S. Nonetheless, the n_β below are well-defined. They are the A-model invariants of local mirror symmetry for S, which is why they are crucial.

Definition 2.2 (See [15, 16, 24, 25]). *Assume (for simplicity of exposition) that β is primitive and let $d \geq 1$. Then the* local BPS state counts *$n_{d\beta} \in \mathbb{Q}$ are defined via the equality of power series:*

$$\sum_{l=1}^{\infty} I_{K_S}(l\beta)\, q^l = \sum_{d=1}^{\infty} n_{d\beta} \sum_{k=1}^{\infty} \frac{1}{k^3}\, q^{dk}.$$

Note that vice versa, one can pass from the n_β to the $I_{K_S}(\beta)$ by a Möbius transform. The following theorem was originally a conjecture attributed to Gopakumar-Vafa and stated in [6]. It was proved in [26] in the case of toric del Pezzo surfaces. These are the del Pezzo surfaces of degree ≥ 6. More generally for toric Calabi-Yau threefolds, a proof was given in [19].

Theorem 1 (Peng in [26]). *If S is a toric del Pezzo surface, then*

$$n_{d\beta} \in \mathbb{Z}$$

for all $d \geq 1$.

We finish our preliminaries with the definition of the geometry that will be relevant in the next sections.

Definition 2.3 (See [17]). *Denote by S a smooth surface, by $D \subseteq S$ a smooth divisor and let $0 \neq \gamma \in \mathrm{H}_2(S, \mathbb{Z})$. The pair (S, D) is said to be* log Calabi-Yau with respect to γ *if*

$$D \cdot \gamma = c_1(S) \cdot \gamma.$$

Provided the above equation holds for all γ, (S, D) is said to be log Calabi-Yau.

A class of examples of log Calabi-Yau surface pairs are the (S, D) for any del Pezzo surface S and any smooth anti-canonical curve D on it.

§3. *A*-model log mirror symmetry for $\mathbb{P}^2$

Log mirror symmetry for the projective plane was explored by N. Takahashi in [27]. Denote by $E \subseteq \mathbb{P}^2$ a smooth cubic, i.e. an elliptic curve, so that $(\mathbb{P}^2, E)$ is a log Calabi-Yau surface pair. According to the conjecture of [27], the log mirror symmetry A-model concerns counts of plane curves satisfying *condition* (AL), which stands for *affine line*:

Definition 3.1. *Let $C \subseteq \mathbb{P}^2$ be a curve. We say that C is an* (AL)-curve *if C is irreducible, reduced and if the normalization of $C \setminus E$ is isomorphic to the affine line $\mathbb{A}^1$.*

Let C be an (AL)-curve of degree d. Then by definition, C is of geometric genus 0 and meets E in exactly one point. If P is that point, then the equation defining C yields the linear equivalence $3d \cdot P \sim 3d \cdot P_0$, where P_0 is any one of the nine inflexion points of E. Recall that the identification of closed points of E with its divisor class group yields, upon choosing a zero element, a group law on the set of closed points of E. Hence by choosing as zero element $0 \in E$ a flex point, it follows that P is a $3d$-torsion point for the resulting group law. Takahashi's proposal for A-model invariants is to count the (AL)-curves that meet E in a chosen point of maximal order:

Definition 3.2. *Choose $0 \in E$ to be a flex point and let P be a point of order $3d$ for the group law induced by the above choice of 0. For $d \geq 1$, define m_d to be the number of* (AL)*-curves of degree d that meet E in P.*

In [27], it is conjectured that the B-model to $(\mathbb{P}^2, E)$ is the same as the B-model to local $\mathbb{P}^2$. We refer to [27] for the precise statements.

Instead, we focus on the conjectured relationship of the A-model invariants m_d to the A-model invariants of local $\mathbb{P}^2$. An equivalent way of stating Takahashi's conjecture then is as follows. Denote by n_d the local BPS state counts of $\mathbb{P}^2$, i.e. we suppress the hyperplane class from the notation.

Conjecture 1 (N. Takahashi in [27]). *For all $d \geq 1$,*

$$3d\, m_d = (-1)^{d+1} n_d.$$

In light of Theorem 2, a more natural statement of the above conjecture is $9d^2\, m_d = (-1)^{d+1}\, 3d\, n_d$. Note that $9d^2$ is the number of $3d$-torsion points on E.

We remarked in Section 2 that an enumerative interpretation of the local BPS state counts remains elusive. The above conjecture would however provide exactly that. According to it, the n_d are counts of rational curves in the projective plane meeting an elliptic curve in a prescribed point of order $3d$.

Remark 1. *As this article is going into press, there has been considerable progress on questions on and related to Conjecture* 1. *A proof of Conjecture* 1 *was announced in* [3, 2]. *An extension of Conjecture* 1 *to all del Pezzo surfaces was formulated and checked up to arithmetic genus* 2 *in* [10, 11, 9]. *At the level of genus* 0 *Gromov-Witten invariants, and for any smooth projective variety with normal crossing divisor whose components are smooth and nef, Conjecture* 1 *translates into the* log-local principle *as formulated in* [12]. *The main result of* [12] *is a proof of this principle when the divisor is smooth. More generally, the log-local principle is proven for $\mathbb{Q}$-factorial toric varieties (with no assumptions on smoothness) in* [5], *when the components of the divisor are very ample in* [23], *and for surfaces and refined invariants in* [4]. *All of these approaches use log enumerative geometry* [18, 7, 1], *see* [13] *for an introduction.*

§4. Relative BPS state counts of del Pezzo surfaces

The pair (S, D) can be thought of as being log Calabi-Yau because the canonical bundle of S is trivial away from the divisor D. The definition of relative BPS state counts mirrors the definition of local BPS state counts of Section 2. While we were considering stable maps then, we now consider relative stable maps, i.e. we additionally prescribe how the maps meet D. Let $\beta \in \mathrm{H}_2(S, \mathbb{Z})$ be the class of a curve and set $w = D \cdot \beta$. Then a generic stable map of degree β meets D in w points. Moreover, the moduli space of stable maps of degree β is of virtual dimension

$$\int_\beta c_1(\mathrm{T}_S) - 1 = w - 1.$$

Each time two points of intersections are identified, the virtual dimension drops by one. Proceeding accordingly with all but one of the intersection points therefore cuts it by $w-1$ and hence, in the minimal case of one intersection point only, yields a virtual dimension of 0.

The virtual counts of such maps are given by relative Gromov-Witten invariants, which were introduced to the algebraic setting by Li in [21]. Denote by $\overline{M}(S/D, w)$ the moduli space of relative genus 0 stable maps of degree β meeting D in one point of maximal tangency w. Since

$$\operatorname{vdim} \overline{M}(S/D, w) = 0,$$

the following definition makes sense.

Definition 4.1. *The* genus 0 degree β relative Gromov-Witten invariant of maximal tangency *is the degree of the corresponding virtual fundamental class:*

$$N_S[w] := \int_{[\overline{M}(S/D,w)]^{vir}} 1 \in \mathbb{Q}.$$

Note that for notational simplicity, β is hidden in the notation. Analogously to the local case, BPS numbers are defined via considering multiple cover contributions over rigid elements. Denote by $\iota : P \to S$ a rigid element of $\overline{M}(S/D, w)$ and for $k \geq 1$, let $M_P[k]$ be the contribution of k-fold multiple covers of P to $N_S[kw]$ (see [17] for precise definitions).

Proposition 1 (Proposition 6.1 in [17])**.**

$$M_P[k] = \frac{1}{k^2}\binom{k(w-1)-1}{k-1}.$$

The next definition exactly mirrors the heuristic discussion of Section 2 with corresponding idealised assumptions for the relative case of one point of intersection.

Definition 4.2 (Paragraph 6.3 in [17])**.** *Let $d \geq 1$. Then the* relative BPS state counts $n_S[dw] \in \mathbb{Q}$ *are defined by means of the equality*

$$\sum_{l=1}^{\infty} N_S[lw]\, q^l = \sum_{d=1}^{\infty} n_S[dw] \sum_{k=1}^{\infty} \frac{1}{k^2}\binom{k(dw-1)-1}{k-1} q^{dk}. \tag{1}$$

As in the local case, a Möbius transform recovers the $N_S[dw]$ from the $n_S[dw]$. Analogously to the local invariants, the relative BPS state counts are conjectured to be integers:

Conjecture 2 (Conjecture 6.2 in [17]). *For $\beta \in \mathrm{H}_2(S,\mathbb{Z})$ a primitive curve class, set $w = \beta \cdot D$. Then, for all $d \geq 1$,*

$$n_S[dw] \in \mathbb{Z}.$$

In the idealised situation where every rational curve meeting D in one point was rigid, and where there are only finitely many such curves in each degree, the relative BPS state counts would correspond to the (AL)-curve counts, except that the point of intersection with D is not prescribed. In that sense, the $n_S[dw]$ are a double generalisation of the m_d: Firstly, any point of any order (for the group law) is allowed as intersection point. Secondly, we consider virtual invariants, i.e. we take the degree of the virtual fundamental class of the relevant moduli stack.

Before proceeding with the linear restatement of Conjecture 1, we introduce some notation. For $n \in \mathbb{Z}$, denote by $\omega(n)$ the number of primes (not counting multiplicities) in the prime factorization of n. Furthermore, let

$$I(n) := \{k \in \mathbb{Z} \,:\, k|n \text{ and } n/k \text{ is square-free}\}.$$

We now introduce the following infinite-dimensional matrix C. If $t|s$, set

$$C_{st} := \frac{(-1)^{sw}}{(s/t)^2} \sum_{k \in I(s/t)} (-1)^{\omega(s/kt)} (-1)^{ktw} \binom{k(tw-1)-1}{k-1}.$$

If $t \nmid s$, set $C_{st} = 0$. Consequently, each row of C has only a finite number of non-zero entries. Therefore, we can apply to C an infinite-dimensional vector without encountering convergence issues.

We are ready to state a virtual version of conjecture [27] by N. Takahashi. Our result applies to all del Pezzo surfaces.

Theorem 2 (See Lemma 12 in [14]).

$$C \cdot [n_S[dw]]_{d\geq 1} = \left[(-1)^{dw+1}\, dw\, n_{d\beta}\right]_{d\geq 1}.$$

In [14], it is proven that the entries of C are integers. Since $\det(C) = 1$, the entries of C^{-1} are integers as well. Hence, the following integrality result follows from Theorem 1.

Corollary 1 (See Corollary 10 in [14]). *For toric del Pezzo surfaces, the relative BPS state counts are integers.*

References

[1] D. Abramovich and Q. Chen, *Stable logarithmic maps to Deligne-Faltings pairs* II, Asian J. Math. 18 (2014), 465–488.

[2] P. Bousseau, *A proof of N. Takahashi's conjecture on genus zero Gromov-Witten theory of* $(\mathbb{P}^2, E)$, arXiv:1909.02992 (2019).

[3] P. Bousseau, *Scattering diagrams, stability conditions, and coherent sheaves on* $\mathbb{P}^2$, arXiv:1909.02985 (2019).

[4] P. Bousseau, A. Brini and M. van Garrel, *On the log-local principle for refined Gromov-Witten theory: log Calabi-Yau surfaces*, in preparation (2019).

[5] P. Bousseau, A. Brini and M. van Garrel, *On the log-local principle for the toric boundary*, arXiv:1908.04371 (2019).

[6] J. Bryan and R. Pandharipande, *BPS states of curves in Calabi-Yau* 3*-folds*, Geom. Topol. 5 (2001), 287–318.

[7] Q. Chen, *Stable logarithmic maps to Deligne-Faltings pairs* I, Ann. of Math. (2) 180 (2014), 455–521.

[8] T.-M. Chiang, A. Klemm, S.-T. Yau and E. Zaslow, *Local mirror symmetry: calculations and interpretations*, Adv. Theor. Math. Phys. 3 (1999), no. 3, 495–565.

[9] J. Choi, M. van Garrel, S. Katz and N. Takahashi, *Contributions of degenerate stable log maps*, arXiv:1908.10906 (2018).

[10] J. Choi, M. van Garrel, S. Katz and N. Takahashi, *Local BPS Invariants: Enumerative Aspects and Wall-Crossing*, Int. Math. Res. Not., (2018).

[11] J. Choi, M. van Garrel, S. Katz and N. Takahashi, *Log BPS numbers of log Calabi-Yau surfaces*, arXiv:1810.02377 (2018).

[12] M. van Garrel, T. Graber and H. Ruddat, *Local Gromov-Witten invariants are log invariants*, Adv. Math. 350 (2019), 860–876.

[13] M. van Garrel, D. P. Overholser and H. Ruddat, *Enumerative aspects of the Gross-Siebert program*, from "Calabi-Yau varieties: arithmetic, geometry and physics", Fields Inst. Monogr. 34, Fields Inst. Res. Math. Sci., Toronto, ON (2015), 337–420.

[14] M. van Garrel, T. W. H. Wong and G. Zaimi, *Integrality of relative BPS state counts of toric del Pezzo surfaces*, Comm. Number Theory Phys. 7 (2013), no. 4, 671–687.

[15] R. Gopakumar and C. Vafa, *M-theory and topological strings* I, arXiv:hep-th/9809187 (1998).

[16] R. Gopakumar and C. Vafa, *M-theory and topological strings* II, arXiv:hep-th/9812127 (1998).

[17] M. Gross, R. Pandharipande and B. Siebert, *The tropical vertex*, Duke Math. J. 153 (2010), no. 2, 297–362.

[18] M. Gross and B. Siebert, *Logarithmic Gromov-Witten invariants*, J. Amer Math. Soc. 26 (2013), 451–510.

[19] Y. Konishi, *Integrality of Gopakumar-Vafa invariants of toric Calabi-Yau threefolds*, Publ. Res. Inst. Math. Sci. 42 (2006), no. 2, 605–648.

[20] Y. Konishi and S. Minabe, *Local B-model and Mixed Hodge Structure*, Adv. Theor. Math. Phys. 14 (2010), no. 4, 1089–1145.

[21] J. Li, *Stable morphisms to singular schemes and relative stable morphisms*, J. Differential Geom. 57 (2001), 509–578.

[22] Y. Manin, *Generating functions in algebraic geometry and sums over trees*, In: The moduli space of curves, Texel Island, 1994, (eds. R. Dijkgraaf, C. Faber and G. van der Geer), Progr. Math. 129, Birkhäuser Boston-Basel-Berlin (1995), 401–417.

[23] N. Nabijou and D. Ranganathan, *Gromov-Witten theory with maximal contacts*, arXiv:1908.04706 (2019).

[24] R. Pandharipande, *Hodge integrals and degenerate contributions*, Comm. Math. Phys. 208 (1999), 489–506.

[25] R. Pandharipande, *Three questions in Gromov-Witten theory*, In: Proceedings of the International Congress of Mathematicians, Beijing (2002), vol. II, Higher Ed. Press., 503–512.

[26] P. Peng, *A Simple Proof of Gopakumar-Vafa Conjecture for Local Toric Calabi-Yau Manifolds*, Commun. Math. Phys. 276 (2007), 551–569.

[27] N. Takahashi, *Log mirror symmetry and local mirror symmetry*, Comm. Math. Phys. 220 (2001), no. 2, 293–299.

[28] C. Voisin, *A mathematical proof of a formula of Aspinwall and Morrison*, Comp. Math. 104 (1996), no. 2, 135–151.

Mathematics Institute,
Zeeman Building,
University of Warwick,
Coventry CV4 7AL, UK
E-mail address: michel.van-garrel@warwick.ac.uk

Advanced Studies in Pure Mathematics 83, 2019
Primitive Forms and Related Subjects — Kavli IPMU 2014
pp. 127–220

Notes on the Hemisphere

Kentaro Hori and Mauricio Romo

Abstract.

In these notes, we provide an introduction to the hemisphere partition function of 2d (2, 2) supersymmetric gauge theories, and discuss its relation to the "D-brane central charge" which were studied in superstring theory, in 2d supersymmetric quantum field theory, and in topological string theory. We also discuss relation to "macroscopic loop" in matrix models. They are mostly reviews of the work by the authors, but contains some new results such as the partition function for a rotated supersymmetry as well as the differential equations.

Contents

Received August 23, 2016.
Revised April 2, 2017.
2010 *Mathematics Subject Classification.* 81T60 (main), 14D05, 14J33, 14N35, 18E30, 32S40, 53D37, 53D45, 81T13, 81T30, 81T40, 81T45.
Key words and phrases. Supersymmetric quantum field theory, gauged linear sigma models, boundary conditions, supersymmetric localization, D-brane central charge, primitive forms, flat structure, Gromov-Witten theory.

1. Introduction

Quantum field theory provides a place where mathematicians and physicists interact with each other, and supersymmetry has been the key in the interaction for more than thirty years since the introduction of Witten index [1]. Having infinite degrees of freedom, quantum field theory so far resisted mathematical definition, and also, even the basic physical property of a given theory is usually difficult to understand. In the presence of supersymmetry, cancellation of infinities happens to a class of observables, and they are sometimes exactly computable and often provide important physical information of the theory. Attempts to define such observables have generated new areas of mathematical research, and certain relations among those observables, which hold for trivial or non-trivial physical reasons, may have dramatic mathematical consequences. For example, as a consequence of mirror symmetry, the number of rational curves in a Calabi-Yau manifold was predicted [2], and that motivated mathematicians to develop the theory of Gromov-Witten invariants and further led to surprising relationship between algebraic geometry and symplectic geometry.

More recently, in the physics side, the class of computable observables are enlarged by using *superconformal* transformations, which become symmetries of the system in special spacetime backgrounds even if the theory is not conformally invariant. Starting from [3] the partition functions of various supersymmetric gauge theories on spheres of dimensions ≤ 5 were computed and certain new information of the theories were obtained. In particular, the partition function of 2d $(2,2)$ supersymmetric gauge theories on the two sphere was computed in [4, 5]. When a certain condition called "the Calabi-Yau condition" is met, it was observed in some examples [6] and later explained in [7, 8] that the partition function determines the Kähler potential of the space of superconformal fixed points of the theory. Motivated by these works, the authors of these notes studied the partition function of 2d $(2,2)$ supersymmetric gauge theories on the hemisphere [9]. (Related works [10, 11] also appeared at the same time.) When the Calabi-Yau condition is met, it was observed in examples that the hemisphere partition function computes the central charge of the D-brane that is placed at the boundary.

The present notes consists of a review of that work [9] and some introductory materials. We also include some new results such as the partition function for a rotated supersymmetry as well as the differential equations. We elaborate on the relation to the central charge in the Calabi-Yau case as well as the discussion of non-Calabi-Yau case. In a non-Calabi-Yau case where the theory is related to non-linear sigam model with a Fano target space X, the hemisphere partition function computes the "central charge" in the Gromov-Witten theory of X. There is also a rather surprising but suggestive relation to "macroscopic loop" in matrix model. We shall discuss these matters in some detail as they were only briefly mentioned in [9].

Notational Remarks

In these notes, we take the following convention. For a compact Lie group G, we write G_0, T and Z_G for the identity component, a maximal torus and the center, respectively. We write the Weyl group of G and G_0 by W and W_0. We write $\mathfrak{g} \supset \mathfrak{t} \supset \mathfrak{z}$ for the Lie algebras of G (or G_0) $\supset T \supset Z_G$ and regard them "pure imaginary". "Reals" are $i\mathfrak{g} \supset i\mathfrak{t} \supset i\mathfrak{z}$ in the complexfied Lie algebras $\mathfrak{g}_{\mathbb{C}} \supset \mathfrak{t}_{\mathbb{C}} \supset \mathfrak{z}_{\mathbb{C}}$. The weight lattice of T is denoted by $\mathrm{P} \subset i\mathfrak{t}^*$. For a $\mathfrak{g}_{\mathbb{C}}$ valued quantity X, we write $X = \mathrm{Re}(X) + i\mathrm{Im}(X)$ where both $\mathrm{Re}(X)$ and $\mathrm{Im}(X)$ are $i\mathfrak{g}$ valued, and write $\overline{X} = \mathrm{Re}(X) - i\mathrm{Im}(X)$.

2. 2d (2, 2) Supersymmetric QFTs

Let us first describe the basics of quantum field theories (QFTs) in two dimensions with $(2, 2)$ supersymmetry, with emphasis on the mathematical aspects.

2.1. 2d (2, 2) supersymmetry

When formulated on the Minkowski spacetime, a 2d $(2, 2)$ supersymmetric QFT has the following symmetry operators: the time translation H (Hamiltonian), the space translation P (momentum), the Lorentz transformation M, the supercharges $Q_+, \overline{Q}_+, Q_-, \overline{Q}_-$, and possibly the vector R-charge F_V and/or the axial R-charge F_A. They act on the Hilbert space of states $\mathcal{H}$ which is $\mathbb{Z}_2$-graded. The supercharges $Q_\pm$ and $\overline{Q}_\pm$ are odd and are the adjoint of each other, while the other operators are even and self-adjoint. With respect to the Lorentz group, H and P form a vector, $i[M, H \pm P] = \mp 2(H \pm P)$, the supercharges are spinors, $i[M, Q_\pm] = \mp Q_\pm$, $i[M, \overline{Q}_\pm] = \mp \overline{Q}_\pm$, and the R-charges are scalars, $[M, F_V] = [M, F_A] = 0$. The supercharges obey[1]

$$\{Q_\pm, \overline{Q}_\pm\} = H \pm P, \tag{2.1}$$

$$\text{all other anticommutators} = 0. \tag{2.2}$$

R-charges are phase rotations of the supercharges

$$[F_V, Q_\pm] = -Q_\pm, \quad [F_V, \overline{Q}_\pm] = \overline{Q}_\pm, \tag{2.3}$$

$$[F_A, Q_\pm] = \mp Q_\pm, \quad [F_A, \overline{Q}_\pm] = \pm \overline{Q}_\pm. \tag{2.4}$$

2.2. A and B

Let us put

$$Q_A := \overline{Q}_+ + Q_-, \qquad Q_B := \overline{Q}_+ + \overline{Q}_-. \tag{2.5}$$

Then, $(Q, F) = (Q_A, F_A)$ or (Q_B, F_V) obey

$$Q^2 = 0, \qquad [F, Q] = Q. \tag{2.6}$$

This means that the space of states forms a complex with differential Q and grading F.[2] The same applies also for the space of local operators. In particular, cohomology classes of local operators form a ring called the

[1]There is a possible modification to (2.2) by central terms which we do not consider in these notes.

[2]In this subsection, we make statements assuming that F_A or F_V is present, but that is not necessary. We can use the $\mathbb{Z}_2$-grading instead.

chiral ring, which we denote by $\mathcal{R}_A$ for (Q_A, F_A) and $\mathcal{R}_B$ for (Q_B, F_V). It is a graded commutative algebra.

When formulated on a half of the Minkowski space, say, where the space coordinate is bounded as $x \leq 0$ and the time t is unbounded, a boundary condition on the fields must be specified at $x = 0$. There are essentially two types of boundary conditions that preserve maximal number of supercharges:

$$\begin{aligned} &\textit{A-type:} \qquad Q_A \text{ and } Q_A^\dagger \text{ conserved.} \\ &\textit{B-type:} \qquad Q_B \text{ and } Q_B^\dagger \text{ conserved.} \end{aligned}$$

Boundary conditions of such types are called *A-branes* and *B-branes* respectively. The pair $(Q, F) = (Q_A, F_A)$ (*resp.* (Q_B, F_V)) acts on local operators inserted on the boundary with an A-type (*resp.* B-type) boundary condition, obeying the same relation as (2.6). The cohomology classes form an algebra, which is non-commutative in general. Two differnt boundary conditions of the same type can be placed on the boundary with a local operator inserted inbetween. The pair (Q, F) acts also on such local operators obeying (2.6),[3] and we may consider the cohomology classes. Then, we have a *category*, which is denoted by $\mathcal{C}_A$ for A-branes and $\mathcal{C}_B$ for B-branes. Objects are boundary conditions and morphisms between boundary conditions are Q-cohomology classes of local operators inserted between them, with the composition represented by the product of operators.

We may also consider combinations of supercharges which are rotated by "the other" R-charge: the vector rotation for the A-type, $\mathrm{e}^{i\alpha F_V} Q_A \,\mathrm{e}^{-i\alpha F_V} = \mathrm{e}^{i\alpha}\overline{Q}_+ + \mathrm{e}^{-i\alpha}Q_-$, and the axial rotation for the B-type, $\mathrm{e}^{i\beta F_A} Q_B \,\mathrm{e}^{-i\beta F_A} = \mathrm{e}^{i\beta}\overline{Q}_+ + \mathrm{e}^{-i\beta}\overline{Q}_-$. D-branes preserving these supercharges and their conjugates are called $A_{\mathrm{e}^{2i\alpha}}$*-branes* and $B_{\mathrm{e}^{2i\beta}}$*-branes* respectively.

2.3. RG flow

In a general QFT, the behaviour of observables depends very much on the energy scale, or inversely, the distance scale. For example, the correlation function of two operators inserted as two points of distance r is in general a complicated function of r. The behaviour at small r i.e. short distance, or equivalently, high energy or ultra-violet (UV), is in general very much different from the behaviour at large r i.e. long distance, or equivalently, low energy or infra-red (IR). If the length scale is increased (i.e. the energy scale is lowered), the behaviour of the theory

[3] There is a potential anomaly to the relations [12].

changes — it may be identified as the behaviour of a different theory before the change of the scale. This change of the theory under the change of the scale is called the *renormalization group flow*, or *RG flow* in short. A theory is scale invariant if it is invariant under the RG flow. The two point correlation function of an operator $\mathcal{O}$ in such a theory depends on the distance as its power,

$$\langle \mathcal{O}(x)\mathcal{O}(y)\rangle = \frac{1}{\mathrm{dist}(x,y)^{2\Delta_{\mathcal{O}}}}. \tag{2.7}$$

The number $\Delta_{\mathcal{O}}$ is called the *dimension* of the operator.

In two-dimensions, scale invariance of a QFT is proven to be equivalent to conformal invariance. For a general QFT, we have conformally invariant field theories (CFTs) in the UV and IR limits. An invariant of a CFT is its central charge c, and it is known that it descreases under the RG flow, $c_{\mathrm{UV}} \geq c_{\mathrm{IR}}$.

The same applies of course to QFTs with supersymmetry. In 2d $(2,2)$ supersymmetric QFTs, there are a class of observables which are invariant under the RG flow, even if the theory is not scale invariant. The chiral ring and the category of branes, which are introduced above, are examples of such observables which are "protected" from renormalization. In a $(2,2)$ superconformal field theory (SCFT), it is convenient to use $\widehat{c} = c/3$ for the central charge.

2.4. Deformations

A QFT can be deformed by adding a local operator $\mathcal{O}$ to its Lagrangian density. If the theory is scale invariant, the deformation is called *irrelevant*, *marginal* and *relevant* if the dimension of the operator minus the dimension of the spacetime, which is $\Delta_{\mathcal{O}} - 2$ in a 2d theory, is positive, zero and negative, respectively. Under the RG flow, an irrelevant deformation decays and a relevant deformation grows. A marginal deformation is called *exactly marginal* if it remains invariant under the RG flow, while it is *marginally irrelevant* (*resp. marginally relevant*) if it decays (*resp.* grows) under the RG. The moduli space of conformal field theories is coordinatized by exactly marginal deformation parameters. It has a natural metric G called the Zamolodchikov metric [13]: Let v_1 and v_2 be tangent vectors at one theory that correspond to exactly marginal operators $\mathcal{O}_1$ and $\mathcal{O}_2$. Then, their inner product $G(v_1, v_2)$ is provided (for 2d) by

$$\langle \mathcal{O}_1(x)\mathcal{O}_2(y)\rangle = \frac{G(v_1,v_2)}{\mathrm{dist}(x,y)^4}. \tag{2.8}$$

In a 2d $(2,2)$ supersymmetric QFT, deformation operators that preserve the supersymmetry are of the following three types,

$$\Delta_D \mathcal{L} = Q_+ Q_- \overline{Q}_- \overline{Q}_+ \mathcal{K}, \tag{2.9}$$

$$\Delta_A \mathcal{L} = Q_+ \overline{Q}_- \mathcal{O}_A \text{ and its adjoint}, \tag{2.10}$$

$$\Delta_B \mathcal{L} = Q_+ Q_- \mathcal{O}_B \text{ and its adjoint}, \tag{2.11}$$

for scalar operators $\mathcal{K}$, $\mathcal{O}_A$ and $\mathcal{O}_B$, where $\mathcal{K}$ is arbitrary, $\mathcal{O}_A$ is *A-chiral*, $\overline{Q}_+ \mathcal{O}_A = Q_- \mathcal{O}_A = 0$, and $\mathcal{O}_B$ is *B-chiral*, $\overline{Q}_\pm \mathcal{O}_B = 0$. We shall call them *D-term*, *A-term*, and *B-term*, respectively.[4] We see from (2.1)–(2.2) that a D-term is an A-term and a B-term at the same time, up to total derivatives. It turns out that A-term deformations modulo D-term deformations are in one to one correspondence with elements of the chiral ring $\mathcal{R}_A$. Similarly, B-term deformations modulo D-term deformations are in one to one correspondence with elements of the chiral ring $\mathcal{R}_B$.

It follows from the algebra (2.1)–(2.2) that D-terms and A-terms are Q_B-exact while D-terms and B-terms are Q_A-exact, up to total derivatives. In particular, the ring $\mathcal{R}_B$ and the category $\mathcal{C}_B$ are invariant under D-term and A-term deformations, while $\mathcal{R}_A$ and the category $\mathcal{C}_A$ are invariant under D-term and B-term deformations.

In a $(2,2)$ SCFT, the D-term deformations are irrelevant, since each supercharge carry dimension $\frac{1}{2}$. Only a part of A-term and B-term deformations are marginal or relevant.

The spaces of parameters of the theory corresponding to A-term deformations and B-term deformations are denoted by $\mathfrak{M}_A$ and $\mathfrak{M}_B$. They have complex structures: tangent vectors of $\mathfrak{M}_A$ (*resp.* $\mathfrak{M}_B$) of type $(1,0)$ correspond to operators of the form $Q_+ \overline{Q}_- \mathcal{O}_A$ (*resp.* $Q_+ Q_- \mathcal{O}_B$) and can naturally be identified as elements of the chiral ring $\mathcal{R}_A$ (*resp.* $\mathcal{R}_B$). For a $(2,2)$ SCFT, the subspaces of exactly marginal parameters, $\mathfrak{M}_A^0 \subset \mathfrak{M}_A$ and $\mathfrak{M}_B^0 \subset \mathfrak{M}_B$, are complex submanifolds. They are also submanifolds of the moduli space of conformal field theories. The Zamolodchikov metric induced on $\mathfrak{M}_A^0$ and $\mathfrak{M}_B^0$ are known to be Kähler [14].

2.5. Topological Twists

The theory can be formulated not just on the (half of) Minkowski space. For example, we can consider the cylinder or the strip, again with the Minkowski metric, which yield closed or open string states. We

---

[4]The standard terminology is: *twisted F-term* instead of A-term, and *F-term* intead of B-term.

may also formulate the theory on these manifolds with Euclidean metric, by Wick rotation of the time line. Furthermore, we may formulate the system on a two-dimensional manifold with an arbitrary metric and spin structure, via the standard covariantization. Does the supersymmetry survive? We can certainly extend the definition of supersymmetry transformation of fields $\mathcal{O}$

$$\delta\mathcal{O} = i\epsilon_+ Q_- \mathcal{O} - i\epsilon_- Q_+ \mathcal{O} - i\overline{\epsilon}_+ \overline{Q}_- \mathcal{O} + i\overline{\epsilon}_- \overline{Q}_+ \mathcal{O}, \tag{2.12}$$

by covariantization of the expressions $Q_\pm\mathcal{O}$ and $\overline{Q}_\pm\mathcal{O}$, and by taking the variational parameters $\epsilon_\pm$ and $\overline{\epsilon}_\pm$ to be sections of the spin bundles $S_\pm$. However, invariance of the covariantized action requires the variational parameters to be covariantly constant, which is impossible on a curved manifold. There are several ways to restore a part of the supersymmetry. One is the *topological twisting* which we now describe. In Section 4, we will consider an alternative way, which yields the main character of the present notes.

Let us assume that $F = F_V$ or F_A is conserved and has *charge integrality*, that is, it generates a $U(1)$ symmetry group under which the non-spinorial and spinorial fields have even and odd charges respectively. The topological twisting is to replace a field of R-charge q with values in a vector bundle E by a field with values in $E\otimes T_\Sigma^{\otimes q/2}$, when we consider the theory on an oriented Riemannian manifold Σ. Here, T_Σ is the holomorphic tangent bundle equipped with the Levi-Civita connection. Note that, due to the constraint on the parity of the R-charge, $E\otimes T_\Sigma^{\otimes q/2}$ makes sense without choice of spin structure of Σ. The same change occurs also for the variational parameters $\epsilon_\pm$ and $\overline{\epsilon}_\pm$, and some of them become scalars. We can take such a parameter to be constant, and the corresponding supercharge is conserved. It is called the *A-twist* for $F = F_V$ and *B-twist* for $F = F_A$. In the A-twisted (*resp.* B-twisted) theory, the supercharges $\overline{Q}_+$ and Q_- and hence their sum $Q = Q_A$ (*resp.* $\overline{Q}_\pm$ and their sum $Q = Q_B$) are conserved. Forthermore, the correlation functions of Q-closed operators depend only on the Q-cohomology classes of the operators and are invariant under deformation of the Riemannian metric on Σ. In particular, they depend only on the topology of Σ — we obtain a topological field theory.

We can further consider *topological string theory* by coupling the twisted theory to a certain theory of 2d gravity called *topological gravity*. A g-loop amplitude is obtained by integration over the moduli space of curves of genus g.

2.6. Examples

Non-linear sigma model

Let X be a compact Kähler manifold. Then, there is a 2d $(2, 2)$ supersymmetric QFT called the *non-linear sigma model* with target X. As a part of the data, we also choose a class $[B] \in \mathrm{H}^2(X, \mathbb{R}/2\pi\mathbb{Z})$ called the *B-field*. The model classically has both vector and axial $U(1)$ R-symmetries with charge integrality, but the axial R-symmetry is anomalous if the first Chern class $c_1(X)$ is non-zero: the axial rotation shifts $[B]$ by $c_1(X)$. The model is classically scale invariant, but the target metric changes under the RG flow. The Kähler class runs according to the Ricci flow: $[\omega] \to [\omega'] = [\omega] + c_1(X)\log(\mu'/\mu)$ for the change $\mu \to \mu'$ of the energy scale.

The sigma model coupling is proportional to the curvature of X which, roughly, is inversely proportional to the size of X or the Kähler class $[\omega]$. Therefore, the energy dependence of the sigma model coupling is controlled by the signs of the eigenvalues of $c_1(X) \simeq$ Ricci curvature. When $c_1(X) > 0$, i.e., when X is a Fano manifold, the Kähler class is larger at higher energies and diverges in the limit $\mu \to +\infty$. That is, the sigma model is free in the ultra-violet limit (asymptotically free and UV complete). At lower energies, the Kähler class is smaller and the coupling is stronger. Finding the infra-red behaviour is a non-trivial problem. When $c_1(X) = 0$, i.e., when X is a Calabi-Yau manifold, the Kähler class does not run under the RG. The theory flows in the infra-red limit to an SCFT with central charge $\widehat{c} = \dim_{\mathbb{C}} X$. When $c_1(X)$ has a negative component, the corresponding component of the Kähler class becomes smaller at higher energies. That is, the coupling partly diverges at some high energy (Landau pole) and the sigma model is not UV complete. In particular, this is the case when $c_1(X) < 0$, i.e., when X is of general type. In that case, the sigma model is free in the infra-red limit.

The chiral ring etc of the model are

$$\begin{aligned}
\mathcal{R}_A &= \mathrm{QH}^*(X) \quad \text{quantum cohomology ring,}\\
\mathcal{R}_B &= \mathrm{H}^*(X, \wedge^* T_X) \quad \text{cohomology ring of polyvector fields,}\\
\mathcal{C}_A &= \mathrm{Fuk}(X) \quad \text{Fukaya category,}\\
\mathcal{C}_B &= \mathrm{D}^b_{Coh}(X) \quad \text{derived category of sheaves with coherent}\\
&\qquad \text{cohomologies,}\\
\mathfrak{M}^{0,c}_A &= \text{the space of complexified Kähler class}\\
&\qquad [\omega - iB] \in \mathrm{H}^2(X, \mathbb{R}/2\pi i\mathbb{Z}),\\
\mathfrak{M}^{0,c}_B &= \text{the moduli space of complex structures of } X.
\end{aligned}$$

$\mathfrak{M}^{0,c}_{A/B}$ is a submanifold of $\mathfrak{M}_{A/B}$ that corresponds to the marginal deformations of the classical system. When $c_1(X) = 0$, the space $\mathfrak{M}^{0,c}_A \times \mathfrak{M}^{0,c}_B$ is identified as an open subset of the moduli space of the IR SCFTs. When $c_1(X) \neq 0$, there is an RG low on $\mathfrak{M}^{0,c}_A$ in the direction of $c_1(X)$, and the shift in the direction of $ic_1(X)$ is absorbed by the axial rotaion.

As the model has vector $U(1)$ R-symmetry with charge integrality, A-twist is always possible. The corresponding topological string theory is known as the *Gromov-Witten theory* in mathematics. B-twist is possible if and only if X is a Calabi-Yau manifold.

Landau-Ginzburg model

Let $W(x)$ be a polynomial of N variables $x = (x_1, \ldots, x_N)$ with complex coefficients, having only isolated critical points. Then, there is a 2d $(2,2)$ supersymmetric QFT called the *Landau-Ginzburg model* with superpotential $W(x)$. It always have an axial $U(1)$ R-symmetry with charge integrality. A vector R-symmetry exists if and only if $W(x)$ is quasi-homogeneous, that is, with a change of coordinates if necessary, there are some numbers $R = (R_1, \ldots, R_N)$ such that $W(\lambda^R x) = \lambda^2 W(x)$, where $\lambda^R x = (\lambda^{R_1} x_1, \ldots, \lambda^{R_N} x_N)$. In that case, the theory flows in the infra-red limit to an SCFT with $\widehat{c} = \mathrm{tr}(1 - R)$. The chiral ring etc of the model are

$$\begin{aligned}
\mathcal{R}_A &= ?,\\
\mathcal{R}_B &= \mathrm{Jac}(W) \quad \text{Jocobi ring } \mathbb{C}[x_1, \ldots, x_N]/(\partial_1 W, \ldots, \partial_N W),\\
\mathcal{C}_A &= \mathrm{Fuk}(W) \quad \text{Fukaya category controled by } W,\\
\mathcal{C}_B &= \mathrm{MF}(W) \quad \text{category of matrix factorizations of } W,\\
\mathfrak{M}_A &= ?,\\
\mathfrak{M}_B &= \text{the moduli space of versal deformations of } W.
\end{aligned}$$

The authors do not know what are $\mathcal{R}_A$ and $\mathfrak{M}_A$ at this moment.

As the model has axial $U(1)$ R-symmetry with charge integrality, B-twist is always possible. The corresponding topological string theory at the tree level (genus zero) is closely related to K. Saito's theory of primitive forms [15–17]. When W is a Morse function, A. Givental proposed a recipe to construct the higher genus amplitudes [18], and C. Teleman proved that they satisfy a mathematical axiom of topological string theory [19].

When W is quasi-homogeneous, there is also a vector R-symmetry. However, it does not possess the charge integrality and the A-twist is not possible. That problem may be cured by orbifolding. Gauge the system by a finite group $\Gamma \subset GL(N,\mathbb{C})$ of symmetries of $W(x)$ that include $e^{\pi i R}$ as its element. Then, the charge integrality holds for gauge invariant fields, and the A-twist becomes possible. The corresponding topological string theory is developed in [20] and is called the *FJRW theory*.

2.7. Mirror Symmetry

The 2d $(2,2)$supersymmetry algebra has an automorphism: $Q_- \leftrightarrow \overline{Q}_-$, $F_V \leftrightarrow F_A$, and the other generators kept intact. A pair of 2d $(2,2)$ supersymmetric QFTs are said to be *mirror* to each other when there is an isomorphism between them under which the supersymmetry generators undergo the above automorphism. There are immediate consequences of the mirror symmetry: the ring $\mathcal{R}_A$ of one theory is isomorphic to the ring $\mathcal{R}_B$ of the mirror, the category $\mathcal{C}_A$ of one theory is equivalent to the category $\mathcal{C}_B$ of the mirror, the parameter space $\mathfrak{M}_A$ of one theory is isomorphic to the parameter space $\mathfrak{M}_B$ of the mirror, and the topological A-model (the A-twisted model or the corresponding topological string theory) of one theory is isomorphic to the topological B-model of the other.

The most famous example of mirror symmetry is the one for the sigma models with Calabi-Yau targets, say X and Y. As a part of the above consequences, we have the relation between the Hodge numbers, $h^{p,q}(X) = h^{n-p,q}(Y)$, where $n = \dim X = \dim Y$. Other well known examples are mirror symmetry between the sigma model with a non-Calabi-Yau target and the Landau-Ginzburg model, and the one between Landau-Ginzburg orbifolds. Some of the consequences in these examples are mathematically proven.

3. Gauged Linear Sigma Models

The main characters of the present notes are a class of 2d $(2, 2)$ supersymmetric gauge theories called gauged linear sigma models (GLSMs). In this section, we provide an introduction to GLSMs.

3.1. The Bulk Theory

A 2d $(2, 2)$ gauge theory is specified by a choice of

- *gauge group* G**:** a compact Lie group,
- *matter representation* V**:** a finite dimensional complex representation of G,
- *superpotential* $W(\phi)$**:** a G invariant polynomial function of $\phi \in V$, and
- *twisted superpotential* $\widetilde{W}(\sigma)$**:** a G invariant polynomial function of $\sigma \in \mathfrak{g}_{\mathbb{C}}$.

As a minor part of the data, we also choose a G-invariant norm $X \in i\mathfrak{g} \mapsto \frac{1}{e^2}(X)^2 \in \mathbb{R}_{\geq 0}$ on $i\mathfrak{g}$ (e is called the *gauge coupling constant*), and a G-invariant hermitian inner product on V. The latter defines a G-invariant symplectic structure on V, and we denote by $\mu : V \to i\mathfrak{g}^*$ the moment map that vanishes at the origin.

A vector $U(1)$ R-symmetry exists when there is a linear map $R : V \to V$ commuting with the G-action such that

$$W(\lambda^R \phi) = \lambda^2 W(\phi). \tag{3.1}$$

The charge integrality holds when $e^{\pi i R} : V \to V$ is the same as the action of a gauge group element, say $J \in G$. An axial $U(1)$ R-symmetry with charge integrality exists at the classical level when $\widetilde{W}(\sigma)$ is linear, and it remains to be a symmetry of the quantum system under *Calabi-Yau condition*: $G \subset SL(V)$. In the present notes, we assume all of the above but the Calabi-Yau condition. We write the linear twisted superpotential as

$$\widetilde{W}(\sigma) = -t(\sigma), \tag{3.2}$$

for an adjoint invariant linear form

$$t = \zeta - i\theta \in \mathfrak{g}_{\mathbb{C}}^{*G}, \tag{3.3}$$

where ζ and θ, both in $i\mathfrak{g}^{*G}$, are called the Fayet-Illiopoulos (FI) parameter and the theta parameter respectively. Note that ζ and θ can also be regarded as elements of $i\mathfrak{t}^{*W}$ or $i\mathfrak{z}^*$ thanks to the natural isomorphisms

$\mathfrak{g}^{*G} \cong \mathfrak{t}^{*\mathrm{W}} \cong \mathfrak{z}^*$. To be precise, the theta parameter is subject to a discrete identification,

$$\theta \equiv \theta + 2\pi n, \tag{3.4}$$

for an image n of a character $G \to U(1)$ under the differential map $\mathrm{Hom}(G, U(1)) \to \mathrm{Hom}(\mathfrak{g}, \mathfrak{u}(1)) = i\mathfrak{g}^*$. Therefore, the space of theta parameter (or theta *angle*) is the compact torus $i\mathfrak{g}^{*G}/2\pi\Lambda_G$, where $\Lambda_G := \mathrm{Image}(\mathrm{Hom}(G, U(1)) \to i\mathfrak{g}^{*G})$. When G is connected, Λ_G is isomorphic to $\mathrm{Hom}(G, U(1))$ and is equal to the lattice P^{W} of Weyl invariant weights of T embedded in $i\mathfrak{g}^{*G}$ via $i\mathfrak{t}^{*\mathrm{W}} \cong i\mathfrak{g}^{*G}$.

The theory consists of two sets of fields called a *matter multiplet* and a *gauge multiplet*, that include scalar fields, ϕ and σ, with values in V and $\mathfrak{g}_{\mathbb{C}}$ respectively. The classical potential for the scalar fields is

$$U(\sigma, \phi) = \frac{1}{8e^2}[\sigma, \overline{\sigma}]^2 + \frac{1}{2}|\sigma\phi|^2 + \frac{1}{2}|\overline{\sigma}\phi|^2 + \frac{e^2}{2}\left(\mu(\phi) - \zeta\right)^2 + |\mathrm{d}W(\phi)|^2. \tag{3.5}$$

Note that each term is non-negative. The space of zero points of U, called *classical vacua*, provides us with a first hint to understand the low energy behaviour of the theory. The vacuum equation $U = 0$ reads

$$[\sigma, \overline{\sigma}] = 0, \quad \sigma\phi = \overline{\sigma}\phi = 0, \quad \mu(\phi) = \zeta, \quad \mathrm{d}W(\phi) = 0. \tag{3.6}$$

The last two equations require ϕ to be in $\mathrm{Crit}(W) \cap \mu^{-1}(\zeta)$ and the first two equations require σ to be in the Cartan subalgebra of the stabilizer subgroup at ϕ. The space of the FI parameter ζ is separated into chambers, called *phases*, according to the topology of the G-space $\mathrm{Crit}(W) \cap \mu^{-1}(\zeta)$. Inside a phase, typically, the stabilizer subgroup is finite at each point of $\mathrm{Crit}(W) \cap \mu^{-1}(\zeta)$, so that σ is forced to vanish — the space of classical vacua is the quotient $X_\zeta = (\mathrm{Crit}(W) \cap \mu^{-1}(\zeta))/G$, called the *Higgs branch*. If that is the case, the theory reduces at low energies to the Landau-Ginzburg model $(\mu^{-1}(\zeta)/G, W_\zeta)$ where W_ζ is the function on $\mu^{-1}(\zeta)/G$ induced from W. If, in addition, W_ζ is Bott-Morse, the theory reduces further to the non-linear sigma model on $\mathrm{Crit}(W_\zeta)$, which is nothing but the Higgs branch X_ζ. Such a phase is called a *geometric phase*. On a wall between chambers (*phase boundary*), there are continuous stabilizer subgroups at some loci of $\mathrm{Crit}(W) \cap \mu^{-1}(\zeta)$. There develops a component of the space of classical vacua, called the *Coulomb branch*, in which σ can take any value in the Cartan subalgebra of the stabilizer subgroup. Emergence of this non-compact space may be regarded as a singularity.

Quantum effects will yield significant modification of this picture. In particular, classical Coulomb branch may be lifted, or quantum Coulomb vacua may emerge even in the absence of classical one. To see this, we explore the region in the field space where σ takes large generic values in a Cartan subalgebra $\mathfrak{t}_{\mathbb{C}}$ of $\mathfrak{g}_{\mathbb{C}}$. Then, the second and the third terms of (3.5) provide masses to many of the components of ϕ, typically all (which we assume for now). Integrating out the massive modes, we obtain the effective theory consisting of the gauge multiplet of the maximal torus T only, with the effective twisted superpotential

$$(3.7)\qquad \widetilde{W}_{\rm eff}(\sigma) = -t(\sigma) + 2\pi i\rho(\sigma) - \sum_i Q_i(\sigma)\left(\log(Q_i(\sigma)/\Lambda) - 1\right).$$

Here, $\rho = \frac{1}{2}\sum_{\alpha>0}\alpha$ is half the sum of positive roots of $\mathfrak{g}$, Q_i's are the weights of V, and Λ is a scale parameter which is needed for renormalization. The gauge coupling constant of the effective theory is a complicated function $e_{\rm eff}(\sigma)$ but it approaches the given value e at $|\sigma/\Lambda| \gg 1$. The effective potential is

$$(3.8)\qquad U_{\rm eff}(\sigma) = \min_{n\in {\rm P}} \frac{e^2_{\rm eff}(\sigma)}{2}\left|{\rm d}\widetilde{W}_{\rm eff}(\sigma) + 2\pi i n\right|^2.$$

Note that the choice of branch of the logarithms in (3.7) has no physical effect — a different choice would shift $\widetilde{W}_{\rm eff}(\sigma)$ by an element of $2\pi i{\rm P}(\sigma)$, but that does not affect (3.8). A point σ_* is a true Coulomb vacuum in the quantum theory when $U_{\rm eff}(\sigma_*) = 0$, that is, ${\rm d}\widetilde{W}_{\rm eff}(\sigma_*) \in 2\pi i{\rm P}$.[1] In particular, a classical Coulomb branch may not survive in the quantum theory, or true Coulomb vacua might appear even in the absence of classical Coulomb branch. For completeness, one should also explore the region in the field space where σ takes large generic values in a Cartan subalgebra for a subgroup $H \subset G$ and large values of H-neutral components of ϕ, and examine whether there are true mixed Coulomb-Higgs vacua.

The character of the theory depends very much on whether the infinitesimal version of the Calabi-Yau condition, $\mathfrak{g} \subset \mathfrak{sl}(V)$, is satisfied or not. That is, whether b_1 defined by

$$(3.10)\qquad b_1(X) := {\rm tr}_V(X) \qquad X \in \mathfrak{g}$$

[1]The vacuum value of the twisted superpotential is

$$(3.9)\qquad \widetilde{W}_{\rm eff}(\sigma_*) = \sum_i Q_i(\sigma_*).$$

This is the value of $\widetilde{W}_{\rm eff}(\sigma)$ on the branch of logarithms on which it is genuinely critical at σ_*, ${\rm d}\widetilde{W}_{\rm eff}(\sigma_*) = 0$. Note that it is not affected by the $2\pi i{\rm P}(\sigma)$ ambiguity of $\widetilde{W}_{\rm eff}(\sigma)$.

is zero or not. Note that b_1 may be regarded as an element of $\mathrm{P}^W \cong \mathrm{P}_{Z_G}$. As an element of P^W, it can also be written as $b_1 = \sum_i Q_i$.

Calabi-Yau case

Suppose it is satisfied, $b_1 = 0$. In this case, the FI parameter is invariant under the renormalization group and the axial $U(1)$ R-symmetry exists in the quantum theory. Accordingly, $\widetilde{W}_{\text{eff}}$ in (3.7) is independent of the parameter Λ. In particular, the vacuum equation $\mathrm{d}\widetilde{W}_{\text{eff}}(\sigma) \in 2\pi i\mathrm{P}$ is invariant under the scaling, $\sigma \to \lambda\sigma$ for $\lambda \in \mathbb{C}^\times$. This means that if σ is a Coulomb vacuum, then, any of its scaling is also. In particular, the space of such vacua, the Coulomb branch, must be non-compact. Also, presence of Coulomb branch imposes a non-trivial constraint on the FI-theta parameter t. In fact, the equation $\mathrm{d}\widetilde{W}_{\text{eff}}(\sigma) \in 2\pi i\mathrm{P}$ produces a parametric representation of t in terms of ratio of σ coordinates, defining a complex hypersurface in the space of t. Let $\Delta \subset \mathfrak{g}_{\mathbb{C}}^{*G}/2\pi i\Lambda_G$ be the discriminant locus on which there is a Coulomb branch and/or mixed branches. It is a union of hyeprsurfaces. When projected to the the ζ space, the discriminant locus Δ projects to an amoeba, in the sense of [21], which asymptotes to the phase boundary. The space of regular values of t is thus

$$\mathfrak{M}_t = \mathfrak{g}_{\mathbb{C}}^{*G}/2\pi i\Lambda_G - \Delta. \tag{3.11}$$

Since $\Delta \subset \mathfrak{g}_{\mathbb{C}}^{*G}/2\pi i\Lambda_G$ has complex codimension one, one can go from one phase to another without meeting it. In particular, there is no sharp transition between different phases.

The theory flows in the infra-red limit to an SCFT with $\widehat{c} = \mathrm{tr}_V(1-R) - \dim G$, and the FI-theta parameter (*resp.* parameters of W) are exactly marginal A-term (*resp.* B-term) deformation parameters of the SCFT. That is, they parameterize submanifolds of the moduli space of SCFTs

$$\mathfrak{M}_t \subset \mathfrak{M}_A^0, \qquad \mathfrak{M}_W \subset \mathfrak{M}_B^0. \tag{3.12}$$

Quite often, the inclusion $\subset$ is equality $=$.

In a geometric phase, $\mathfrak{M}_t$ and $\mathfrak{M}_W$ are respectively (parts of) the moduli space of complexified Kähler class and the moduli space of complex structures of the Higgs branch, respectively.

Non Calabi-Yau case

Suppose the condition is violated, $b_1 \neq 0$. In this case, the FI parameter runs under the renormalization group — for a change of energy scale $\mu \to \mu'$ it changes as

$$\zeta \to \zeta' = \zeta + \log(\mu'/\mu) b_1, \tag{3.13}$$

and the classical axial $U(1)$ R-symmetry is anomalous — the axial rotation $\mathrm{e}^{i\beta} \in U(1)_A$ shifts the theta angle as

$$\theta \to \theta + 2\beta b_1. \tag{3.14}$$

Accordingly, $\widetilde{W}_{\text{eff}}$ in (3.7) depends non-trivially on Λ. The parameter t in $\widetilde{W}_{\text{eff}}$ is the FI-theta parameter at the scale Λ. Since the vacuum equation $\mathrm{d}\widetilde{W}_{\text{eff}}(\sigma) \in 2\pi i \mathrm{P}$ has no scaling invariance, the space of Coulomb vacua does not have to be non-compact. Quite often, there are isolated Coulomb vacua. Such a vacuum cannot be found by the classical analysis of $U(\sigma, \phi)$ but should be taken into account as a sound vacuum of the quantum theory. Of course, there can be vacua at special values of σ, such as $\sigma = 0$, which can be found by the classical analysis.

The theory flows in the infra-red limit to one of the isolated Coulomb vacua, which is typically a massive vacuum, or to the Higgs branch theory $(\mu^{-1}(\zeta_{\text{IR}})/G, W_{\zeta_{\text{IR}}})$ at $\sigma = 0$ where ζ_{IR} is the IR value of the FI parameter, or to a mixture of these two types. Some of the Higgs branch theory can be a non-trivial SCFT.

One should be careful for the use of the term "phase" for two reasons; one is that the FI parameter runs under the renormalization group and another is that there are other vacua at different regions of the field space, such as Coulomb vacua. When we say "phase", we mean the theory at certain range of energy scales where ζ is in a certain chamber *and* in the region of the field space where the gauge symmetry is broken to a finite group and the classical analysis is valid. When we want to make it clear, we shall sometimes use the term "regime" instead. In a geometric regime, the flow of ζ corresponds to the flow of the Kähler class and the axial shift of θ corresponds to that of the B-field.

Example $\mathbf{T}^{U(1)}_{N,d}$

Let us consider a model $\mathrm{T}^{U(1)}_{N,d}$ labelled by two positive integers N and d, with the following data

$$\begin{aligned} G &= U(1), \\ V &= \mathbb{C}(-d) \oplus \mathbb{C}(1)^{\oplus N} \ni (p, x_1, \ldots, x_N), \\ W &= p f(x_1, \ldots, x_N), \\ \widetilde{W} &= -t\sigma. \end{aligned}$$

$\mathbb{C}(i)$ is the representation of $U(1)$ of weight i, $f(x_1,\ldots,x_N)$ is a degree d polynomial which is generic in the sense that $\partial f/\partial x_i = 0$ for all i implies $x_1 = \cdots = x_N = 0$, and $t = \zeta - i\theta \in \mathbb{C}/2\pi i\mathbb{Z}$.

The R-charge is unique up to gauge, $R = (2 - d\epsilon, \epsilon, \ldots, \epsilon)$, and satisfies the charge integrality condition with $J = \mathrm{e}^{\pi i\epsilon}$. The space of FI-parameter $i\mathfrak{z}^* \cong \mathbb{R}$ is separated into two phases — the geometric phase $\zeta > 0$ and the Landau-Ginzburg phase $\zeta < 0$. In $\zeta > 0$, the classical vacuum equation $U = 0$ forces x to have a non-zero value which breaks the gauge group completely. The theory reduces to the sigma model whose target space is the degree d hypersurface $X_f \subset \mathbb{CP}^{N-1}$ defined by $f = 0$. The Kahler and the B-field classes are approximately given by

$$(3.15)\quad [\omega] \simeq \zeta H \in \mathrm{H}^2(X_f,\mathbb{R}), \quad [B] \simeq [(\theta + \pi d)H] \in \mathrm{H}^2(X_f,\mathbb{R}/2\pi\mathbb{Z}),$$

where H is the hyperplane class of $\mathbb{P}^{N-1}$ restricted on X_f. More precisely, the correction is exponentially small in the $\zeta \to +\infty$ limit, $[\omega - iB] = [(t - d\pi i)H] + O(\mathrm{e}^{-t})$. In $\zeta < 0$, the classical vacuum equation $U = 0$ forces p to have a non-zero value which breaks the gauge group to the subgroup $\mathbb{Z}_d \subset U(1)$ consisting of d-th roots of unity. The theory reduces to the Landau-Ginzburg orbifold $(\mathbb{C}^N/\mathbb{Z}_d, W = f)$ at low energies. The FI parameter runs as $\zeta' = \zeta + (N - d)\log(\mu'/\mu)$, except when the Calabi-Yau condition $d = N$ is satisfied. The effective twisted superpotential is

$$(3.16)\quad \widetilde{W}_{\mathrm{eff}}(\sigma) = -t\sigma + d\sigma(\log(-d\sigma/\Lambda) - 1) - N\sigma(\log(\sigma/\Lambda) - 1),$$

and the equation for the Coulomb vacuum is $\partial_\sigma \widetilde{W}_{\mathrm{eff}} \equiv 0 \bmod 2\pi i\mathbb{Z}$, or

$$(3.17)\qquad (\sigma/\Lambda)^{N-d} = (-d)^d\,\mathrm{e}^{-t}.$$

When $d = N$, we have a family of superconformal field theories with $\widehat{c} = N - 2$ parametrized by t as well as the parameters for f. Since the equation (3.17) has solutions (i.e. arbitrary $\sigma \neq 0$) only for $\mathrm{e}^t = (-N)^N$, the discriminant locus Δ is one point at $t \equiv N\log N + N\pi i$,

$$(3.18)\qquad \mathfrak{M}_t = \mathbb{C}/2\pi i\mathbb{Z} - \{[N\log N + N\pi i]\}.$$

We see that the non-linear sigma model on the Calabi-Yau manifold X_f is continuously connected to the Landau-Ginzburg orbifold $(\mathbb{C}^N/\mathbb{Z}_N, f)$. This is the basic example of *CY/LG correspondence*. In the present model the inclusions in (3.12) are both equalities, $\mathfrak{M}_t = \mathfrak{M}_A^0$ and $\mathfrak{M}_W = \mathfrak{M}_B^0$.

When $d < N$, the FI parameter ζ runs from positive to negative. The high energy theory is the non-linear sigma model on the Fano manifold X_f whose size decreases as the energy scale is lowered. At lower energies, the sigma model description is no longer valid. There are $(N-d)$ Coulomb vacua with mass gap at $\sigma^{N-d} = \Lambda^{N-d}(-d)^d\,\mathrm{e}^{-t}$, as well as one Higgs branch theory at $\sigma = 0$ (for $d > 1$) which is the Landau-Ginzburg orbifold $(\mathbb{C}^N/\mathbb{Z}_d, f)$. When $3 \leq d < N$, the Higgs branch theory further flows in the infra-red limit to a superconformal field theory with $\widehat{c} = N(1-2/d)$. When $d = 2$, the Higgs branch theory has two (*resp.* one) supersymmetric ground states with a mass gap for even (*resp.* odd) N.

When $d > N$, the FI parameter ζ runs from negative to positive. The theory can be regarded as the superconformal field theory with $\widehat{c} = N(1-2/d)$ corrsponding to the Landau-Ginzburg orbifold $(\mathbb{C}^N/\mathbb{Z}_d, f)$ which is perturbed by a relevant operator of dimension $2N/d$. There are $(d-N)$ Coulomb vacua with mass gap at $\sigma^{d-N} = \Lambda^{d-N}(-d)^{-d}\,\mathrm{e}^{t}$ as well as one Higgs branch theory at $\sigma = 0$ which is the non-linear sigma model on the hypersurface X_f of general type.

3.2. Boundary Conditions

Our main interests are B-branes in GLSM and their low energy behaviour. The classical data for a B-brane is

- *Chan-Paton vector space* $M = M^{\mathrm{ev}} \oplus M^{\mathrm{od}}$**:** a $\mathbb{Z}_2$-graded representation of G,
- *matrix factorization* $Q(\phi)$**:** a G-equivariant polynomial function of $\phi \in V$ with values in $\mathrm{End}^{\mathrm{od}}(M)$ satisfying $Q(\phi)^2 = W(\phi)\mathrm{id}_M$, and
- $\gamma \subset \mathfrak{t}_{\mathbb{C}}$**:** a Weyl invariant Lagrangian submanifold which is a deformation of the real locus $i\mathfrak{t} \subset \mathfrak{t}_{\mathbb{C}}$

G-equivariance reads $g^{-1}Q(g\phi)g = Q(\phi)$. We assume that the vector $U(1)$ R-symmetry extends to the boundary: M is also a representation of $U(1)_V$, commuting with G, such that $Q(\phi)$ has R-charge 1. That is, there is a linear map $\mathbf{r} : M \to M$ commuting with the G-action such that

$$\lambda^{\mathbf{r}} Q(\lambda^R \phi)\lambda^{-\mathbf{r}} = \lambda Q(\phi). \tag{3.19}$$

We further assume that the charge integrality is maintained: $\mathrm{e}^{\pi i \mathbf{r}} J$ agrees with the $\mathbb{Z}_2$ grading on M, which is $+1$ (*resp.* -1) on even (*resp.* odd) elements of M. When the Lie algebra $\mathfrak{g}$ has a non-zero center $\mathfrak{z}$, there is an unphysical ambiguity in the R-charge: for each $\varepsilon \in i\mathfrak{z}$, we may do

$$R \to R + \varepsilon, \qquad J \to J\,\mathrm{e}^{\pi i \varepsilon}, \qquad \mathbf{r} \to \mathbf{r} - \varepsilon. \tag{3.20}$$

Given this data, we have a boundary condition on the fields as well as interaction terms at the boundary [22, 9]. Neumann boundary condition is imposed on the scalar ϕ in the matter multiplet while the scalar σ in the gauge multiplet is required to have boundary values in the adjoint orbit of γ. Also, we have boundary interaction determined by the data (M, Q), called the *Chan-Paton factor*: For $\Sigma = (-\infty, 0] \times \mathbb{R}$ with coordinate (x, t), it is $\mathrm{Pexp}\left(-i \int_{\partial\Sigma} \mathcal{A}_t \mathrm{d}t\right) : M \to M$ where[2]

$$\mathcal{A}_t = v_t - \mathrm{Re}(\sigma) - \frac{1}{2\sqrt{2\pi}} \psi^i \partial_i Q(\phi) + \frac{1}{2\sqrt{2\pi}} \overline{\psi}^{\bar{\imath}} \partial_{\bar{\imath}} Q(\phi)^\dagger + \frac{1}{4\pi} \{Q(\phi), Q(\phi)^\dagger\}. \tag{3.21}$$

Here, $v_t dt$ is the gauge potential on the boundary, while ψ^i and $\overline{\psi}^{\bar{\imath}}$ are the boundary values of the fermionic components of the matter multiplet. Of course, $v_t - \mathrm{Re}(\sigma)$ in (3.21) should be understood as its representation on M.

In the presence of boundary, the theta parameter is no longer periodic: the shift $\theta \to \theta + 2\pi n$ for $n \in \Lambda_G$ generates a boundary term $-\oint_{\partial\Sigma} n(v)$ in the action. This, however, can be compensated by a change of M: since $n \in \Lambda_G$ is the infinitesimal version of a character $f_n \in \mathrm{Hom}(G, U(1))$, the shift is compensated if the Chan-Paton representation M, $m \mapsto gm$, is replaced by a new representation $M(f_n^{-1})$, $m \mapsto f_n(g)^{-1} gm$. In other words,

$$\theta \to \theta + 2\pi n \quad \text{is equivalent to} \quad M \to M(f_n). \tag{3.22}$$

We may also consider $\mathrm{B}_{\mathrm{e}^{2i\beta}}$-branes in the model. Since $\mathrm{B}_{\mathrm{e}^{2i\beta}}$-type supersymmetry is obtained from the B-type supersymmetry by the classical axial rotation $\mathrm{e}^{i\beta F_A}$, we may obtain a $\mathrm{B}_{\mathrm{e}^{2i\beta}}$-brane from the B-brane (M, Q, γ) by performing a change of variable $X \to \mathrm{e}^{i\beta F_A} X$. In particular, σ in (3.21) is replaced by $\mathrm{e}^{2i\beta}\sigma$, and the submanifold $\gamma \subset \mathfrak{t}_{\mathbb{C}}$ is rotated as

$$\gamma \to \mathrm{e}^{-2i\beta}\gamma. \tag{3.23}$$

There is a possible mathematical application. The general principle of supersymmetry dictates that the category of B-branes is invariant under A-term deformations as well as under the renormalization group.

[2]Somewhat unusual appearance of powers of $\sqrt{2\pi}$ in (3.21) is due to our convention that the entire bulk action is divided by 2π. See (A.10).

When the infinitesimal Calabi-Yau condition is satisfied $\mathfrak{g} \subset \mathfrak{sl}(V)$, the GLSM defines a family of SCFTs over the parameter space $\mathfrak{M}_t \times \mathfrak{M}_W$. Since the FI-theta parameter is an A-term parameter, the category of B-branes in the SCFT is locally constant on $\mathfrak{M}_t$. In particular, if we draw a path in $\mathfrak{M}_t$, there must be an equivalence of categories at two different points on the path. If the path connects two different phases, there must be an equivalence of the categories of B-branes in the two phases, and the equivalence should not change under deformation of the path. If we draw a loop in $\mathfrak{M}_t$ around the discriminant locus Δ, we must have an autoequivalence of the category at any point of the loop. When the infinitesimal Calabi-Yau condition is violated $\mathfrak{g} \not\subset \mathfrak{sl}(V)$, the GLSM defines an RG flow from a UV theory to an IR theory. From the RG-invariance, the category of B-branes in the UV theory must be equivalent to the category of B-branes in the IR theory.

A natural problem is to learn about these equivalences, both in the Calabi-Yau and non Calabi-Yau cases, by studying B-branes in GLSM. The hemisphere partition function will play an important rôle in this problem, as will be discussed in later sections.

At this stage, let us mention that there is a category $\mathfrak{D}_{\rm LSM} = {\rm MF}_G(V, W)$ whose objects are B-brane data (M, Q). A morphism from (M_1, Q_1) to (M_2, Q_2) is an $\mathbb{C}[\phi]$-module map $a : M_1 \otimes \mathbb{C}[\phi] \to M_2 \otimes \mathbb{C}[\phi]$ compatible with the $U(1)_V \times G$ action obeying $Q_2 a = a Q_1$. This may be regarded as the category of branes of the theory where the gauge coupling constant is turned off, $e \searrow 0$. In a regime where the gauge symmetry is broken to a finite subgroup by the non-zero values of the matter fields and the theory reduces to a Higgs branch theory $(\mu^{-1}(\zeta)/G, W_\zeta)$, the rôle of the gauge multiplet is expected to be unimportant, and the data (M, Q) alone, without reference to $\gamma \subset \mathfrak{t}_{\mathbb{C}}$, should be enough to determine a D-brane. In such a regime, there is a functor $\pi_\zeta : \mathfrak{D}_{\rm LSM} \to {\rm MF}(\mu^{-1}(\zeta)/G, W_\zeta)$ that represents the reduction.[3] Since $e \searrow 0$ is a singular limit, unlike in the usual RG flow, this is far from being an equivalence.

Example $\mathbf{T}^{U(1)}_{N,d}$

The reduction of a B-brane data $\mathfrak{B} = (M, Q)$ in the geometric and LG regimes of the theory $\mathrm{T}^{U(1)}_{N,d}$ are described as follows. Suppose the

[3]In a geometric regime where $W_\zeta : \mu^{-1}(\zeta)/G \to \mathbb{C}$ is a Bott-Morse function with a smooth critical locus $X_\zeta = {\rm Crit}(W_\zeta)$, the target category ${\rm MF}(\mu^{-1}(\zeta)/G, W_\zeta)$ is expected to be equivalent to the derived category $D^b_{\rm Coh}(X_\zeta)$. This is a global version of Knörrer periodicity [23]. Proofs in various set ups have been given in [24–26].

Chan-Paton vector space is

$$M = \bigoplus_j \mathbb{C}(r_j, q_j) \tag{3.24}$$

where $\mathbb{C}(r_i, q_i)$ stands for a component of R-charge r_i and $U(1)$ gauge charge q_i. In the geometric regime, it is convenient to set $\epsilon = 0$ so that x_i's have R-charge zero, and we write r_j^0 for r_j at $\epsilon = 0$. Then, the geometric image of $\mathfrak{B}$ is the complex of vector bundles on X_f,

$$\begin{aligned} \pi_+(\mathfrak{B}) = (\mathcal{E}, d) : \mathcal{E} = \bigoplus_{n=0}^{\infty} \bigoplus_j \mathcal{O}_{X_f}(q_j + dn)[r_j^0 + 2n], \\ d = Q(p_+, x), \end{aligned} \tag{3.25}$$

where p_+ stands for the shift operator $\otimes \mathcal{O}_{X_f}(d)[2]$. To elaborate on "$d = Q(p_+, x)$", if $Q(p, x)$ has a term $p^m x^\alpha$ in the component that sends $\mathbb{C}(r_{j_2}^0, q_{j_2})$ to $\mathbb{C}(r_{j_1}^0, q_{j_1})$, which is possible only if $-q_{j_1} + q_{j_2} - dm + |\alpha| = 0$ and $r_{j_1}^0 - r_{j_2}^0 + 2m = 1$, the corresponding term in d is multiplication by x^α in the components that send $\mathcal{O}_{X_f}(q_{j_2} + dn)[r_{j_2}^0 + 2n]$ to $\mathcal{O}_{X_f}(q_{j_1} + d(n + m))[r_{j_2}^0 + 2(n + m)]$ for $n = 0, 1, \ldots$. This $(\mathcal{E}, d)$ is a complex of vector bundles on X_f. It is infinitely long to the right but is exact beyond a certain degree, and hence defines a bounded complex of coherent sheaves on X_f. We should note that the B-field class is no longer periodic in the presence of boundary, just like the theta parameter. The *absolute* B-field class is approximately

$$[B] \simeq (\theta + \pi d) H \in \mathrm{H}^2(X_f, \mathbb{R}). \tag{3.26}$$

In the LG regime, it is convenient to set $\epsilon = 2/d$ so that p has R-charge zero, and we write r_j^{LG} for r_j at $\epsilon = 2/d$. Then, the LG image of $\mathfrak{B}$ is simply

$$\begin{aligned} \pi_-(\mathfrak{B}) = (M_-, Q_-) : M_- = \bigoplus_j \mathbb{C}(r_j^{\mathrm{LG}}, q_j) \ \text{ as } U(1)_V \times \mathbb{Z}_d\text{-module}, \\ Q_-(x) = Q(1, x). \end{aligned} \tag{3.27}$$

This is indeed a brane data in the LG orbifold $(\mathbb{C}^N/\mathbb{Z}_d, f)$, that is, the GLSM with gauge group $\mathbb{Z}_d$, matter representation V and superpotential $f(x)$.

Let us present two examples of B-brane data, $\mathfrak{B}_1 = (M_1, Q_1)$ and $\mathfrak{B}_2 = (M_2, Q_2)$, with

$$M_1 = \mathbb{C}(0,0) \oplus \mathbb{C}(1-d\epsilon, d), \tag{3.28}$$

$$Q_1 = \begin{pmatrix} 0 & p \\ f(x) & 0 \end{pmatrix}, \tag{3.29}$$

and

$$M_2 = \bigoplus_{j=0}^{N} \mathbb{C}(j - j\epsilon, j)^{\oplus \binom{N}{j}}, \tag{3.30}$$

$$Q_2 = \sum_{i=1}^{N} \left(x_i \overline{\eta}_i + \frac{1}{d} p \frac{\partial f(x)}{\partial x_i} \eta_i \right). \tag{3.31}$$

In the second data, M_2 is a representation of the Clifford algebra $\{\eta_i, \overline{\eta}_j\} = \delta_{i,j}$, $\{\eta_i, \eta_j\} = \{\overline{\eta}_i, \overline{\eta}_j\} = 0$, in such a way that $\overline{\eta}_i$'s raise the (R, gauge) charge by $(1-\epsilon, 1)$ while η_i's lower it by the same amount. We may also consider $\mathfrak{B}_1(i,q)$ and $\mathfrak{B}_2(i,q)$ where the (R, gauge) charge of each component of M_1 and M_2 is shifted by (i,q). For example, $\mathfrak{B}_1(i,q)$ has $M = \mathbb{C}(i,q) \oplus \mathbb{C}(1 - d\epsilon + i, d+q)$ with everything else unchanged.

In fact, these data play crucial rôles in the study of low energy behaviour of D-branes in the geometric regime or in the LG regime. For $Q = Q_1$ and Q_2, let us compute $\{Q, Q^\dagger\}$ that enters into (3.21) as the boundary potential $\frac{1}{4\pi}\{Q, Q^\dagger\}$:

$$\{Q_1, Q_1^\dagger\} = \left(|p|^2 + |f(x)|^2\right) \mathrm{id}_{M_1}, \tag{3.32}$$

$$\{Q_2, Q_2^\dagger\} = \sum_{i=1}^{N} \left(|x_i|^2 + \frac{1}{d^2} \left| p \frac{\partial f(x)}{\partial x_i} \right|^2 \right) \mathrm{id}_{M_2}. \tag{3.33}$$

In the geometric regime where $x \neq 0$, $\{Q_2, Q_2^\dagger\}$ is positive definite which means that $\mathfrak{B}_2$ and any of its shifts are empty at low energies. In the Landau-Ginzburg regime where $p \neq 0$, $\{Q_1, Q_1^\dagger\}$ is positive definite which means that $\mathfrak{B}_1$ and its shifts are empty at low energies. We can also see the emptiness by looking at the reductions, (3.25) and (3.27), $\pi_+(\mathfrak{B}_2(i,q)) \cong 0$ and $\pi_-(\mathfrak{B}_1(i,q)) \cong 0$.

A side remark: In the geometric regime, $\mathfrak{B}_1$ and its shifts can be non-trivial, as $\{Q_1, Q_1^\dagger\}$ vanishes at $p = f(x) = 0$. Indeed, $\mathfrak{B}_1(i,q)$ descends under the reduction (3.25) to the line bundle $\mathcal{O}(q)[j]$ over the hypersurface X_f. In the LG regimes, $\mathfrak{B}_1$ and its shifts can be non-trivial, as $\{Q_2, Q_2^\dagger\}$ vanishes at $x = 0$. Indeed, $\mathfrak{B}_2(i,q)$ descends under the reduction (3.27) to a non-trivial B-brane in the LG orbifold, which further flows to a famous B-brane called "Recknagel-Schomerus brane" in Gepner model (when f is Fermat polynomial) [27].

Presence of empty branes makes it manifest that different D-branes in GLSM can lead to the same D-brane at low energies. In the geometric regime, the branes $\mathfrak{B}_2(i,q)$ can be added to any D-brane $\mathfrak{B}$ without changing its low energy behaviour. This holds even when we bind $\mathfrak{B}_2(i,q)$'s and $\mathfrak{B}$ with a non-trivial map between them (cone construction). The resulting brane data can be simplified by cancelling a pair of components. In this way, $\mathfrak{B}$ can be replaced by another brane which bahaves in the same way at low energies, but with a different set of Chan-Paton gauge charges. If you wish, repeating this process if necessary, you can increase the minimum gauge charge or lower the maximal gauge charge. Since the set of gauge charges of $\mathfrak{B}_2(j,q)$ is $\{q,q+1,\ldots,q+N\}$, the given brane $\mathfrak{B}$ can be replaced by another brane $\mathfrak{B}'$ whose gauge charges are in a set of N consecutive integers, say $\{0,1,\ldots,N-1\}$. The same holds in the LG regime, where empty branes are $\mathfrak{B}_1(i,q)$ whose set of gauge charges is $\{q,q+d\}$. A given brane can be replaced by another brane whose gauge charges are in a set of d consecutive integers, say $\{1,2,\ldots,d\}$, without changing the low energy behaviour.

This result has a categorical interpretation. The functors $\pi_+ : \mathfrak{D}_{\mathrm{LSM}} \to D^b_{\mathrm{Coh}}(X_f)$ and $\pi_- : \mathfrak{D}_{\mathrm{LSM}} \to \mathrm{MF}_{\mathbb{Z}_d}(f)$ that represent the reduction in the geometric and LG regimes have huge kernels — many many different objects are sent to the same object. But there are nice *slices*. For a subset $I \subset \mathbb{Z}$, we write $\mathcal{T}_I \subset \mathfrak{D}_{\mathrm{LSM}}$ for the full subcategory consisting of data whose gauge charges are in the set I. Then, for sets $I_+ \subset \mathbb{Z}$ and $I_- \subset \mathbb{Z}$ of N and d consecutive integers respectively, the functors

$$\mathcal{T}_{I_+} \hookrightarrow \mathfrak{D}_{\mathrm{LSM}} \xrightarrow{\pi_+} D^b_{\mathrm{Coh}}(X_f), \tag{3.34}$$

$$\mathcal{T}_{I_-} \hookrightarrow \mathfrak{D}_{\mathrm{LSM}} \xrightarrow{\pi_-} \mathrm{MF}_{\mathbb{Z}_d}(f), \tag{3.35}$$

are equivalences of categories. A proof can be found in [28] with a combination of [24–26]. This was extended to a more general situation in [29, 30].

4. The Hemisphere Partition Function

In this section, we present the partition function of gauged linear sigma model on the hemisphere, and describe some of its properties, such as the behaviour at large values of the radius r of the hemisphere, expressions in the geometric and the Landau-Ginzburg regimes, and the differential equations with respect to the radius r as well as the FI-thtea parameters.

4.1. Supersymmetry on the sphere and the hemisphere

We first present the supersymmetry on the sphere and the hemisphere. The key is to employ the *superconformal* transformations. A superconformal transformation has conformal Killing spinors as the variational parameters: holomorphic sections ϵ_+ and $\overline{\epsilon}_+$ of $S_+ \cong \sqrt{T_\Sigma}$ and anti-holomorphic sections ϵ_- and $\overline{\epsilon}_-$ of $S_- \cong \sqrt{\overline{T}_\Sigma}$. It reduces to the supersymmetry transformation (2.12) for constant variational parameters on a flat Euclidean or Minkowski space. The transformation is specified by a choice of vector and axial R-transformations of the fields. The commutator of the superconformal transformations $\delta_1^{\rm sc}$ and $\delta_2^{\rm sc}$ with different variational parameters is the sum of conformal and R transformations

$$[\delta_1^{\rm sc}, \delta_2^{\rm sc}] = \delta_{X_{12}}^{\rm conformal} + \delta_{\Theta_{V,12}}^{\rm vector} + \delta_{\Theta_{A,12}}^{\rm axial}. \tag{4.1}$$

Let us first consider the sphere. It is covered by two charts, the z-plane and the w-plane, which are related by $zw = 1$. We give it an $O(3)$ symmetric round metric of radius r, which is ${\rm d}s^2 = 4r^2|{\rm d}z|^2/(1+|z|^2)^2$ on the z-plane. There are four conformal Killing spinors

$$\mathbf{s}_{-\frac{1}{2}} = \sqrt{\frac{\partial}{\partial z}}, \quad \mathbf{s}_{\frac{1}{2}} = z\sqrt{\frac{\partial}{\partial z}} \text{ and } \widetilde{\mathbf{s}}_{-\frac{1}{2}} = \sqrt{\frac{\partial}{\partial \overline{z}}}, \quad \widetilde{\mathbf{s}}_{\frac{1}{2}} = \overline{z}\sqrt{\frac{\partial}{\partial \overline{z}}}. \tag{4.2}$$

Thus, we may consider the superconformal transformations $\delta^{\rm sc}(\epsilon_+, \epsilon_-, \overline{\epsilon}_+, \overline{\epsilon}_-)$ where the variational parameters are chosen from the four. However, we can accept only the combinations whose commutators are symmetries of the system. In particular, since we consider theories which are not necessarily conformally invariant, like GLSM, the vector field X_{12} that appear in (4.1) must be an $\mathfrak{o}(3)$ isometry. There are essentially two types of such combinations — A-type and B-type:

A-type supercharges are the four combinations

$$\begin{aligned} Q_{(+)}^{A+} &= \delta^{\rm sc}(0,0,\mathbf{s}_{\frac{1}{2}},\widetilde{\mathbf{s}}_{-\frac{1}{2}}), \quad & Q_{(+)}^{A-} &= \delta^{\rm sc}(\mathbf{s}_{-\frac{1}{2}},\widetilde{\mathbf{s}}_{\frac{1}{2}},0,0), \\ Q_{(-)}^{A+} &= \delta^{\rm sc}(0,0,\mathbf{s}_{-\frac{1}{2}},-\widetilde{\mathbf{s}}_{\frac{1}{2}}), \quad & Q_{(-)}^{A-} &= \delta^{\rm sc}(\mathbf{s}_{\frac{1}{2}},-\widetilde{\mathbf{s}}_{-\frac{1}{2}},0,0). \end{aligned} \tag{4.3}$$

The $\mathfrak{o}(3)$ rotations and the vector R-transformation appear as their anti-commutators.

B-type supercharges are the four combinations

$$(4.4)\qquad \begin{aligned} Q^{B+}_{(+)} &= \delta^{\rm sc}(\mathbf{s}_{\frac{1}{2}},0,0,\widetilde{\mathbf{s}}_{-\frac{1}{2}}), & Q^{B-}_{(+)} &= \delta^{\rm sc}(0,\widetilde{\mathbf{s}}_{\frac{1}{2}},\mathbf{s}_{-\frac{1}{2}},0),\\ Q^{B+}_{(-)} &= \delta^{\rm sc}(\mathbf{s}_{-\frac{1}{2}},0,0,-\widetilde{\mathbf{s}}_{\frac{1}{2}}), & Q^{B-}_{(-)} &= \delta^{\rm sc}(0,-\widetilde{\mathbf{s}}_{-\frac{1}{2}},\mathbf{s}_{\frac{1}{2}},0). \end{aligned}$$

The $\mathfrak{o}(3)$ rotations and the axial R-transformation appear as their anti-commutators.

To be precise, there are variants obtained by the other R-rotations. $\mathrm{A}^{e^{2i\beta}}$-type is obtained from A-type by the axial R-rotation: $\epsilon_\pm \to e^{\pm i\beta}\epsilon_\pm$ and $\overline{\epsilon}_\pm \to e^{\mp i\beta}\overline{\epsilon}_\pm$; $\mathrm{B}^{e^{2i\alpha}}$-type is obtained from B-type by the vector R-rotation: $\epsilon_\pm \to e^{-i\alpha}\epsilon_\pm$ and $\overline{\epsilon}_\pm \to e^{i\alpha}\overline{\epsilon}_\pm$. Together with the $\mathfrak{o}(3)$ rotations and the R-charge (F_V for $\mathrm{A}^{e^{2i\beta}}$-type and F_A for $\mathrm{B}^{e^{2i\alpha}}$-type), they form a Lie super-algebra isomorphic to $\mathfrak{osp}(2|2)$.

Next, let us consider the hemisphere, which is realized as a half of the sphere, say, the region $|z|\leq 1$ ("southern" hemisphere) in the z-plane. We again give it the round metric of radius r. There are infinitely many conformal Killing spinors, $\mathbf{s}_r = z^{r+\frac{1}{2}}\sqrt{\frac{\partial}{\partial z}}$ and $\widetilde{\mathbf{s}}_r = \overline{z}^{r+\frac{1}{2}}\sqrt{\frac{\partial}{\partial \overline{z}}}$ with $r\in\mathbb{Z}_{\geq 0}-\frac{1}{2}$, but the boundary condition at $|z|=1$ can admit only the pairs $(\mathbf{s}_{\frac{1}{2}},\widetilde{\mathbf{s}}_{-\frac{1}{2}})$ and $(\mathbf{s}_{-\frac{1}{2}},\widetilde{\mathbf{s}}_{\frac{1}{2}})$ as the variational parameters. Note also that the hemisphere only has $O(2)$ isometry, and hence we can accept only the combinations whose anti-commutators (4.1) have the $\mathfrak{o}(2)$ isometry as X_{12}. These constraints leave us with only four possibilities: $\mathrm{A}_{(+)}$, $\mathrm{A}_{(-)}$, $\mathrm{B}_{(+)}$ and $\mathrm{B}_{(-)}$ where

$\mathrm{A}_{(\pm)}$-type supercharges are the two combinations

$$(4.5)\quad Q^{A+}_{(\pm)} = \delta^{\rm sc}(0,0,\mathbf{s}_{\pm\frac{1}{2}},\pm\widetilde{\mathbf{s}}_{\mp\frac{1}{2}}) \text{ and } Q^{A-}_{(\pm)} = \delta^{\rm sc}(\mathbf{s}_{\mp\frac{1}{2}},\pm\widetilde{\mathbf{s}}_{\pm\frac{1}{2}},0,0).$$

Their anticommutator is the sum of the $\mathfrak{o}(2)$ rotation and the vector R-rotation.

$\mathrm{B}_{(\pm)}$-type supercharges are the two combinations

$$(4.6)\quad Q^{B+}_{(\pm)} = \delta^{\rm sc}(\mathbf{s}_{\pm\frac{1}{2}},0,0,\pm\widetilde{\mathbf{s}}_{\mp\frac{1}{2}}) \text{ and } Q^{B-}_{(\pm)} = \delta^{\rm sc}(0,\pm\widetilde{\mathbf{s}}_{\pm\frac{1}{2}},\mathbf{s}_{\mp\frac{1}{2}},0).$$

Their anticommutator is the sum of the $\mathfrak{o}(2)$ rotation and the axial R-rotation.

Again, there are variants, $\mathrm{A}^{e^{2i\beta}}_{(\pm)}$ and $\mathrm{B}^{e^{2i\alpha}}_{(\pm)}$, by applying the other R-rotation. The commutator algebra for $(Q,\overline{Q},F) = (e^{-\frac{\pi i}{4}}Q^{A\pm}_{(\pm)}, e^{-\frac{\pi i}{4}}Q^{A\mp}_{(\pm)}, \pm F_V)$ or $(e^{-\frac{\pi i}{4}}Q^{B\pm}_{(\pm)}, e^{-\frac{\pi i}{4}}Q^{B\mp}_{(\pm)}, \pm F_A)$ and the $\mathfrak{o}(2)$ rotation generator L

is

$$Q^2 = \overline{Q}^2 = 0, \quad \{Q, \overline{Q}\} = -2L + F,$$
$$(4.7) \qquad [L, Q] = \tfrac{1}{2}Q, \quad [L, \overline{Q}] = -\tfrac{1}{2}\overline{Q},$$
$$[F, Q] = Q, \quad [F, \overline{Q}] = -\overline{Q}.$$

If we zoom in to a point on the boundary $|z| = 1$ and take the flat space limit $r \to \infty$, we see that the supercharges become

$$(4.8) \qquad Q^{A+}_{(\pm)} \propto \mathrm{e}^{i\beta}\overline{Q}_+ \pm \mathrm{e}^{-i\beta}\overline{Q}_-, \quad Q^{A-}_{(\pm)} \propto \mathrm{e}^{-i\beta}Q_+ \pm \mathrm{e}^{i\beta}Q_-,$$

$$(4.9) \qquad Q^{B+}_{(\pm)} \propto \mathrm{e}^{i\alpha}\overline{Q}_+ \mp \mathrm{e}^{-i\alpha}Q_-, \quad Q^{B-}_{(\pm)} \propto \mathrm{e}^{-i\alpha}Q_+ \mp \mathrm{e}^{i\alpha}\overline{Q}_-.$$

Therefore, the boundary condition at $|z| = 1$ must be a $\mathrm{B}_{\pm\,\mathrm{e}^{2i\beta}}$-brane (*resp.* $\mathrm{A}_{\mp\,\mathrm{e}^{2i\alpha}}$-brane) for the $\mathrm{A}^{\mathrm{e}^{2i\beta}}_{(\pm)}$-type (*resp.* $\mathrm{B}^{\mathrm{e}^{2i\alpha}}_{(\pm)}$-type) supersymmetry.

Supergravity approach

A general approach for supersymmetry on a curved space is to couple the system to the supergravity and to choose a supergravity background that is invariant under a part of the supersymmetry. Topological twisting, introduced in Section 2.5, is an example of such a procedure — turn on the $U(1)$ connection for an R-symmetry so that some of the variational parameters can be constant scalars. The supersymmetry on the (hemi)sphere discussed above is another example where a different component of the supergravity multiplet is turned on. Certain linear combinations, ϵ and $\overline{\epsilon}$, of the variational parameters satisfy the following equations for some $\omega \in \mathbb{R}/2\pi\mathbb{Z}$ [1]

$$(4.10) \qquad \nabla_\mu \epsilon = \frac{i}{2r}\gamma_\mu\, \mathrm{e}^{i\omega\gamma_3}\epsilon, \qquad \nabla_\mu \overline{\epsilon} = \frac{i}{2r}\gamma_\mu\, \mathrm{e}^{-i\omega\gamma_3}\overline{\epsilon},$$

where γ_μ are the gamma matrices, γ_3 is the chirality operator, $\gamma_3 = \pm 1$ on $S_\mp$. This is indeed the condition for supersymmetry in a certain supergravity background [31].

4.2. Formulation

We consider the GLSM on the sphere and the hemisphere preserving A-type supersymmetry. In what follows, we describe the hemisphere with $\mathrm{A}_{(\pm)}$-type supersymmetry and B-branes at the boundary. The

---

[1] $\epsilon = \epsilon_- + \epsilon_+$, $\overline{\epsilon} = \overline{\epsilon}_- + \overline{\epsilon}_+$, $\omega = 2\beta - \frac{\pi}{2}$ for $\mathrm{A}^{\mathrm{e}^{2i\beta}}$; and $\epsilon = \epsilon_- + \overline{\epsilon}_+$, $\overline{\epsilon} = \overline{\epsilon}_- + \epsilon_+$, $\omega = 2\alpha - \frac{\pi}{2}$ for $\mathrm{B}^{\mathrm{e}^{2i\alpha}}$.

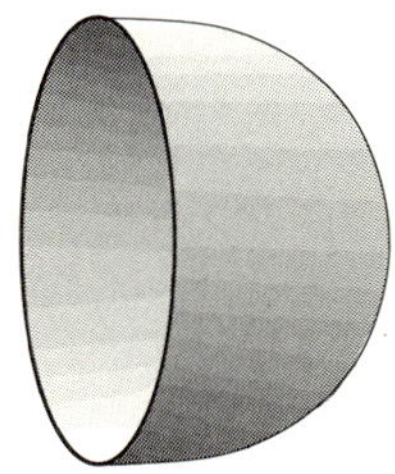

Fig. 1. The hemisphere

case of the sphere can be obtained from that easily. The supersymmetry is specified for a choice of vector R-transformation of the matter field ϕ, and for this we use the R-symmetry $R : V \to V$ of the system. We shall write $Q^- = Q^{A-}_{(\pm)}$ and $Q^+ = Q^{A+}_{(\pm)}$, to simplify the notations.

The action consists of four scalar terms

$$S = S_{\rm g} + S_{\rm m} + S_W + S_t, \tag{4.11}$$

and a matrix term $\int_{\partial D^2} \mathcal{A}$ that enters into the Chan-Paton factor

$$\mathrm{tr}_M \mathrm{Pexp}\left(-\int_{\partial D^2} \mathcal{A}\right). \tag{4.12}$$

These are obtained by modifying the covariantized version of the Lagrangian by terms that depend on the radius r as well as the bulk and the boundary R-charges R and $\mathbf{r}$. See Appendix A.4 for the detail. $S_{\rm g}$ and $S_{\rm m}$ are respectively the kinetic terms for the gauge multiplet and the matter multiplet. They are individually supersymmetric and Q-exact,

$$S_{\rm g} = Q^- Q^+(\cdots), \qquad S_{\rm m} = Q^- Q^+(\cdots). \tag{4.13}$$

S_W is the superpotential term. It is not supersymetric by itself, but the combination $\mathrm{e}^{-S_W}\mathrm{tr}_M \mathrm{Pexp}\left(-\int_{\partial D^2} \mathcal{A}\right)$ is. If we deform the superpotential $W \to W + \Delta W$ and the matrix factorization $Q \to Q + \Delta Q$ while maintaining the condition of supersymmetry, $\Delta Q Q + Q \Delta Q = \Delta W \mathrm{id}_M$, then, the combination changes by a Q-exact term

$$\Delta\left[\mathrm{e}^{-S_W}\mathrm{tr}_M \mathrm{Pexp}\left(-\int_{\partial D^2} \mathcal{A}\right)\right] = (Q^+ + Q^-)(\cdots). \tag{4.14}$$

Finally, S_t depends on the FI-theta parameter and is supersymmetric by itself. It is the sum of terms which are holomorphic and anti-holomorphic in t, $S_t = S_t^{\rm hol} + S_t^{\rm antihol}$. For the $\mathrm{A}_{(+)}$-type (*resp.* $\mathrm{A}_{(-)}$-type) supersymmetry, the latter (*resp.* former) is Q-exact,

$$\begin{aligned} \mathrm{A}_{(+)} &: S_t^{\rm antihol} = Q^- Q^+(\cdots), \\ \mathrm{A}_{(-)} &: S_t^{\rm hol} = Q^- Q^+(\cdots). \end{aligned} \tag{4.15}$$

The partition function is defined to be the path-integral over the fields on the hemisphere

$$Z_{D^2} = \int \mathcal{D}\,[\text{fields}]\, \exp\left(-S\right) \mathrm{tr}_M \mathrm{Pexp}\left(-\int_{\partial D^2} \mathcal{A}\right). \tag{4.16}$$

Due to the exactness (4.14), the partition function is invariant under the deformation of the superpotential W and the matrix factorization Q that maintain the condition $Q^2 = W \mathrm{id}_M$. Also, for the system preseving the $\mathrm{A}_{(+)}$-type (*resp.* $\mathrm{A}_{(-)}$-type) supersymmetry, due to the exactness (4.15), the partition function is annihilated by the antiholomorphic (*resp.* holomorphic) derivative with respect to the FI-theta parameter t, that is, it depends holomorphically (it resp. antiholomorphically) on t.

The exactness of the kinetic terms (4.13) is relevant for the computation. This means that the result does not change when these terms are scaled up. In particular, the path integral localizes on the supersymmetric locus where the bosonic part of the kinetic terms vanish — the one-loop approximation around such locus yields the exact answer.

4.3. The result

Let us write down the result. Let

$$V = \bigoplus_i \mathbb{C}(R_i, Q_i) \tag{4.17}$$

be the weight decomposition of the matter representation V of $U(1)_V \times G$. We assume that the R-charges are chosen in the range

$$0 < R_i < 2. \tag{4.18}$$

The partition function of the system preserving the $\mathrm{A}_{(+)}$-type supersymmetry on the hemisphere with the B-brane (M, Q, γ) at the boundary is

$$\begin{aligned} Z_{D^2}^{\mathrm{A}_{(+)}}(M,Q,\gamma) = C(r\Lambda)^{\widehat{c}/2} \int_{\gamma'} \mathrm{d}^\ell \sigma' \\ \times \prod_{\alpha>0} \alpha(\sigma')\sinh\left(\pi\alpha(\sigma')\right) \prod_i \Gamma\Big(iQ_i(\sigma') + \frac{R_i}{2}\Big) \\ \times \exp\Big(it_R(\sigma')\Big)\, f_M(\sigma'), \end{aligned} \tag{4.19}$$

with

$$f_M(\sigma') := \mathrm{tr}_M \exp\Big(\pi i \mathbf{r} + 2\pi\sigma'\Big). \tag{4.20}$$

For the system preserving the $\mathrm{A}_{(-)}$-type supersymmetry on the hemisphere with the B_{-1}-brane $(M, Q, -\gamma)$ at the boundary (see (3.23) for the replacement $\gamma \to -\gamma$), it is its complex conjugate,

$$Z_{D^2}^{\mathrm{A}_{(-)}}(M, Q, -\gamma) = \Big(Z_{D^2}^{\mathrm{A}_{(+)}}(M, Q, \gamma)\Big)^*. \tag{4.21}$$

In the above expressions, r is the radius of the (hemi)sphere, Λ is the energy scale that is needed to renormalize the theory, and $\widehat{c} = \mathrm{tr}_V(1 - R) - \dim G$. The integration variable σ' takes values in $\mathfrak{t}_{\mathbb{C}}$ and comes from the Cartan zero mode of the field σ times the radius r. The contour $\gamma' \subset \mathfrak{t}_{\mathbb{C}}$ in (4.19) is γ times r, and $\mathrm{d}^\ell\sigma'$ is the flat holomorphic volume form

$$\mathrm{d}^\ell\sigma' = \mathrm{d}\sigma'_1 \wedge \cdots \wedge \mathrm{d}\sigma'_\ell. \tag{4.22}$$

The product $\prod_{\alpha>0}$ is over positive roots of the gauge group. t_R is the renormalized FI-theta parameter defined by

$$t_R = t - b_1 \log(r\Lambda). \tag{4.23}$$

Information of the D-brane enters into $f_M(\sigma')$ which we call the brane factor. Note that it depends only on M as the representation of $U(1)_V \times G$. The detail of the matrix factorization $Q(\phi)$ does not matter.

The above result is first derived for the case where γ is the real locus $i\mathfrak{t}$ and then for a deformed γ using holomorphy. The integrand has poles at the hyperplanes

$$iQ_i(\sigma') + \frac{R_i}{2} = 0, -1, -2, \ldots \tag{4.24}$$

which misses the real locus $i\mathfrak{t}$ under the condition (4.18). We propose that γ is acceptable only when (i) it can be continuously deformed to the real locus $i\mathfrak{t}$ without meeting the pole hyperplanes (4.24). Since the integral (4.19) is over a non-compact contour, whether it is convergent is a non-trivial question. The asymptotic behaviour of the integrand can be found easily using Stirling's formula for the Gamma function. For the irreducible decomposition $M = \oplus_j U_j$ with respect to the identity component G_0 of G, the brane factor $f_M(\sigma')$ decomposes into the sum $\sum_j \mathrm{tr}_{U_j} \mathrm{e}^{\pi i \mathbf{r} + 2\pi\sigma'}$. Stirling's formula says that the integrand for the term corresponding to the representation U_j of highest weight λ_j behaves as

$$\text{a power of } \sigma \times \exp\Big(-ir\widetilde{W}_{\mathrm{eff},\lambda_j}(\sigma)\Big), \tag{4.25}$$

with[2]
(4.26)
$$\widetilde{W}_{\mathrm{eff},\lambda_j}(\sigma) = -t(\sigma) + 2\pi i(\lambda_j + \rho)(\sigma) - \sum_i Q_i(\sigma)\left(\log\left(\frac{Q_i(\sigma)}{-i\Lambda}\right) - 1\right).$$

We see that the integrand may grow exponentially depending on the direction of $\mathfrak{t}_{\mathbb{C}}$. This leads us to another condition — (ii) the integrand must decay at infinity of γ. We shall call γ *admissible* for (M, Q) when both of the conditions (i) and (ii) are satisfied. We may also say that γ is admissible with respect to a representation U of G_0 when the decay condition holds if the brane factor is replaced by $\mathrm{tr}_U\, \mathrm{e}^{2\pi\sigma'}$.

We propose that *the classical data (M, Q, γ) defines a B-brane in the quantum theory when γ is admissible for (M, Q).*

The case of Landau-Ginzburg model

Let us also present the partition functions for the Landau-Ginzburg model of N variables $x_1, \ldots, x_N$ with superpotential $W(x_1, \ldots, x_N)$, on the hemisphere with $\mathrm{B}_{(\pm)}$-type supersymmetry. For the hemisphere with $\mathrm{B}_{(\pm)}$-type supersymmetry, we need to put an $\mathrm{A}_{\mp 1}$-type boundary condition at $|z| = 1$ which is specified by a Lagrangian submanifold $L_\pm \subset \mathbb{C}^N$ such that $\mp\mathrm{Im}(W)$ are bounded below on $L_\pm$. The partition function is

$$Z_{D^2}^{\mathrm{B}_{(+)}}(L_+) = (r\Lambda)^{\frac{N}{2}} \int_{L_+} \exp\Big(-irW(x)\Big)\mathrm{d}x_1 \wedge \cdots \wedge \mathrm{d}x_N, \tag{4.27}$$

$$Z_{D^2}^{\mathrm{B}_{(-)}}(L_-) = (r\Lambda)^{\frac{N}{2}} \int_{L_-} \exp\Big(-ir\overline{W(x)}\Big)\mathrm{d}\overline{x}_1 \wedge \cdots \wedge \mathrm{d}\overline{x}_N. \tag{4.28}$$

These are absolutely convergent or convergent oscillatory integrals.

If $W(x)$ is quasihomogeneous, $W(\lambda^R x) = \lambda^2 W(x)$ for some linear map $x \mapsto Rx$, and if the class of L_+ is invariant under $x \mapsto \lambda^R x$, we see from a change of variables that the partition function (4.27) depends on the radius r simply as an overall power

$$Z_{D^2}^{\mathrm{B}_{(+)}}(L_+) = (r\Lambda)^{\widehat{c}/2} \cdot Z_{D^2}^{\mathrm{B}_{(+)}}(L_+)|_{r\Lambda=1}, \tag{4.29}$$

---

[2]Each logarithm is required to have imaginary parts in the interval $(-\pi, \pi)$. Note that $\widetilde{W}_{\mathrm{eff},\lambda}(\sigma)$ looks similar to the effective twisted superpotential $\widetilde{W}_{\mathrm{eff}}(\sigma)$ on the Coulomb branch (3.7). There is a differnce though — While $\widetilde{W}_{\mathrm{eff}}(\sigma)$ is defined only modulo shifts by $2\pi i\mathrm{P}(\sigma)$, $\widetilde{W}_{\mathrm{eff},\lambda}(\sigma)$ is defined absolutely by the specific choice of branch of the logarithms. Nevertheless, there is a reason for the similarity, as will be explained momentarily.

where $\widehat{c} = \mathrm{tr}(1 - R)$ is one third of the central charge of the SCFT to which the LG model flows in the IR limit. The other function (4.28) has the same dependence. This power behaviour is expected to be a characteristic of the partition function of a superconformal field theory.

On the other hand, if $W(x)$ is a Morse function and if L_+ passes through one of the critical points, the partition function (4.27) in the large radius limit is dominated by the Gaussian integral near the critical point x_*,

$$Z_{D^2}^{\mathrm{B}(+)}(L_+) \sim C_* \exp\Big(-irW(x_*)\Big), \qquad r \to \infty. \tag{4.30}$$

The other function (4.28) has a similar behaviour. This exponential behaviour is expected to be a characteristic of the partition function of a theory with massive vacua.

Rotated supersymmetry

We may also consider the partition functions on the hemisphere with the rotated supersymmetry, $\mathrm{A}_{(\pm)}^{\mathrm{e}^{2i\beta}}$ or $\mathrm{B}_{(\pm)}^{\mathrm{e}^{2i\alpha}}$.

The effect of rotation is easiest to state in the LG model with B-type supersymmetry. It is simply to make the replacement

$$W(x) \to \mathrm{e}^{2i\alpha}W(x), \qquad \overline{W(x)} \to \mathrm{e}^{-2i\alpha}\overline{W(x)} \tag{4.31}$$

in the result, such as (4.27)–(4.28) and (4.30). Note that the brane $L_\pm$ must be an $\mathrm{A}_{\mp\,\mathrm{e}^{2i\alpha}}$-brane on which $\mp\mathrm{Im}(\mathrm{e}^{2i\alpha}W)$ is bounded below, so that the result remains to be an absolutely convergent or convergent oscillatory integral.

For GLSM with A-type supersymmetry, we may employ the classical axial R-symmetry $\mathrm{e}^{i\beta F_A}$. Then, the effect is simply the change of variables $X \to \mathrm{e}^{i\beta F_A}X$ and the rotation of the contour $\gamma \to \mathrm{e}^{-2i\beta}\gamma$, which does not seem to change anything: we may simply denote $r\,\mathrm{e}^{2i\beta}\sigma$ again by σ'. However, this change of variables is possibly anomalous, and will shift the theta angle as $\theta \to \theta + 2\beta\mathrm{tr}_V$. The precise effect is executed by the replacement

$$\Lambda \to \mathrm{e}^{2i\beta}\Lambda, \quad \sigma \to \mathrm{e}^{2i\beta}\sigma, \quad \gamma \to \mathrm{e}^{-2i\beta}\gamma, \tag{4.32}$$

with σ' and γ' unchanged, in all of the formula above for GLSM within this subsection. This in particular changes the asymptotic behaviour of the integrand (4.25) to

$$\text{a power of } \sigma \times \exp\left(-ir\,\mathrm{e}^{2i\beta}\widetilde{W}_{\mathrm{eff},\lambda_j}(\sigma)\right), \tag{4.33}$$

with the expression (4.26) for $\widetilde{W}_{\mathrm{eff},\lambda_j}(\sigma)$ unchanged.

Remarks

(i) Let us write down the expression for the partition functions on the sphere. Let $\mathrm{Q}^\vee \subset i\mathfrak{t}$ be the cocharacter lattice, that is, the set of elements that takes integer values on the weight lattice $\mathrm{P} \subset i\mathfrak{t}^*$. Then, the partition function of the GLSM on the sphere with the A-type supersymmetry is

$$\begin{aligned} Z^{\mathrm{A}}_{S^2} = C|r\Lambda|^{\widehat{c}} \sum_{m\in \mathrm{Q}^\vee} \int_{i\mathfrak{t}} \mathrm{d}^\ell \sigma' \\ \times \prod_{\alpha>0} \left(\frac{\alpha(m)^2}{4} + \alpha(\sigma')^2\right) \prod_i \frac{\Gamma\left(iQ_i(\sigma') - \frac{Q_i(m)}{2} + \frac{R_i}{2}\right)}{\Gamma\left(1 - iQ_i(\sigma') - \frac{Q_i(m)}{2} - \frac{R_i}{2}\right)} \\ \times \exp\Big(2i\zeta_R(\sigma') + i(\theta_R + 2\pi\rho)(m)\Big). \end{aligned} \tag{4.34}$$

Here $t_R = \zeta_R - i\theta_R$. The integrand grows as a power in magnitude but it rapidly oscillates along the contour $i\mathfrak{t} \subset \mathfrak{t}_{\mathbb{C}}$ of integration. Thus, it is convergent though not absolutely. The partition function of the LG model on the sphere with B-type supersymmetry is

$$Z^{\mathrm{B}}_{S^2} = |r\Lambda|^N \int_{\mathbb{C}^N} \exp\left(-ir(W(x) + \overline{W(x)})\right) \mathrm{d}^{2N}x. \tag{4.35}$$

This is also a convergent oscillatory integral.

(ii) The similarity between $\widetilde{W}_{\mathrm{eff},\lambda_j}(\sigma)$ in (4.26) and $\widetilde{W}_{\mathrm{eff}}(\sigma)$ in (3.7) is not a coincidence. If $\widetilde{W}_{\mathrm{eff}}(\sigma)$ has a non-degenerate critical point σ_*, there is a Coulomb vacuum with a mass gap. If γ passes through σ_*, the hemisphere partition function should behave in the large radius limit $r \to \infty$ as (4.30), that is, $\sim \exp\left(-ir\widetilde{W}_{\mathrm{eff}}(\sigma_*)\right)$. Note that the critical value is $\widetilde{W}_{\mathrm{eff}}(\sigma_*) = b_1(\sigma_*)$ and is not affected by the $2\pi i\mathrm{P}(\sigma)$ ambiguity of $\widetilde{W}_{\mathrm{eff}}(\sigma)$ — see the discussion around (3.9). This is indeed the behaviour of the integral (4.19) provided $\widetilde{W}_{\mathrm{eff},\lambda_j}(\sigma) \equiv \widetilde{W}_{\mathrm{eff}}(\sigma) \bmod 2\pi i\mathrm{P}(\sigma)$. This comparison yields the precise relationship between the two Λ's, the one in (3.7) for $\widetilde{W}_{\mathrm{eff}}(\sigma)$ and the one for the hemisphere partition function which appears in (4.26):

$$\Lambda|_{(3.7)} = -i\Lambda|_{(4.26)}. \tag{4.36}$$

(iii) Recall that there is an unphysical ambiguity (3.20) in the R-charge when $\mathfrak{g}$ has a non-zero center $\mathfrak{z}$. Let us see the effect of this change, $R \to R + \varepsilon$ and $\mathbf{r} \to \mathbf{r} - \varepsilon$, on the hemisphere partition function (4.19).

This affects the exponent $\widehat{c}$ of the prefactor as $\widehat{c} \to \widehat{c} - b_1(\varepsilon)$ as well as the Gamma function factors and the brane factor. The effect of these functions is absorbed by the change of variables $\sigma' \to \sigma' + \frac{i}{2}\varepsilon$. The net effect is an overall factor $\mathrm{e}^{-t(\varepsilon)/2}$ and the shift of the contour $\gamma' \to \gamma' - \frac{i}{2}\varepsilon$. As long as the bound (4.18) is not violated by the shift $R \to R + \varepsilon$, we can bring the contour back to γ' without hitting the poles. Thus, the effect is simply

$$Z_{D^2}^{\mathrm{A}(+)} \longrightarrow \mathrm{e}^{-\frac{1}{2}t(\varepsilon)} Z_{D^2}^{\mathrm{A}(+)}. \tag{4.37}$$

The effect on the other functions are $Z_{D^2}^{\mathrm{A}(-)} \to \mathrm{e}^{-\frac{1}{2}\bar{t}(\varepsilon)} Z_{D^2}^{\mathrm{A}(-)}$ and $Z_{S^2}^{\mathrm{A}} \to \mathrm{e}^{-\frac{1}{2}t(\varepsilon) - \frac{1}{2}\bar{t}(\varepsilon)} Z_{S^2}^{\mathrm{A}}$.

(iv) The expression (4.19) can be simplified a little. Let $M = \oplus_j U_j(r_j)$ be the irreducible decomposition with respect to the identity component G_0 of G. If U is the irreducible representation of G_0 with highest weight λ, we have the Weyl character formula

$$\mathrm{tr}_U\, \mathrm{e}^{2\pi\sigma'} = \frac{\sum_{w \in W_0} (-1)^{\ell(w)}\, \mathrm{e}^{2\pi w(\lambda+\rho)(\sigma')}}{\prod_{\alpha>0} \left(\mathrm{e}^{\pi\alpha(\sigma')} - \mathrm{e}^{-\pi\alpha(\sigma')}\right)}. \tag{4.38}$$

Notice that the denominator is nothing but the factor $\prod_{\alpha>0} \sinh(\pi\alpha(\sigma'))$ in (4.19) up to a factor of 2. Note also that $\prod_{\alpha>0} \alpha(\sigma')$ is Weyl odd. Using these as well as the Weyl invariance of γ, we see that the partition function can be written as

(4.39)

$$Z_{D^2}^{\mathrm{A}(+)}(M, Q, \gamma) = C \frac{|W_0|}{2^{|\Delta^+|}} (r\Lambda)^{\widehat{c}/2} \sum_j \mathrm{e}^{\pi i r_j} \int_{\gamma'} \mathrm{d}^\ell \sigma'$$
$$\times \prod_{\alpha>0} \alpha(\sigma') \prod_i \Gamma\left(iQ_i(\sigma') + \frac{R_i}{2}\right) \exp\Bigl(it_R(\sigma') + 2\pi(\lambda_j + \rho)(\sigma')\Bigr),$$

where $|\Delta^+|$ is the number of positive roots. In fact, this has been used in finding the behaviour (4.25).

(v) The gamma function has an integral expression

$$\Gamma(z) = \int_0^\infty \mathrm{e}^{-t} t^{z-1} \mathrm{d}t, \tag{4.40}$$

when the real part of z is positive. The same holds for other range of z provided the contour of integration is chosen appropriately. Applying this for each gamma function in (4.39), we find

$$Z_{D^2}^{\mathrm{A}_{(+)}}(M,Q,\gamma)=C\frac{|W_0|}{2^{|\Delta^+|}\Lambda^{\ell+|\Delta^+|}}(r\Lambda)^{\frac{\ell+d_V}{2}}\sum_j \mathrm{e}^{\pi i r_j} \times\int_\Gamma \mathrm{d}^\ell\sigma\,\mathrm{d}^{d_V}y\prod_{\alpha>0}\alpha(\sigma)\prod_i \mathrm{e}^{-\frac{R_i}{2}y_i}\exp\Big(-ir\widetilde{W}(\sigma,y)\Big), \tag{4.41}$$

with $d_V=\dim_{\mathbb{C}}V$ and

$$\widetilde{W}(\sigma,y)=\left(\sum_i Q_iy_i-t+2\pi i(\lambda_j+\rho)\right)(\sigma)-i\Lambda\sum_i \mathrm{e}^{-y_i}, \tag{4.42}$$

for a contour $\Gamma\subset\mathfrak{t}_{\mathbb{C}}\times\mathbb{C}^{d_V}$ that projects onto $\gamma\subset\mathfrak{t}_{\mathbb{C}}$. Compared to (4.27), the formula (4.41) looks to be the hemisphere partition function $Z_{D^2}^{\mathrm{B}}$ for an A-brane in the LG model with superpotential (4.42). In fact, (4.41) is nothing but the formula for the central charge of the B-brane in the mirror theory [32], except that the dependence on the brane is made more precise and that the mirror superpotential modulo $2\pi i\mathrm{P}(\sigma)$ is corrected by a shift by $2\pi i\rho(\sigma)$.

4.4. The example $\mathbf{T}_{N,d}^{U(1)}$

Let us look at the hemisphere partition function $Z_{D^2}=Z_{D^2}^{\mathrm{A}_{(+)}}$ in the example $\mathrm{T}_{N,d}^{U(1)}$. We assign R-charge $2-d\epsilon$ on p and ϵ on x_i's with $0<\epsilon<2/d$, so that the bound (4.18) is satisfied. Then, we have

$$Z_{D^2}(M,Q,\gamma)=(r\Lambda)^{\widehat{c}/2}\int_\gamma \mathrm{d}\sigma'\,\Gamma\left(-di\sigma'+1-\tfrac{d\epsilon}{2}\right)\Gamma\left(i\sigma'+\tfrac{\epsilon}{2}\right)^N \times\mathrm{e}^{it_R\sigma'}f_M(\sigma') \tag{4.43}$$

where $\widehat{c}=N-2-(N-d)\epsilon$, $t_R=t-(N-d)\log(r\Lambda)$, and

$$f_M(\sigma')=\mathrm{tr}_M\,\mathrm{e}^{\pi i\mathbf{r}+2\pi\sigma'}=\sum_j \mathrm{e}^{\pi i r_j+2\pi q_j\sigma'}, \tag{4.44}$$

for the $U(1)_V\times U(1)$ weight decomposition

$$M=\bigoplus_j \mathbb{C}(r_j,q_j). \tag{4.45}$$

The Gamma function factor of the integrand has order N poles at $\sigma'=i\left(n_x+\frac{\epsilon}{2}\right)$ with $n_x=0,1,2,\ldots$ and simple poles at $\sigma'=i\left(-\frac{n_p+1}{d}+\frac{\epsilon}{2}\right)$ with $n_p=0,1,2,\ldots$. We shall call them *x-poles* and *p-poles* respectively.

The x-poles are on the positive imaginary axis and the p-poles are on the negative imaginary axis. By Stirling's formula, the j-th term of the integrand behaves as $\mathrm{e}^{-A_{q_j}(\sigma')}$ up to a power factor, where

$$(4.46)\quad A_q(\sigma') = \mathrm{Im}(\sigma')\left(\zeta - d\log d + (N-d)\left(\log\left|\frac{\sigma'}{r\Lambda}\right| - 1\right)\right) + |\mathrm{Re}(\sigma')|\left(\frac{N+d}{2}\pi + (N-d)\arctan\left[\frac{\mathrm{Im}(\sigma')}{|\mathrm{Re}(\sigma')|}\right]\right) - \mathrm{Re}(\sigma')(\theta + 2\pi q).$$

The contour γ is admissible with repect to the charge q when A_q goes to positive infinity at the ends of γ. We see that the behaviour of A_q depends very much on whether $(N-d)$ is zero, positive or negative. Below, we shall discuss these cases separately.

Important rôles will be played by the brane data $\mathfrak{B}_1$ and $\mathfrak{B}_2$ introduced earlier. They have the following brane factors

$$(4.47)\quad f_{M_1}(\sigma') = 1 - \mathrm{e}^{-\pi i d\epsilon}\,\mathrm{e}^{2\pi d\sigma'},$$

$$(4.48)\quad f_{M_2}(\sigma') = \sum_{j=0}^{N}\binom{N}{j}\mathrm{e}^{\pi i j(1-\epsilon)}\,\mathrm{e}^{2\pi j\sigma'} = (1 - \mathrm{e}^{-\pi i\epsilon}\,\mathrm{e}^{2\pi\sigma'})^N.$$

We see that $f_{M_1}(\sigma')$ has simple zero at $\sigma' = i\left(\frac{n}{d} + \frac{\epsilon}{2}\right)$, while $f_{M_2}(\sigma')$ has N-th order zero at $\sigma' = i\left(n + \frac{\epsilon}{2}\right)$, both for $n \in \mathbb{Z}$. In particular, the integrand for $\mathfrak{B}_1$ has vanishing residue at the p-poles but has non-zero residues at the x-poles, while the integrand for $\mathfrak{B}_2$ has vanishing residue at the x-poles but has non-zero residues at some of the p-poles (four out of the five series). Such a property does not change under the shift $\mathfrak{B} \mapsto \mathfrak{B}(i,q)$ which does $f_M(\sigma') \mapsto f_M(\sigma')(-1)^i\,\mathrm{e}^{2\pi q\sigma'}$.

$\underline{d = N}$ Family of conformal field theories

Recall that the model with $d = N$ defines a family of SCFTs parameterized by $t = \zeta - i\theta$, with $\widehat{c} = N - 2$. In the presence of boundary, the 2π periodicity of θ is lost, and we should consider the unwrapped moduli space $\widetilde{\mathfrak{M}}_t = \mathbb{C} - \widetilde{\Delta}$, where $\widetilde{\Delta}$ is the set of singular points, $\zeta = N\log N$ and $\theta \in N\pi + 2\pi\mathbb{Z}$. $\zeta \gg 0$ is the geometric phase and $\zeta \ll 0$ is the Landau-Ginzburg orbifold phase. By the general principle of supersymmetry, we expect to have an equivalence $D^b_{\mathrm{Coh}}(X_f) \cong \mathrm{MF}_{\mathbb{Z}_N}(f)$ for each homotopy class of paths in $\widetilde{\mathfrak{M}}_t$ between the two phases, as well as an autoequivalence for each loop around a singular point.

The dependence of Z_{D^2} on the radius r is simply the overall power

$$(4.49)\quad Z_{D^2}(\mathfrak{B},\gamma) = (r\Lambda)^{\frac{N-2}{2}} \cdot Z_{D^2}(\mathfrak{B},\gamma)|_{r\Lambda=1}.$$

This is indeed the characteristic behaviour of the partition function for a superconformal field theory with $\widehat{c} = N - 2$. The growth function A_q also simplifies for $d = N$ to

(4.50) $$A_q(\sigma') = (\zeta - N\log N)\mathrm{Im}(\sigma') + N\pi|\mathrm{Re}(\sigma')| - (\theta + 2\pi q)\mathrm{Re}(\sigma').$$

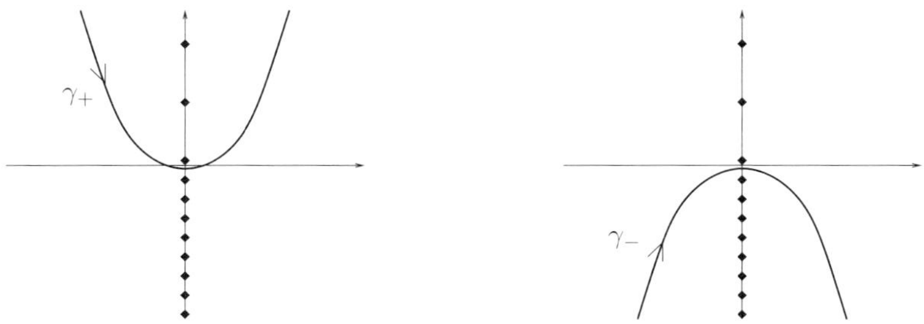

Fig. 2. Contours γ_+ and γ_-

In the geometric phase $\zeta \gg 0$, the function A_q for any q blows up linearly and the integrand decays expotentially in the direction where $\mathrm{Im}(\sigma')/|\mathrm{Re}(\sigma')|$ is large positive. Therefore, if we take $\gamma = \gamma_+$ as in Fig 2, it is admissible with respect to any q and hence for any (M, Q). The contour γ_+ can be deformed further up toward the positive imaginary direction, so that the integral becomes the sum of residues at the x-poles. This shows $Z_{D^2}(\mathfrak{B}_2(i,q), \gamma_+) = 0$ and $Z_{D^2}(\mathfrak{B}_1(i,q), \gamma_+) \neq 0$. This is consistent with the fact in the geometric regime that $\mathfrak{B}_2(i,q)$ is empty but $\mathfrak{B}_1(i,q)$ is not. In particular, the hemisphere partition function does not change under the brane replacement $\mathfrak{B} \rightsquigarrow \mathfrak{B}'$ introduced in Section 3.2, $Z_{D^2}(\mathfrak{B}, \gamma_+) = Z_{D^2}(\mathfrak{B}', \gamma_+)$, since $\mathfrak{B}'$ is obtained from $\mathfrak{B}$ by binding $\mathfrak{B}_2$ and its shifts and by cancelling identical pairs.

In the Landau-Ginzburg phase $\zeta \ll 0$, the function A_q for any q blows up linearly and the integrand decays expotentially in the direction where $\mathrm{Im}(\sigma')/|\mathrm{Re}(\sigma')|$ is large negative. Therefore, if we take $\gamma = \gamma_-$ as in Fig 2, it is admissible with respect to any q and hence for any (M, Q). The contour γ_- can be deformed further down toward the negative imaginary direction, so that the integral becomes the sum of residues at the p-poles. This shows $Z_{D^2}(\mathfrak{B}_1(i,q), \gamma_-) = 0$ and $Z_{D^2}(\mathfrak{B}_2(i,q), \gamma_-) \neq 0$. This is consistent with the fact in the Landau-Ginzburg regime that $\mathfrak{B}_1(i,q)$ is empty but $\mathfrak{B}_2(i,q)$ is not. In particular, the hemisphere partition function does not change under the brane replacement, $Z_{D^2}(\mathfrak{B}, \gamma_-) = Z_{D^2}(\mathfrak{B}', \gamma_-)$, since $\mathfrak{B}'$ is obtained from $\mathfrak{B}$ by binding $\mathfrak{B}_1$ and its shifts and by cancelling identical pairs.

On the line $\zeta = N\log N$, the dependence on $\mathrm{Im}(\sigma')$ disappears from (4.50),

$$A_q(\sigma') = N\pi|\mathrm{Re}(\sigma')| - (\theta + 2\pi q)\mathrm{Re}(\sigma'). \tag{4.51}$$

Note that $\theta+2\pi q$ never coinsides with $\pm N\pi$ unless t is one of the singular points. If $|\theta+2\pi q| > N\pi$, the function A_q goes down to negative infinity for either $\mathrm{Re}(\sigma') \to +\infty$ or $\mathrm{Re}(\sigma') \to -\infty$; No γ is admissible with respect to q. On the other hand, if $|\theta + 2\pi q| < N\pi$, it goes up to positive infinity for both $\mathrm{Re}(\sigma') \to \pm\infty$; Admissibles are the real locus $\gamma = \mathbb{R}$ as well as any of its deformation as long as it remains to extend to $\mathrm{Re}(\sigma') \to \pm\infty$. To summarize, at $\zeta = N\log N$, there is a unique homotopy class of admissible contours if and only if

$$-\frac{N}{2} < q + \frac{\theta}{2\pi} < \frac{N}{2}. \tag{4.52}$$

This yields the *grade restriction rule* concerning D-brane transport along paths in $\widetilde{\mathfrak{M}}_t$ from one phase to another. Such a path must go through a window on the interface $\zeta = N\log N$, $-N\pi + 2\pi n < \theta < -N\pi + 2(n+1)\pi$ for some $n \in \mathbb{Z}$. For each window $\mathbf{w}$, the bound (4.52) defines a set $[\mathbf{w}]$ of N consecutive integers. For example, if $\mathbf{w}$ is $(-N-2)\pi < \theta < -N\pi$ at $\zeta = N\log N$, the set is $[\mathbf{w}] = \{1, 2, \ldots, N\}$. If $\mathbf{w}'$ is the next one on the right, $-N\pi < \theta < (-N+2)\pi$, the set is $[\mathbf{w}'] = \{0, 1, 2, \ldots, N-1\}$. We shall say that a brane data $\mathfrak{B} = (M, Q)$ is *grade restricted with respect to* $\mathbf{w}$ when all the gauge charges of M belong to the set $[\mathbf{w}]$.

Suppose a brane data $\mathfrak{B}$ is grade restricted with respect to a window $\mathbf{w}$. Then, one can find a family of contours γ along a path through $\mathbf{w}$, interpolating γ_+ at $\zeta \gg 0$ and γ_- at $\zeta \ll 0$, so that it is admissible for $\mathfrak{B}$ all the way. This defines a family of quantum B-branes along the path. This is the rule of D-brane transport in the grade restricted case. Since the integral (4.43) is absolutely convergent all the way, $Z(\mathfrak{B}, \gamma_+)$ at $\zeta \gg 0$ and $Z(\mathfrak{B}, \gamma_-)$ at $\zeta \ll 0$ are related by the analytic continuation along the path.

Suppose, on the other hand, $\mathfrak{B}$ is not grade restricted with respect to $\mathbf{w}$. Then, a family of admissible contours does not exist along any path through $\mathbf{w}$. The brane $(\mathfrak{B}, \gamma_+)$ does make sense in the geometric phase $\zeta \gg 0$, but it cannot go to the LG phase $\zeta \ll 0$ through the window $\mathbf{w}$. In such a situation, we employ the idea of *brane replacement* introduced in Section 3.2. While in the geometric phase $\zeta \gg 0$, we can replace $\mathfrak{B}$ by another data $\mathfrak{B}'$ whose gauge charges belong to a set of N consecutive integers without changing the low energy behaviour

— without changing the hemisphere partition function $Z_{D^2}(\mathfrak{B},\gamma_+) = Z_{D^2}(\mathfrak{B}',\gamma_+)$ in particular. If we choose the set to be $[\mathbf{w}]$, then, $\mathfrak{B}'$ is grade restricted with respect to $\mathbf{w}$. Then, we can find a family of contours γ along a path through $\mathbf{w}$, starting from γ_+ in $\zeta \gg 0$ and ending with γ_- in $\zeta \ll 0$, which is admissible for $\mathfrak{B}'$ all the way. This defines a family of quantum B-branes along the path, starting with $(\mathfrak{B},\gamma_+) \cong (\mathfrak{B}',\gamma_+)$ in the geometric phase and ending with $(\mathfrak{B}',\gamma_-)$ in the LG phase. This is the rule of D-brane tranport. In particular, $Z_{D^2}(\mathfrak{B},\gamma_+)$ in $\zeta \gg 0$ analytically continues along the path to $Z_{D^2}(\mathfrak{B}',\gamma_-)$ in $\zeta \ll 0$. The same works for the transport backward. If we start from a brane $(\mathfrak{B},\gamma_-)$ in the LG phase $\zeta \ll 0$, we first find a grade restricted representative $(\mathfrak{B}'',\gamma_-)$ before the transport, and then go through the window $\mathbf{w}$. The analytic continuation of $Z_{D^2}(\mathfrak{B},\gamma_-)$ in $\zeta \ll 0$ ends with $Z_{D^2}(\mathfrak{B}'',\gamma_+)$ in $\zeta \gg 0$.

This rule allows us to find the monodromy. Let us consider a loop around one of the singular points, say, $t_* = N\log N + N\pi i$, with the base point in the geometric phase $\zeta \gg 0$. If the loop goes counter-clockwise, it may be regarded as a concatenation of a path from $\zeta \gg 0$ to $\zeta \ll 0$ through the window $\mathbf{w}$ on the left of t_* and a path from $\zeta \ll 0$ to $\zeta \gg 0$ through the window $\mathbf{w}'$ on the right of t_*. Let us start from a brane $(\mathfrak{B},\gamma_+)$ at $\zeta \gg 0$. While in the geometric phase, we replace $\mathfrak{B}$ by $\mathfrak{B}'$ which is grade restricted with respect to $\mathbf{w}$ and then move $(\mathfrak{B}',\gamma)$ through $\mathbf{w}$ to $\zeta \ll 0$. While in the LG phase $\zeta \ll 0$, we replace $\mathfrak{B}'$ by $\mathfrak{B}''$ which is grade restricted with respect to $\mathbf{w}'$, and then move $(\mathfrak{B}'',\gamma)$ through $\mathbf{w}'$ back to $\zeta \gg 0$. Thus, we end up with $(\mathfrak{B}'',\gamma_+)$.

By now it should be clear how to describe the (auto)equivalences of categories. Recall the definition of $\mathcal{T}_I \subset \mathfrak{D}_{\mathrm{LSM}}$ and the equivalences (3.34) and (3.35) in the two phases. The transport along a path through a window $\mathbf{w}$ results in the equivalence

$$D^b_{\mathrm{Coh}}(X_f) \xrightarrow{\pi_+^{-1}} \mathcal{T}_{[\mathbf{w}]} \xrightarrow{\pi_-} \mathrm{MF}_{\mathbb{Z}_N}(f). \tag{4.53}$$

The monodromy along the loop in the previous paragraph is

$$D^b_{\mathrm{Coh}}(X_f) \xrightarrow{\pi_+^{-1}} \mathcal{T}_{[\mathbf{w}]} \xrightarrow{\pi_-} \mathrm{MF}_{\mathbb{Z}_N}(f) \xrightarrow{\pi_-^{-1}} \mathcal{T}_{[\mathbf{w}']} \xrightarrow{\pi_+} D^b_{\mathrm{Coh}}(X_f). \tag{4.54}$$

$d < N$ Flow from sigma model

The model with $d < N$ describes an RG flow from the sigma mode with target X_f to the LG orbifold $W = f(x)/\mathbb{Z}_d$ or one of the $(N-d)$ massive vacua at

$$\sigma_k = -i\widetilde{\Lambda}\exp\left(i\,\frac{\theta+\pi d+2\pi k}{N-d}\right), \qquad k \in \mathbb{Z}/(N-d)\mathbb{Z}, \tag{4.55}$$

where $\widetilde{\Lambda} = \Lambda d^{\frac{d}{N-d}} \mathrm{e}^{-\frac{\zeta}{N-d}}$. Here Λ is the scale parameter for the hemisphere partition function which we assume to be real positive. See (3.17) and (4.36). The value of the twisted superpotential at these vacua are $\widetilde{W}_{\text{eff}}(\sigma_k) = (N-d)\sigma_k$. Note that the UV limit has $\widehat{c}_{\text{LV}} = N-2$ while the LG orbifold flows further to an SCFT with $\widehat{c}_{\text{LG}} = N(1-2/d)$. We expect that $D^b_{\text{Coh}}(X_f)$ is equivalent to a category that includes $\text{MF}_{\mathbb{Z}_d}(f)$ and $(N-d)$ exceptional objects.

Let us study the behaviour of the growth function (4.46). It can be written as

$$A_q(\sigma') = \zeta_{\text{eff}}\text{Im}(\sigma') + N_{\text{eff}}\pi|\text{Re}(\sigma')| - (\theta + 2\pi q)\text{Re}(\sigma'), \tag{4.56}$$

where

$$\begin{aligned} \zeta_{\text{eff}} &= (N-d)\left(\log\left|\frac{\sigma'}{r\widetilde{\Lambda}}\right| - 1\right), \\ N_{\text{eff}} &= \frac{N+d}{2} + \frac{N-d}{\pi}\arctan\left[\frac{\text{Im}(\sigma')}{|\text{Re}(\sigma')|}\right]. \end{aligned} \tag{4.57}$$

ζ_{eff} is positive outside the circle $|\sigma'| = r\widetilde{\Lambda}\,\mathrm{e}$ and negative inside. N_{eff} is bounded as $d < N_{\text{eff}} < N$ — the lower (*resp.* upper) bound is approached in the direction of the negative (*resp.* positive) imaginary axis.

The fact that ζ_{eff} is large positive for large enough $|\sigma'|$ and that N_{eff} is bounded as $d < N_{\text{eff}} < N$ means that γ_+ in Fig. 2 is admissible with respect to any q. Thus for any brane data $\mathfrak{B} = (M, Q)$, γ_+ is admissible and $(\mathfrak{B}, \gamma_+)$ defines a D-brane in the quantum theory. If we wish, we can deform γ_+ further up in the positive imaginary direction, and the integral (4.19) is written as the sum of residues at the x-poles. This shows

$$Z_{D^2}(\mathfrak{B}_2(i,q), \gamma_+) = 0. \tag{4.58}$$

This vanishing allows us to employ the brane replacement using $\mathfrak{B}_2$ and its shifts, so that the brane data $\mathfrak{B}$ can be taken from $\mathcal{T}_{I_+}$ for a set I_+ of N consecutive integers. In what follows, we study the behaviour of $Z_{D^2}(\mathfrak{B}, \gamma_+)$ in the small radius limit $r\Lambda \searrow 0$ (UV limit) as well as in the large radius limit $r\Lambda \nearrow \infty$ (IR limit).

When $r\Lambda \ll 1$, ζ_{eff} quickly becomes large positive as σ' goes away from the origin, and the above deformation of contour is the best choice, resulting in the sum of residues at the x-poles. Since t_R is large positive for $r\Lambda \ll 1$, the dominant is the residue at the first pole, $\sigma' = i\epsilon/2$, whose depenence on the radius is simply

$$Z_{D^2} \sim (r\Lambda)^{\widehat{c}/2}\,\mathrm{e}^{it_R\cdot i\epsilon/2} \sim (r\Lambda)^{\frac{N-2}{2}}. \tag{4.59}$$

This is indeed the characteristic behaviour for an SCFT with $\widehat{c}_{\mathrm{LV}} = N-2$. We will find the precise expression of the residue in Section 4.5 which will show that it is non-zero whenever the geometric image of the brane data $\mathfrak{B}$ is non-trivial.

When $r\Lambda \gg 1$, ζ_{eff} is negative over an extended region $|\sigma'| < r\widetilde{\Lambda}\,\mathrm{e}$ inside which A_q is positive (*resp.* negative) in a neighborhood of the negative (*resp.* positive) imaginary axis. Therefore, we may deform the contour γ_+ to $\widetilde{\gamma}_+$ as in Fig. 3. The new contour can be separated into two

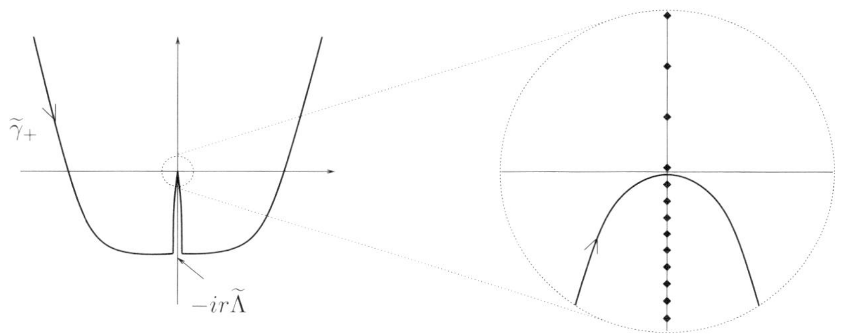

Fig. 3. Deformed contour $\widetilde{\gamma}_+$

parts, $\widetilde{\gamma}_+ = \widetilde{\gamma}_{\mathrm{cent}} + \widetilde{\gamma}_{\mathrm{rest}}$, where $\widetilde{\gamma}_{\mathrm{cent}}$ is the central part that encircles a large number of p-poles. Integration over $\widetilde{\gamma}_{\mathrm{cent}}$ yields the sum of residues at the p-poles. Since t_R is large negative for $r\Lambda \gg 1$, the dominant is the residue at the first pole, $\sigma' = i(-1/d + \epsilon/2)$, whose dependence on the radius is simply

$$Z_{D^2}|_{\mathrm{central}} \sim (r\Lambda)^{\widehat{c}/2}\,\mathrm{e}^{it_R\cdot i(-1/d+\epsilon/2)} \sim (r\Lambda)^{\frac{N(1-2/d)}{2}}. \tag{4.60}$$

This is the characteristic behaviour for a brane in an SCFT with $\widehat{c}_{\mathrm{LG}} = N(1-2/d)$. To see what the rest $\widetilde{\gamma}_{\mathrm{rest}}$ gives, let us examine whether the integrand has critical points. When $|\sigma'| \sim r\widetilde{\Lambda} \gg 1$ and $\mathrm{Re}(\sigma') \neq 0$, we can use the asymptotic behaviour (4.25) and look for the critical points of $\widetilde{W}_{\mathrm{eff},q}(\sigma)$. That is to find solutions to $\partial_\sigma \widetilde{W}_{\mathrm{eff},q} = 0$ literally, not modulo $2\pi i\mathbb{Z}$. It turns out that this equation is equivalent to

$$|\sigma'| = r\widetilde{\Lambda}, \qquad (\theta + 2\pi q)\mathrm{sgn}(\mathrm{Re}(\sigma')) = N_{\mathrm{eff}}\pi. \tag{4.61}$$

We see that there is a solution only when $d\pi < |\theta + 2\pi q| < N\pi$. In such a case, the solution is unique and is nothing but one of the Coulomb vacua in (4.55) — it is $\sigma_{k(q)}$ with $k(q) = q$ (*resp.* $q - d$) when $\theta + 2\pi q$ is in $(-N\pi, -d\pi)$ (*resp.* $(d\pi, N\pi)$). The contour $\widetilde{\gamma}_{\text{rest}}$ can be chosen to go through it, and the saddle point approximation to the integral gives

$$\text{const} \times \exp\left(-ir\widetilde{W}_{\text{eff},q}(\sigma_{k(q)})\right). \tag{4.62}$$

This is the characteristic behaviour for a brane supported at the massive vacuum $\sigma_{k(q)}$. When $|\theta + 2\pi q| < d\pi$, there is no critical point. In this case, we can choose $\widetilde{\gamma}_{\text{rest}}$ along which the integrand is very small entirely, and can show that the integral vanishes exponentially as $r\Lambda \to \infty$, much faster than any of $\mathrm{e}^{-ir\widetilde{W}_{\text{eff}}(\sigma_k)}$'s. When $|\theta+2\pi q| > N$, there is no solution again. In this case, one cannot avoid $\widetilde{\gamma}_{\text{rest}}$ to go through a region where the integrand is large, and it is hard to estimate the integral. However, we may avoid the case $|\theta+2\pi q| > N$ to begin with, via brane replacement using $\mathfrak{B}_2$ and its shifts. The special case $|\theta+2\pi q| = d\pi$ (*resp.* $N\pi$) arizes when $\theta \equiv d\pi$ (*resp.* $N\pi$) modulo $2\pi\mathbb{Z}$. Then, one of the Coulomb vacua σ_k is on the negative (*resp.* positive) imaginary axis where Stirling's formula breaks down. In order to avoid possible complications, we shall assume $\theta \not\equiv d\pi, N\pi$ modulo $2\pi\mathbb{Z}$.

Under this assumption, $[\theta; d]$ and $[\theta; N]$, with

$$[\theta; m] := \left\{ q \in \mathbb{Z} \,\middle|\, -\frac{m}{2} < q + \frac{\theta}{2\pi} < \frac{m}{2} \right\}, \tag{4.63}$$

are sets of d and N consecutive integers. The above analysis leads us to the following claim concerning the low energy behaviour of a brane $(\mathfrak{B}, \gamma_+)$. If $\mathfrak{B}$ belongs to $\mathcal{T}_{[\theta;d]}$, the brane flows purely to a brane in the LG orbifold. If $\mathfrak{B}$ belongs to $\mathcal{T}_{[\theta;N]}$ but not to $\mathcal{T}_{[\theta;d]}$, there is at least one gauge charge q such that $d\pi < |\theta + 2\pi q| < N\pi$. Let q_* be the one that maximizes $\mathrm{Im}\widetilde{W}_{\text{eff},q}(\sigma_{k(q)})$ among such q's. Then, the brane at low energies has a component supported at the Coulomb vacuum $\sigma_{k(q_*)}$.

Let us examine the case of $\mathfrak{B} = \mathfrak{B}_1(i, q)$ which reduces to the shifted line bundle $\mathcal{O}(q)[i]$ in the high energy sigma model. For any q, it does not belong to $\mathcal{T}_{[\theta;d]}$ since the set of gauge charges, which is $\{q, q + d\}$, cannot fit into $[\theta; d]$. It belongs to $\mathcal{T}_{[\theta;N]}$ when q is one of the $(N - d)$ integers satisfying $-N\pi < \theta + 2\pi q < (N - 2d)\pi$. In this case, q_* is q if $\theta + 2\pi q$ is in $(-N\pi, -d\pi)$ and $q + d$ if $\theta + 2\pi q$ is in $(-d\pi, (N - 2d)\pi)$. In either case, $k(q_*)$ is q and hence the brane has a component supported at σ_q. In fact, one can show that it is supported purely at σ_q. Therefore, for q in this range, the low energy limit of the line bundle $\mathcal{O}(q)$ on X_f is a brane supported purely at the Coulomb vacuum σ_q. When q is outside this range, we need to use its replacement in $\mathcal{T}_{[\theta;N]}$, obtained by binding $\mathfrak{B}_2$ and its shifts. Since $\mathfrak{B}_2$ and its shifts have non-zero residues at the p-poles, the low energy brane may have an SCFT component as well as components supported at Coulomb vacua. For example, let us consider $\mathfrak{B}_1$ in the theory with $\theta = -N\pi + \delta$ for a small negative δ, for which $[\theta; N] = \{1, 2, \ldots, N\}$. Obviously $\mathfrak{B}_1$ is not in $\mathcal{T}_{[\theta;N]}$ since its set of gauge charges is $\{0, d\}$. Its representative in $\mathcal{T}_{[\theta;N]}$ can be obtained by binding $\mathfrak{B}_1$ with $\mathfrak{B}_2(-1, 0)$,[3] which gives

(4.64)

$$\mathbb{C}(1,d) \underset{f'}{\overset{px}{\rightleftarrows}} \mathbb{C}(0,1)^N \underset{pf'}{\overset{x}{\rightleftarrows}} \mathbb{C}(1,2)^{\binom{N}{2}} \underset{pf'}{\overset{x}{\rightleftarrows}} \cdots \underset{pf'}{\overset{x}{\rightleftarrows}} \mathbb{C}(N-1,N)$$

where $x = \sum_i x_i\overline{\eta}_i$ and $f' = \sum_i \partial_i f \eta_i/d$. Note that the last component $\mathbb{C}(N-1, N)$ yields a brane supported at σ_0 which has the highest value of $\mathrm{Im}\widetilde{W}_{\mathrm{eff}}(\sigma_k)$. This shows that the low energy limit of the structure sheaf $\mathcal{O}$ of X_f includes the Recknagel-Shomerus brane $\pi_-(\mathfrak{B}_2(-1,0))$ of the SCFT and is also supported at some of the Coulomb vacua including σ_0.

What happens when θ crosses the special values, $d\pi$ and $N\pi$ modulo $2\pi\mathbb{Z}$? As remarked above, as θ crosses $d\pi$ (*resp.* $N\pi$) modulo $2\pi\mathbb{Z}$, one of the Coulomb vacua crosses the negative (*resp.* positive) imaginary axis on which there are p-poles (*resp.* x-poles). In fact, this is where there is a zero mode for p (*resp.* x) localized near the boundary and something special can happen. The example in the last paragraph can be used to illustrate what happens when θ crosses $N\pi$ (mod 2π), say, from $\theta > -N\pi$ to $\theta < -N\pi$, under which the vacuum σ_0 crosses the positive imaginary axis. Under this, $[\theta; N]$ changes from $\{0, 1, \ldots, N-1\}$ to $\{1, 2, \ldots, N\}$, and $\mathfrak{B}_1$ moves from inside $\mathcal{T}_{[\theta;N]}$ to the outside. Before the crossing, the brane $(\mathfrak{B}_1, \gamma_+)$ is supported purely at the vacuum σ_0 at low energies, but after the crossing, it flows to an SCFT brane as well as branes supported at some of the Coulomb vacua including σ_0. To see what happens when θ crosses $d\pi$ (mod 2π), let us look at a brane whose low energy limit is supported purely in the SCFT. To be specific, we consider a move from $\theta < -d\pi$ to $\theta > d\pi$, under which $[\theta; d]$ changes from $\{1, 2, \ldots, d\}$ to $\{0, 1, \ldots, d-1\}$. We are looking at a brane $(\mathfrak{B}, \gamma_+)$ with $\mathfrak{B} \in \mathcal{T}_{\{1,\ldots,d\}}$. Note that at each component of charge d, $\mathfrak{B}$ is of the form

$$\mathbb{C}(i,d) \underset{b}{\overset{pa}{\rightleftarrows}} \mathfrak{B}_{\mathrm{rest}}, \tag{4.65}$$

[3]We take the limit $\epsilon \searrow 0$ for a while, to simplify the expressions.

for some $a : \mathbb{C}(i-2,0) \to \mathfrak{B}_{\rm rest}$ such that $b \cdot a$ is $\mathbb{C}(i-2,0) \overset{f(x)\times}{\longrightarrow} \mathbb{C}(i,d)$. This can be expressed as

$$
\begin{array}{ccc}
\mathbb{C}(i-2,0) & \underset{pb}{\overset{a}{\rightleftarrows}} & \mathfrak{B}_{\rm rest} \\
{\scriptstyle -\mathrm{id}}\downarrow & & \downarrow{\scriptstyle b} \\
\mathbb{C}(i-1,0) & \underset{p}{\overset{f}{\rightleftarrows}} & \mathbb{C}(i,d)
\end{array}
\tag{4.66}
$$

We see that $\mathbb{C}(i,d)$ can be replaced by $\mathbb{C}(i-2,0)$ by forming a bound state with $\mathfrak{B}_1(i-1,0)$. Applying this for each component of charge d, $\mathfrak{B} \in \mathcal{T}_{\{1,\ldots,d\}}$ can be expressed as a bound state of $\mathfrak{B}' \in \mathcal{T}_{\{0,\ldots,d-1\}}$ and a number of copies of $\mathfrak{B}_1(j,0)$'s. Note that the LG images of $\mathfrak{B}$ and $\mathfrak{B}'$ are equivalent, $\pi_-(\mathfrak{B}) \cong \pi_-(\mathfrak{B}')$, and also that $\mathfrak{B}_1(j,0)$'s are supported purely at the Coulomb vacuum σ_0 that crosses the negative imaginary axis under the move. To summarize, as θ crosses $-d\pi$, a brane supported purely in the SCFT at low energies acquires components supported at the Coulomb vacuum that crossed the negative imaginary axis while the SCFT component remains the same. These effects of the move can be regarded as the "brane creation" as discussed in [33]. See Fig. 4.

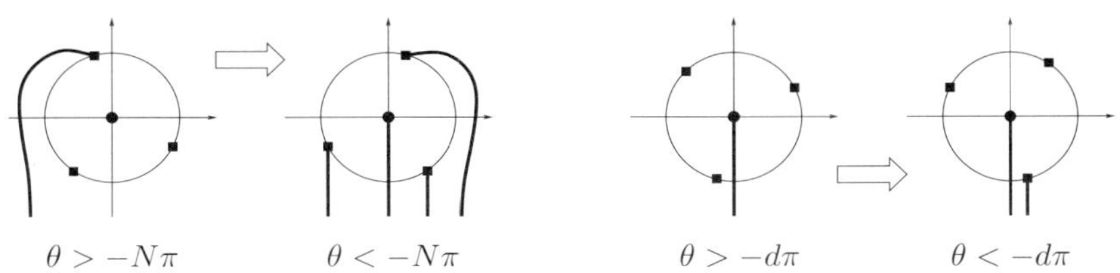

Fig. 4. Brane creations

Under the expected equivalence between $D^b_{\rm Coh}(X_f)$ and a category including $\mathrm{MF}_{\mathbb{Z}_d}(f)$ and $(N-d)$ exceptional objects, $\mathrm{MF}_{\mathbb{Z}_d}(f)$ is included in $D^b_{\rm Coh}(X_f)$ by

$$
\mathrm{MF}_{\mathbb{Z}_d}(f) \cong \mathcal{T}_{[\theta;d]} \hookrightarrow \mathcal{T}_{[\theta;N]} \cong D^b_{\rm Coh}(X_f). \tag{4.67}
$$

As θ crosses $d\pi$ (*resp.* $N\pi$) mod 2π, the category $\mathcal{T}_{[\theta;d]}$ (*resp.* $\mathcal{T}_{[\theta;N]}$) moves inside $\mathfrak{D}_{\rm LSM}$, and accordingly, the inclusion (4.67) changes. This change yields the brane creation discussed above.

$d > N$ Flow from Landau-Ginzburg model

The model with $d > N$ describes a deformation of an SCFT corresponding to the LG orbifold $W = f(x)/\mathbb{Z}_d$, which flows to the sigma model with target X_f or one of the $(d-N)$ massive vacua at

$$\sigma_k = i\widetilde{\Lambda}\exp\left(i\,\frac{\theta+\pi N+2\pi k}{N-d}\right), \qquad k\in\mathbb{Z}/(d-N)\mathbb{Z}, \tag{4.68}$$

where $\widetilde{\Lambda} = \Lambda d^{\frac{d}{N-d}}\,\mathrm{e}^{-\frac{\zeta}{N-d}}$, with $\widetilde{W}_{\rm eff}(\sigma_k) = (N-d)\sigma_k$. Note that the UV SCFT has $\widehat{c}_{\rm LG} = N(1-2/d)$ while the IR sigma model has $\widehat{c}_{\rm LV} = N-2$. We expect that $\mathrm{MF}_{\mathbb{Z}_d}(f)$ is equivalent to a category that includes $D^b_{\rm Coh}(X_f)$ and $(d-N)$ exceptional objects. The analysis of the model goes similarly to the case $d < N$ and hence we shall only highlight the main points.

For any brane data $\mathfrak{B} = (M, Q)$, γ_- as in Fig. 2 is admissible and $(\mathfrak{B}, \gamma_-)$ defines a D-brane in the quantum theory. The hemisphere partition function $Z_{D^2}(\mathfrak{B}, \gamma_-)$ may be written as the sum of residues at the p-poles. In particular, we have

$$Z_{D^2}(\mathfrak{B}_1(i,q), \gamma_-) = 0. \tag{4.69}$$

This vanishing allows us to assume $\mathfrak{B} \in \mathcal{T}_{I_-}$ for a set I_- of d consecutive integers.

When $r\Lambda \ll 1$, t_R is large negative and the series of the p-pole residues is dominated by the first term. Its radius dependence is simply

$$Z_{D^2} \sim (r\Lambda)^{\widehat{c}/2}\,\mathrm{e}^{it_R\cdot i(-1/d+\epsilon/2)} \sim (r\Lambda)^{\frac{N(1-2/d)}{2}}, \tag{4.70}$$

which is indeed the characteristic behaviour for an SCFT with $\widehat{c}_{\rm LG} = N(1-2/d)$. This leading term is non-vanishing whenever the LG image of $\mathfrak{B}$ is non-trivial.

When $r\Lambda \gg 1$, the integrand is very small (*resp.* very large) in a neighborhood of the positive (*resp.* negative) imaginary axis within $|\sigma'| < r\widetilde{\Lambda}\,\mathrm{e}$. Therefore, we may deform the contour γ_- to $\widetilde{\gamma}_-$ as in Fig. 5. The new contour can be separated into two parts, $\widetilde{\gamma}_- = \widetilde{\gamma}_{\rm cent} + \widetilde{\gamma}_{\rm rest}$, where $\widetilde{\gamma}_{\rm cent}$ is the central part that encircles a large number of x-poles. Integration over $\widetilde{\gamma}_{\rm cent}$ yields the sum of residues at the x-poles. Since t_R is large positive for $r\Lambda \gg 1$, the dominant is the residue at the first pole, $\sigma' = i\epsilon/2$, whose dependence on the radius is simply

$$Z_{D^2}|_{\rm central} \sim (r\Lambda)^{\widehat{c}/2}\,\mathrm{e}^{it_R\cdot i\epsilon/2} \sim (r\Lambda)^{\frac{N-2}{2}}. \tag{4.71}$$

This is the characteristic behaviour for a brane in an SCFT with $\widehat{c}_{\rm LV} = N-2$. The bahaviour of the contribution from $\widetilde{\gamma}_{\rm rest}$ depends on the range

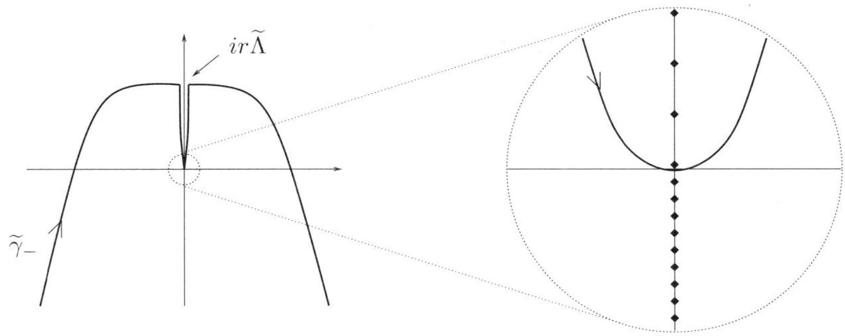

Fig. 5. Deformed contour $\widetilde{\gamma}_-$

of the gauge charge q. We assume that $\theta \not\equiv d\pi, N\pi$ modulo $2\pi\mathbb{Z}$ so that the Coulomb vacua σ_k are all away from the imaginary axis. When $N\pi < |\theta+2\pi q| < d\pi$, the integrand has a unique critical point at the Coulomb vacuum $\sigma_{k(q)}$ with $k(q) = q$ (*resp.* $q-N$) when $\theta+2\pi q$ is in $(-d\pi, -N\pi)$ (*resp.* $(N\pi, d\pi)$), and the integral behaves as $\sim \mathrm{e}^{-ir\widetilde{W}_{\mathrm{eff},q}(\sigma_{k(q)})}$ as $r \to \infty$. When $|\theta+2\pi q| < N\pi$, the integral on $\widetilde{\gamma}_{\mathrm{rest}}$ is much smaller than any of $\mathrm{e}^{-ir\widetilde{W}_{\mathrm{eff}}(\sigma_k)}$'s. The case $|\theta + 2\pi q| > N\pi$ where it is hard to estimate the integral can be avoided by a choice of $\mathfrak{B}$. From this analysis, we may make the following claim concerning the low energy behaviour of the brane $(\mathfrak{B}, \gamma_+)$: If $\mathfrak{B}$ belongs to $\mathcal{T}_{[\theta;N]}$, the brane flows purely to a brane in the sigma model on X_f. If $\mathfrak{B}$ belongs to $\mathcal{T}_{[\theta;d]}$ but not to $\mathcal{T}_{[\theta;N]}$, some of the gauge charges q are in $N\pi < |\theta + 2\pi q| < d\pi$. Let q_* be the one that maximizes $\mathrm{Im}\widetilde{W}_{\mathrm{eff},q}(\sigma_{k(q)})$ among such q's. Then, the brane at low energies has a component supported at the Coulomb vacuum $\sigma_{k(q_*)}$.

Let us examine the case of $\mathfrak{B} = \mathfrak{B}_2(i,q)$ which reduces to the Recknagel-Schomerus brane in the UV SCFT. For any q, it does not belong to $\mathcal{T}_{[\theta;N]}$ since the set of gauge charges $\{q, q+1, \ldots, q+N\}$ cannot fit into $[\theta; N]$. When $-d\pi < \theta+2\pi q < (d-2N)\pi$, it belongs to $\mathcal{T}_{[\theta;d]}$, and the brane $(\mathfrak{B}_2(i,q), \gamma_-)$ is supported purely at the vacuum σ_q at low energies. When q is outside this range, we need to use a replacement of in $\mathcal{T}_{[\theta;d]}$ to find the low energy behaviour of $(\mathfrak{B}_2(i,q), \gamma_-)$.

When θ crosses the special values, $d\pi$ and $N\pi$ modulo $2\pi\mathbb{Z}$, one (or two) of the Coulomb vacua crosses the imaginary axis. This is where there is a matter zero mode localized near the boundary, and something special can happen. When θ crosses $d\pi$ (mod 2π), a brane supported at the vacuum that crosses the negative imaginary axis will acuire a component supported at the sigma model on X_f. When θ croses $N\pi$

(mod 2π), a brane that flows to the sigma model brane will acquire a component supported at the vacuum that crosses the positive imaginary axis.

Under the expected equivalence between $\mathrm{MF}_{\mathbb{Z}_d}(f)$ and a category including $D^b_{\mathrm{Coh}}(X_f)$ and $(d-N)$ exceptional objects, $D^b_{\mathrm{Coh}}(X_f)$ is included in $\mathrm{MF}_{\mathbb{Z}_d}(f)$ by

$$\mathrm{MF}_{\mathbb{Z}_d}(f) \cong \mathcal{T}_{[\theta;d]} \hookleftarrow \mathcal{T}_{[\theta;N]} \cong D^b_{\mathrm{Coh}}(X_f). \tag{4.72}$$

As θ crosses $d\pi$ (*resp.* $N\pi$) mod 2π, the category $\mathcal{T}_{[\theta;d]}$ (*resp.* $\mathcal{T}_{[\theta;N]}$) moves inside $\mathfrak{D}_{\mathrm{LSM}}$, and accordingly, the inclusion (4.72) changes. This change yields the creation of new components as discussed above.

4.5. Geometric and LG expressions

In a regime where the gauge symmetry is broken to a finite subgroup, the GLSM reduces to the Higgs branch theory $(\mu^{-1}(\zeta)/G, W_\zeta)$. From the general result (4.19), we may attempt to extract the hemisphere partition function of such a Higgs branch theory. The result for a brane data $\mathfrak{B}$ should be expressed in terms of the image $\pi_\zeta(\mathfrak{B})$ in the reduced theory.

In this subsection, we present the partition function of the theory in the geometrc and the LG regimes of the model $\mathrm{T}^{U(1)}_{N,d}$. As is clear from the previous subsection, the one in the geometric regime is the sum of residues at the x-poles, while the one in the LG regime is the sum of residues at the p-poles. They may or may not capture the partition function of the full theory. In the Calabi-Yau case $d=N$, the geometric and the LG expressions are the limiting behaviour of the full partition function in the two phases. In the non-Calabi-Yau case $d\neq N$, one is the full partition function in the short distance limit $r\Lambda\to 0$, while the other is a part of the full partition function in the long distance limit $r\Lambda\to\infty$.

To simplify some of the expressions, in what follows, we shall take the limit $\epsilon\searrow 0$ in which we write $f^0_M(\sigma')$ for the brane factor. If we wish, we can always bring back the ganeral ϵ using (4.37).

Geometric regime

We first consider the hemisphere partition function in the geometric regime. The geometric image of a brane data $\mathfrak{B}=(M,Q)$ is $\pi_+(\mathfrak{B})$ given in (3.25), with $[B]\simeq(\theta+d\pi)H$. In particular, its Chern character is

$$\mathrm{ch}(\pi_+(\mathfrak{B})) = \sum_{n=0}^{\infty}\sum_j (-1)^{r^0_j}\,\mathrm{e}^{(q_j+dn)H} = \frac{f^0_M(\frac{1}{2\pi}H)}{1-\mathrm{e}^{dH}}. \tag{4.73}$$

Since the x-poles are at $\sigma' = in$ with $n = 0, 1, 2, \ldots$, the partition function in the geometric regime is

$$(4.74)\quad Z_{D^2}^{\rm LV}(\mathfrak{B}) = (r\Lambda)^{\frac{\hat{c}_{\rm LV}}{2}} \sum_{n=0}^{\infty} \oint_0 \frac{\mathrm{d}z}{2\pi} \Gamma\left(dn + \tfrac{dz}{2\pi i} + 1\right) \Gamma\left(-n - \tfrac{z}{2\pi i}\right)^N \times \mathrm{e}^{-t_R n + \frac{i}{2\pi} t_R z} f_M^0\left(in + \tfrac{z}{2\pi}\right).$$

This is a power series in e^{-t_R} which is a small parameter when $t_R \gg 1$. Note that $f_M^0(in + \frac{z}{2\pi}) = f_M^0(\frac{z}{2\pi})$ since $\mathrm{e}^{2\pi q_j(in)} = 1$. Using this and $\Gamma(x)\Gamma(1-x) = \pi/\sin(\pi x)$, it can be written as

(4.75)

$$\begin{aligned} Z_{D^2}^{\rm LV}(\mathfrak{B}) &= -C(r\Lambda)^{\frac{\hat{c}_{\rm LV}}{2}} \sum_{n=0}^{\infty} \oint_0 \frac{\mathrm{d}z}{2\pi i} \left(\frac{(-1)^n}{2\sinh\left(\frac{z}{2}\right)}\right)^N \frac{\Gamma\left(1 + \frac{dz}{2\pi i} + dn\right)}{\Gamma\left(1 + \frac{z}{2\pi i} + n\right)^N} \\ &\qquad \times \mathrm{e}^{-nt_R + \frac{i}{2\pi} t_R z} f_M^0\left(\tfrac{z}{2\pi}\right) \\ &= C(r\Lambda)^{\frac{\hat{c}_{\rm LV}}{2}} \sum_{n=0}^{\infty} \oint_0 \frac{\mathrm{d}z}{2\pi i} \frac{1}{z^N} \cdot d \cdot z \cdot \frac{z^{N-1} 2\sinh\left(\frac{dz}{2}\right)}{d\left(2\sinh\left(\frac{z}{2}\right)\right)^N} \frac{\Gamma\left(1 + \frac{dz}{2\pi i} + dn\right)}{\Gamma\left(1 + \frac{z}{2\pi i} + n\right)^N} \\ &\qquad \times (-1)^{Nn} \mathrm{e}^{-nt_R} \mathrm{e}^{\frac{i}{2\pi}(t_R - d\pi i)z} \cdot \frac{f_M^0\left(\frac{z}{2\pi}\right)}{1 - \mathrm{e}^{dz}}, \end{aligned}$$

with $C = -i(-2\pi i)^N$. Using the identity

$$(4.76)\qquad \oint_0 \frac{\mathrm{d}z}{2\pi i} \frac{1}{z^N} \cdot d \cdot z \cdot g(z) = \int_{\mathbb{P}^{N-1}} d \cdot H \cdot g(H) = \int_{X_f} g(H),$$

for a power series $g(z)$ in z, and the expression (4.73) for the Chern character of $\pi_+(\mathfrak{B})$, we find

$$(4.77)\quad Z_{D^2}^{\rm LV}(\mathfrak{B}) = C(r\Lambda)^{\frac{\hat{c}_{\rm LV}}{2}} \sum_{n=0}^{\infty} (-1)^{Nn} \mathrm{e}^{-nt_R} \int_{X_f} \widehat{\Gamma}_{X_f}(n) \times \exp\left(\frac{i}{2\pi}(t_R - d\pi i)H\right) \mathrm{ch}(\pi_+(\mathfrak{B})),$$

with

$$(4.78)\qquad \widehat{\Gamma}_{X_f}(n) := \frac{H^{N-1} 2\sinh\left(\frac{dH}{2}\right)}{d\left(2\sinh\left(\frac{H}{2}\right)\right)^N} \cdot \frac{\Gamma\left(1 + d\left(\frac{H}{2\pi i} + n\right)\right)}{\Gamma\left(1 + \frac{H}{2\pi i} + n\right)^N}.$$

Note that

$$(4.79)\qquad \widehat{\Gamma}_{X_f}(0) = \widehat{\mathrm{A}}_{X_f} \cdot \frac{\Gamma(1 + \frac{d}{2\pi i} H)}{\Gamma(1 + \frac{1}{2\pi i} H)^N} = \widehat{\mathrm{A}}_{X_f} \cdot \frac{1}{\widehat{\Gamma}_{X_f}^*} = \widehat{\Gamma}_{X_f},$$

where $\widehat{\mathrm{A}}_{X_f}$, $\widehat{\Gamma}_{X_f}$ and $\widehat{\Gamma}^*_{X_f}$ are the characteristic classes of the holomorphic tangent bundle of X_f obtained by inserting the roots of the total Chern class

$$c(X_f) = \frac{(1+H)^N}{(1+dH)} \tag{4.80}$$

into the functions $\widehat{\mathrm{A}}(x) = \frac{x/2}{\sinh(x/2)}$, $\widehat{\Gamma}(x) = \Gamma\left(1 - \frac{x}{2\pi i}\right)$ and $\widehat{\Gamma}^*(x) = \Gamma\left(1 + \frac{x}{2\pi i}\right)$. We said earlier that the complexified Kähler class is related to the FI-theta parameter by $[\omega - iB] = (t - d\pi i)H + O(\mathrm{e}^{-t})$, but this should be understood to be the relation between the renormalized parameters, that is, $[\omega_R - iB]$ versus t_R, where

$$\omega_R := \omega - \underbrace{c_1(X_f)}_{(N-d)H} \log(r\Lambda). \tag{4.81}$$

We see that the asymptotic behaviour at $\zeta_R \to \infty$ is

$$\begin{aligned} &Z^{\mathrm{LV}}_{D^2}(\mathfrak{B}) \\ &= C(r\Lambda)^{\frac{\hat{c}_{\mathrm{LV}}}{2}} \left[\int_{X_f} \widehat{\Gamma}_{X_f} \exp\left(\frac{1}{2\pi}(B + i\omega_R)\right) \mathrm{ch}(\pi_+(\mathfrak{B})) + O(\mathrm{e}^{-\omega_R + iB}) \right], \end{aligned} \tag{4.82}$$

where $O(\mathrm{e}^{-\omega_R+iB})$ stands for a correction of the form $\sum_{n=1}^{\infty} c_n\, \mathrm{e}^{-n\langle \omega_R - iB, l\rangle}$ where l is a second homology class of X_f such that $\langle H, l\rangle = 1$.

As seen in the previous subsection, in the case $d < N$ (*resp.* $d = N$), this is the expansion of the full partition function in the ultra-voilet limit $r\Lambda \to 0$ (*resp.* the large target volume limit $\zeta \to +\infty$), while in the case $d > N$, this is the expansion of a part of the full partition function in the infra-red limit $r\Lambda \to \infty$. Note the relation $\mathrm{e}^{-t_R} = \mathrm{e}^{-t}(r\Lambda)^{N-d}$. This difference can also be seen from the growth of the power series (4.77). The n-th term is roughly $\frac{\Gamma(1+dn)}{\Gamma(1+n)^N}\,\mathrm{e}^{-nt_R}$ times a constant. By Stirling's formula, it behaves for $n \gg 1$ as

$$\frac{\Gamma(1+dn)}{\Gamma(1+n)^N}\,\mathrm{e}^{-nt_R} \sim \mathrm{e}^{n(-t_R + N - d + d\log d)} n^{(d-N)n - \frac{\hat{c}_{\mathrm{LV}}+1}{2}} \left(1 + O\left(\tfrac{1}{n}\right)\right). \tag{4.83}$$

We see that the series is abolutely convergent for any e^{-t_R} if $d < N$ and for $|\,\mathrm{e}^{-t}| < N^{-N}$ if $d = N$, while it is divergent if $d > N$. In the case $d \leq N$, the series can be summed up to an analytic function of e^{-t_R} in an appropriate domain and may be used as a definition of the full partition function there. In the case $d > N$, on the other hand, one needs to resort to some summation technique to define an analytic function, but that usually involves choices. This reflects the fact that the sigma model is ultra-violet complete for $d \leq N$ but not for $d > N$. In the latter case, to say something about short distance $r\Lambda \ll 1$, the theory needs to be embedded into a ultra-violet complete theory, but things can depend on the choice of the completion. Of course, the GLSM provides one choice of UV completion, and the full partition function is the fully analytic completion of the divergent series (4.77).

Landau-Ginzburg regime

We next consider the hemisphere partition function in the LG regime. The LG image of a brane data $\mathfrak{B} = (M, Q)$ is $\pi_-(\mathfrak{B}) = (M_-, Q_-)$ given in (3.27). From the charge integrality of $\mathfrak{B}$ and $\pi_-(\mathfrak{B})$, we see that $(-1)^{\mathbf{r}^0}$ plays the rôle of the $\mathbb{Z}_2$-grading operator of M_-.

Since the p-poles are simple poles at $\sigma' = -i(n+1)/d$ with $n = 0, 1, 2, \ldots$, the partition function in the LG regime is

$$(4.84)\quad Z^{\rm LG}_{D^2}(\mathfrak{B}) = \frac{2\pi}{d}(r\Lambda)^{\frac{N-2}{2}} \sum_{n=0}^{\infty} \frac{(-1)^n}{n!}\Gamma\left(\tfrac{n+1}{d}\right)^N \mathrm{e}^{t_R\frac{n+1}{d}} f^0_M\left(-i\tfrac{n+1}{d}\right).$$

Note that $(r\Lambda)^{\frac{N-2}{2}}\,\mathrm{e}^{t_R/d} = \mathrm{e}^{t/d}(r\Lambda)^{\frac{N(1-2/d)}{2}} = \mathrm{e}^{t/d}(r\Lambda)^{\frac{\hat{c}_{\rm LG}}{2}}$ and

$$(4.85)\qquad f^0_M\left(-i\tfrac{n+1}{d}\right) = \mathrm{tr}_M\left(\mathrm{e}^{\pi i \mathbf{r}^0}\,\mathrm{e}^{-2\pi i\frac{n+1}{d}}\right) = \mathrm{Str}_{M_-}\left(\omega_1^{-n-1}\right),$$

where ω_1 is the generator $\mathrm{e}^{2\pi i/d}$ of $\mathbb{Z}_d$ and $\mathrm{Str}_{M_-}(?)$ is the supertrace $\mathrm{tr}_{M_-}((-1)^{\mathbf{r}^0}?)$. Thus, we find

$$(4.86)\quad Z^{\rm LG}_{D^2}(\mathfrak{B})$$
$$= \frac{2\pi}{d}\,\mathrm{e}^{t/d}(r\Lambda)^{\frac{\hat{c}_{\rm LG}}{2}} \sum_{n=0}^{\infty}\frac{(-1)^n}{n!}\,\mathrm{e}^{nt_R/d}\Gamma\left(\tfrac{n+1}{d}\right)^N \mathrm{Str}_{M_-}\left(\omega_1^{-n-1}\right)$$
$$= \frac{2\pi}{d}\,\mathrm{e}^{t/d}\Gamma\left(\tfrac{1}{d}\right)^N\cdot\underbrace{(r\Lambda)^{\frac{\hat{c}_{\rm LG}}{2}}\mathrm{Str}_{M_-}\left(\omega_1^{-1}\right)} + \mathrm{e}^{t/d}O(\mathrm{e}^{t_R/d}).$$

The underbraced factor of the leading term in the limit $t_R \to -\infty$ is indeed the hemisphere partition function for the brane $\pi_-(\mathfrak{B})$ of the LG orbifold $(\mathbb{C}^N/\mathbb{Z}_d, f)$, obtained by applying the result (4.19) to the theory with gauge group $\mathbb{Z}_d$, matter representation V and superpotential $f(x)$.

In the case $d > N$ (*resp.* $d = N$), this is the expansion of the full partition function in the ultra-voilet limit $r\Lambda \to 0$ (*resp.* the LG limit $\zeta \to -\infty$), while in the case $d < N$, this is the expansion of a part of the full partition function in the infra-red limit $r\Lambda \to \infty$. Note the relation $\mathrm{e}^{t_R/d} = \mathrm{e}^{t/d}(r\Lambda)^{1-N/d}$. This difference can also be seen from the growth of the power series (4.86). The n-th term behaves for $n \gg 1$ as

$$\frac{\mathrm{e}^{nt_R/d}}{n!}\Gamma\left(\tfrac{n+1}{d}\right)^N \sim \mathrm{e}^{n(t_R-N+d-N\log d)/d} n^{(N-d)n/d-\frac{\widehat{c}_{\mathrm{LG}}+1}{2}}\left(1+O\left(\tfrac{1}{n}\right)\right). \tag{4.87}$$

We see that the series is abolutely convergent for any $\mathrm{e}^{t_R/d}$ if $d > N$ and for $|\,\mathrm{e}^{t/N}| < N$ if $d = N$, while it is divergent if $d < N$. In the case $d \geq N$, the series can be summed up to an analytic function of $\mathrm{e}^{t_R/d}$ in an appropriate domain and may be used as a definition of the full partition function there. In the case $d < N$, on the other hand, one needs to resort to some summation technique to define an analytic function, but that usually involves choices. This reflects the fact that the theory under question is a deformation of the LG orbifold by a relevant, exactly marginal and irrelevant operator respectively for the case $d > N$, $d = N$ and $d < N$. In the latter case, to say something about short distances $r\Lambda \ll 1$, the theory needs to be embedded into a ultra-violet complete theory, but things can depend on the choice of the completion. Of course, the GLSM provides one choice of UV completion, and the full partition function is the fully analytic completion of the divergent series (4.86).

4.6. Differential Equations

As a function of the radius r and the FI-theta parameters t, the hemisphere partition function of the GLSM satisfies certain differential equations. We shall write down such equations for $Z_{D^2} = Z_{D^2}^{\mathrm{A}(+)}$.

Renormalization group eqaution

Let us first write down the renormalization group equation that shows how the partition function Z_{D^2} changes as the distance scale is varied. That is, we look at the responce to the differentiation with respect to the radius r of the hemisphere. The radius enters into the prefactor $(r\Lambda)^{\widehat{c}/2}$ and possibly also in the factor $\mathrm{e}^{it_R(\sigma')}$ of the integrand as $t_R(\sigma') = t(\sigma) - b_1(\sigma')\log(r\Lambda)$, where we recall $b_1 = \sum_i Q_i$ from (3.10). Noting $i\sigma'_a\,\mathrm{e}^{it_R(\sigma')} = \frac{\partial}{\partial t^a}\,\mathrm{e}^{it_R(\sigma')}$, we find

$$r\frac{\partial}{\partial r}Z_{D^2} = \left(\frac{\widehat{c}}{2} - b_1\left(\frac{\partial}{\partial t}\right)\right)Z_{D^2}, \tag{4.88}$$

where $b_1\left(\frac{\partial}{\partial t}\right) := \sum_a b_1^a \frac{\partial}{\partial t^a}$. The equation is invariant under the unphysical shift of R-charge, which does $\widehat{c} \to \widehat{c} - b_1(\varepsilon)$ and $Z_{D^2} \to \mathrm{e}^{-t(\varepsilon)/2}Z_{D^2}$.

For example, in the model $\mathrm{T}^{U(1)}_{N,d}$ (with $\epsilon \searrow 0$), the equation is

$$r\frac{\partial}{\partial r}Z_{D^2} = \left(\frac{N-2}{2} - (N-d)\frac{\partial}{\partial t}\right) Z_{D^2}. \tag{4.89}$$

Picard-Fuchs equations

The partition function Z_{D^2} satisfies another set of differential equations associated to relations of local operators. We may consider inserting an operator $\mathcal{O}$, say, at the center $z = 0$ of the hemisphere. The

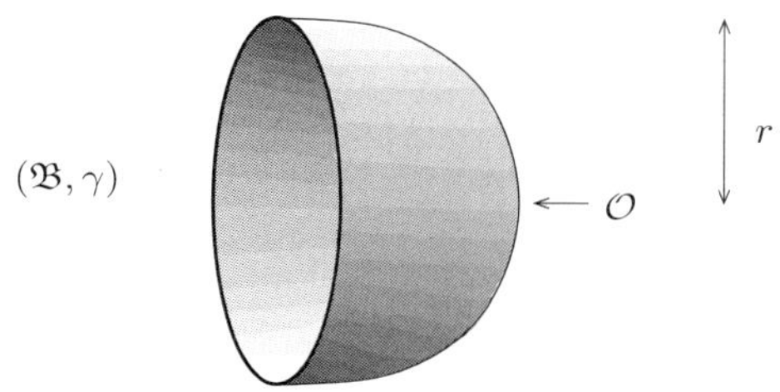

Fig. 6. The hemisphere with an insertion of an operator

supersymmetric localization still work if $\mathcal{O}$ is invariant under the $\mathrm{A}_{(+)}$-type supersymmetry, $Q^{A+}_{(+)}\mathcal{O} = Q^{A-}_{(+)}\mathcal{O} = 0$. For example, an adjoint invariant polynomial $\Phi(\sigma)$ of the scalar σ in the gauge multiplet, inserted at $z = 0$, satisfies this condition. With insertion of such an operator, the computation of the partition function goes through as before and the result is simply (4.19) with an insertion of $\Phi(\sigma'/r)$ in the integrand. If the operator is $Q^{A+}_{(+)}$ or $Q^{A-}_{(+)}$ exact, then the path-integral should vanish. In particular, if there is an operator relation at $z = 0$, $F(t; \Phi_1(\sigma), \Phi_2(\sigma), \ldots) \equiv 0$ modulo $Q^{A+}_{(+)}$ or $Q^{A-}_{(+)}$ exact operators, the partition function must vanish. Note that inserting a central component of σ' in the integrand is implemented by differentiation with respect to it. Therefore, for each relation $F(t; \sigma_{z_1}, \sigma_{z_2}, \ldots) \equiv 0$ among the central components $\sigma_{z_1}, \sigma_{z_2}, \ldots$, we have a differential equation

$$F\left(t; \frac{1}{ir}\frac{\partial}{\partial t^{z_1}}, \frac{1}{ir}\frac{\partial}{\partial t^{z_2}}, \ldots\right) Z_{D^2} = 0. \tag{4.90}$$

Note that $Q^{A+}_{(+)}$ and $Q^{A-}_{(+)}$ at $z = 0$ approach the supercharges $\overline{Q}_+$ and Q_- in the limit $r \to \infty$ where we get back the standard flat background. This implies that the operator relation $F(t; \Phi_1(\sigma), \Phi_2(\sigma), \ldots) \equiv 0$ approaches in this limit to a relation of the chiral ring $\mathcal{R}_A$. In other words, it should be a deformation of a chiral ring relation by terms that vanish as $r \to \infty$.

Let us illustrate this in the model $\mathrm{T}^{U(1)}_{N,d}$. The equation can be found by using the Gamma function identity

$$z\Gamma(z) = \Gamma(z+1), \tag{4.91}$$

and the property that the brane factor vanishes at $\sigma' = 0$,

$$f^0_M(0) = \mathrm{tr}_M(-1)^{\mathbf{r}_0} = \dim M^{\mathrm{ev}} - \dim M^{\mathrm{od}} = 0, \tag{4.92}$$

which follows from the matrix factorization equation $Q^2 = W\mathrm{id}_M$. Indeed, repeatedly using (4.91), we find

$$\begin{aligned}(i\sigma')^{N-1} \cdot \Gamma(-id\sigma'+1)\Gamma(i\sigma')^N &= -d\Gamma(-id\sigma')\Gamma(i\sigma'+1)^N \\ &\overset{\sigma'\to\sigma'+i}{\longrightarrow} -d\Gamma(-id\sigma'+d)\Gamma(i\sigma')^N \\ &= -d(-id\sigma'+d-1)\cdots(-id\sigma'+1)\cdot\Gamma(-id\sigma'+1)\Gamma(i\sigma')^N.\end{aligned} \tag{4.93}$$

Because of $f^0_M(0)=0$, the function $(i\sigma')^{N-1}\Gamma(-id\sigma'+1)\Gamma(i\sigma')^N\,\mathrm{e}^{it_R\sigma}f^0_M(\sigma')$ has no pole between γ' and $\gamma'+i$. Note also that $f^0_M(\sigma'+i) = f^0_M(\sigma')$. Therefore, for an admissible $(\mathfrak{B},\gamma)$ we have

$$\begin{aligned}0 &= \left[\int_{\gamma'} - \int_{\gamma'+i}\right] \mathrm{d}\sigma'(i\sigma')^{N-1}\cdot\Gamma(-id\sigma'+1)\Gamma(i\sigma')^N\,\mathrm{e}^{it_R\sigma}f^0_M(\sigma') \\ &\overset{(4.93)}{=} \int_{\gamma'} \mathrm{d}\sigma'\mathcal{D}(\sigma')\Gamma(-id\sigma'+1)\Gamma(i\sigma')^N\,\mathrm{e}^{it_R\sigma}f^0_M(\sigma')\end{aligned} \tag{4.94}$$

with

$$\mathcal{D}(\sigma') = (i\sigma')^{N-1} + d\,\mathrm{e}^{-t_R}(-id\sigma'+d-1)\cdots(-id\sigma'+1). \tag{4.95}$$

The admissibility is used to ensure that the contour $\gamma' - (\gamma'+i)$ can be closed at infinity. This shows that the partition function is annihilated by $\mathcal{D}(-i\frac{\partial}{\partial t})$, that is, it satisfies the differential equation

$$\left[\left(\frac{\partial}{\partial t}\right)^{N-1} - (-d)^d\,\mathrm{e}^{-t_R}\prod_{n=1}^{d-1}\left(\frac{\partial}{\partial t} - \frac{n}{d}\right)\right] Z_{D^2} = 0. \tag{4.96}$$

For $N = d = 5$, this is the famous Picard-Fuchs equation for the periods of the mirror quintic [2]. The corresponding operator relation must be $\mathcal{D}(r\sigma) = 0$. Using $\mathrm{e}^{-t_R} = \mathrm{e}^{-t}(r\Lambda)^{N-d}$, it reads

$$\sigma^{N-1} - (-d)^d\,\mathrm{e}^{-t}(-i\Lambda)^{N-d}\prod_{n=1}^{d-1}\left(\sigma + i\frac{n}{dr}\right) = 0. \tag{4.97}$$

In the limit $r \to \infty$, this becomes

$$\sigma^{N-1} - (-d)^d \, \mathrm{e}^{-t} (-i\Lambda)^{N-d} \sigma^{d-1} = 0. \tag{4.98}$$

This is indeed the chiral ring relation of the model that follows partially from $\partial_\sigma \widetilde{W}_{\text{eff}}(\sigma) \equiv 0 \bmod 2\pi i \mathbb{Z}$ for (3.16) with (4.36). See [34]. The relation can also be derived using the mirror description [32]. The relation in the sigma model for the case $d < N$ is proved mathematically in [35].

We may try to find the equation in other models, say, in a model with $U(1)$ gauge group. Let us put

$$\mathcal{D}_\pm(\sigma') := \prod_{\pm Q_i > 0} \prod_{n=0}^{|Q_i|-1} \left(iQ_i\sigma' + \frac{R_i}{2} + n \right). \tag{4.99}$$

Using (4.91), we find

$$\begin{aligned} &\mathcal{D}_\pm(\sigma') \prod_i \Gamma\left(iQ_i\sigma' + \frac{R_i}{2} \right) \\ &\quad = \prod_{\pm Q_i < 0} \Gamma\left(iQ_i\sigma' + \frac{R_i}{2} \right) \prod_{\pm Q_i > 0} \Gamma\left(iQ_i(\sigma' \mp i) + \frac{R_i}{2} \right), \end{aligned} \tag{4.100}$$

which means that

$$\mathcal{D}_+(\sigma') \prod_i \Gamma\left(iQ_i\sigma' + \frac{R_i}{2} \right) \overset{\sigma' \to \sigma' + i}{\longrightarrow} \mathcal{D}_-(\sigma') \prod_i \Gamma\left(iQ_i\sigma' + \frac{R_i}{2} \right). \tag{4.101}$$

Note that the left hand side of (4.101) has no pole between γ' and $\gamma' + i$ as long as the bound $0 < R_i < 2$ is satisfied. Therefore, for an admissible $(\mathfrak{B}, \gamma)$ we have

$$\begin{aligned} 0 &= \left[\int_{\gamma'} - \int_{\gamma'+i} \right] \mathrm{d}\sigma' \mathcal{D}_+(\sigma') \prod_i \Gamma\left(iQ_i\sigma' + \frac{R_i}{2} \right) \mathrm{e}^{it_R\sigma'} f_M(\sigma') \\ &= \int_{\gamma'} \mathrm{d}\sigma' \left(\mathcal{D}_+(\sigma') - \mathrm{e}^{-t_R} \mathcal{D}_-(\sigma') \right) \\ &\qquad \times \prod_i \Gamma\left(iQ_i\sigma' + \frac{R_i}{2} \right) \mathrm{e}^{it_R\sigma'} f_M(\sigma'), \end{aligned} \tag{4.102}$$

which yields the differential equation

(4.103)

$$\left[\prod_{Q_i>0}\prod_{n=0}^{Q_i-1}\left(Q_i\frac{\partial}{\partial t}+\frac{R_i}{2}+n\right)-\mathrm{e}^{-t_R}\prod_{Q_i<0}\prod_{n=0}^{|Q_i|-1}\left(Q_i\frac{\partial}{\partial t}+\frac{R_i}{2}+n\right)\right]Z_{D^2} = 0.$$

This holds in any model, with or without superpotential W. In the model $\mathrm{T}^{U(1)}_{N,d}$, this equation is one order higher than (4.96). In fact, it is nothing but $\frac{\partial}{\partial t}$(4.96). To get (4.96), we needed to use the property (4.92) of the brane factor. Similarly, if the model has a non-zero superpotential W, the brane factor $f_M(\sigma')$ has an extra property by which we may be able to find a lower order equation.

Let us show this in the model $\mathrm{T}^{U(1)}_{\vec{w},d}$ labelled by $\vec{w}=(w_1,\ldots,w_N)\in(\mathbb{Z}_{>0})^N$ and $d\in\mathbb{Z}_{>0}$. The model is similar to $\mathrm{T}^{U(1)}_{N,d}$, but is different just in that $\mathbb{C}(1)^{\oplus N}$ is replaced by $\oplus_{i=1}^N\mathbb{C}(w_i)$ and $f(x)$ is quasihomogeneous, $f(c^{w_1}x_1,\ldots,c^{w_N}x_N)=c^d f(x_1,\ldots,x_N)$. The R-charge is $R=(2-d\epsilon,w_1\epsilon,\ldots,w_N\epsilon)$, and we shall work again in the limit $\epsilon\searrow 0$. $\zeta\gg 0$ is the geometric phase with the target space being the hypersurface X_f in the weighted projective space $\mathbb{P}(w_1,\ldots,w_N)$, and $\zeta\ll 0$ is the phase where we have the LG orbifold $(\oplus_{i=1}^N\mathbb{C}(w_i)/\mathbb{Z}_d,f)$. We assume that X_f misses the orbifold loci. This requires that f contains a term $c_i x_i^{d_i}$ ($c_i\neq 0$) for each i with $w_i>1$, which means $d=d_iw_i$, and that w_i's are pairwise coprime. Then, for each i with $w_i>1$, for a brane data $\mathfrak{B}=(M,Q)$, if Q_i denotes Q with all x_j with $j\neq i$ set equal to zero, (M,Q_i) is a matrix factorization of $c_ipx_i^{d_i}$. By the gauge invariance, Q_i commutes with the gauge action of $g_i=\mathrm{e}^{2\pi i/w_i}$ on M and hence preserves each subspace $M_q^{(i)}\subset M$ of weight $q\in\mathbb{Z}/w_i\mathbb{Z}$. That is, the matrix factorization (M,Q_i) decomposes into the sum of $(M_q^{(i)},Q_i)$'s. Since the vanishing (4.92) holds for each component $(M_q^{(i)},Q_i)$, we find that, for any $l\in\mathbb{Z}$,

$$f_M^0(il/w_i)=\mathrm{tr}_M(-1)^{\mathbf{r}_0}g_i^l=\sum_{q\in\mathbb{Z}/w_i\mathbb{Z}}\mathrm{e}^{2\pi ilq/w_i}\mathrm{tr}_{M_q^{(i)}}(-1)^{\mathbf{r}_0}=0. \tag{4.104}$$

In the present case, $\mathcal{D}_\pm$ are

$$\mathcal{D}_+(\sigma')=\prod_{i=1}^N\prod_{n_i=0}^{w_i-1}(iw_i\sigma'+n_i),\quad \mathcal{D}_-(\sigma')=\prod_{n=0}^{d-1}(-id\sigma'+1+n). \tag{4.105}$$

Note that the factor $C(\sigma') = i(\sigma'+i)\prod_{w_i>1}\prod_{n_i=1}^{w_i-1}(iw_i(\sigma'+i)+n_i)$ of $\mathcal{D}_+(\sigma'+i)$ divides $\mathcal{D}_-(\sigma')$. Therefore, if we write $\mathcal{D}_+(\sigma'+i) = \widetilde{\mathcal{D}}_+(\sigma'+i)C(\sigma')$ and $\mathcal{D}_-(\sigma') = \widetilde{\mathcal{D}}_-(\sigma')C(\sigma')$, we find

(4.106)

$$\widetilde{\mathcal{D}}_+(\sigma')\Gamma(-id\sigma'+1)\prod_i \Gamma(iw_i\sigma') \overset{\sigma'\to\sigma'+i}{\longrightarrow} \widetilde{\mathcal{D}}_-(\sigma')\Gamma(-id\sigma'+1)\prod_i \Gamma(iw_i\sigma').$$

Between γ' and $\gamma'+i$, the left hand side of (4.106) has simple poles at $\sigma'=0$ and $\sigma'=il_i/w_i$ for $w_i>1$ and $l_i=1,\ldots,w_i-1$, but the brane factor $f_M^0(\sigma')$ vanishes at each of these points, thanks to (4.104). Therefore, for an admissible $(\mathfrak{B},\gamma)$ we have

(4.107)

$$\begin{aligned} 0 &= \left[\int_{\gamma'} - \int_{\gamma'+i}\right] \mathrm{d}\sigma' \widetilde{\mathcal{D}}_+(\sigma')\Gamma(-id\sigma'+1)\prod_i \Gamma(iw_i\sigma')\mathrm{e}^{it_R\sigma'} f_M^0(\sigma') \\ &= \int_{\gamma'} \mathrm{d}\sigma' \left(\widetilde{\mathcal{D}}_+(\sigma') - \mathrm{e}^{-t_R}\widetilde{\mathcal{D}}_-(\sigma')\right) \\ &\qquad\qquad \times \Gamma(-id\sigma'+1)\prod_i \Gamma(iw_i\sigma')\mathrm{e}^{it_R\sigma'} f_M^0(\sigma'), \end{aligned}$$

which yields $\left[\widetilde{D}_+(-i\frac{\partial}{\partial t}) - \mathrm{e}^{-t_R}\widetilde{\mathcal{D}}_-(-i\frac{\partial}{\partial t})\right] Z_{D^2} = 0$, that is,

$$\left[\left(\frac{\partial}{\partial t}\right)^{N-1} - \frac{(-d)^d}{\prod_i w_i^{w_i}}\mathrm{e}^{-t_R}\prod_{\substack{1\le n\le d-1\\ \frac{n}{d}\ne\frac{n_i}{w_i}}}\left(\frac{\partial}{\partial t} - \frac{n}{d}\right)\right] Z_{D^2} = 0. \tag{4.108}$$

The corresponding operator relation is

$$\sigma^{N-1} - \frac{(-d)^d}{\prod_i w_i^{w_i}}\mathrm{e}^{-t}(-i\Lambda)^{\sum_i w_i - d}\prod_{\substack{1\le n\le d-1\\ \frac{n}{d}\ne\frac{n_i}{w_i}}}\left(\sigma + i\frac{n}{dr}\right) = 0. \tag{4.109}$$

It reduces in the flat space limit $r\to\infty$ to the chiral ring relation.

For example, for the cases $(\vec{w};d) = (1,1,1,1,2;6), (1,1,1,1,4;8), (1,1,1,2,5;10)$, the above equations agree with the Picard-Fuchs equations for the periods in the mirror of the Calabi-Yau three-folds X_f with $h^{1,1}(X_f)=1$ [36–38]. The analysis can be extended to more general models. For example, if we apply it to models for Calabi-Yau hypersurfaces in toric varieties, we obtain the system of Picard-Fuchs equations as studied in, say, [39].

5. D-Brane Central Charges

In this section, we discuss what the hemisphere partition function is computing. It turns out that it is related to what is known as the central charge of the brane. The central charge (in this context) is originally defined in a particular situation in superstring theory, but is generalized to other areas of reserach. The hemisphere is related both to the original definition and to the generalizations. We will also find a surprising and suggestive relation to "macroscopic loop" in matrix models.

5.1. The Central Charge in 4d $\mathcal{N} = 2$ Compactifications

As is clear from the formulation and manifest in the resulting expression, the hemisphere partition function is invariant under continuous deformations of the brane. In this sense, it may be regarded as a "topological invariant" of the brane. In particle mechanics, the basic examples of invariant quantity of a particle are its electric and magnetic charges. In superstraing theory, D-branes are dynamical objects and are charged under Ramond-Ramond (RR) gauge potentials [40],[1] just like the electron is charged under the electro-magnetic gauge potential. Therefore, the first guess is that the hemisphere partition function is determined by the RR-charges of the brane. Indeed, we have seen in the model $\mathrm{T}_{N,d}^{U(1)}$ that the geometric expression depends on the brane $\mathcal{E} = \pi_+(\mathfrak{B})$ via its Chern character $\mathrm{ch}(\mathcal{E})$, while the LG expression depends on the brane $(M_-, Q_-) = \pi_-(\mathfrak{B})$ via $\mathrm{Str}_{M_-}(\omega)$, $\omega \in \mathbb{Z}_d \setminus \{1\}$. They are known to determine the RR charges of the branes in the sigma model [41, 42] and in the LG orbifold [43] respectively.

There are several RR potentials in general and hence each D-brane $\mathcal{B}$ has several RR charges. When we compactify Type II superstring theory on a three-dimensional Calabi-Yau manifold, we obtain an $\mathcal{N} = 2$ supersymmetric theory in the remaining four dimensions, and there is a distiguished linear combination of the RR charges, denoted by $Z(\mathcal{B})$ and called the *central charge*. The name is after the fact that it is the eigenvalue of the central element of the 4d $\mathcal{N} = 2$ supersymmetry algebra.[2] The central charge plays a crucial rôle in determining the stability of D-brane states, and motivated the mathematical study of the stability condition of the category of branes. For Type IIA (*resp.* Type IIB) string on a Calabi-Yau three-fold X, the relevant D-branes are B-branes (*resp.* A-branes) in the sigma model with target X. For

[1] See Appendix B for what "RR" means.

[2] This should not be confused with the central charge c (or $\widehat{c}$) which is the central part of the 2d (super)conformal symmetry algebra.

Type IIB, the central charge of the A-brane wrapped on a Lagrangian submanifold L is the period integral $Z(L) = \int_L \Omega$ of a holomorphic volume form Ω of X. The corresponding state is stable (or BPS) when L is a special Lagrangian submanifold of X so that $|Z(L)|$ is equal to the volume of L times $|\int_X \Omega \wedge \overline{\Omega}|^{1/2}$.

The theory $\mathrm{T}_{5,5}^{U(1)}$ yields a family of SCFTs on $\mathfrak{M}_A^0 = \{\mathrm{e}^t\} = \mathbb{C} \setminus \{(-5)^5\}$ — a compactification of (3.18) at the LG point $\mathrm{e}^t = 0$ — that can be used for a 4d $\mathcal{N} = 2$ compactification, say, of Type IIA string theory. By mirror symmetry, it is equivalent to Type IIB string theory on a family of Calabi-Yau three-folds Y_t, and a detailed study of the theory is performed by Candelas, de la Ossa, Green and Parkes in [2]. In particular, they computed the period integrals of Y_t, that is, the central charges of A-branes in Y_t. Later, using mirror symmetry, the results are rephrased as the formula for the central charge for B-branes in the quintic three-fold X_f [44]. Comparing these results with the results obtained in the previous section, we see that the central charge agrees with the hemisphere partition function,

$$Z_{D^2}(\mathcal{B}) = Z(\mathcal{B}) \quad \text{in } \mathrm{T}_{5,5}^{U(1)}, \tag{5.1}$$

up to the factor $(r\Lambda)^{3/2}$ of radius dependence. Indeed, the integral formula for the periods given in [2] matches exactly our integral formula for the hemisphere (4.19), more specifically (4.43) with $d = N = 5$. For example, see Eqn (3.14) in [2],

$$\varpi_0(\psi) = \frac{1}{2\pi i}\int_C \mathrm{d}s \frac{\Gamma(-s)\Gamma(5s+1)}{\Gamma(s+1)^4}\, \mathrm{e}^{\pi i s}(5\psi)^{-5s}, \tag{5.2}$$

where $0 < \mathrm{Arg}(\psi) < 2\pi/5$ and C is a contour parallel to the imaginary axis located in $-1/5 < \mathrm{Re}(s) < 0$, as depicted in Fig. 5 of [2]. The parameter ψ is related to our t via $\mathrm{e}^t = -(5\psi)^5$. Using $1/\Gamma(s+1) = 2i\sin(\pi s)\Gamma(-s)/(-2\pi i)$, after the change of integration variables $s = -i\sigma'$, we see that this integral is nothing but the integral (4.43) with $\epsilon \searrow 0$ and $f_M(\sigma') = (\mathrm{e}^{\pi\sigma'} - \mathrm{e}^{-\pi\sigma'})^4$ for $-\pi < \theta < \pi$, up to an overall constant and $(r\Lambda)^{3/2}$, where the contour γ' is parallel to the real axis located in $-1/5 < \mathrm{Im}(\sigma') < 0$. If this $f_M(\sigma')$ comes from a brane data $\mathfrak{B}$, its Chan-Paton charges would be from $\{0, \pm 1, \pm 2\}$. Thus, this would be grade restricted with respect to the window $-\pi < \theta < \pi$, and the contour is the same as the one we proposed for the hemisphere for any value of ζ. In the geometric phase, the corresponding Chern character would be

$$\text{“ch}(\pi_+(\mathfrak{B}))\text{”} = \frac{f_M(\frac{H}{2\pi})}{1 - \mathrm{e}^{5H}} = -\frac{1}{5}H^3 = \text{a generator of } \mathrm{H}^6(X_f, \mathbb{Z}), \tag{5.3}$$

which is the Chern character of the skyscraper sheaf at a point pf X_f (a D0-brane). Another decisive support of the relation (5.1) can be found in the geometric regime. The formula for the central charge for a general B-brane in the quintic X_f is extracted in [44] from [2] via mirror symmetry, and it agrees with the large volume expansion (4.77) for the hemisphere partition function, up to $(r\Lambda)^{3/2}$.

The relation (5.1) is supported also by the differential equation. It was found in [2] that the period integrals in Y_t satisfy the Picard-Fuchs equation, which is the same as the equation (4.96) with $d = N = 5$ satisfied by the hemisphere partition function, as noted earlier. This extends also to other models. Therefore, it is very plausible that the relation between the hemisphere and the central charge holds in a general model for 4d $\mathcal{N} = 2$ compactification. In fact, the relation can be extended to a wider range of models.

5.2. The Central Charge in 2d $(2, 2)$ Supersymmetric QFTs

From the viewpoint of the string worldsheet, the RR charges of a D-brane $\mathcal{B}$ are measured by taking the overlap of the boundary state of the D-brane ${}_{\rm RR}\langle\mathcal{B}|$ and the RR ground states. See Appendix B for the definition of boundary states. And the central charge is the overlap with a distingusihed ground state $|0\rangle_{\rm RR}$,

$$Z(\mathcal{B}) = {}_{\rm RR}\langle\mathcal{B}|0\rangle_{\rm RR}\,. \tag{5.4}$$

If $\mathcal{B}$ is a B-brane (*resp.* an A-brane), $|0\rangle_{\rm RR}$ is the state that corresponds to the identity operator under the A-type (*resp.* B-type) spectral flow. That is, the overlap (5.4) can be represented in path-integral as the partition function on the semi-infinite cigar, with the D-brane $\mathcal{B}$ as the boundary condition and A-twist (*resp.* B-twist) in the curved region. See Fig. 7. This formulation allows us to generalize the definition of

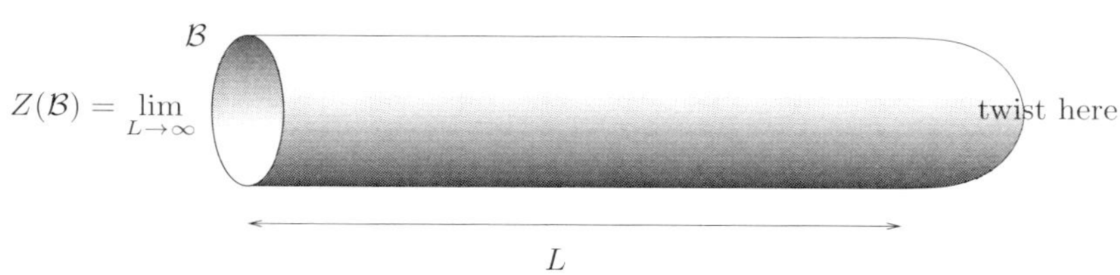

Fig. 7. The CV Central Charge

the central charge in much wider range of theories than the SCFTs for 4d $\mathcal{N} = 2$ string compactifications. The overlap (5.4) or the path-integral Fig. 7 makes sense for a B-brane (*resp.* A-brane) in a 2d (2, 2) supersymmetric QFT as long as the theory admits an A-twist (*resp.* a B-twist). This is so in SCFTs with an arbitrary $\widehat{c}$ and even in models which are not necessarily conformal. The only thing we need is A-twist (*resp.* B-twist). We shall call it the *CV central charge*, after the work by Cecotti and Vafa [45] who studied the inner products of RR ground states using worldsheets like Fig. 7. We may then ask whether the hemisphere partition function is equal to the CV central charge in general

$$Z_{D^2}(\mathcal{B}) \stackrel{?}{=} Z_{\rm CV}(\mathcal{B}). \tag{5.5}$$

The space of RR ground states form a vector bundle over the parameter space $\mathfrak{M} = \mathfrak{M}_A$ (*resp.* $\mathfrak{M}_B$), which we shall call the *vacuum bundle* $\mathcal{H}_{\rm v}$. To be specific, we consider states defined on the circle of circumference $\boldsymbol{\beta}$. This bundle is equipped with the hermitian inner product g induced from the one on the full space of states, and also with a holomorphic structure; the worldsheet like Fig. 7 with an insertion of a chiral operator ϕ at the tip defines a RR ground state $|\phi\rangle_{\rm RR}$, and if $\phi(t)$ is a holomorphic family of chiral operaors over $U \subset \mathfrak{M}$, then $|\phi(t)\rangle_{\rm RR}$ is a holomorphic section of $\mathcal{H}_{\rm v}$ over U. We denote the associated hermitian connection by D. For a local operator $\mathcal{O}$, we denote by $C_{\mathcal{O}}$ the operator on $\mathcal{H}_{\rm v}$ defined by multiplication by $\mathcal{O}$ followed by the projection to $\mathcal{H}_{\rm v}$. In [45], it was shown that these structures satisfies a system of equations called the tt^* equations, which amounts to the flatness of the connection ∇ of $\mathcal{H}_{\rm v}$ defined by

$$\nabla_{\phi} := D_{\phi} + \boldsymbol{\beta}\,{\rm e}^{iv} C_{\phi}, \qquad \nabla_{\overline{\phi}} := D_{\overline{\phi}} + \boldsymbol{\beta}\,{\rm e}^{-iv} C_{\overline{\phi}}, \tag{5.6}$$

for an arbitrary phase ${\rm e}^{iv}$. Here, ϕ and $\overline{\phi}$ are A-chiral and anti-A-chiral operators (*resp.* B-chiral and anti-B-chiral operators) and may be identified respectively as the holomorphic and anti-holomorphic tangent vectors of $\mathfrak{M} = \mathfrak{M}_A$ (*resp.* $\mathfrak{M}_B$) that correspond to the deformation of the (Minkowski) action by $-Q_+\overline{Q}_-\phi$ and $\overline{Q}_+Q_-\overline{\phi}$ (*resp.* $-Q_+Q_-\phi$ and $\overline{Q}_+\overline{Q}_-\overline{\phi}$).

Properties of $Z_{\rm CV}(\mathcal{B}) = {}_{\rm RR}\langle\mathcal{B}|0\rangle_{\rm RR}$, or more generally, of the overlap of the boundary state and all the RR ground states are studied in [47, 33]. It was found that ${}_{\rm RR}\langle\mathcal{B}|$ defines a parallel section of $\mathcal{H}_{\rm v}^*$ with respect to the connection ∇ for the phase ${\rm e}^{iv} = -i$.[3] In terms of local holomorphic

[3]More generally, the phase is ${\rm e}^{iv} = -i\,{\rm e}^{iu}$ if $\mathcal{B}$ is a ${\rm B}_{{\rm e}^{iu}}$-brane (*resp.* ${\rm A}_{{\rm e}^{iu}}$-brane).

coordinates t^α, this is rephrased as follows. Let ϕ_α be the chiral operator corresponding to $\frac{\partial}{\partial t^\alpha}$ and put $\Pi^{\mathcal{B}}_\alpha = {}_{\rm RR}\langle \mathcal{B}|\phi_\alpha\rangle_{\rm RR}$, which is obtained by the path-integral as in Fig. 7 with ϕ_α inserted at the tip of the cigar. Then,

$$(5.7)\qquad \frac{\partial}{\partial t^\alpha}\Pi^{\mathcal{B}}_\beta = A^\gamma_{\alpha\beta}\Pi^{\mathcal{B}}_\gamma - i\beta C^\gamma_{\alpha\beta}\Pi^{\mathcal{B}}_\gamma, \qquad \frac{\partial}{\partial \overline{t^\alpha}}\Pi^{\mathcal{B}}_\beta = i\beta C^\gamma_{\overline{\alpha}\beta}\Pi^{\mathcal{B}}_\gamma.$$

Here $A^\gamma_{\alpha\beta}$, $C^\gamma_{\alpha\beta}$ and $C^\gamma_{\overline{\alpha}\beta}$ are the representating matrices for D_{ϕ_α}, C_{ϕ_α} and $C_{\overline{\phi_\alpha}}$ with respect to the holomorphic frame $\{|\phi_\alpha\rangle_{\rm RR}\}$. $C^\gamma_{\alpha\beta}$ is the same as the structure function $\phi_\alpha\phi_\beta = \phi_\gamma C^\gamma_{\alpha\beta}$ of the chiral ring. Using the tt^* metric $g_{\overline{\alpha}\beta} := {}_{\rm RR}\langle\phi_\alpha|\phi_\beta\rangle_{\rm RR}$, the other two elements can also be written as $A^\gamma_{\alpha\beta} = g^{\gamma\overline{\lambda}}\partial_\alpha g_{\overline{\lambda}\beta}$ and $C^\gamma_{\overline{\alpha}\beta} = (g_{\overline{\beta}\nu}C^\nu_{\alpha\lambda}g^{\lambda\overline{\gamma}})^*$. It was also found in [47, 33] that $\Pi^{\mathcal{B}}_\alpha$'s are independent of the B-chiral (*resp.* A-chiral) parameters.

When we consider a family of superconformal field theories over a subspace of $\mathfrak{M}^0_A$ (*resp.* $\mathfrak{M}^0_B$), it follows from (5.7) that the CV central charge $Z_{\rm CV}(\mathcal{B}) = \Pi^{\mathcal{B}}_0$ depends holomorphically on the parameters $(t^i)^\ell_{i=1}$ of the family and that it obeys holomorphic differential equations which take the form of deformation of the chiral ring relation among the corresponding marginal operators $\{\phi_i\}^\ell_{i=1}$.

This can be shown as folows. Recall that the vector (*resp.* axial) $U(1)$ R-symmetry with charge integrality is used for the twist. A superconformal field theory also have an axial (*resp.* vector) $U(1)$ R-symmetry, which we shall call R'-symmetry for the moment. The chiral ring $\mathcal{R}_A$ (*resp.* $\mathcal{R}_B$) is graded by the R'-charge. The identity is the unique operator of the minimum R'-charge zero, and there is a unique operator of the maximal R'-charge $2\widehat{c}$. The R'-charge of the state $|\phi_\alpha\rangle_{\rm RR}$ is equal to the R'-charge of the operator ϕ_α minus $\widehat{c}$. This yields the selection rule: $g_{\overline{\alpha}\beta} = 0$ unless ϕ_α and ϕ_β have the same R'-charge, and in particular $g_{\overline{0}\alpha} = \delta_{\alpha,0}g_{\overline{0}0}$. Therefore, $C^\gamma_{\overline{i}0} = (g_{\overline{0}\nu}C^\nu_{i\lambda}g^{\lambda\overline{\gamma}})^* = (g_{\overline{0}0}C^0_{i\lambda}g^{\lambda\overline{\gamma}})^*$. By the R'-grading of the chiral ring, the product of a marginal operator ϕ_i with any operator cannot include the identity component, $C^0_{i\lambda} = 0$ for any λ. This shows $C^\gamma_{\overline{i}0} = 0$ for any γ and hence the second equations of (5.7) read $\partial_{\overline{i}}\Pi^{\mathcal{B}}_0 = 0$. That is, $\Pi^{\mathcal{B}}_0$ is a holomorphic function of the marginal parameters t^i. Next, let us look at the first equations in (5.7) for the components $\Pi^{\mathcal{B}}_\nu$ where ν ranges over the basis of the subspace of the chiral ring generated by the marginal operators under question, including the identity. Let us align the components in the order of increasing R'-charge, starting from $\Pi^{\mathcal{B}}_0$, and put them in a column vector $\vec{\Pi}$. Then, the equations takes the form

$$(5.8)\qquad \partial_i\vec{\Pi} = (A_i - i\beta C_i)^T\vec{\Pi},$$

where A_i are R'-block diagonal and non-holomorphic matrices and C_i are lower-triangular, holomorphic, and mutually commutative matrices. By definition, all the components of $\vec{\Pi}$ can be written as linear combinations of holomorphic derivatives of the first component $\Pi_0^{\mathcal{B}}$, where the coefficients are not necessarily holomorphic functions. Suppose there is a chiral ring relation $R(t;\phi_1,\ldots,\phi_\ell)=0$ which we may assume homogeneous. It implies a matrix identity $R(t;C_1,\ldots,C_\ell)=0$ or its transposed version. Then, using (5.8), we find an equation of the form $R\left(t;\frac{\partial}{\partial t^1},\ldots,\frac{\partial}{\partial t^\ell}\right)\vec{\Pi}=M\vec{\Pi}$, for some matrix M which is the sum of products of A_i^T's, C_i^T's and their derivatives. Expressing $\vec{\Pi}$ on the right hand side in terms of the derivatives of Π_0 and taking the first component, we obtain a differential equation

$$\left[R\left(t;\frac{\partial}{\partial t^1},\ldots,\frac{\partial}{\partial t^\ell}\right)+\delta D_R\right]\Pi_0^{\mathcal{B}}=0, \tag{5.9}$$

for some differential operator δD_R of lower order than $R\left(t;\frac{\partial}{\partial t^1},\ldots,\frac{\partial}{\partial t^\ell}\right)$. The differentials in δD_R are holomorphic ones by construction, but the coefficients are not a priori holomorphic. However, the flatness of ∇ guarantees that they must be holomorphic.

This argument holds even when the theories are not exactly conformal — the only requirement is the existence of R'-symmetry with the properties quoted at the begining of the previous paragraph. For example, the sigma model with Calabi-Yau target X over the moduli space of complexified Kähler class (*resp.* complex structure), the GLSM obeying the Calabi-Yau condition over $\mathfrak{M}_t$ (*resp.* $\mathfrak{M}_W$), and LG model with a family of quasi-homogenious superpotentials. All these models are expected to flow in the infra-red limit to a family of superconformal field theories over the same parameter space. In the present discussion, we shall call such theories "conformal" and the parameters "marginal" by a slight abuse of language.

In the conformal case, we have seen that the CV central charge $Z_{\rm CV}(\mathcal{B})$ has all the properties that the hemisphere partition function $Z_{D^2}(\mathcal{B})$ has, concerning the dependence on the marginal parameters: no dependence on the B-chiral (*resp.* A-chiral) parameters, holomorphic dependence on the A-chiral (*resp.* B-chiral) parameters, and a differential equation for each relation among marginal A-chiral (*resp.* B-chiral) operators. This is a strong support for the relation (5.5).

Another support for (5.5) can be obtained through the sphere partition function Z_{S^2}. Right after Z_{S^2} for GLSM was computed in [4, 5], it was observed in [6] in some Calabi-Yau examples including $T_{5,5}^{U(1)}$ that it is related to the Kähler potential K for the Zamolodchikov metric on

$\mathfrak{M}_t \subset \mathfrak{M}_A^0$ via

$$Z_{S^2} = \mathrm{e}^{-K}. \tag{5.10}$$

Later, this was shown to hold in general [7, 8]. In [45], it was shown that the Kähler potential K on $\mathfrak{M}_A^0$ (*resp.* $\mathfrak{M}_B^0$) is related to the RR-ground state $|0\rangle_{\mathrm{RR}}$ obtained via A-twist (*resp.* B-twist) by

$$\mathrm{e}^{-K} = {}_{\mathrm{RR}}\langle 0|0\rangle_{\mathrm{RR}}. \tag{5.11}$$

Combining (5.10) and (5.11) we have

$$Z_{S^2} = {}_{\mathrm{RR}}\langle 0|0\rangle_{\mathrm{RR}}. \tag{5.12}$$

This was observed also in the Landau-Ginzburg model with quasihomogeneous superpotential: The inner product ${}_{\mathrm{RR}}\langle 0|0\rangle_{\mathrm{RR}}$ is determined from the tt^* equation in [45], and it was found to be given by the same integral as in (4.35) for Z_{S^2}. The relation (5.12) results in an important consequence concerning (5.5). Note that, for any pair of basis $\{|i\rangle_{\mathrm{RR}}\}$ and $\{|j'\rangle_{\mathrm{RR}}\}$ of the space $\mathcal{H}_{\mathrm{v}}^0$ of RR ground states with vector (*resp* axial) R-charge zero[4] we have

$$ {}_{\mathrm{RR}}\langle 0|0\rangle_{\mathrm{RR}} = \sum_{i,j'} {}_{\mathrm{RR}}\langle 0|j'\rangle_{\mathrm{RR}}\, g^{j'\bar{\imath}}\, {}_{\mathrm{RR}}\langle i|0\rangle_{\mathrm{RR}}, \tag{5.13}$$

where $g^{j'\bar{\imath}}$ is the inverse matrix to $g_{\bar{\imath}j'} = {}_{\mathrm{RR}}\langle i|j'\rangle_{\mathrm{RR}}$. If we can take basis of $\mathcal{H}_{\mathrm{v}}^0$ corresponding to sets of D-branes, this can be regarded as a "bilinear identity" expressing the sphere partition function as a sum of products of the hemisphere partition functions, in view of the fact (5.12) and supposing that the conjectural relation (5.5) is indeed true. Indeed, such a relation seems to hold. Let $\{\mathcal{B}_i\}$ be a set of B-branes (*resp.* A_--branes) such that the ground state projections of their RR boundary states, $\{P_{\mathrm{v}}|\mathcal{B}_i\rangle_{\mathrm{RR}}\}$, form a basis of $\mathcal{H}_{\mathrm{v}}^0$. This is the case when $\mathcal{B}_i$'s represent a basis of the Grothendieck group of the category of B-branes (*resp.* A_--branes). Let $\{\mathcal{B}'_i\}$ be a set of B_--branes (*resp.* A-branes) with the same properties. Then, we claim the factorization formula

$$Z_{S^2} \stackrel{!}{=} \sum_{i,j} Z_{D^2}^{(-)}(\mathcal{B}'_j)\cdot \mathrm{I}^{\mathcal{B}_i,\overline{\mathcal{B}}'_j}\cdot Z_{D^2}^{(+)}(\mathcal{B}_i), \tag{5.14}$$

[4]Note that the state $|0\rangle_{\mathrm{RR}}$ obtained by the A-twist (*resp.* B-twist) as well as the boundary state of a brane preserving the vector (*resp.* axial) R-symmetry have vanishing vector (*resp.* axial) R-charge.

where $\mathrm{I}^{\mathcal{B}_i,\overline{\mathcal{B}}'_j}$ is the inverse matrix of

$$\mathrm{I}_{\overline{\mathcal{B}}'_j,\mathcal{B}_i} := \mathrm{Tr}_{\mathcal{H}_{\overline{\mathcal{B}}'_j,\mathcal{B}_i}}(-1)^F\,\mathrm{e}^{-\beta H}, \tag{5.15}$$

which is the Witten index for the open string with the boundary conditions $\overline{\mathcal{B}}'_j$ on the left and the boundary condition $\mathcal{B}_i$ on the right. See Appendix B for the definition of the conjugation $\mathcal{B} \mapsto \overline{\mathcal{B}}$ of boundary conditions. Here $Z^{(\pm)}_{D^2}$ is the partition function on the hemisphere with $\mathrm{A}_{(\pm)}$-type (*resp.* $\mathrm{B}_{(\pm)}$-type) supersymmetry. (Recall that we had been looking at $Z_{D^2} = Z^{(+)}_{D^2}$ for a while.) Note that the conjecture (5.5) is $Z^{(+)}_{D^2}(\mathcal{B}_i) = {}_{\mathrm{RR}}\langle \mathcal{B}_i|0\rangle_{\mathrm{RR}}$ and extends to $Z^{(-)}_{D^2}(\mathcal{B}'_i) = {}_{\mathrm{RR}}\langle \mathcal{B}'_i|\overline{0}\rangle_{\mathrm{RR}}$, where ${}_{\mathrm{RR}}\langle \mathcal{B}'_i|\overline{0}\rangle_{\mathrm{RR}}$ is obtained from the path-integral on the worldsheet as Fig. 7 but with *anti*-topological twist in the curved region. By rotation, we can simply identify ${}_{\mathrm{RR}}\langle \mathcal{B}'_i|\overline{0}\rangle_{\mathrm{RR}}$ with ${}_{\mathrm{RR}}\langle 0|\mathcal{B}'_i\rangle_{\mathrm{RR}}$. Note also that the open string Witten index is given by $\mathrm{I}_{\overline{\mathcal{B}}'_j,\mathcal{B}_i} = {}_{\mathrm{RR}}\langle \mathcal{B}_i|P_{\mathrm{v}}|\mathcal{B}'_j\rangle_{\mathrm{RR}}$. Therefore, if the conjecture is true, the factorization formula (5.14) is nothing but the identity (5.13) and hence must hold. In [9], the formula was indeed shown to hold for GLSM in the geometric regime where we can use the Riemann-Roch formula for the index, $\mathrm{I}_{\overline{\mathcal{E}}_j,\mathcal{E}_i} = \int_X \mathrm{ch}(\mathcal{E}_j^*)\mathrm{ch}(\mathcal{E}_i)\widehat{\mathrm{A}}_X$. The formula also holds in the LG model due to Poincaré-Lifshetz duality: Let $B_\pm \subset \mathbb{C}^n$ be the subset defined by $\pm\mathrm{Im}W > R$ for a large positive R. Then, $\mathrm{H}^n(\mathbb{C}^n, B_\pm)$ is the group of RR-charges of $\mathrm{A}_\pm$-branes. Let us choose a basis of these groups, $\{\gamma_i\} \subset \mathrm{H}^n(\mathbb{C}^n, B_-)$ and $\{\gamma'_i\} \subset \mathrm{H}^n(\mathbb{C}^n, B_+)$. Then, the intersection matrix $\mathrm{I}_{\overline{\jmath},i} = \#(\gamma'_j \cap \gamma_i)$ has an inverse $\mathrm{I}^{i,\overline{\jmath}}$ and the folowing identity holds

$$\int_{\mathbb{C}^n} \mathrm{e}^{-ir(W+\overline{W})}\overline{\Omega}\wedge\Omega = \sum_{i,j}\int_{\gamma'_j}\mathrm{e}^{-ir\overline{W}}\overline{\Omega}\cdot \mathrm{I}^{i,\overline{\jmath}}\cdot\int_{\gamma_i}\mathrm{e}^{-irW}\Omega, \tag{5.16}$$

for a holomorphic volume form Ω such as $\mathrm{d}x_1\wedge\cdots\wedge\mathrm{d}x_n$. The intersection number $\#(\gamma'_j\cap\gamma_i)$ is identified as the Witten index for the open string between $\overline{\gamma}'_j$ and γ_i, and hence the identity (5.16) is nothing but (5.14).

The perfect similarity bewteen Z_{D^2} and Z_{CV} goes away if we consider non-conformal theoies, such as a non-Calabi-Yau GLSM or a LG model with a non-quasi-homogeneous superpotential. While they are both independent of B-chiral (*resp.* A-chiral) parameters, they differ in the dependence on the A-chiral (*resp.* B-chiral) parameters: $Z_{D^2}(\mathcal{B})$ is holomorphic but $Z_{\mathrm{CV}}(\mathcal{B})$ is not — the proof requires R'-symmetry which is absent in a non-conformal theory. However, it is true that $Z_{D^2}(\mathcal{B})$ depends on the brane $\mathcal{B}$ only through its RR charge. Therefore, we may still say $Z_{D^2}(\mathcal{B}) = {}_{\mathrm{RR}}\langle\mathcal{B}|\Omega\rangle_{\mathrm{RR}}$ for some RR ground state $|\Omega\rangle_{\mathrm{RR}}$. On the other hand, it was shown in [48] that the sphere partition function is invariant under a one parameter family of deformation of the background, and the limit of the deformation can be interpreted as the inner product of a pair of RR ground states, $Z_{S^2} = {}_{\rm RR}\langle\Omega'|\Omega''\rangle_{\rm RR}$. If the ground states $|\Omega\rangle_{\rm RR}$, $|\Omega'\rangle_{\rm RR}$, etc that appear in the above relations are related to each other, then, via the identity like (5.13), we should again have a factorization formula for the sphere partition function. Indeed, the formula (5.14) was shown in [9] to hold in GLSM with a geometric regime with Fano targets space as well, such as $\mathrm{T}^{U(1)}_{N,d}$ with $d < N$. Also, the identity (5.16) holds whether or not W is quasihomogeneous. The formula suggests a more precise version of the relations: There are families of RR ground states $|\Omega^{e^{iu}}_{(\pm)}\rangle_{\rm RR} = |\Omega, \mathrm{A}^{e^{iu}}_{(\pm)}\rangle_{\rm RR}$ (*resp.* $|\Omega, \mathrm{B}^{e^{iu}}_{(\pm)}\rangle_{\rm RR}$) parametrized by $e^{iu} \in U(1)$, such that the partition function on the sphere with the A-type (*resp.* B-type) supersymmetry is given by

$$Z_{S^2} = {}_{\rm RR}\langle\Omega^{-1}_{(+)}|\Omega^{1}_{(+)}\rangle_{\rm RR} = {}_{\rm RR}\langle\Omega^{-1}_{(-)}|\Omega^{1}_{(-)}\rangle_{\rm RR}, \tag{5.17}$$

and the partition function on the hemisphere with the $\mathrm{A}_{(\pm)}$-type (*resp.* $\mathrm{B}_{(\pm)}$-type) supersymmetry and a $\mathrm{B}_{\pm 1}$-brane (*resp.* $\mathrm{A}_{\mp 1}$-brane) $\mathcal{B}_\pm$ at the boundary is given by

$$Z^{(+)}_{D^2}(\mathcal{B}_+) = {}_{\rm RR}\langle\mathcal{B}_+|\Omega^{1}_{(+)}\rangle_{\rm RR} = {}_{\rm RR}\langle\Omega^{-1}_{(-)}|\mathcal{B}_+\rangle_{\rm RR}, \tag{5.18}$$

$$Z^{(-)}_{D^2}(\mathcal{B}_-) = {}_{\rm RR}\langle\mathcal{B}_-|\Omega^{1}_{(-)}\rangle_{\rm RR} = {}_{\rm RR}\langle\Omega^{-1}_{(+)}|\mathcal{B}_-\rangle_{\rm RR}. \tag{5.19}$$

For the sphere or hemisphere with $\mathrm{A}^{e^{iu}}$ or $\mathrm{A}^{e^{iu}}_{(\pm)}$-type (*resp.* $\mathrm{B}^{e^{iu}}$ or $\mathrm{B}^{e^{iu}}_{(\pm)}$-type) supersymmetry, we replace $\Omega^{\pm 1}_{(\pm)}$ by $\Omega^{\pm e^{iu}}_{(\pm)}$ in the above expressions.

5.3. The Central Charge in Topological String Theory

The hemisphere partition function can also be related to a natural observable in topological field/string theory. A central element in this relation is what is known as Dubrovin's connection of a vector bundle on the space $\mathfrak{M} \times \mathbb{C}^*$, where $\mathfrak{M}$ is the space of supersymmetry preserving parameters of the theory, $\mathfrak{M}_A$ or $\mathfrak{M}_B$, and $\mathbb{C}^*$ is the space of an additional parameter. It turns out that this additional parameter is related to the radius r of the hemisphere. We will also find a relation to what is known as *macroscopic loops* in the theory of 2d quantum gravity. In this connection, the radius r is related to the length ℓ of the loop.

In the literature, several different symbols are used for Dubrovin's parameter. For convenience, we list below these symbols and the parameter z used in this section as well as the corresponding parameters in the 2d gravity and the hemisphere:

Reference	A	B	C	D	E	this section	2d gravity	hemisphere
Notation	δ	$\hbar^{-1}$	z	z^{-1}	u^{-1}	$-z^{-1}$	$-\ell$	$-r\Lambda$

The references are: (A) Kyoji Saito's original papers [15–17]; (B) Givental's earlier paper [49]; (C) Dubrovin's paper [51]; (D) Givental's later paper [52] and Iritani [53]; (E) Katzarkov et al [54].

Frobenius Manifold

Recall that a 2d $(2,2)$ supersymmetric field theory with a vector (*resp.* an axial) $U(1)$ R-symmetry with charge integrality admits an A-twist (*resp.* a B-twist). Local operators of the resulting topological field theory are elements of the chiral ring $\mathcal{R} = \mathcal{R}_A$ (*resp.* $\mathcal{R}_B$), and the sphere three point functions $\langle \mathcal{O}_1\mathcal{O}_2\mathcal{O}_3\rangle_{S^2}$ define a structure of *Frobenius algebra* on $\mathcal{R}$: it has a symmetric non-degenerate bilinear form $\eta(\mathcal{O}_1,\mathcal{O}_2) = \langle \,\mathrm{id}\; \mathcal{O}_1\mathcal{O}_2\rangle_{S^2}$ under which the product is self-adjoint, $\eta((\mathcal{O}_1\cdot\mathcal{O}_2),\mathcal{O}_3) = \eta(\mathcal{O}_1,(\mathcal{O}_2\cdot\mathcal{O}_3))$. Symmetry and self-adjointness follow from the properties of topological correlation functions, and non-degeneracy can be shown, say, by the spectral flow to RR ground states. Recall that elements of $\mathcal{R}$ are naturally identified as tangent vectors of type $(1,0)$ of the deformation space $\mathfrak{M} = \mathfrak{M}_A$ (*resp.* $\mathfrak{M}_B$), and hence we have an isomorphism $T_{\mathfrak{M}} \cong \mathcal{R}$ of holomorphic vector bundles on $\mathfrak{M}$. In particular, each fiber of $T_{\mathfrak{M}}$ is equipped with a structure of Frobenuius algebra. It varies holomorphically on $\mathfrak{M}$ and the bilinear form η can be regarded as a holomorphic metric on $\mathfrak{M}$, called the *topological metric.*

In [56], Dijkgraaf, Verline and Verlinde studied properties of the sphere three point functions of the family of topological field theories obtained by twisting a 2d $(2,2)$ superconformal field theory of central charge $c = 3\,\widehat{c}$ and its deformations. Later in [50, 51], Dubrovin extracted the finding and summarized into the structure of *Frobenius manifold* on the space $\mathfrak{M}$, which turned out to be the same as the *flat structure* found by K. Saito a decade earlier [15–17]. In the formulation of [51], it is stated as follows:

(i) The topological metric η is flat, i.e. the Levi-Civita connection ∇^η has zero curvature.

(ii) The vector field corresponding to the identity is ∇^η-parallel.

(iii) There is a holomorphic vectore field, called *Euler vector field* E, such that $L_E\eta = (2-\widehat{c})\eta$ and that $\nabla^\eta E : T_{\mathfrak{M}} \to T_{\mathfrak{M}}$ defined by $v \mapsto \nabla^\eta_v E$ is diagonalizable.

(iv) The connection $\widetilde{\nabla}$ of the vector bundle $T_{\mathfrak{M}}$ over $\mathfrak{M}\times\mathbb{C}^*_z$ defined below is flat,

$$\widetilde{\nabla}_v := \nabla^\eta_v - z^{-1}C_v, \tag{5.20}$$

$$\widetilde{\nabla}_{z\frac{\partial}{\partial z}} := z\frac{\partial}{\partial z} + z^{-1}C_E + \mu \tag{5.21}$$

where C_v is the multiplication by $v\in T_{\mathfrak{M}}$ and

$$\mu := 1 - \frac{\widehat{c}}{2} - \nabla^\eta E. \tag{5.22}$$

We emphasize that the connections ∇^η and $\widetilde{\nabla}$ of $T_{\mathfrak{M}}$ are holomorphic connections and are not the same as the C^∞ connections D and ∇ of $\mathcal{H}_{\rm v}$ introduced in the previous subsection, even though $T_{\mathfrak{M}}\cong\mathcal{R}$ and $\mathcal{H}_{\rm v}$ are isomorphic as holomorphic bundles via the spectral flow. The relation between them was discussed in [57] (see also [58]).

We can find *flat coordinates* $(\mathrm{t}^\alpha)_{\alpha=0}^{N-1}$ of $(\mathfrak{M},\eta)$, such that $\frac{\partial}{\partial \mathrm{t}^0}$ corresponds to the identity operator, the Euler vector field is of the form $E=\sum_\alpha\left(E_0^\alpha+(1-q_\alpha)\mathrm{t}^\alpha\right)\frac{\partial}{\partial\mathrm{t}^\alpha}$, and $\eta_{\alpha\beta}$ is non-zero only if $q_\alpha+q_\beta=\widehat{c}$, for some constants E_0^α and q_α. In particular $\mu^\alpha{}_\beta=\left(q_\alpha-\frac{\widehat{c}}{2}\right)\delta^\alpha_\beta$. If the theory at the origin $\mathrm{t}=0$ is a superconformal field theory, then E_0^α are zero and $2q_\alpha$ is the R'-charge, that is, the axial (*resp.* vector) R-charge, of the operator ϕ_α corresponding to $\frac{\partial}{\partial\mathrm{t}^\alpha}$. In particular, q_0 is zero and t^i with $q_i=1$ are marginal parameters. In the flat coordinate system, a section $\xi=\xi_\alpha d\mathrm{t}^\alpha$ of $T^*_{\mathfrak{M}}$ over $\mathfrak{M}\times\mathbb{C}^*$ is $\widetilde{\nabla}$-parallel when

$$\frac{\partial}{\partial\mathrm{t}^\beta}\xi_\alpha = -z^{-1}C^\gamma_{\beta\alpha}\xi_\gamma, \tag{5.23}$$

$$z\frac{\partial}{\partial z}\xi_\alpha = \left(-E^\beta\frac{\partial}{\partial\mathrm{t}^\beta}+q_\alpha-\frac{\widehat{c}}{2}\right)\xi_\alpha. \tag{5.24}$$

The identity component satisfies

$$z\frac{\partial}{\partial z}\xi_0 = \left(-E^\beta\frac{\partial}{\partial\mathrm{t}^\beta}-\frac{\widehat{c}}{2}\right)\xi_0. \tag{5.25}$$

For each chiral ring relation $R(\mathrm{t};\phi_{i_1},\ldots,\phi_{i_s})=0$, following the same procedure as in the derivation of (5.9), we find a differential equation of the form

$$\left[R\left(\mathrm{t};z\frac{\partial}{\partial\mathrm{t}^{i_1}},\ldots,z\frac{\partial}{\partial\mathrm{t}^{i_s}}\right)+\delta\mathcal{D}_R\right]\xi_0=0, \tag{5.26}$$

where $\delta\mathcal{D}_R$ is a holomorphic differential operator of lower degree than R.

The equation (5.25) looks similar to the renormalization group equation (4.88) on the hemisphere partition function if we identify z with r^{-1} and the Euler vector field E as $-b_1\left(\frac{\partial}{\partial t}\right)$. The fact that we have an equation (5.26) for each chiral ring relation R is also similar to the fact that we have an equation (4.90) for each deformed relation among the central components of σ (see (4.96) and (4.108) for concrete examples). This suggests that the hemisphere partition function $Z_{D^2}(\mathcal{B})$ for the GLSM may be regarded as the identity component of a $\widetilde{\nabla}$-parallel section of $T^*_{\mathfrak{M}}$ over $\mathfrak{M}_t \times \mathbb{C}^*$, where the coordinate z of $\mathbb{C}^*$ is identified as a complexification of the inverse radius r^{-1}.

We next confirm this in GLSM for the example $\mathrm{T}^{U(1)}_{N,d}$ for $d \leq N$ where the full partition function can be captured in the geometric regime.

Gromov-Witten Theory

For the non-linear sigma model on a Kähler manifold X, as quoted earlier, the chiral ring $\mathcal{R}_A$ is the quantum cohomology ring $(\mathrm{H}^*(X;\mathbb{C}),)$ and the space $\mathfrak{M}_A$ may be identified as an open subset of $\mathrm{H}^*(X;\mathbb{C})$. The topological metric is simply the Poincaré pairing $\eta(\omega_1,\omega_2) = \int_X \omega_1 \wedge \omega_2$ and therefore affine coordinates of $\mathfrak{M}_A \subset \mathrm{H}^*(X;\mathbb{C})$ are flat. The sigma model is not conformal in general, but may be regarded as a "deformation" of a conformal theory with $\widehat{c} = \dim_{\mathbb{C}} X =: D$, which is the model with a fine-tuned metric when X is Calabi-Yau and the model with infinite volume when X is not Calabi-Yau. The conformal limit has an axial R-symmetry where the axial R-charge of a chiral ring element is the degree of the corresponding cohomology class.

Let $\{\phi_\alpha\}_{\alpha=0}^{N-1} \subset \mathrm{H}^*(X;\mathbb{C})$ be a basis consisting of classes of definite degrees, $\phi_\alpha \in \mathrm{H}^{2q_\alpha}(X;\mathbb{C})$, and let $(\mathrm{t}^\alpha)_{\alpha=0}^{N-1}$ be the corresponding affine coordinates. We take $\phi_0 = 1 \in \mathrm{H}^0(X;\mathbb{C})$ and $\{\phi_i\}_{i=1}^{\ell}$ to be the basis of $\mathrm{H}^{1,1}(X)$. Note that $\eta_{\alpha\beta} = \int_X \phi_\alpha \wedge \phi_\beta$ is non-zero only when $q_\alpha + q_\beta = D$. The Euler vector field is given by

$$E = \sum_{i=1}^{\ell} c_1(X)^i \frac{\partial}{\partial \mathrm{t}^i} - \sum_{\alpha=0}^{N-1} (q_\alpha - 1)\mathrm{t}^\alpha \frac{\partial}{\partial \mathrm{t}^\alpha} \tag{5.27}$$

where $\sum c_1(X)^i \phi_i = c_1(X)$ is the first Chern class of X.

$\widetilde{\nabla}$-parallel sections of $T^*_{\mathfrak{M}_A}$ over $\mathfrak{M}_A \times \mathbb{C}^*$ can be constructed using genus zero amplitudes in the associated topological string theory, that is, the Gromov-Witten theory. In this theory, each element $\mathcal{O} \in \mathrm{H}^*(X;\mathbb{C}) = \mathcal{R}_A$ defines a *primary operator* denoted by the same symbol (the one for $\mathcal{O} = \phi_0 = 1$ is called the *puncture operator* P), as well as operators $\tau_n\mathcal{O} = \psi^n\mathcal{O}$ called the *gravitational descendants*, for

$n = 1, 2, \ldots$. See for example [59–63] for the definition and properties.[5] For each $\mathcal{O} \in \mathrm{H}^*(X; \mathbb{C})$, let us put

$$w[\mathcal{O}] := \frac{\mathcal{O}}{1 + z^{-1}\psi} = \sum_{n=0}^{\infty} (-1)^n z^{-n} \tau_n \mathcal{O}, \tag{5.28}$$

and

$$\xi_\alpha[\mathcal{O}] := \left\langle\, P\,\phi_\alpha\, w\left[z^{-\mu}\left(z^{c_1(X)}\mathcal{O}\right)\right] \right\rangle, \tag{5.29}$$

where $\langle \mathcal{O}_1 \cdots \mathcal{O}_s \rangle$ is the genus zero amplitude in the background t. Then, $\xi[\mathcal{O}] = \xi_\alpha[\mathcal{O}]\mathrm{dt}^\alpha$ is a $\widetilde{\nabla}$-parallel section of $T^*_{\mathfrak{M}_A}$. The construction is due to Givental [49] and Dubrovin [51]. Givental introduced the combination $w[\mathcal{O}]$ and showed that it defines parallel sections in the $\mathfrak{M}_A$ direction, and Dubrovin put $z^{-\mu} z^{c_1(X)}$ to make it parallel also in the z-direction. In view of its importance, let us give a proof of the assertion that $\xi[\mathcal{O}]$ is $\widetilde{\nabla}$-parallel, that is, $\xi_\alpha[\mathcal{O}]$ satisfy (5.23) and (5.24). Equation (5.23) is an immediate consequence of the topological recursion relation [59],

$$\begin{aligned} \frac{\partial}{\partial \mathrm{t}_\beta} \xi_\alpha[\mathcal{O}] &= \sum_{n=0}^{\infty} (-1)^n z^{-n} \langle\, \phi_\beta\, P\, \phi_\alpha\, \tau_n(\mathcal{O}_z)\,\rangle \\ &= \sum_{n=1}^{\infty} (-1)^n z^{-n} \langle\, \tau_{n-1}(\mathcal{O}_z)\, P\, \phi_\gamma\,\rangle \langle\, \phi^\gamma \phi_\beta \phi_\alpha\,\rangle \\ &= -z^{-1} C^\gamma_{\beta\alpha} \xi_\gamma[\mathcal{O}], \end{aligned} \tag{5.30}$$

with $\mathcal{O}_z := z^{-\mu}(z^{c_1(X)}\mathcal{O})$ where we used $\langle \phi^\gamma P \phi_\beta \phi_\alpha \rangle = 0$. To see (5.24), first note that

$$\begin{aligned} z\frac{\partial}{\partial z} \xi_\alpha[\mathcal{O}] = &\sum_{n=1}^{\infty} (-1)^n (-n) z^{-n} \langle\, P\, \phi_\alpha\, \tau_n(\mathcal{O}_z)\,\rangle \\ &+ \sum_{n=0}^{\infty} (-1)^n z^{-n} \langle\, P\, \phi_\alpha\, \tau_n(-\mu\mathcal{O}_z + z^{-1} c_1(X) \wedge \mathcal{O}_z)\,\rangle. \end{aligned} \tag{5.31}$$

[5] $\sigma_n \mathcal{O}$ in [59, 61] is equal to $n! \times \tau_n \mathcal{O}$.

On the other hand, the divisor equation [64] for $c_1(X)$ and the selection rule (degree-matching) yield

(5.32)

$$\begin{aligned} & c_1(X)^i \frac{\partial}{\partial \mathrm{t}^i}\xi_\alpha[\mathcal{O}] = \\ & \sum_{n=0}^{\infty}(-1)^n z^{-n}\Bigg[\left(-\frac{D}{2}+q_\alpha+n+\sum_\beta (q_\beta-1)\mathrm{t}^\beta\frac{\partial}{\partial \mathrm{t}^\beta}\right)\langle\, P\,\phi_\alpha\,\tau_n(\mathcal{O}_z)\,\rangle \\ & \qquad + \langle\, P\,\phi_\alpha\,\tau_n(\mu\mathcal{O}_z)\,\rangle\Bigg] + \sum_{n=1}^{\infty}(-1)^n z^{-n}\langle\, P\,\phi_\alpha\,\tau_{n-1}(c_1(X)\wedge\mathcal{O}_z)\,\rangle. \end{aligned}$$

Summing the above two, we find that (5.24) holds.

For a B-brane in the sigma model, $\mathcal{E}\in \mathrm{D}^b_{Coh}(X)$, following [53], we define its GW central charge as the identity component of $\xi[\mathcal{O}]$ for a particular $\mathcal{O}$ determined by the brane,

(5.33)

$$\begin{aligned} & Z_{\mathrm{GW}}(\mathcal{E}) \\ & \quad := \frac{1}{(-2\pi i)^D}\left\langle P\,P\,w\left[z^{-\mu}\left(z^{c_1(X)}(-2\pi i)^{\frac{\deg}{2}}\left(\widehat{\Gamma}_X\,\mathrm{ch}(\mathcal{E})\right)\right)\right]\right\rangle\Big|_{\mathfrak{M}_A^{0,c}\times\mathbb{C}^*} \\ & \quad = \frac{1}{(-2\pi i)^D}\xi_0\left[(-2\pi i)^{\frac{\deg}{2}}\left(\widehat{\Gamma}_X\,\mathrm{ch}(\mathcal{E})\right)\right]\Big|_{\mathfrak{M}_A^{0,c}\times\mathbb{C}^*}, \end{aligned}$$

where $\mathfrak{M}_A^{0,c}\subset\mathfrak{M}_A$ is the subspace in which only the Kähler parameters $-\mathrm{t}^i$ are turned on.[6] The equation (5.25) reads

$$z\frac{\partial}{\partial z}Z_{\mathrm{GW}}(\mathcal{E}) = \left(-c_1(X)^i\frac{\partial}{\partial \mathrm{t}^i}-\frac{D}{2}\right)Z_{\mathrm{GW}}(\mathcal{E}), \tag{5.34}$$

which looks a lot closer to the renormalization group equation (4.88), epsecially to (4.89), with $z\to r^{-1}$. This means that $Z_{\mathrm{GW}}(\mathcal{E})$ is $z^{-\frac{D}{2}}$ times a function of

$$\mathrm{t}_R := \mathrm{t} - c_1(X)\log(z). \tag{5.35}$$

Using the divisor equation repeatedly, we find

$$Z_{\mathrm{GW}}(\mathcal{E}) = z^{-\frac{D}{2}}\left[\int_X \exp\left(-\frac{i}{2\pi}\,\mathrm{t}_R\right)\widehat{\Gamma}_X\,\mathrm{ch}(\mathcal{E}) + O(\mathrm{e}^{\mathrm{t}_R})\right], \tag{5.36}$$

[6] Z_{GW} is related to the central charge Z_{Iritani} of [53] by $Z_{\mathrm{GW}}(\mathcal{E}) = (-1)^D z^{-\frac{D}{2}}Z_{\mathrm{Iritani}}(\mathcal{E}^\vee)|_{\mathfrak{M}_A^{0,c}\times\mathbb{C}^*}$.

where $O(\mathrm{e}^{\mathrm{t}_R})$ is a power series of the form $\sum c_\beta\,\mathrm{e}^{\langle \mathrm{t}_R,\beta\rangle}$ where the sum is over the effective classes of X. This matches with the large volume behaviour of the hemisphere partition function (4.82) up to the numerical constant C, under

$$z=(r\Lambda)^{-1},\qquad \mathrm{t}_R=-\omega_R+iB. \tag{5.37}$$

Let us examine the relation more closely, for the model $\mathrm{T}^{U(1)}_{N,d}$ with $d\leq N$ where we have a large volume formula (4.77) that captures the full partition function. For convenience, we copy the formula here with $r\Lambda\to z^{-1}$

$$\begin{aligned} Z^{\mathrm{LV}}_{D^2}(\mathfrak{B})=Cz^{-\frac{D}{2}}\sum_{n=0}^{\infty}(-1)^{Nn}\,\mathrm{e}^{-nt_R}\int_{X_f}\widehat{\Gamma}_{X_f}(n)\\ \times\exp\left(\frac{i}{2\pi}(t_R-d\pi i)H\right)\mathrm{ch}(\mathcal{E}), \end{aligned} \tag{5.38}$$

where $t_R=t+(N-d)\log(z)$, $\mathcal{E}=\pi_+(\mathfrak{B})$ and

$$\begin{aligned} \widehat{\Gamma}_{X_f}(n)&=\frac{\Gamma\left(1+\frac{H}{2\pi i}\right)^N}{\Gamma\left(1+\frac{H}{2\pi i}+n\right)^N}\frac{\Gamma\left(1+d\left(\frac{H}{2\pi i}+n\right)\right)}{\Gamma\left(1+d\frac{H}{2\pi i}\right)}\cdot\widehat{\Gamma}_{X_f}(0)\\ &=\frac{\prod_{a=1}^{dn}\left(a+\frac{dH}{2\pi i}\right)}{\prod_{b=1}^{n}\left(b+\frac{H}{2\pi i}\right)^N}\cdot\widehat{\Gamma}_{X_f}. \end{aligned} \tag{5.39}$$

The integral remains the same if we operate $\left(-\frac{2\pi i}{z}\right)^{\frac{\deg}{2}}$ to the integrand, which does $H\to-\frac{2\pi i}{z}H$ for example, and then divide the result by $\left(-\frac{2\pi i}{z}\right)^{D}$. This transforms the expression to

(5.40)
$$Z^{\mathrm{LV}}_{D^2}(\mathfrak{B})=\frac{C}{(-2\pi i)^D}\int_{X_f}I_{X_f}(\widetilde{t},z)\cdot z^{-\mu}\left(z^{c_1(X_f)}(-2\pi i)^{\frac{\deg}{2}}\left(\widehat{\Gamma}_{X_f}\mathrm{ch}(\mathcal{E})\right)\right),$$

where

$$\widetilde{t}:=t-d\pi i \tag{5.41}$$

and

$$I_{X_f}(\widetilde{t},z):=\mathrm{e}^{\widetilde{t}H/z}\sum_{n=0}^{\infty}\mathrm{e}^{-n\widetilde{t}}\cdot\frac{\prod_{a=1}^{dn}(dH-za)}{\prod_{b=1}^{n}(H-zb)^N}. \tag{5.42}$$

At this point, we quote Givental's mirror theorem [49] on the relationship between this $I_{X_f}(\widetilde{t},z)$ and the Gromov-Witten theory of X_f. Its states that (see also [62])

$$\int_{X_f} I_{X_f}(\widetilde{t},z)\alpha = \mathcal{Y}(\widetilde{t},z)\,\langle\, P\,P\,w[\alpha]\,\rangle\Big|_{\mathrm{t}=-\mathcal{T}(\widetilde{t})}, \tag{5.43}$$

for any $\alpha \in \mathrm{H}^*(X_f,\mathbb{C})$, where

(5.44)

	$d \leq N$	$d = N-1$	$d = N$
$\mathcal{Y}(\widetilde{t},z)$	1	$\exp\left(-d!\,\mathrm{e}^{-\widetilde{t}}/z\right)$	$C^{-1}z^{\frac{D}{2}}Z(\mathrm{D0})$
$\mathcal{T}(\widetilde{t})$	$\widetilde{t}H$	$\widetilde{t}H$	$-2\pi i\dfrac{Z(\mathrm{D2})}{Z(\mathrm{D0})}H$

in which $Z(\mathrm{D0})$ and $Z(\mathrm{D2})$ are $Z_{D^2}^{\mathrm{LV}}$ for the brane data $\mathfrak{B}$ such that $\mathrm{ch}(\pi_+(\mathfrak{B}))$ is the point class [pt] and the line class l respectively. Applying this mirror theorem to (5.40), we find the relationship between the hemisphere partition function of $\mathrm{T}_{N,d}^{U(1)}$ and the GW central charge of X_f for the case $d \leq N$:

$$Z_{D^2}^{\mathrm{LV}}(\mathfrak{B})\Big|_{r\Lambda=z^{-1}} = C\cdot\mathcal{Y}(\widetilde{t},z)\cdot Z_{\mathrm{GW}}(\pi_+(\mathfrak{B}))|_{\mathrm{t}=-\mathcal{T}(\widetilde{t})}\,. \tag{5.45}$$

That is, they are equal up to a multiplicative renormalization and a change of parameter. As a consistency test, let us compute Z_{GW} for the D0-brane in a Calabi-Yau manifld X:

$$Z_{\mathrm{GW}}(\mathrm{D0}) = z^{-\frac{D}{2}}\,\langle P\,P\,w\,[[\mathrm{pt}]]\rangle = z^{-\frac{D}{2}}\sum_{n=0}^{\infty}(-z)^{-n}\langle P\,P\,\tau_n[\mathrm{pt}]\rangle. \tag{5.46}$$

On $\mathfrak{M}_A^0\times\mathbb{C}^*$, it follows from the selection rule that only the $n=0$ term is non-zero, and from the puncture equation [61] that only the degree zero map without Kähler perturbation contributes, finding

$$Z_{\mathrm{GW}}(\mathrm{D0})|_{\mathfrak{M}_A^0\times\mathbb{C}^*} = z^{-\frac{D}{2}}. \tag{5.47}$$

Therefore, for $d = N$ we indeed have $Z_{D^2}^{\mathrm{LV}}(\mathrm{D0}) = C\cdot C^{-1}z^{\frac{D}{2}}Z(\mathrm{D0})\cdot z^{-\frac{D}{2}} = Z(\mathrm{D0})$.

Macroscopic Loop

The development of topological string in [59–61], especially the introduction of gravitational descendants, is motivated by the solution of two-dimensional quantum gravity via matrix models [65]. The partition function of the one matrix model is

$$Z(N,V)=\int_{H_N}\mathrm{d}M\exp\Big(-N\,\mathrm{tr}\,V(M)\Big) \tag{5.48}$$

where H_N is the space of $N\times N$ hermitian matrices, $\mathrm{d}M$ is the flat measure and V is a polynomial with a quadratic and higher order terms. The Feynman diagram expansion of this integral can be organized into the sum over triangulated surfaces [66]. In a certain limit where the size N is sent to infinity and the potential V is tuned at the same time (double scaling limit), the expansion can be regarded as the sum over closed oriented 2d Riemannian manifolds, including the sum over the genus, with the weight given by the partition function of a conformal field theory. That is, the theory can be regarded as the 2d quantum gravity coupled to a CFT. There is a series of double scaling limits, labelled by $m=1,2,3,\ldots$, where the CFT is the $(p,q)=(2m-1,2)$ minimal model. The $m=2$ theory, where the CFT is empty, is the pure gravity. The $m=1$ theory is identified [67] with the pure topological gravity, that is, the Gromov-Witten theory of a point. The model with ν matrix variables yields the theory of gravity coupled to the $(p,\nu+1)$ minimal CFT, and the $(1,\nu+1)$ theory is the topological string associated with the LG B-model with superpotential $W=x^{\nu+1}$, or equivalently, the FJRW theory for $(\mathbb{C}/\mathbb{Z}_{\nu+1},x^{\nu+1})$.

Insertion of $\mathrm{tr}M^j$ into the integrand of (5.48) creates a j-gonal hole in the triangulated surfaces. See Fig. 8. In the double scaling limit, certain linear combinations of $\mathrm{tr}M^j$ with finite j's become local operators $\sigma_0=:P,\sigma_1,\sigma_2,\ldots$ while a certain limit of $\mathrm{tr}M^j$ with $j\to\infty$ yields a *macroscopic loop operator* $w(\ell)$ that creates a hole of length ℓ in the 2d manifold [68]. The puncture operator and the gravitational descendants, $\tau_n=\tau_n\phi_0$ with $n=0,1,2,\ldots,$ in the topological gravity are the local operators in the $m=1$ theory. Correlation functions involving macroscopic loops have been studied in [68–70] (see [71] for a review). For example, at genus zero

$$\langle P\,P\,w(\ell)\rangle=\mathrm{const}\cdot\sqrt{\ell}\,\mathrm{e}^{-\ell\langle PP\rangle}\langle PPP\rangle, \tag{5.49}$$

$$\langle P\,w(\ell_1)w(\ell_2)\rangle=\mathrm{const}\cdot\sqrt{\ell_1\ell_2}\,\mathrm{e}^{-(\ell_1+\ell_2)\langle PP\rangle}\langle PPP\rangle. \tag{5.50}$$

The macroscopic loop $w(\ell)$ for small ℓ can be expanded as a power series in ℓ with coefficients given by the local operators. In the $m=1$ theory

Fig. 8. Insertion of $\mathrm{tr}M^9$ creates a nine-gonal hole.

that corresponds to topological gravity, the series is

$$w(\ell) = \mathrm{const} \cdot \sqrt{\ell} \sum_{n=0}^{\infty} (-1)^n \ell^n \tau_n. \tag{5.51}$$

Indeed, under this, the results (5.49) and (5.50) can be reproduced from the constitutive relations [61], $n!m!\langle \tau_n \tau_m \rangle = \langle PP \rangle^{n+m+1}/(n+m+1)$. We would also like to quote that the Virasoro constraint in the double scaled theory [73, 74] can be written via (5.51) as a loop equation on $w(\ell)$ or its Laplace transform $\widehat{w}(\zeta)$, and that it can be stated as regularity of the energy-momentum tensor of a free chiral CFT on the ζ-space [74]. For the ν matrix model, there are ν different macroscopic loop operators $w_1(\ell), \ldots, w_\nu(\ell)$, and it was suggested [72] that the small ℓ expansion of $w_i(\ell)$ in the $(1, \nu+1)$ theory is given by (5.51) where τ_n is replaced by $\tau_n O_i$ for some primary O_i.

We see that $w[P]$ (5.28) in the Gromov-Witten theory of a point is nothing but the small ℓ expansion (5.51) of the macroscopic loop $w(\ell)$ with $\ell = z^{-1}$, up to a normalization factor including $z^{-\frac{1}{2}}$. This suggests that we may view Givental's operator $w[\mathcal{O}]$ in the Gromov-Witten theory of a general target X, or in a general topological string theory, as an expansion of the macroscopic loop operator labelled by $\mathcal{O}$ that creates a hole of length $\ell = z^{-1}$. Furthermore, if the relation of the type (5.45) holds generally, then, the partition function on the hemisphere of radius r is proportional to a genus zero correlation function involving a macroscopic loop of length r in the topological string,

$$Z_{D^2(r)}(\mathcal{B}) \sim \langle P\,P\,w_{[\mathcal{B}]}(r)\rangle. \tag{5.52}$$

This would be a strikingly simple relation. It is a very interesting problem to find a precise formulation of macroscopic loops in the topological string theory that leads to (5.28) in the small length expansion, and to see whether such a relation can be derived naturally.

Acknowledgement

We would like to thank the organizer of the meeting *Primitive forms and related subjects*, Changzheng Li, Si Li and Kyoji Saito, for suggesting to write these notes and for patience to wait for the completion. KH would also like to thank Jaume Gomis for the conversation in 2013 from which the idea to relate the differential equation to the deformed chiral ring relation emerged. We would also like to thank the referee for suggestions to improve the note.

Appendix

A. Lagrangian and supersymmetry

In this appendix, we describe the supersymmetric action on the Minkowski spacetime as well as on the sphere and the hemisphere for GLSMs and LG models. For the GLSMs on the (hemi)sphere, we describe the one that preserves A (or $\mathrm{A}_{(\pm)}$) type supercharges. For the LG models on the (hemi)sphere, we describe the one that preserves B (or $\mathrm{B}_{(\pm)}$) type supercharges.

A.1. Minkowski space

We describe the supersymmetry and the Lagrangian of GLSM on the Minkowski space, with time and space coordinates x^0 and x^1 and the metric $\mathrm{d}s^2 = -(\mathrm{d}x^0)^2 + (\mathrm{d}x^1)^2$. We often use the light-cone coordinates $x^{\pm} = x^0 \pm x^1$.

The gauge multiplet with gauge group G consists of a G connection v_μ, as well as a scalar σ, a Dirac fermion λ and a scalar D with values in $\mathfrak{g}_{\mathbb{C}}$, $\mathfrak{g}_{\mathbb{C}}$ and $i\mathfrak{g}$. The matter multplet in representation V consists of a scalar ϕ, a Dirac fermion ψ and a scalar F, all with values in V. The gauge connection v_μ is "real-valued" so that the curvature is $v_{\mu\nu} = \partial_\mu v_\nu - \partial_\nu v_\mu + i[v_\mu, v_\nu]$ and the gauge covariant derivative is $D_\mu\phi = \partial_\mu\phi + iv_\mu\phi$.

The supersymmetry transformation $\delta = i\epsilon_+ Q_- - i\epsilon_- Q_+ - i\overline{\epsilon}_+\overline{Q}_- + i\overline{\epsilon}_-\overline{Q}_+$ is

$$
\begin{aligned}
\delta v_\pm &= \frac{i}{2}\overline{\epsilon}_\pm\lambda_\pm + \frac{i}{2}\epsilon_\pm\overline{\lambda}_\pm,\\
\delta\sigma &= -i\overline{\epsilon}_+\lambda_- - i\epsilon_-\overline{\lambda}_+,\\
\delta\lambda_\pm &= i\epsilon_\pm\left((D \pm iv_{01}) \pm \frac{1}{2}[\sigma,\overline{\sigma}]\right) + \epsilon_\mp(D_0 \pm D_1)\sigma_\mp,\\
\delta D &= \left[\frac{1}{2}\epsilon_+\big((D_0 - D_1)\overline{\lambda}_+ + i[\sigma,\overline{\lambda}_-]\big)\right.\\
&\quad \left. + \frac{1}{2}\epsilon_-\big((D_0 + D_1)\overline{\lambda}_- + i[\overline{\sigma},\overline{\lambda}_+]\big)\right] + c.c.,
\end{aligned}
\tag{A.1}
$$

and

$$\begin{aligned}\text{(A.2)}\qquad \delta\phi &= \epsilon_+\psi_- - \epsilon_-\psi_+,\\ \delta\psi_\pm &= \pm i\overline{\epsilon}_\mp(D_0 \pm D_1)\phi \mp \overline{\epsilon}_\pm\sigma_\mp\phi + \epsilon_\pm F,\\ \delta F &= -i\overline{\epsilon}_+(D_0 - D_1)\psi_+ - i\overline{\epsilon}_-(D_0 + D_1)\psi_-\\ &\quad + \overline{\epsilon}_+\overline{\sigma}\psi_- + \overline{\epsilon}_-\sigma\psi_+ + i(\overline{\epsilon}_-\overline{\lambda}_+ - \overline{\epsilon}_+\overline{\lambda}_-)\phi.\end{aligned}$$

where we use the notation $\sigma_+ = \sigma$ and $\sigma_- = \overline{\sigma}$ just in here. The transformation of the complex conjugates are obtained from these by the rule $(AB)^\dagger = B^\dagger A^\dagger$. For example, $\delta\overline{\sigma} = i\overline{\lambda}_-\epsilon_+ + i\lambda_+\overline{\epsilon}_- = -i\epsilon_+\overline{\lambda}_- - i\overline{\epsilon}_-\lambda_+$. Note that σ is A-chiral and ϕ is B-chiral

$$\text{(A.3)}\qquad \overline{Q}_+\sigma = Q_-\sigma = 0, \qquad \overline{Q}_+\phi = \overline{Q}_-\phi = 0.$$

Before writing down the Lagrangian, we choose an adjoint invariant innder product on $i\mathfrak{g}$ and G-invariant hermitian inner product on V,

$$\text{(A.4)}\ \ (X, Y) \in i\mathfrak{g} \times i\mathfrak{g} \longmapsto \frac{1}{e^2}XY \in \mathbb{R},\ \ (\phi_1, \phi_2) \in V \times V \longmapsto \overline{\phi}_1\phi_2 \in \mathbb{C},$$

which are both positive definite. We also write $\frac{1}{e^2}XX = \frac{1}{e^2}X^2$, $\overline{\phi}\phi = |\phi|^2$, etc. The supersymmetric Lagrangian is

$$\begin{aligned}\text{(A.5)}\qquad \mathcal{L} &= Q_+Q_-\overline{Q}_+\overline{Q}_-\left(-\frac{1}{2e^2}|\sigma|^2 + |\phi|^2\right)\\ &\quad + \mathrm{Re}\,Q_+Q_-W(\phi) + \mathrm{Re}\,Q_+\overline{Q}_-\Big(-t(\sigma)\Big)\\ &\quad + \text{total derivative}\\ &= \mathcal{L}_{\mathrm{g}} + \mathcal{L}_{\mathrm{m}} + \mathcal{L}_W + \mathcal{L}_t,\end{aligned}$$

where

(A.6)

$$\begin{aligned}\mathcal{L}_{\mathrm{g}} = \frac{1}{2e^2}\Big[&|D_0\sigma|^2 - |D_1\sigma|^2 + i\overline{\lambda}_-(D_0 + D_1)\lambda_- + i\overline{\lambda}_+(D_0 - D_1)\lambda_+\\ &+ (v_{01})^2 + D^2 - \frac{1}{4}[\sigma, \overline{\sigma}]^2 - \lambda_+[\sigma, \overline{\lambda}_-] + [\overline{\sigma}, \lambda_-]\overline{\lambda}_+\Big],\end{aligned}$$

$$\begin{aligned}\text{(A.7)}\quad \mathcal{L}_{\mathrm{m}} &= |D_0\phi|^2 - |D_1\phi|^2 + i\overline{\psi}_-(D_0 + D_1)\psi_- + i\overline{\psi}_+(D_0 - D_1)\psi_+\\ &\quad + |F|^2 + \overline{\phi}D\phi - \frac{1}{2}|\sigma\phi|^2 - \frac{1}{2}|\overline{\sigma}\phi|^2 - \overline{\psi}_-\sigma\psi_+ - \overline{\psi}_+\overline{\sigma}\psi_-\\ &\quad - i\overline{\phi}\lambda_-\psi_+ + i\overline{\phi}\lambda_+\psi_- + i\overline{\psi}_+\overline{\lambda}_-\phi - i\overline{\psi}_-\overline{\lambda}_+\phi,\end{aligned}$$

(A.8) $$\mathcal{L}_W = \mathrm{Re}\left[F^i \frac{\partial W}{\partial \phi^i} - \psi_+^i \psi_-^j \frac{\partial^2 W}{\partial \phi^i \partial \phi^j}\right],$$

(A.9) $$\mathcal{L}_t = -\zeta(D) + \theta(v_{01}).$$

The Lagrangian (A.5) is manifestly supersymmetric since σ is A-chiral and ϕ is B-chiral (A.3).

The fields D and F are "auxiliary fields" — they do not have the kinetic terms. They can be eliminated by the equations of motion. After doing that, we obtain the scalar potential (3.5).

In the present notes, we take the convention that the action is the integral of Lagragian density *divided by* 2π:

(A.10) $$S = \frac{1}{2\pi}\int_{\mathbb{R}^2} \mathrm{d}^2x\, \mathcal{L}.$$

In particular, the theta term enters into the action as $\frac{1}{2\pi}\int_{\mathbb{R}^2} \mathrm{d}^2x\, \theta(v_{01}) = \int_{\mathbb{R}^2} \theta(\frac{i}{2\pi}F_v)$ where $F_v = iv_{01}\mathrm{d}x^0 \wedge \mathrm{d}x^1$ is the $\mathfrak{g}$-valued ("imaginary") curvature two-form.

A.2. Wick rotation and covariantization

The Lagrangian and supersymmetry on the Euclidean space is obtained by Wick rotation $x^0 \to -ix^2$ and $\mathcal{L} \to -\mathcal{L}_E$. The auxiliary fields are also rotated as

(A.11) $$D \to iD_E, \quad D_E \in i\mathfrak{g},$$

(A.12) $$(F, \overline{F}) \to (if, i\overline{f}).$$

The fermion pairs, $(\lambda_\pm, \overline{\lambda}_\pm)$ and $(\psi_\pm, \overline{\psi}_\pm)$, are no longer related by complex conjugation.

It is starightforward to formulate the theory on a general two-manifold Σ with a Riemannian metric h and a spin structure, by the standard covariantization. Note that choice of a principal G bundle P on Σ is a part of the variables. A scalar with values in a representation U of G should now be regraded as a section of the vector bundle $P \times_G U$ on Σ. A Dirac fermion $\xi = (\xi_+, \xi_-)$ is a section of the spinor bundle $S = S_+ \oplus S_-$. The positive and negative chirality spinor bundles can also be regarded as the square roots of the (holomorphic or anti-holomorphic) tangent or cotangent bundles:

(A.13) $$S_+ \cong \sqrt{\overline{K}_\Sigma} \cong \sqrt{T_\Sigma}, \qquad S_- \cong \sqrt{K_\Sigma} \cong \sqrt{\overline{T}_\Sigma}.$$

The Lagrangian and the supersymmetry transformation on a general surface (Σ, h) can then be found straightforwardly, by covariantization.

The variational parameters $\epsilon_\pm$ and $\overline{\epsilon}_\pm$ should now be regarded as sections of $S_\pm$. The variation of the Lagrangian under this is
(A.14)
$$\delta\mathcal{L} = \nabla_\mu\epsilon_+ G^\mu_- - \nabla_\mu\epsilon_- G^\mu_+ - \nabla_\mu\overline{\epsilon}_+\overline{G}^\mu_- + \nabla_\mu\overline{\epsilon}_-\overline{G}^\mu_+ + \text{total derivative},$$

where $G^\mu_\pm$ and $\overline{G}^\mu_\pm$ are the supercurrents corresponding to $Q_\pm$ and $\overline{Q}_\pm$. This shows that the condition for supersymmetry is $\nabla_\mu\epsilon_\pm = \nabla_\mu\overline{\epsilon}_\pm = 0$, which is impossible on a curved manifold.

A.3. Superconformal transformation

The superconformal transformation of the fields are obtained from the Weyl covariantization. For the fields in GLSM, this is specified by a choice vector R-charge for the B-chiral scalar ϕ. For this, we use the one $R : V \to V$ that determines the vector R-symmetry of the system. Given this choice, the superconformal transformation is

$$\begin{gathered}\delta^{\rm sc}v_\pm = \delta v_\pm,\\ \delta^{\rm sc}\sigma = \delta\sigma,\qquad \delta^{\rm sc}\overline{\sigma} = \delta\overline{\sigma},\\ \delta^{\rm sc}\lambda = \delta\lambda + i\sigma\,\nabla\!\!\!\!/\,\overline{\widetilde{\epsilon}},\qquad \delta^{\rm sc}\overline{\lambda} = \delta\overline{\lambda} - i\overline{\sigma}\,\nabla\!\!\!\!/\,\widetilde{\epsilon},\\ \delta^{\rm sc}D_E = \delta D_E - \tfrac{1}{2}\langle\nabla\!\!\!\!/\,\widetilde{\epsilon},\overline{\lambda}\rangle - \tfrac{1}{2}\langle\nabla\!\!\!\!/\,\overline{\widetilde{\epsilon}},\lambda\rangle,\end{gathered} \tag{A.15}$$

and

$$\begin{gathered}\delta^{\rm sc}\phi = \delta\phi,\qquad \delta^{\rm sc}\overline{\phi} = \delta\overline{\phi},\\ \delta^{\rm sc}\psi = \delta\psi + \tfrac{i}{2}R\phi\,\nabla\!\!\!\!/\,\overline{\epsilon},\qquad \delta^{\rm sc}\overline{\psi} = \delta\overline{\psi} - \tfrac{i}{2}R\overline{\phi}\,\nabla\!\!\!\!/\,\epsilon,\\ \delta^{\rm sc}f = \delta f - \tfrac{1}{2}\langle\nabla\!\!\!\!/\,\overline{\epsilon},R\psi\rangle,\qquad \delta^{\rm sc}\overline{f} = \delta\overline{f} + \tfrac{1}{2}\langle R\overline{\psi},\nabla\!\!\!\!/\,\epsilon\rangle,\end{gathered} \tag{A.16}$$

where δ is the transformation obtained from (A.1) and (A.2) via Wick rotation and covariantization. In the above expressions, we use $\lambda = -i\lambda_- + i\overline{\lambda}_+$, $\overline{\lambda} = i\overline{\lambda}_- - i\lambda_+$, $\psi = \psi_- + \psi_+$, $\overline{\psi} = \overline{\psi}_- + \overline{\psi}_+$, $\epsilon = \epsilon_- + \epsilon_+$, $\overline{\epsilon} = \overline{\epsilon}_- + \overline{\epsilon}_+$, $\widetilde{\epsilon} = \epsilon_- + \overline{\epsilon}_+$ and $\overline{\widetilde{\epsilon}} = \overline{\epsilon}_- + \epsilon_+$; $\nabla\!\!\!\!/$ is the Dirac operator; and $\langle\epsilon,\eta\rangle$ is the invariant pairing of spinors, which reduces to $\epsilon_+\eta_- - \epsilon_-\eta_+$ on the flat Minkowski space.

The commutator of the superconformal transformations $\delta^{\rm sc}_1$ and $\delta^{\rm sc}_2$ with different variational parameters is

$$[\delta^{\rm sc}_1,\delta^{\rm sc}_2] = L_{Z_{12}} + \widetilde{L}_{\widetilde{Z}_{12}} + J_{\Theta_{12}} + \widetilde{J}_{\widetilde{\Theta}_{12}}, \tag{A.17}$$

where L_Z (*resp.* $\widetilde{L}_{\widetilde{Z}}$) is the conformal transfomation by a holomorphic vector field Z (*resp.* anti-holomorphic vector field $\widetilde{Z}$) and J_Θ (*resp.* $\widetilde{J}_{\widetilde{\Theta}}$) is the right-handed R-rotation $\frac{1}{2}(F_V - F_A)$ (*resp.* left-handed R-rotation $\frac{1}{2}(F_V + F_A)$) with the variational parameter Θ (*resp.* $\widetilde{\Theta}$). $Z_{12}, \widetilde{Z}_{12}, \Theta_{12}$

and $\widetilde{\Theta}_{12}$ are the bilinears of the two superconformal variational parameters $\epsilon_{i+} = \epsilon_i(z)\sqrt{\frac{\partial}{\partial z}}$, $\overline{\epsilon}_{i+} = \overline{\epsilon}_i(z)\sqrt{\frac{\partial}{\partial z}}$, $\epsilon_{i-} = \epsilon_i(\overline{z})\sqrt{\frac{\partial}{\partial \overline{z}}}$, $\overline{\epsilon}_{i-} = \overline{\epsilon}_i(\overline{z})\sqrt{\frac{\partial}{\partial \overline{z}}}$ $(i = 1, 2)$,

$$Z_{12} = -2i\epsilon_{[1}(z)\overline{\epsilon}_{2]}(z)\frac{\partial}{\partial z}, \qquad \widetilde{Z}_{12} = 2i\epsilon_{[1}(\overline{z})\overline{\epsilon}_{2]}(\overline{z})\frac{\partial}{\partial \overline{z}},$$
$$\Theta_{12} = -2i\epsilon_{[1}(z)\overleftrightarrow{\partial_z}\overline{\epsilon}_{2]}(z), \qquad \widetilde{\Theta}_{12} = 2i\epsilon_{[1}(\overline{z})\overleftrightarrow{\partial_{\overline{z}}}\overline{\epsilon}_{2]}(\overline{z}),$$

where $A\overleftrightarrow{\partial_x}B := \frac{1}{2}A\partial_x B - \frac{1}{2}(\partial_x A)B$.

A.4. Action on the (hemi)sphere

Let $\mathcal{L}_{\rm cov}$ be the Lagrangian obtained from the one (A.5) on the Minsowski spacetime by Wick rotation and covariantization. The action $\int_\Sigma \mathcal{L}_{\rm cov}\sqrt{h}\mathrm{d}^2x$ is not invariant under any of the superconformal tranformations in general. However, on the sphere or on the hemisphere, there is a way to modify the action so that it is (mostly) invariant under a part of the superconformal transformations. The modified action is of the form

$$S_{S^2} = \int_{S^2} (\mathcal{L}_{\rm cov} + \Delta\mathcal{L})\sqrt{h}\mathrm{d}^2x, \tag{A.18}$$

$$S_{D^2} = \int_{D^2} (\mathcal{L}_{\rm cov} + \Delta\mathcal{L})\sqrt{h}\mathrm{d}^2x + \int_{\partial D^2} \mathcal{L}_{\rm bdry}\mathrm{d}\tau, \tag{A.19}$$

where $\Delta\mathcal{L}$ and $\mathcal{L}_{\rm bdry}$ are given below, h is the round (hemi)sphere metric of radius r and $\mathrm{d}\tau$ is the line element of ∂D^2. $\Delta\mathcal{L}$ is the sum of

$$\Delta\mathcal{L}_{\rm g} = \frac{1}{2e^2}\left[\frac{2}{r}\left(D_E\mathrm{Re}(\sigma) + \frac{v_{12}}{\sqrt{h}}\mathrm{Im}(\sigma)\right) + \frac{1}{r^2}\overline{\sigma}\sigma\right], \tag{A.20}$$

$$\Delta\mathcal{L}_{\rm m} = \overline{\phi}\left[\frac{2R^2 - R^2}{4r^2} - i\frac{R}{r}\mathrm{Re}(\sigma)\right]\phi, \tag{A.21}$$

and $\mathcal{L}_{\rm bdry}$ is the sum of

$$\mathcal{L}_{\rm g,\,bdry} = -\frac{1}{4e^2}\left[\partial_n(\overline{\sigma}\sigma) \pm 2i\left(D_E\mathrm{Im}(\sigma) - \frac{v_{12}}{\sqrt{h}}\mathrm{Re}(\sigma)\right)\right], \tag{A.22}$$

$$\mathcal{L}_{\rm m,\,bdry} = \pm\frac{i}{2}\langle\overline{\psi}, \psi\rangle \mp \overline{\phi}\,\mathrm{Im}(\sigma)\phi, \tag{A.23}$$

$$\mathcal{L}_{\rm t,\,bdry} = \pm\mathrm{Im}\left(\frac{1}{2\pi}t(\sigma)\right), \tag{A.24}$$

where ∂_n is the outward normal derivative. Then, S_{S^2} is invariant under A-type supercharges as defined in Section 4.1. For the hemisphere,

$S_{D^2,\mathrm{g}}$, $S_{D^2,\mathrm{m}}$, and $S_{D^2,t}$ (whose definition is hopefully obvious) are invariant under $\mathrm{A}_{(\pm)}$-type supercharges, but $S_{D^2,W}$ transforms as

$$\delta S_{D^2,\mathrm{W}} = \mp\frac{i}{2}\int_{\partial D^2}\left[\langle\overline{\epsilon},\psi^i\rangle\partial_i W + \langle\epsilon,\overline{\psi}^{\bar{\imath}}\rangle\partial_{\bar{\imath}}\overline{W}\right]\mathrm{d}\tau. \tag{A.25}$$

This failure of invariance is compensated by the transformation of the Chan-Paton factor (4.12), $\mathrm{tr}_M\mathrm{Pexp}\left(-\int_{\partial D^2}\mathcal{A}\right)$, with $\mathcal{A}=\mathcal{A}_\tau\mathrm{d}\tau$ given by

$$\mathcal{A}_\tau = iv_\tau \mp \mathrm{Re}(\sigma) - \frac{1}{2}\psi^i\partial_i Q + \frac{1}{2}\overline{\psi}^{\bar{\imath}}\partial_{\bar{\imath}}Q^\dagger + \frac{1}{2}\{Q,Q^\dagger\} \mp \frac{i}{2r}\mathbf{r}. \tag{A.26}$$

In this expression,

$$\psi := \frac{1}{\sqrt{r}}\left[z^{\frac{1}{2}}\psi_-^{\{z\}} \pm \overline{z}^{\frac{1}{2}}\psi_+^{\{z\}}\right],\quad \overline{\psi} := \frac{1}{\sqrt{r}}\left[z^{\frac{1}{2}}\overline{\psi}_-^{\{z\}} \pm \overline{z}^{\frac{1}{2}}\overline{\psi}_+^{\{z\}}\right], \tag{A.27}$$

where the superscript $\{z\}$ is put for the component with respect to the "z frames". For example $\psi_+ = \psi_+^{\{z\}}\sqrt{\mathrm{d}\overline{z}}$ and $\psi_- = \psi_-^{\{z\}}\sqrt{\mathrm{d}z}$. Also, in the last term of (A.26), r in the denominator is the radius of the hemisphere while $\mathbf{r}$ in the numerator is the vector R-charge on the Chan-Paton vector space M that appears, say, in Eqn (3.19). Note that ψ and $\overline{\psi}$ are fermionic and antiperiodic along ∂D^2. At first sight, it appears strange to add the terms $-\frac{1}{2}\psi^i\partial_i Q + \frac{1}{2}\overline{\psi}^{\bar{\imath}}\partial_{\bar{\imath}}Q^\dagger$ and the rest in (A.26). In fact, the Chan-Paton factor $\mathrm{tr}_M\mathrm{Pexp}\left(-\int_{\partial D^2}\mathcal{A}\right)$ is defined so that $\partial_i Q$ and $\partial_{\bar{\imath}}Q^\dagger$ are treated as fermionic, and it make a perfect sense only when ψ and $\overline{\psi}$ are antiperiodic. See Appendix B of [9] for more details.

A.5. The case of Landau-Ginzburg model

The Landau-Ginzburg model associated with a polynomial $W(x_1,\ldots,x_N)$ can be regarded as the GLSM with the trivial gauge group $G=\{1\}$, the matter representation $V=\mathbb{C}^N$ and the superpotential $W(\phi)=W(x_1,\ldots,x_N)$ for $\phi^i=x_i$.[1] In particular, the Lagrangian, the supersymmetry transformation and the superconformal transformation in the theory can be obtained from the above formulae by setting (v,σ,λ,D) all zero (we shall denote this process by "$(-)|$"). That is, the Minkowski Lagrangian is $\mathcal{L}=\mathcal{L}_{\mathrm{m}}+\mathcal{L}_W$ given by (A.7)$|$ and (A.8) and the supersymmetry transformation is given by (A.2)$|$. Wick rotation

[1]The Landau-Ginzburg orbifold associated with a pair (W,Γ) is obtained likewise by taking $G=\Gamma$ and V to be the representation corresponding to the given Γ action on the variables $x_1,\ldots,x_N$.

and covariantization are done in the standard fashion, though we follow (A.12). The superconformal transformation with a vector R-charge $R : \mathbb{C}^N \to \mathbb{C}^N$ is given by (A.16)|.

Let us describe the actions on the sphere and on the hemisphere of radius r that are invariant under the B-type and the $\mathrm{B}_{(\pm)}$-type supercharges. B-type supersymmetry does not require the theory to have a vector $U(1)$ R-symmetry, and we may pick an arbitrary R. In fact, the B-type supercharge action on (ϕ, ψ, f) with R-charge R is the same as the B-type supercharge action on $(\phi, \psi, f_!)$ with the vanishing R-charge, with

$$f_! = f + \frac{R}{2r}\phi. \tag{A.28}$$

The B-type supersymmetric Lagrangian on the sphere is $\mathcal{L} = \mathcal{L}_{\rm m} + \mathcal{L}_W$ with

$$\mathcal{L}_{\rm m} = \mathcal{L}_{\rm m,\,cov}|_{f \to f_!}, \tag{A.29}$$

$$\mathcal{L}_W = \mathcal{L}_{W,\,\rm cov}|_{f \to f_!} + \frac{i}{r}\mathrm{Re}(W). \tag{A.30}$$

When W is not quasihomogeneous, it is the simplest to choose $R = 0$, so that $f \to f_!$ is trivial. When W is quasihomogeneous, we may take R to the the one that makes $W(\lambda^R\phi) = \lambda^2 W(\phi)$. Then, we have $\mathcal{L}_W = \mathcal{L}_{W,\,\rm cov}$. On the hemisphere, the action

$$S_{D^2} = \int_{D^2} \mathcal{L}\sqrt{h}\mathrm{d}^2x + \int_{\partial D^2} \mathcal{L}_{\rm bdry}\mathrm{d}\tau, \tag{A.31}$$

in invariant under the $\mathrm{B}_{(\pm)}$-type supercharges with

$$\mathcal{L}_{\rm m,\,bdry} = -\frac{1}{2}\partial_n(\overline{\phi}\phi) \pm (\overline{f}\phi - \overline{\phi}f), \tag{A.32}$$

$$\mathcal{L}_{W,\,\rm bdry} = \mp\mathrm{Im}(W). \tag{A.33}$$

B. Boundary States Etc

In this Appendix, we describe the definition of the boundary states and related matters in order to set the convention used in Section 5.2.

Let us consider a unitary quantum field theory in 1+1 dimensions. If we quantize the system on a Minkowski cylinder, with space S^1 and time $\mathbb{R}$, we have the Hilbert space $\mathcal{H}_{S^1}$ of states on S^1 and the Hamiltonian H_c ("c" for "closed" string). We employ the standard notation where the conjugation $|a\rangle \mapsto \langle a|$ is defined by $\langle a|b\rangle = (|a\rangle, |b\rangle)$ for the hermitian inner product $(-,-)$ of $\mathcal{H}_{S^1}$. Of course, $(\langle a|b\rangle)^* = \langle b|a\rangle$. Let us consider the Euclidean path integral on the finite cylinder $[-\ell, 0] \times S^1$ where we impose the boundary condition $\mathcal{B}$ at the end $\{0\} \times S^1$ and put a state $|a\rangle$ at the beginning $\{-\ell\} \times S^1$. We define the boundary state $\langle \mathcal{B}|$ by the property that this path integral is equal to $\langle \mathcal{B}| \,\mathrm{e}^{-\ell H_c} |a\rangle$. (To be precise, $\langle \mathcal{B}|$ is not normalizable but $\langle \mathcal{B}| \,\mathrm{e}^{-\ell H_c}$ is.) By definition, $\langle a| \,\mathrm{e}^{-\ell H_c} |\mathcal{B}\rangle$ is the complex conjugate of this $\langle \mathcal{B}| \,\mathrm{e}^{-\ell H_c} |a\rangle$ and can be regarded as the Euclidean path integral on the finite cylinder $[0, \ell] \times S^1$ where we put the conjugate state $\langle a|$ at the end $\{\ell\} \times S^1$ and impose a "conjugate boundary condition" $\overline{\mathcal{B}}$ at the beggining $\{0\} \times S^1$.

For example, the path integral on the cylinder $[0, L] \times S^1$ with the boundary condition $\mathcal{B}_f$ at the end $\{L\} \times S^1$ and the boundary condition $\overline{\mathcal{B}}_i$ at the beginning $\{0\} \times S^1$ is given by $\langle \mathcal{B}_f| \,\mathrm{e}^{-LH_c} |\mathcal{B}_i\rangle$. This can also be regarded as the partition sum of the open string states on $[0, L]$ with the boundary condition $\overline{\mathcal{B}}_i$ at the left end $\{0\}$ and the boundary condition $\mathcal{B}_f$ at the right end $\{L\}$. If the circumference of the circle is β, then, it is $\mathrm{Tr}_{\mathcal{H}_{\overline{\mathcal{B}}_i, \mathcal{B}_f}} \mathrm{e}^{-\beta H_o}$ where $\mathcal{H}_{\overline{\mathcal{B}}_i, \mathcal{B}_f}$ is the Hilbert space of open string states and H_o is the Hamiltonian ("o" for "open" string). We obtained a relation

$$ {}_{\beta}\langle \mathcal{B}_f| \,\mathrm{e}^{-LH_c} |\mathcal{B}_i\rangle_{\beta} = \mathrm{Tr}_{\mathcal{H}_{\overline{\mathcal{B}}_i, \mathcal{B}_f}(L)} \mathrm{e}^{-\beta H_o}, \tag{B.1} $$

where the dependence of the boundary states and the Hilbert space on the lengths β and L is made explicit.

More generally, let Σ be a surface with an outgoing boundary circle $\partial\Sigma \cong S^1$, having a neighborhood isomorphic to $(-\epsilon, 0] \times S^1$ for some $\epsilon > 0$. The path integral on the fields on the interior of Σ defines a state $|\Sigma\rangle \in \mathcal{H}_{S^1}$. Then, the partition function on Σ with the boundary condition $\mathcal{B}$ at $\partial\Sigma$ is given by

$$ Z(\mathcal{B}|\Sigma) = \langle \mathcal{B}|\Sigma\rangle. \tag{B.2} $$

Its complex conjugate can be regarded as the partition function on a "conjugate surface" $\overline{\Sigma}$ with the conjugate boundary condition $\overline{\mathcal{B}}$,

$$ Z(\overline{\Sigma}|\overline{\mathcal{B}}) = \langle \Sigma|\mathcal{B}\rangle. \tag{B.3} $$

The surface $\overline{\Sigma}$ has an incoming boundary circle $\partial\overline{\Sigma} \cong -S^1$, with a neighborhood isomorphic to $[0, \epsilon) \times S^1$. Note that it can be regarded as the outgoing boundary by the sign flip of the coordinates, which involves the orientation reversal of the boundary. The conjugate boundary condition $\overline{\mathcal{B}}$ at the incoming boundary is sometimes equal to the original boundary condition $\mathcal{B}$ when the boundary is regarded as outgoing. For example, for the sigma model with target X, the D-brane $\mathcal{B}$ supporting a hermitian vector bundle on X with a unitary connection (E, A) is obtained by putting the Chan-Paton factor $\mathrm{tr}_E \mathrm{Pexp}\left(-i\int_{\partial\Sigma}\phi^* A\right)$ in the path integral weight. The complex conjugate of this factor is

$$\text{(B.4)} \qquad \left[\mathrm{tr}_E \mathrm{Pexp}\left(-i\int_{\partial\Sigma}\phi^* A\right)\right]^* = \mathrm{tr}_{E^\vee}\mathrm{Pexp}\left(i\int_{-\partial\overline{\Sigma}}\phi^* A^T\right) = \mathrm{tr}_E \mathrm{Pexp}\left(-i\int_{\partial\overline{\Sigma}}\phi^* A\right)$$

where we used the unitarity of the connection, $A^* = A^T$, in the first equality. The above equation means that the conjugate boundary condition $\overline{\mathcal{B}}$ supports the dual bundle with the dual connection $(E^\vee, -A^T)$, and that $\overline{\mathcal{B}}$ at the incoming boundary $-\partial\overline{\Sigma}$ is equal to the original $\mathcal{B}$ at the outgoing boundary $\partial\overline{\Sigma}$. In particular, we have $Z(\overline{\Sigma}|\overline{\mathcal{B}}) = Z(\mathcal{B}|\overline{\Sigma})$, that is,

$$\text{(B.5)} \qquad (\langle\mathcal{B}|\Sigma\rangle)^* = \langle\Sigma|\mathcal{B}\rangle = \langle\mathcal{B}|\overline{\Sigma}\rangle.$$

A boundary condition is said to be *unitary* when it has this property. All the boundary conditions we consider in this paper are unitary.

When the system has spinors among the fields, we need to specify the spin structure on the domain surface. On the flat cylinder, for each of the two chiralities, there are two choices — Ramond (R) sector and Neveu-Schwarz (NS) sector in which the parallel transport along the non-trivial circle is the identity and the sign flip respectively. In total, there are four sectors on the cylinder, (R,R), (NS,NS), (R,NS) and (NS,R), or RR, NSNS, RNS and NSR for short.[1] At the boundary of a surface, we need to specify an identification of the spin bundles of the opposite chiralities (which is also a part of the spin structure). Therefore, a boundary circle has two spin structures — RR sector and NSNS sector. Accordingly, for each boundary condition $\mathcal{B}$, there are two boundary states — $|\mathcal{B}\rangle_{\mathrm{RR}}$ and $|\mathcal{B}\rangle_{\mathrm{NSNS}}$. In particular, there are two versions of the identity (B.1),

$$\text{(B.6)} \qquad {}_{\mathrm{RR}}\langle\mathcal{B}_f|\,\mathrm{e}^{-LH_c}|\mathcal{B}_i\rangle_{\mathrm{RR}} = \mathrm{Tr}_{\mathcal{H}_{\overline{\mathcal{B}}_i,\mathcal{B}_f}}(-1)^F\,\mathrm{e}^{-\beta H_o},$$

$$\text{(B.7)} \qquad {}_{\mathrm{NSNS}}\langle\mathcal{B}_f|\,\mathrm{e}^{-LH_c}|\mathcal{B}_i\rangle_{\mathrm{NSNS}} = \mathrm{Tr}_{\mathcal{H}_{\overline{\mathcal{B}}_i,\mathcal{B}_f}}\mathrm{e}^{-\beta H_o}.$$

In a supersymmetric system, when $\mathcal{B}_f$ and $\overline{\mathcal{B}}_i$ preserve a common supersymmetry on the strip $[0, L]\times\mathbb{R}$, then, the right hand side of (B.6)

[1]The "RR gauge potentials" mentioned in Section 5.1 are degrees of freedom coming from the closed string states in the RR sector.

is the open string Witten index $\mathrm{I}_{\overline{\mathcal{B}}_i,\mathcal{B}_f}$. Since the index does not depend on the lengths, it must be the same as the value in the limit $L \to \infty$. Since e^{-LH_c} approaches the projection P_{v} to the subspace of zero energy states (supersymmetric ground states) as $L \to \infty$, we find

$$\text{(B.8)} \qquad \mathrm{I}_{\overline{\mathcal{B}}_i,\mathcal{B}_f} = {}_{\mathrm{RR}}\langle \mathcal{B}_f | P_{\mathrm{v}} | \mathcal{B}_i \rangle_{\mathrm{RR}}.$$

Let us now consider a $(2,2)$ supersymmetric quantum field theory. Let $\mathcal{B}$ be a B-brane boundary condition. If it is imposed at the right boundary $x = 0$ of the left-half Minkoswki spacetime $\{(x,t)|\, x \leq 0\}$, then the supercharges $\overline{Q}_+ + \overline{Q}_-$ and $Q_+ + Q_-$ are preserved. This means that the space components of the supercurrents vanishes at the boundary, $(\overline{G}^x_+ + \overline{G}^x_-)|_{x=0} = (G^x_+ + G^x_-)|_{x=0} = 0$. In order to translate this into the condition on the boundary state, we need to perform the Wick rotation $t \to -i\tau$ first, and then the 90° rotation $(x', \tau') = (-\tau, x)$ in order to trade the space and time. Note that $z = x + i\tau$ and $z' = x' + i\tau'$ are related by $z' = iz$. Since $\overline{G}_+\sqrt{\mathrm{d}\overline{z}} = \overline{G}_{+'}\sqrt{\mathrm{d}\overline{z}'}$, $\overline{G}_-\sqrt{\mathrm{d}z} = \overline{G}_{-'}\sqrt{\mathrm{d}z'}$, $G_+\sqrt{\mathrm{d}\overline{z}} = G_{+'}\sqrt{\mathrm{d}\overline{z}'}$ and $G_-\sqrt{\mathrm{d}z} = G_{-'}\sqrt{\mathrm{d}z'}$, the equations of the current become the condition on the boundary states

$$\text{(B.9)} \qquad \langle \mathcal{B}|(\overline{G}^{\tau'}_{+'} + i\overline{G}^{\tau'}_{-'}) = \langle \mathcal{B}|(G^{\tau'}_{+'} + iG^{\tau'}_{-'}) = 0.$$

By conjugation, they become

$$\text{(B.10)} \qquad (\overline{G}^{\tau'}_{+'} - i\overline{G}^{\tau'}_{-'})|\mathcal{B}\rangle = (G^{\tau'}_{+'} - iG^{\tau'}_{-'})|\mathcal{B}\rangle = 0.$$

These are both in RR and NSNS sectors. Undoing the 90° rotation and the Wick rotation, this means that the conjugate boundary condition $\overline{\mathcal{B}}$ at the left boundary $x = 0$ of the right-half Minkowski spacetime $\{(x,t)|\, x \geq 0\}$ preserves the supercharges $\overline{Q}_+ - \overline{Q}_-$ and $Q_+ - Q_-$. That is, $\overline{\mathcal{B}}$ is a B_{-1}-brane boundary condition. Likewise, if $\mathcal{B}$ is a $\mathrm{B}_{\mathrm{e}^{iu}}$-brane (*resp.* $\mathrm{A}_{\mathrm{e}^{iu}}$-brane), then $\overline{\mathcal{B}}$ is a $\mathrm{B}_{-\mathrm{e}^{iu}}$-brane (*resp.* $\mathrm{A}_{-\mathrm{e}^{iu}}$-brane). As noted earlier, (B.6) can be interpreted as an open string Witten index and (B.8) holds, provided $\mathcal{B}_f$ and $\overline{\mathcal{B}}_i$ preserve the same set of supercharges. This is when $\mathcal{B}_f$ and $\mathcal{B}_i$ preserve the same types of supercharges but with the opposite relative signs. For example, when $\mathcal{B}_f$ is a B-brane and $\mathcal{B}_i$ is a B_{-1}-brame.

The hemisphere partition function is the partition function on D^2 with a particular background, having an outgoing boundary $\partial D^2 = S^1$ in the NSNS sector. For example, the one preserving $\mathrm{A}_{(\pm)}$-type supersymmetry with a $\mathrm{B}_{\pm 1}$-type boundary condition $\mathcal{B}$ may be denoted by

$$\text{(B.11)} \qquad Z^{\mathrm{A}_{(\pm)}}_{D^2}(\mathcal{B}) = Z(\mathcal{B}|D^2, \mathrm{A}_{(\pm)}) = {}_{\mathrm{NSNS}}\langle \mathcal{B}|D^2, \mathrm{A}_{(\pm)}\rangle_{\mathrm{NSNS}}.$$

To see what the complex conjugate is, we first note that the conjugate hemisphere is the northern hemiphere D^2_∞ (if D^2 is a southern hemisphere) obtained by $z \to \overline{z}^{-1}$. This transforms the conformal Killing spinors as $\mathbf{s}_{\pm\frac{1}{2}} \to \mp i\widetilde{\mathbf{s}}_{\mp\frac{1}{2}}$ and $\widetilde{\mathbf{s}}_{\pm\frac{1}{2}} \to \pm i\mathbf{s}_{\mp\frac{1}{2}}$. This implies that the conjugate background is $\overline{(D^2, \mathrm{A}_{(\pm)})} = (D^2_\infty, \mathrm{A}^{-1}_{(\pm)})$ with an incoming boundary, giving $Z(\mathcal{B}|D^2, \mathrm{A}_{(\pm)})^* = Z(D^2_\infty, \mathrm{A}^{-1}_{(\pm)}|\overline{\mathcal{B}})$. Note that $\overline{\mathcal{B}}$ is an $\mathrm{B}_{\mp 1}$-brane and that is consistent with the fact that the background $D^2_\infty, \mathrm{A}^{-1}_{(\pm)}$ preserves the $\mathrm{B}_{\mp 1}$-type supersymmetry at the boundary. We can regard the northern hemisphre as the southern hemisphere by the map $z \to z^{-1}$. Under this, the conformal Killing spinors transform as $\mathbf{s}_{\pm\frac{1}{2}} \to \pm i\mathbf{s}_{\mp\frac{1}{2}}$ and $\widetilde{\mathbf{s}}_{\pm\frac{1}{2}} \to \mp i\widetilde{\mathbf{s}}_{\mp\frac{1}{2}}$. This means that $(D^2_\infty, \mathrm{A}^{-1}_{(\pm)})$ with incoming boundary can be regarded as $(D^2, \mathrm{A}^{-1}_{(\mp)})$ with outgoing boundary. If $\mathcal{B}$ is unitary, we have $Z(D^2_\infty, \mathrm{A}^{-1}_{(\pm)}|\overline{\mathcal{B}}) = Z(\mathcal{B}|D^2, \mathrm{A}^{-1}_{(\mp)})$. Combining what we have seen, we find

$$Z(\mathcal{B}|D^2, \mathrm{A}_{(\pm)})^* = Z(D^2_\infty, \mathrm{A}^{-1}_{(\pm)}|\overline{\mathcal{B}}) = Z(\mathcal{B}|D^2, \mathrm{A}^{-1}_{(\mp)}), \tag{B.12}$$

that is, ${}_{\mathrm{NSNS}}\langle D^2, \mathrm{A}_{(\pm)}|\mathcal{B}\rangle_{\mathrm{NSNS}} = {}_{\mathrm{NSNS}}\langle \mathcal{B}|D^2, \mathrm{A}^{-1}_{(\mp)}\rangle_{\mathrm{NSNS}}$. The same holds when $(\mathrm{A}_{(\pm)}, \mathrm{A}^{-1}_{(\pm)})$ is replaced by $(\mathrm{A}^{\mathrm{e}^{iu}}_{(\pm)}, \mathrm{A}^{-\mathrm{e}^{iu}}_{(\pm)})$ or $(\mathrm{B}^{\mathrm{e}^{iu}}_{(\pm)}, \mathrm{B}^{-\mathrm{e}^{iu}}_{(\pm)})$. Note that the expressiond for the hemisphere partition functions, (4.21) and (4.32) for GLSMs and (4.27)–(4.28) and (4.31) for LG models, obey this rule.

C. Additional References

After the first version of the present note is submitted for refereeing process, we noticed a few relevant references on which we make some comments.

C.1. Hemisphere versus central charge

A proof of the relation (5.5), $Z_{D^2}(\mathcal{B}) = {}_{\mathrm{RR}}\langle \mathcal{B}|0\rangle_{\mathrm{RR}} =: \Pi_0^{\mathcal{B}}$ up to a constant multiple, between the hemisphere partition function and the CV central charge was proposed by Bachas and Plencner in [75] in the case where the theory is a $(2,2)$ superconformal field theory with charge integrality and the boundary condition $\mathcal{B}$ is superconformal and "stable". A superconformal boundary condition $\mathcal{B}$ is said to be stable when there is a phase $\mathrm{e}^{i\varphi}$ such that

$${}_{\mathrm{RR}}\langle \mathcal{B}|\phi_\alpha\rangle_{\mathrm{RR}} = \mathrm{e}^{i\varphi}\,{}_{\mathrm{NSNS}}\langle \mathcal{B}|\phi_\alpha\rangle_{\mathrm{NSNS}}, \tag{C.1}$$

for any chiral primary field ϕ_α where the states $|\phi_\alpha\rangle_{\mathrm{RR}}$ and $|\phi_\alpha\rangle_{\mathrm{NSNS}}$ are related by the spectral flow. Let us briefly review their logic. To be specific, we consider the hemisphere for a B-brane $\mathcal{B}$, so that the focus will be the dependence on the moduli space $\mathfrak{M}^0 = \mathfrak{M}^0_A$ of superconformal field theories parametrized by the exactly marginal A-term deformation parameters.

The argument closely follows the lines of [8] for the proof of the relation (5.10), $Z_{S^2} = \mathrm{e}^{-K}$, between the sphere partition function and the Kähler potential of the moduli space $\mathfrak{M}^0$. First, they formulated the system on a surface with boundary, coupled to a supergravity background, and wrote down the "anomaly formula" that determines the possible form of the response of the partition function to a change of the background. The formula includes the Kähler potential K and a holomorphic function $h^{\mathcal{B}}$ on the moduli space $\mathfrak{M}^0$, where the latter depends on the boundary condition $\mathcal{B}$. Integrating the formula, they found that the hemisphere partition function is given by

$$Z_{D^2}(\mathcal{B}) = \text{constant} \times \mathrm{e}^{h^{\mathcal{B}}}. \tag{C.2}$$

By conformal perturbation theory, K and $h^{\mathcal{B}}$ are related to the one point function of the A-term marginal operators $\mathcal{O}_i$ on the left half-space $\mathbb{C}_{\leq 0} = \{w \in \mathbb{C} \,|\, \mathrm{Re}(w) \leq 0\}$ as[1]

$$\frac{1}{4\pi}\partial_i(K + h^{\mathcal{B}}) = \langle \mathcal{O}_i(-1)\rangle^{\mathcal{B}}_{\mathbb{C}_{\leq 0}}. \tag{C.3}$$

The operator $\mathcal{O}_i$ is related to an A-chiral primary operator ϕ_i of conformal weight $(\frac{1}{2}, \frac{1}{2})$ by the descent relation $\mathcal{O}_i = Q_+\overline{Q}_-\phi_i$. By the supersymmetric Ward identity, this one point function is proportional to the one for ϕ_i, and by conformal Ward identity for $w \mapsto z = \frac{1+w}{1-w}$, it is proportional to the one point function on the disc $D^2 = \{z \in \mathbb{C} \,|\, |z| \leq 1\}$:

$$\begin{aligned} \langle \mathcal{O}_i(-1)\rangle^{\mathcal{B}}_{\mathbb{C}_{\leq 0}} &= -i\langle \phi_i(-1)\rangle^{\mathcal{B}}_{\mathbb{C}_{\leq 0}} = -\frac{i}{2}\langle \phi_i(0)\rangle^{\mathcal{B}}_{D^2} \\ &= -\frac{i}{2}\frac{{}_{\mathrm{NSNS}}\langle \mathcal{B}|\phi_i\rangle_{\mathrm{NSNS}}}{{}_{\mathrm{NSNS}}\langle \mathcal{B}|\,\mathbf{1}\,\rangle_{\mathrm{NSNS}}} = -\frac{i}{2}\frac{{}_{\mathrm{RR}}\langle \mathcal{B}|\phi_i\rangle_{\mathrm{RR}}}{{}_{\mathrm{RR}}\langle \mathcal{B}|\,\mathbf{1}\,\rangle_{\mathrm{RR}}} \\ &= -\frac{i}{2}\frac{\Pi_i^{\mathcal{B}}}{\Pi_0^{\mathcal{B}}}. \end{aligned} \tag{C.4}$$

In the second line, we used the stability condition (C.1). On the other hand, the first equation in (5.7) for the circumference $\beta = 2\pi$ and for

[1] As noted in Section 5.2, we normalize the deformation as $\Delta S_E = \int_\Sigma \mathrm{d}^2x(\Delta t^i\mathcal{O}_i + \Delta\bar{t}^{\bar{i}}\overline{\mathcal{O}}_{\bar{i}})$. In this normalization, Zamolodchikov metric is $G(\partial_i, \partial_{\bar{j}}) = \langle \mathcal{O}_i(1)\overline{\mathcal{O}}_{\bar{j}}(0)\rangle_{\mathbb{C}} = \frac{1}{\pi^2}\partial_i\partial_{\bar{j}}K$.

the operators $\phi_\alpha = \phi_i$ and $\phi_\beta = \mathbf{1}$ reads $\partial_i \Pi_0^{\mathcal{B}} = -\partial_i K \Pi_0^{\mathcal{B}} - 2\pi i \Pi_i^{\mathcal{B}}$ since $A_{i0}^0 = g^{0\bar{0}} \partial_i g_{\bar{0}0} = -\partial_i K$, where we used $g_{\bar{0}0} = \mathrm{e}^{-K}$ [45]. That is, $\partial_i(K + \log \Pi_0^{\mathcal{B}}) = -2\pi i \Pi_i^{\mathcal{B}} / \Pi_0^{\mathcal{B}}$. Thus, we find

$$\langle \mathcal{O}_i(-1) \rangle_{\mathbb{C}_{\leq 0}}^{\mathcal{B}} = \frac{1}{4\pi} \partial_i (K + \log \Pi_0^{\mathcal{B}}). \tag{C.5}$$

Comparing (C.3) and (C.5), we find that $h^{\mathcal{B}} = \log \Pi_0^{\mathcal{B}}$ up to a constant addition. Combining this with (C.2), we obtain the wanted relation, $Z_{D^2}(\mathcal{B}) = \Pi_0^{\mathcal{B}}$ up to a constant multiple.

Since the hemisphere partition function and the CV central charge are both invariant under continuous deformation of the boundary condition, the relation $Z_{D^2}(\mathcal{B}) = \Pi_0^{\mathcal{B}}$ holds for a general, not necessarily superconformal nor stable, boundary condition $\mathcal{B}$ provided there exists a superconformal and stable boundary condition in the same homotopy class.

C.2. Mathematical works related to the renormalization group flow

In [9] (see also Section 4.4), taking the example $\mathrm{T}_{N,d}^{U(1)}$ with $N \neq d$ for illustration, we examined the behaviour of the hemisphere partition function in the ultra-violet ($r \to 0$) and infra-red ($r \to \infty$) limits, and the result was used to learn which brane in the UV theory flows to which part of the IR theory. In [76–78], analogous problems were studied in the context of the Gromov-Witten theory of a Fano manifold X.

In [76] (see also [77]), Galkin, Golyshev and Iritani studied the behaviour of solutions to the quantum differential equation $\widetilde{\nabla} s = 0$ in the limit $z \to 0$ and presented two conjectures, *Gamma conjectures I* and *II*. Recall from Section 5.3 that the hemisphere partition function $Z_{D^2}^{\mathrm{LV}}$ of the sigma model is proportional to a component Z_{GW} of a solution to the quantum differential equation, when the parameter z in the latter is set to be the inverse radius of the hemisphere, $z \propto r^{-1}$. Therefore, $z \to 0$ corresponds to the infra-red limit $r \to \infty$. When translated into the sigma model language via this connection, the conjectures become assertions concerning the IR behaviour of B-branes in the sigma model.

Let h be the Fano index of X, the largest integer such that $c_1(X)/h$ is an integral class. Then, the group G_A of anomaly free axial R-symmetries is isomorphic to $\mathbb{Z}_{2h}$. It includes the fermion sign flip $(-1)^F$ which can never be spontaneously broken. We consider the sigma model with vanishing B-field, $B = 0$. We assume that the set of vacua that maximize $|\widetilde{W}|$ is a single G_A orbit with stabilizer $\{1, (-1)^F\}$ including one with $\widetilde{W} = |\widetilde{W}|$. Then, the conjectures read as follows.

Gamma conjecture I

The structure sheaf $\mathcal{O}_X$ descends at low energies to a brane supported purely at the vacuum with $\widetilde{W} = |\widetilde{W}|_{\max}$.

Gamma conjecture II

Suppose that all the vacua, say $\mathbf{v}_1, \dots \mathbf{v}_N$, have mass gaps and that $D^b_{\mathrm{Coh}}(X)$ has a full exceptional collection. Then, for each phase $\mathrm{e}^{i\phi}$ such that any non-zero $\widetilde{W}(\mathbf{v}_i) - \widetilde{W}(\mathbf{v}_j)$ is not parallel to $\mathrm{e}^{i\phi}\mathbb{R}$, there are branes $E_1^\phi, \dots, E_N^\phi$ in the sigma model that descend at low energies to branes supported purely at the vacua $\mathbf{v}_1, \dots \mathbf{v}_N$ respectively. Moreover, if $\mathrm{Im}(\mathrm{e}^{-i\phi}\widetilde{W}(\mathbf{v}_{\sigma(i)})) \leq \mathrm{Im}(\mathrm{e}^{-i\phi}\widetilde{W}(\mathbf{v}_{\sigma(j)}))$ for a permutation $\sigma \in \mathfrak{S}_N$, $E^\phi_{\sigma(1)}, \dots, E^\phi_{\sigma(N)}$ form a full exceptional collection of $D^b_{\mathrm{Coh}}(X)$.[2]

The model $\mathrm{T}^{U(1)}_{N,d}$ with $d < N$ corresponds to the Fano sigma model where the target space X is the degree d hypersurface $f = 0$ in the projective space $\mathbb{P}^{N-1}$, with Fano index $h = N - d$. The theory has $N - d$ massive vacua at $(\sigma/\Lambda)^{N-d} = (-d)^d\,\mathrm{e}^{-t}$ (see (3.17)) with $\widetilde{W} = (N-d)\sigma$, breaking the axial R-symmetry $G_A = \mathbb{Z}_{2(N-d)}$ to $\mathbb{Z}_2$, and, for $d \geq 2$, axial symmetry preserving vacua with $\widetilde{W} = 0$ of an SCFT with $\widehat{c} = N(1-2/d)$, the infra-red fixed point of the LG orbifold $(\mathbb{C}^N/\mathbb{Z}_d, f)$. In the case $d = 2$, the sector at $\widetilde{W} = 0$ has $\widehat{c} = 0$ and consists of one massive vacuum (*resp.* two massive vacua) when N is odd (*resp.* even). Note that B-field vanishes when $\theta = -\pi d$ (see (3.15)), in which case the G_A-breaking massive vacua are at $\sigma_k = \Lambda(d^d\,\mathrm{e}^{-\zeta})^{\frac{1}{(N-d)}}\,\mathrm{e}^{\frac{2\pi i k}{N-d}}$ for $k \in \mathbb{Z}/(N-d)\mathbb{Z}$ and have $\widetilde{W} = (N-d)\sigma_k$. Indeed, when Λ is real, the vacuum σ_0 has $\widetilde{W} = |\widetilde{W}|_{\max}$. In comparison with the results with the hemisphere, we should rotate the twisted superpotential as $\widetilde{W} \to -i\widetilde{W}$ due to the change $\Lambda \to -i\Lambda$ (4.36). We have seen in Section 4.4 that the line bundle $\mathcal{O}_X(q)$ for $-N\pi < \theta + 2\pi q < (N-2d)\pi$ descends to a brane supported purely at σ_q. Note that the bound on q for $\theta = -\pi d$ (for $B = 0$) reads $-(N-d)/2 < q < (N-d)/2$, and $q = 0$ satisfies it. In particular, the structure sheaf $\mathcal{O}_X$ descends to a brane at the vacuum σ_0 that has $\widetilde{W} = -i|\widetilde{W}|_{\max}$. Gamma conjecture I indeed holds. In the case $d = 1$ ($X \cong \mathbb{P}^{N-2}$) and $d = 2$ (X is a quadric in $\mathbb{P}^{N-1}$), all the vacua are massive and $D^b_{\mathrm{Coh}}(X)$ has an exceptional collection. Thus, the set-up of Gamma conjecture II is satisfied. The phase $\mathrm{e}^{i\phi}$ may be identified with the phase $\mathrm{e}^{-2i\beta}$ that determines the preserved supersymmetry. The case $d = 1$ is as discussed in [76] and the conjecture

---

[2]The inequality appears to be opposite to the one in [76, 77], but this is due to the duality $\mathcal{E} \to \mathcal{E}^\vee$ that appears in the footnote in page 195.

holds. For the case $d = 2$ we need to consider the vacua (vacuum) with $\widetilde{W} = 0$ that preserve(s) the axial R-symmetry. It would be interesting to explicitly contruct the branes in the sigma model that descend prely to these (this) vacua (vacuum). In the case $d \geq 3$, the theory has a non-trivial SCFT as an IR fixed point and the set-up of Gamma conjecture II is not satisfied. However, we certainly have a (non-full) exceptional collection in $D^b_{\mathrm{Coh}}(X)$ that descend to branes supported at the massive vacua, and $D^b_{\mathrm{Coh}}(X)$ admits a semi-orthogonal decomposition by these objects and the category $\mathrm{MF}_{\mathbb{Z}_d}(f)$ from the SCFT at $\widetilde{W} = 0$, where the latter sits in the "middle".

In [78], Acosta considered the Picard-Fuchs differential equation (4.108) for the model $\mathrm{T}^{U(1)}_{\vec{w},d}$ and studied the behaviour of solutions at $\mathrm{e}^{-t_R} \to 0$ and at $\mathrm{e}^{-t_R} \to \infty$. Let us restrict the description of this work to the model $\mathrm{T}^{U(1)}_{N,d}$ which is familiar to us, though extension to the model with general $\vec{w}$ is very easy. The focus of [78] is the comparison between the solutions at $\mathrm{e}^{-t_R} \sim 0$ determined by the Gromov-Witten theory of the hypersurface X_f and the solutions at $\mathrm{e}^{-t_R} \sim \infty$ determined by the FJRW theory of $(\mathbb{C}^N/\mathbb{Z}_d, f)$. When $d < N$ where the UV theory is the sigma model, the former is analytic while the latter is formal, and when $d > N$ where the UV theory is the LG orbifold, the former is formal while that latter is analytic. In [78], it was shown that a part of the the analytic solution, when continued to the opposite regime, has the formal solution as its asymptotic expansion. This is relevant to our question: which branes in the UV theory descend to branes in the non-trivial IR fixed point.

C.3. Picard-Fuchs equations as deformed chiral ring relations

In Section 4.6, we presented a picture of Picard-Fuchs equations as a consequence of operator relations at the origin of the hemisphere, which approach the chiral ring relations in the flat space limit $r \to \infty$.

Similar pictures were presented in [79] in a different but related context (two point functions on the Ω-deformed sphere) and also in [80] in a closer context (two point function on the sphere with the background of [4, 5]); Picard-Fuchs equations are regarded as relations among correlators and they are interpreted as consequences of the operator relations, the Ω-deformed chiral ring relations in [79] and the relations identical to ours in [80].

References

[1] E. Witten, "Constraints on Supersymmetry Breaking," Nucl. Phys. B **202** (1982) 253.

[2] P. Candelas, X. C. De La Ossa, P. S. Green and L. Parkes, "A Pair of Calabi-Yau manifolds as an exactly soluble superconformal theory," Nucl. Phys. B **359** (1991) 21.

[3] V. Pestun, "Localization of gauge theory on a four-sphere and supersymmetric Wilson loops," Commun. Math. Phys. **313** (2012) 71 [arXiv:0712.2824 [hep-th]].

[4] F. Benini and S. Cremonesi, "Partition Functions of $\mathcal{N} = (2,2)$ Gauge Theories on S^2 and Vortices," Commun. Math. Phys. **334** (2015) no. 3, 1483 [arXiv:1206.2356 [hep-th]].

[5] N. Doroud, J. Gomis, B. Le Floch and S. Lee, "Exact Results in D=2 Supersymmetric Gauge Theories," JHEP **1305** (2013) 093 [arXiv:1206.2606 [hep-th]].

[6] H. Jockers, V. Kumar, J. M. Lapan, D. R. Morrison and M. Romo, "Two-Sphere Partition Functions and Gromov-Witten Invariants," Commun. Math. Phys. **325** (2014) 1139, doi:10.1007/s00220-013-1874-z [arXiv:1208.6244 [hep-th]].

[7] E. Gerchkovitz, J. Gomis and Z. Komargodski, "Sphere Partition Functions and the Zamolodchikov Metric," JHEP **1411** (2014) 001 [arXiv:1405.7271 [hep-th]].

[8] J. Gomis, P. S. Hsin, Z. Komargodski, A. Schwimmer, N. Seiberg and S. Theisen, "Anomalies, Conformal Manifolds, and Spheres," JHEP **1603** (2016) 022 [arXiv:1509.08511 [hep-th]].

[9] K. Hori and M. Romo, "Exact Results In Two-Dimensional (2, 2) Supersymmetric Gauge Theories With Boundary," arXiv:1308.2438 [hep-th].

[10] S. Sugishita and S. Terashima, "Exact Results in Supersymmetric Field Theories on Manifolds with Boundaries," JHEP **1311** (2013) 021 [arXiv:1308.1973 [hep-th]].

[11] D. Honda and T. Okuda, "Exact results for boundaries and domain walls in 2d supersymmetric theories," JHEP **1509** (2015) 140 [arXiv:1308.2217 [hep-th]].

[12] K. Fukaya, Y-G. Oh, H. Ohta and K. Ono, "Lagrangian intersection Floer theory: anomaly and obstruction, Part I and II" (American Mathematical Society, 2009).

[13] A. B. Zamolodchikov, "Irreversibility of the Flux of the Renormalization Group in a 2D Field Theory," JETP Lett. **43** (1986) 730 [Pisma Zh. Eksp. Teor. Fiz. **43** (1986) 565].

[14] V. Periwal and A. Strominger, "Kahler Geometry of the Space of $N = 2$ Superconformal Field Theories," Phys. Lett. B **235** (1990) 261.

[15] K. Saito, "On the periods of primitive integrals I," Harvard preprint (1980).

[16] K. Saito, "Period Mapping Associated to a Primitive Form," Publ. RIMS **19** (1983) 1231–1264.

[17] K. Saito, "Primitive forms for a universal unfolding of a function with an isolated critical point," J. Fac. Sci. Univ. Tokyo Sec. IA Math. **28** (1982) 775–792.

[18] A. Givental, "Semisimple Frobenius structures at higher genus," Int. Math. Res. Notices (2001) 1265–1286 [math/0008067].

[19] C. Teleman, "The structure of 2D semi-simple field theories," Invent. Math. **188** (2012) 525–588 [arXiv:0712.0160 [math.AT]].

[20] H. Fan, T. Jarvis and Y. Ruan, "The Witten equation, mirror symmetry, and quantum singularity theory," Ann. Math. **178** (2013) 1–106 [arXiv:0712.4021 [math.AG]]

[21] I.M. Gelfand, M.M. Kapranov and A.V. Zelevinsky, *Discriminants, Resultants and Multidimensional Determinants*, (Birkhäuser, 1994).

[22] M. Herbst, K. Hori and D. Page, "Phases Of N=2 Theories In 1+1 Dimensions With Boundary," arXiv:0803.2045 [hep-th].

[23] H. Knörrer, "Cohen-Macaulay modules on hypersurface singularities. I.," Invent. Math. **88** (1987) 153–164.

[24] M.U. Isik, "Equivalence of the derived category of a variety with a singularity category," Int. Math. Res. Notices (2013) 2787–2808 [arXiv:1011.1484 [math.AG]].

[25] I. Shipman, "A Geometric approach to Orlov's theorem," Compos. Math. **148** (2012) 1365–1389 [arXiv:1012.5282 [math.AG]].

[26] A. Preygel, "Thom-Sebastiani & duality for matrix factorizations," MIT Thesis (2012) [arXiv:1101.5834 [math.AG]].

[27] A. Recknagel and V. Schomerus, "D-branes in Gepner models," Nucl. Phys. B **531** (1998) 185 [hep-th/9712186].

[28] E. Segal, "Equivalences between GIT quotients of Landau-Ginzburg B-models," Commun. Math. Phys. **304** (2011) 411–432; `arXiv:0910.5534`.

[29] D. Halpern-Leistner, "The derived category of a GIT quotient," J. Amer. Math. Soc. **28** (2015) 871–912 [arXiv:1203.0276 [math.AG]].

[30] M. Ballard, D. Favero and L. Katsarkov, "Variation of geometric invariant theory quotients and derived categories," arXiv:1203.6643 [math.AG].

[31] C. Closset and S. Cremonesi, "Comments on $\mathcal{N} = (2,2)$ supersymmetry on two-manifolds," JHEP **1407** (2014) 075 [arXiv:1404.2636 [hep-th]].

[32] K. Hori and C. Vafa, "Mirror symmetry," hep-th/0002222.

[33] K. Hori, A. Iqbal and C. Vafa, "D-branes and mirror symmetry," hep-th/0005247.

[34] E. Witten, Lecture 15 in *Quantum Fields and Strings: A Course for Mathematicians*, Vol. 2, P. Deligne et al, Eds, (AMS 2000).

[35] M. Jinzenji, "On Quantum cohomology rings for hypersurfaces in CP**(N-1)," J. Math. Phys. **38** (1997) 6613 [hep-th/9511206]; A. Collino and M. Jinzenji, "On the structure of small quantum cohomology rings for projective hypersurfaces," Commun. Math. Phys. **206** (1999) 157 [hep-th/9611053].

[36] D. R. Morrison, "Picard-Fuchs equations and mirror maps for hypersurfaces," In *Yau, S.T. (ed.): Mirror symmetry I* 185–199 [hep-th/9111025].

[37] A. Klemm and S. Theisen, "Considerations of one modulus Calabi-Yau compactifications: Picard-Fuchs equations, Kahler potentials and mirror maps," Nucl. Phys. B **389** (1993) 153 [hep-th/9205041].

[38] A. Font, "Periods and duality symmetries in Calabi-Yau compactifications," Nucl. Phys. B **391** (1993) 358 [hep-th/9203084].

[39] S. Hosono, A. Klemm, S. Theisen and S. T. Yau, "Mirror symmetry, mirror map and applications to Calabi-Yau hypersurfaces," Commun. Math. Phys. **167** (1995) 301 [hep-th/9308122].

[40] J. Polchinski, "Dirichlet Branes and Ramond-Ramond charges," Phys. Rev. Lett. **75** (1995) 4724 [hep-th/9510017].

[41] R. Minasian and G. W. Moore, "K theory and Ramond-Ramond charge," JHEP **9711** (1997) 002 [hep-th/9710230].

[42] Y. K. E. Cheung and Z. Yin, "Anomalies, branes, and currents," Nucl. Phys. B **517** (1998) 69 [hep-th/9710206].

[43] J. Walcher, "Stability of Landau-Ginzburg branes," J. Math. Phys. **46** (2005) 082305 [hep-th/0412274].

[44] S. Hosono, "Central charges, symplectic forms, and hypergeometric series in local mirror symmetry," hep-th/0404043.

[45] S. Cecotti and C. Vafa, "Topological antitopological fusion," Nucl. Phys. B **367** (1991) 359.

[46] S. Cecotti and C. Vafa, "On classification of N=2 supersymmetric theories," Commun. Math. Phys. **158** (1993) 569 [hep-th/9211097].

[47] H. Ooguri, Y. Oz and Z. Yin, "D-branes on Calabi-Yau spaces and their mirrors," Nucl. Phys. B **477** (1996) 407 [hep-th/9606112].

[48] J. Gomis and S. Lee, "Exact Kahler Potential from Gauge Theory and Mirror Symmetry," JHEP **1304** (2013) 019 [arXiv:1210.6022 [hep-th]].

[49] A. Givental, "Equivariant Gomov-Witten Invariants," In. Math. Res. Notices **13** (1996) 613–663; [math/9603021].

[50] B. Dubrovin, "Geometry of 2-D topological field theories," Lect. Notes Math. **1620** (1996) 120 [hep-th/9407018].

[51] B. Dubrovin, "Painleve transcendents and two-dimensional topological field theory," [math/9803107].

[52] A. Givental, "Gromov-Witten invariants and quantization of quadratic hamiltonians," Mosc. Math. J. **1** (2001) 551-568. [arXiv:math/0108100].

[53] H. Iritani, "An integral structure in quantum cohomology and mirror symmetry for toric orbifolds," Adv. Math. **222** (2009), no.3, 1016-1079 arXiv:0903.1463 [math.AG].

[54] L. Katzarkov, M. Kontsevich and T. Pantev, "Hodge theoretic aspects of mirror symmetry," arXiv:0806.0107 [math.AG].

[55] H. Iritani, "Quantum Cohomology and Periods," arXiv:1101.4512 [math.AG].

[56] R. Dijkgraaf, H. L. Verlinde and E. P. Verlinde, "Topological strings in $d < 1$," Nucl. Phys. B **352** (1991) 59.
[57] C. Hertling, "tt^* geometry, Frobenius manifolds, their connections, and the construction for singularities," J. Reine Angew. Math. **555** (2003) 77-161; [arXiv:math/0203054].
[58] H. Iritani, "tt^* geometry and quantum cohomology," [arXiv:0906.1307].
[59] E. Witten, "On the Structure of the Topological Phase of Two-dimensional Gravity," Nucl. Phys. B **340** (1990) 281.
[60] E. Witten, "'Two-dimensional gravity and intersection theory on moduli space," Surveys Diff. Geom. **1** (1991) 243.
[61] R. Dijkgraaf and E. Witten, "Mean Field Theory, Topological Field Theory, and Multimatrix Models," Nucl. Phys. B **342** (1990) 486.
[62] D. Cox and S. Katz, *Mirror Symmetry and Algebraic Geometry*, Mathematical Surveys and Monographs **68** (American Mathematical Society, 1991).
[63] K. Hori, S. Katz, A. Klemm, R. Pandharipande, R. Thomas, C. Vafa, R. Vakil and E. Zaslow, *Mirror Symmetry*, Clay Mathematics Monographs **1** (American Mathematical Society, 2003).
[64] K. Hori, "Constraints for topological strings in $D \geq 1$," Nucl. Phys. B **439** (1995) 395 [hep-th/9411135].
[65] E. Brezin and V. A. Kazakov, "Exactly Solvable Field Theories of Closed Strings," Phys. Lett. B **236** (1990) 144; M. R. Douglas and S. H. Shenker, "Strings in Less Than One-Dimension," Nucl. Phys. B **335** (1990) 635; D. J. Gross and A. A. Migdal, "Nonperturbative Two-Dimensional Quantum Gravity," Phys. Rev. Lett. **64** (1990) 127.
[66] G. 't Hooft, "A Planar Diagram Theory for Strong Interactions," Nucl. Phys. B **72** (1974) 461.
[67] J. Distler, "2-D Quantum Gravity, Topological Field Theory and the Multicritical Matrix Models," Nucl. Phys. B **342** (1990) 523.
[68] T. Banks, M. R. Douglas, N. Seiberg and S. H. Shenker, "Microscopic and Macroscopic Loops in Nonperturbative Two-dimensional Gravity," Phys. Lett. B **238** (1990) 279.
[69] J. Ambjorn, J. Jurkiewicz and Y. M. Makeenko, "Multiloop correlators for two-dimensional quantum gravity," Phys. Lett. B **251** (1990) 517.
[70] G. W. Moore, N. Seiberg and M. Staudacher, "From loops to states in 2-D quantum gravity," Nucl. Phys. B **362** (1991) 665.
[71] P. H. Ginsparg and G. W. Moore, "Lectures on 2-D gravity and 2-D string theory," (based on lectures at TASI Summer School 1992) [hep-th/9304011].
[72] R. Dijkgraaf, H. L. Verlinde and E. P. Verlinde, "Notes on topological string theory and 2-D quantum gravity," Lectures at Trieste and Cargèse, 1990.
[73] M. Fukuma, H. Kawai and R. Nakayama, "Continuum Schwinger-dyson Equations and Universal Structures in Two-dimensional Quantum Gravity," Int. J. Mod. Phys. A **6** (1991) 1385.

[74] R. Dijkgraaf, H. L. Verlinde and E. P. Verlinde, "Loop equations and Virasoro constraints in nonperturbative 2-D quantum gravity," Nucl. Phys. B **348** (1991) 435.

[75] C. Bachas and D. Plencner, "Boundary Weyl anomaly of $\mathcal{N} = (2,2)$ superconformal models," JHEP **1703** (2017) 034 [arXiv:1612.06386 [hep-th]].

[76] S. Galkin, V. Golyshev and H. Iritani, "Gamma classes and quantum cohomology of Fano manifolds: gamma conjectures," Duke Math. J. **165** (2016) 2005–2077; [arXiv:1404.6407 [math.AG]].

[77] S. Galkin and H. Iritani, "Gamma conjecture via mirror symmetry," this volume; [arXiv:1508.00719 [math.AG]].

[78] P. Acosta, "Asymptotic expansion and the LG/(Fano, general type) correspondence," [arXiv:1411.4162 [math.AG]].

[79] C. Closset, S. Cremonesi and D. S. Park, "The equivariant A-twist and gauged linear sigma models on the two-sphere," JHEP **1506** (2015) 076 [arXiv:1504.06308 [hep-th]].

[80] K. Hosomichi, S. Lee and T. Okuda, work in progress; talk "Vortex defects in 2D SUSY gauge theories," given by K. Hosomichi at Symposium *Frontiers in Mathematical Physics*, Rikkyo University, January 6-9, 2017.

Kavli Institute for the Physics and Mathematics of the Universe,
University of Tokyo,
Kashiwa, Chiba 277-8583, Japan

Institute for Advanced Study, Einstein Drive, Princeton,
NJ 08540, USA

Advanced Studies in Pure Mathematics 83, 2019
Primitive Forms and Related Subjects — Kavli IPMU 2014
pp. 221–236

On Dubrovin's Frobenius structures on Hurwitz spaces

Sergei K. Lando

Abstract.

Hurwitz spaces are spaces of meromorphic functions on algebraic curves. B. Dubrovin introduced Frobenius structures on Hurwitz spaces, which serve as one of the most spectacular examples of such structures. On the other hand, Hurwitz spaces admit natural compactification by stable maps. The main goal of the present paper is to formulate a question concerning the behavior of Dubrovin's Frobenius structures on Hurwitz spaces on the boundary of the compactification. Sample computations justifying the question are given.

Contents

§1. Introduction

According to B. Dubrovin [2], a *Frobenius structure* on a manifold M consists of the following data:

- a commutative associative algebra structure on each tangent space T_xM; we will denote the multiplication in this algebra by $\circ$;
- a unity vector field denoted by e, which possesses the property $e \circ e_1 = e_1$ for any vector field e_1 on M;

Received November 11, 2015.
2010 *Mathematics Subject Classification.* 14H70, 14D22, 30F20.
Key words and phrases. Frobenius structure, Riemann surfaces, moduli space, Lyashko–Looijenga map, Hurwitz spaces.

- an Euler vector field denoted by E;
- a flat metric $(\cdot,\cdot)$, which is invariant with respect to multiplication, meaning that $(e_1 \circ e_2, e_3) = (e_1, e_2 \circ e_3)$ for any three vector fields e_1, e_2, e_3 on M.

These data must satisfy certain compatibility conditions; we refer the reader to [2] for precise definitions.

Among the numerous important examples of Frobenius structures, Dubrovin considers Frobenius structures on Hurwitz spaces. A Hurwitz space is the space of rational functions on genus g complex curves, with prescribed ramification type over infinity. Dubrovin's Frobenius structures on Hurwitz spaces are, in fact, Frobenius structures on some coverings of the Hurwitz spaces, defined by picking up a symplectic basis of cycles in the complex curve, that depends continuously on the moduli of the curve.

The main goal of the present paper is to formulate a question concerning a possible behavior of Dubrovin's Frobenius structures on the boundary of compactified Hurwitz spaces. A compactification of Hurwitz spaces was defined in [3] for the needs of proving a formula, which generalizes Hurwitz's formula for the number of ramified coverings of the sphere by the sphere to coverings by surfaces of higher genera. Then it was shown in [4] that this compactification coincides with that by stable mappings.

Our sample computations show that at least some representatives of Frobenius structures on Hurwitz spaces have rather nice behavior on the boundary (can be extended holomorphically or have poles). The question we are asking is

Whether there exists a natural Frobenius structure on Hurwitz spaces whose extension to the boundary has at most poles as singularities?

The exposition below makes this question explicit.

§2. Saito's Frobenius structure on the space of polynomials

In [6, 7] K. Saito suggested a flat metric on the spaces of orbits of Coxeter groups. In particular, for groups of series A this gives a flat metric on spaces of polynomials in one variable. Below, we describe Dubrovin's interpretation of this metric and explain the additional required data producing, together with the metric, the Frobenius structure on the space of polynomials.

2.1. Saito's metric

The group A_{k-1} is the symmetric group of permutations of k elements. It acts on the vector space $\mathbb{C}^k$ by permuting the coordinates.

This action is reducible and decomposable into a direct sum of a trivial 1-dimensional representation and an irreducible $(k-1)$-dimensional representation of the symmetric group on the hyperplane $z_1 + \cdots + z_k = 0$. The quotient space of the latter action is naturally identified with the space of polynomials

$$P_k = \{z^k + a_2 z^{k-1} + \cdots + a_k\},$$

the coordinates $(z_1, \ldots, z_k)$ being the roots of the polynomials. The space $\mathcal{P}_k$ is the $(k-1)$-dimensional vector space $\mathbb{C}^{k-1}$ with the coordinates $(a_2, \ldots, a_k)$ (up to signs, the elementary symmetric functions in the roots $(z_1, \ldots, z_k)$).

The tangent space $T_f\mathcal{P}_k$ to $\mathcal{P}_k$ at any point $f \in \mathcal{P}_k$ is naturally identified with the vector space of polynomials in z of degree at most $k-2$. The *scalar product* (p_1, p_2) of two such polynomials *in the Saito metric* at a point $f \in \mathcal{P}_k$ is defined by the formula

$$(p_1, p_2) = -\mathrm{Res}_{z=\infty} \frac{p_1 p_2 dz}{f'}.$$

This metric is polynomial in the coordinates $a_2, \ldots, a_k$ and flat. For example, for $k = 4$, the matrix of the Saito metric in the basis $\partial/\partial a_i$, $i = 2, 3, 4$, looks like

$$\begin{pmatrix} 0 & 0 & \frac{1}{4} \\ 0 & \frac{1}{4} & 0 \\ \frac{1}{4} & 0 & -\frac{1}{8}a_2 \end{pmatrix}$$

A coordinate change leading to an anti-diagonal constant matrix of the Saito metric, whence to flat coordinates, now can be obtained easily.

2.2. Multiplication, the unity field and the Euler field

The multiplication in each tangent space $T_f\mathcal{P}_k$ is nothing but the multiplication in the local ring of the polynomial f:

$$p_1 \circ p_2 = p_1 p_2 \mod f',$$

that is, $p_1 \circ p_2$ is the polynomial remainder of the product $p_1 p_2$ divided by f'.

The unity field and the Euler field correspond to the additive and the multiplicative subgroup in the affine group of automorphisms of the target curve $\mathbb{C}P^1$, respectively:

$$e = \frac{\partial}{\partial a_k}, \qquad E = \frac{1}{k}\left(2a_2 \frac{\partial}{\partial a_2} + \cdots + k a_k \frac{\partial}{\partial a_k}\right).$$

2.3. Canonical coordinates

Canonical coordinates are well defined for the domain in $\mathcal{P}_k$ consisting of Morse polynomials, that is, of polynomials with $k-1$ pairwise distinct critical values. Choosing a numbering $t_1, \dots, t_{k-1}$ of these critical values, we obtain a system of local coordinates in a neighborhood of a Morse point in $\mathcal{P}_k$. In these coordinates, the multiplication $\circ$ looks like

$$\frac{\partial}{\partial t_i} \circ \frac{\partial}{\partial t_j} = \delta_{ij} \frac{\partial}{\partial t_i}.$$

Indeed, when shifting along a tangent vector $p \in T_f\mathcal{P}_k$ at a point $f \in \mathcal{P}_k$ we shift the i th critical value with the velocity proportional to the value of p at the i th critical point of f, while multiplication $p_1 \circ p_2$ of two tangent vectors simply multiplies the values of p_1 and p_2 at each critical point of f. In other words, numbering of the critical points of $f \in \mathcal{P}_k$ establishes an identification of the tangent space $T_f\mathcal{P}_k$ with the vector space $\mathbb{C}^{k-1}$ of functions on the set of critical points of f.

The elementary symmetric functions in the critical values of polynomials (= in the canonical coordinates) can be computed as follows.

Lemma 1 ([8]). *The critical values of a polynomial $f \in \mathcal{P}_k$ are the roots of the polynomial*

$$\operatorname{discrim}_z(f - t)$$

in the variable t; here $\operatorname{discrim}_z$ *denotes the discriminant of a polynomial in the variable z.*

Due to the Frobenius property, the flat metric in the canonical coordinates has a diagonal form

$$\sum_{i=1}^{k-1} \eta_{ii} dt_i \otimes dt_i.$$

Here η_{ii} is inverse to the second derivative of f at the i th critical point. The coordinate a_2 serves as the *potential* for this metric, which means that

$$\eta_{ii} = \frac{\partial a_2}{\partial t_i}, \qquad i = 1, \dots, k-1.$$

§3. Hurwitz spaces

Hurwitz spaces are spaces of rational (meromorphic) functions on complex curves. Such a function is a holomorphic mapping $f : C \to \mathbb{C}P^1$ of a smooth compact complex curve C to the projective line $\mathbb{C}P^1$. We will always assume that there is a distinguished complex coordinate on the projective line; this means, in particular, that there is a distinguished point $\infty \in \mathbb{C}P^1$, the infinity of the distinguished coordinate.

Two meromorphic functions $f_1 : C_1 \to \mathbb{C}P^1$ $f_2 : C_2 \to \mathbb{C}P^1$ are said to be equivalent if there is a biholomorphic mapping $h : C_1 \to C_2$ such that $f_1 = f_2 \circ h$, and we make no distinction between equivalent functions.

We start with giving examples of Hurwitz spaces, and then proceed with general definitions. We restrict ourselves with Hurwitz spaces consisting of meromorphic functions defined on curves of a given genus and having a prescribed ramification over the point $\infty \in \mathbb{C}P^1$. Such a space is characterized by the value g of the genus of the curves and the orders $k_1, \dots, k_n$ of the poles of the functions.

3.1. Examples of Hurwitz spaces

Due to complicated topology, Hurwitz spaces rarely have global coordinate descriptions. However, certain genus 0 Hurwitz spaces admit nice explicit presentations.

Example 1 (space of polynomials). *The Hurwitz space $\mathcal{H}_{0;k^1}$ of rational functions on genus 0 curves having a single pole of order k is, essentially, a space of polynomials. Indeed, by an appropriate choice of a coordinate z in the source curve $\mathbb{C}P^1$, each such function can be given the form*

$$f(z) = z^k + a_2 z^{k-2} + \cdots + a_k. \tag{1}$$

However, the Hurwitz space does not coincide with this space of polynomials: the coordinate transformation $z \mapsto \varepsilon_k z$, where ε_k is a primitive root of unity of degree k, takes each polynomial into an equivalent one. Thus, the Hurwitz space of rational functions with a single pole of order k is the $(k-1)$-dimensional quotient space of the space of polynomials of the form (1) modulo the action of the cyclic group $\mathbb{Z}_k$. The action is defined by the action of the generator

$$\varepsilon_k : (a_2, \dots, a_{k-1}, a_k) \mapsto (\varepsilon_k^2 a_2, \dots, \varepsilon_k^{k-1} a_{k-1}, a_k).$$

The line $(0, \dots, 0, a_k)$ consists of the orbifold points of this quotient space.

Example 2 (space of trigonometric polynomials). *A coordinate presentation of the Hurwitz space of rational functions on genus 0 curves with two poles, of orders k_1 and k_2, was suggested by Arnold in [1]. He calls this space the space of trigonometric polynomials, since the restriction of such a function to the unit circle is a finite linear combination* *of sine, cosine and their powers. By an appropriate choice of a coordinate z in the source curve $\mathbb{C}P^1$, each rational function can be given the form*

(2)
$$f(z) = z^{k_1} + a_1 z^{k_1-1} + \dots + a_{k_1-1} z + c + b_{k_2-1}\frac{u}{z} + \dots + b_1\left(\frac{u}{z}\right)^{k_2-1} + \left(\frac{u}{z}\right)^{k_2},$$

where $u \neq 0$. This form is preserved by the action of the finite group with two generators $\varepsilon_{k_1} : (z,u) \mapsto (\varepsilon_{k_1} z, \varepsilon_{k_1} u)$ and $\varepsilon_{k_2} : u \mapsto \varepsilon_{k_2} u$. These generators act on the coefficients as follows:

$$\varepsilon_{k_1} : (a_1, \dots, a_{k_1-1}, b_1, \dots, b_{k_2-1}, c, u) \mapsto (\varepsilon_{k_1} a_1, \dots, \varepsilon_{k_1}^{k_1-1} a_{k_1-1}, b_1, \dots, b_{k_2-1}, c, \varepsilon_{k_1} u);$$
$$\varepsilon_{k_2} : (a_1, \dots, a_{k_1-1}, b_1, \dots, b_{k_2-1}, c, u) \mapsto (a_1, \dots, a_{k_1-1}, \varepsilon_{k_2} b_1, \dots, \varepsilon_{k_2}^{k_2-1} b_{k_2-1}, c, \varepsilon_{k_2} u).$$

Thus, the corresponding Hurwitz space is the complement to the hyperplane $u = 0$ in the space of coordinates $(a_1, \dots, a_{k_1-1}, b_1, \dots, b_{k_2-1}, c, u)$ modulo the product $\mathbb{Z}_{k_1} \times \mathbb{Z}_{k_2}$ action.

These two examples have an immediate generalization to the spaces of rational functions with three poles, of orders k_1, k_2, k_3, which we leave to the reader. The case of poles of order 1 is considered below.

3.2. General Hurwitz spaces and their compactifications

In the description of general Hurwitz spaces below we follow the approach of [3].

Let $\mathcal{M}_{g;n}$ denote the moduli space of smooth genus g curves with n marked points, $n \geq 1$. For a smooth genus g curve C and a point $x \in C$, consider two meromorphic functions $f_1, f_2 : (C,x) \to (\mathbb{C}P^1, \infty)$ having poles of order k at x. We say that f_1, f_2 *have the same principal parts at $x \in C$* if the difference $f_1 - f_2$ has no pole at x.

For an arbitrary local coordinate z at a point $x \in C$, any principal part of order k at x admits a presentation of the form

$$\left(\frac{u}{z}\right)^k + a_1\left(\frac{u}{z}\right)^{k-1} + \dots + a_{k-1}\frac{u}{z}.$$

This presentation is unique up to the action of the cyclic group $\mathbb{Z}_k$ defined by $\varepsilon_k : u \mapsto \varepsilon_k u$, where $\varepsilon_k \in \mathbb{C}$ is a primitive k th root of unity. Principal parts of orders $k_1, \dots, k_n$ at the marked points $x_1, \dots, x_n$, respectively, form cone bundles $\mathcal{P}_1, \dots, \mathcal{P}_n$ over $\mathcal{M}_{g;n}$.

Taking a meromorphic function in a Hurwitz space $\mathcal{H}_{g;k_1,\dots,k_n}$ to the n-tuple of its principal parts at the marked points, we obtain a mapping from this space to the direct sum of the cone bundles $\mathcal{P}_1 \oplus \dots \oplus \mathcal{P}_n$ over $\overline{\mathcal{M}}_{g;n}$, which is a cone bundle in its own turn. This mapping takes two functions that differ by an additive constant to the same n-tuple of principal parts. One can make this mapping into an embedding by adding a trivial line bundle $\mathbb{C}$ over $\overline{\mathcal{M}}_{g;n}$ to the direct sum of cone bundles and associating to each function the sum of its values at its critical points, see details in [3]. By definition, the space $\overline{\mathcal{H}}_{g;k_1,\dots,k_n}$ is the closure of the image of this latter mapping in $\mathcal{P}_1 \oplus \cdots \oplus \mathcal{P}_n \oplus \mathbb{C}$. It is shown in [4] that this closure coincides with the space of relative stable mappings of genus g curves with n marked points to $(\mathbb{C}P^1, \infty)$.

Although $\overline{\mathcal{H}}_{g;k_1,\dots,k_n}$ is not a compact space, its projectivization $P\overline{\mathcal{H}}_{g;k_1,\dots,k_n}$ consisting of all nonzero functions modulo multiplications by nonzero complex numbers is. By abuse of language, we will call the space $\overline{\mathcal{H}}_{g;k_1,\dots,k_n}$ the *stable compactification* of the space $\mathcal{H}_{g;k_1,\dots,k_n}$.

§4. Frobenius structures on Hurwitz spaces

On Hurwitz spaces, there is a unique natural choice of the multiplication, the unity vector field, and the Euler field, while the choice of a flat metric is ambiguous. That is why we discuss these parts of data separately.

4.1. Algebra structure, unity, and Euler field

Each Hurwitz space can be realized as a ramified covering of a space of polynomials in one variable. Namely, the *Lyashko–Looijenga mapping* takes a meromorphic function to the monic polynomial whose roots are the critical values of the function. A generic function in the Hurwitz space $\overline{\mathcal{H}}_{g;k_1,\dots,k_n}$ has

$$\mu = \dim \overline{\mathcal{H}}_{g;k_1,\dots,k_n} = 2g - 2 + n + k_1 + \cdots + k_n$$

pairwise distinct critical values, and the Lyashko–Looijenga mapping takes this function to the polynomial

$$\mathrm{LL} : f \mapsto (t - t_1)\dots(t - t_\mu) = t^\mu + b_1 t^{\mu-1} + \cdots + b_\mu.$$

The target polynomial, clearly, is well-defined, that is, it does not depend on the chosen numbering of the critical values. Similarly to the case of polynomials above, for rational functions this polynomial can be computed explicitly due to the following statement.

Lemma 2 ([5]). *The critical values of a rational function $f(z) = \frac{P(z)}{Q(z)}$, P, Q being coprime polynomials, are the roots of the polynomial*

$$\mathrm{discrim}_z(P - tQ)$$

in the variable t; here $\mathrm{discrim}_z$ *denotes the discriminant of a polynomial in the variable z.*

It is shown in [3] that the Lyashko–Looijenga mapping admits an extension to the whole space $\overline{\mathcal{H}}_{g;k_1,\dots,k_n}$ that is an algebraic morphism.

The algebra structure, the unity and the Euler field on the Hurwitz space are inherited from the space of polynomials. Namely,

- the multiplication comes from the rule

$$\frac{\partial}{\partial t_i}\circ\frac{\partial}{\partial t_j}=\delta_{ij}\frac{\partial}{\partial t_i},\qquad i,j=1,\dots,\mu;$$

- the unity field is

$$e=\frac{\partial}{\partial t_1}+\dots+\frac{\partial}{\partial t_\mu}=\frac{\partial}{\partial b_1}$$

  and corresponds to the 1-parameter group of adding a constant to the function f;
- the Euler field is

$$E=t_1\frac{\partial}{\partial t_1}+\dots+t_\mu\frac{\partial}{\partial t_\mu}=b_1\frac{\partial}{\partial b_1}+2b_2\frac{\partial}{\partial b_2}+\dots+\mu b_\mu\frac{\partial}{\partial b_\mu}$$

  and corresponds to the 1-parameter group of multiplying the function f by a nonzero constant.

4.2. Primitive forms and flat metrics

The situation with the flat metric is more complicated. In Chapter 5 of [2] Dubrovin suggested a mechanism to introduce flat metrics on Hurwitz spaces that mimics the situation on the space of polynomials P_k. Essentially, this mechanism looks like follows.

The tangent space to a Hurwitz space at some point $f: C\to\mathbb{C}P^1$ is identified with a space of meromorphic functions on C having the same poles as f itself. Then the notion of a primitive form is introduced whose role is to make functions into differential 1-forms on C. We do not reproduce the definition of a *primitive form* here just mentioning that it is a section of a certain Hodge bundle over an appropriate covering of the moduli space of curves with marked poles, that is flat with respect to the Gauss–Manin connection. In the case of polynomials, the primitive form is dz, which is the only 1-form that has poles at the poles of the functions and no zeroes.

Then, having a primitive form ω, one can introduce the scalar product (p_1,p_2) of two tangent vectors $p_1,p_2\in T_f\mathcal{H}$ by the formula

$$(p_1,p_2)=\sum_{df=0}\mathrm{Res}\frac{p_1\omega\otimes p_2\omega}{df}.$$

Summation here is carried over all critical points of the function f. Note that since the sum of residues of any meromorphic 1-form on a compact curve is 0, the same formula can be written as

$$(p_1, p_2) = -\sum_{f=\infty} \mathrm{Res}\frac{p_1\omega \otimes p_2\omega}{df},$$

where this time the summation is carried over all poles of f. The latter formula is in a sense more convenient, since computing the critical points of a function is a complicated task, while its poles are made explicit in advance. Of course, the scalar product thus defined depends on the choice of the primitive form ω.

Furthermore, Dubrovin suggested several series of primitive forms producing flat metrics on open domains in Hurwitz spaces. Dubrovin has shown that these metrics are compatible with the other data of the Frobenius structure, and that the open domains where the metrics are non-degenerate cover the whole Hurwitz space. These flat metrics together define the Forbenius structure on the covering of the Hurwitz space in question.

4.3. Tangent spaces to Hurwitz spaces

Let $\overline{\mathcal{H}}_{g;k_1,\dots,k_n}$ be a Hurwitz space, $\mu = 2g - 2 + n + k_1 + \cdots + k_n$ being its dimension. The Hurwitz space is fibered over the moduli space $\overline{\mathcal{M}}_{g;n}$ of genus g stable curves with a fiber being a cone; in the case of $k_1 = \cdots = k_n = 1$, the fiber over each point is a vector space, so that $\overline{\mathcal{H}}_{g;1^n}$ is the total space of a vector bundle over $\overline{\mathcal{M}}_{g;n}$ provided n is sufficiently large.

For an arbitrary vector bundle $\pi : E \to M$, the tangent space at a point $f \in E$ is subject to the short exact sequence

$$0 \longrightarrow T_f\pi^{-1}\pi(f) \longrightarrow T_fE \longrightarrow \pi^*TM_{\pi(f)} \longrightarrow 0$$

where $T_f\pi^{-1}\pi(f)$ is the tangent space to the fiber of the bundle containing f and $\pi^*TM_{\pi(f)}$ is the pull-back of the fiber of the tangent bundle to the base M. In the case of the Hurwitz space $\overline{\mathcal{H}}_{g;k_1,\dots,k_n}$, the fiber $T^\vee_{(C;x_1,\dots,x_n)}\overline{\mathcal{M}}_{g;n}$ of the cotangent bundle to the base space $\overline{\mathcal{M}}_{g;n}$ at a point $(C; x_1, \dots, x_n)$ is naturally identified with the vector space of quadratic differentials on the curve C having poles of order at most k_i at the points x_i. It is unclear, however, how to define the "value of a dual quadratic differential at a critical point of f", which is necessary to identify the tangent space at f with the space of functions on the set of critical points of f.

4.4. Explicit formulas for $\overline{\mathcal{H}}_{0;1^3}$

Denote the Hurwitz space of rational functions with n poles of order 1 by $\mathcal{H}_{0;1^n}$, $\overline{\mathcal{H}}_{0;1^n}$ being its stable compactification. Below, we give explicit formulas for the Frobenius structures on the Hurwitz space $\mathcal{H}_{0;1^3}$ and discuss their extension to $\overline{\mathcal{H}}_{0;1^3}$.

The Hurwitz space $\overline{\mathcal{H}}_{0;1^3}$ can be identified with the space of rational functions of the form

$$\left\{\frac{a}{z-1}+\frac{b}{z}+\frac{c}{z+1}+d\right\},$$

while the space $\mathcal{H}_{0;1^3}$ is the complement in $\overline{\mathcal{H}}_{0;1^3}$ to the union of the three hyperplanes $a=0$, $b=0$, $c=0$.

Since $\overline{\mathcal{H}}_{0;1^3}$ is a vector space, the tangent space $T_f\overline{\mathcal{H}}_{0;1^3}$ at any point $f\in\overline{\mathcal{H}}_{0;1^3}$ can be identified with the space itself.

The unity vector field

$$e=\frac{\partial}{\partial d}$$

and the Euler field

$$E=\frac{\partial}{\partial a}+\frac{\partial}{\partial b}+\frac{\partial}{\partial c}+\frac{\partial}{\partial d}$$

correspond to the group of shifts and the group of dilations of the target projective line, respectively.

The algebra structure on the tangent plane $T_f\overline{\mathcal{H}}_{0;1^3}$ can be described as follows.

Proposition 3. *In the basis* $\frac{\partial}{\partial a}, \frac{\partial}{\partial b}, \frac{\partial}{\partial c}, \frac{\partial}{\partial d}$, *multiplication* $\circ$ *at a point* $f=(a,b,c,d)\in\overline{\mathcal{H}}_{0;1^3}$ *is given by the following multiplication table*:

$\circ$	$\frac{\partial}{\partial a}$	$\frac{\partial}{\partial b}$	$\frac{\partial}{\partial c}$	$\frac{\partial}{\partial d}$
$\frac{\partial}{\partial a}$	$-\frac{3}{2}\frac{\partial}{\partial a}-\frac{b}{2a}\frac{\partial}{\partial b}+\frac{c}{2a}\frac{\partial}{\partial c}-\frac{a+b+c}{2a}\frac{\partial}{\partial d}$	$\frac{\partial}{\partial a}-\frac{\partial}{\partial b}$	$\frac{1}{2}\left(\frac{\partial}{\partial a}-\frac{\partial}{\partial c}\right)$	$\frac{\partial}{\partial a}$
$\frac{\partial}{\partial b}$	$\frac{\partial}{\partial a}-\frac{\partial}{\partial b}$	$\frac{2a}{b}\frac{\partial}{\partial a}-\frac{2c}{b}\frac{\partial}{\partial c}+\frac{a+b+c}{b}\frac{\partial}{\partial d}$	$\frac{\partial}{\partial b}-\frac{\partial}{\partial c}$	$\frac{\partial}{\partial b}$
$\frac{\partial}{\partial c}$	$\frac{1}{2}\left(\frac{\partial}{\partial a}-\frac{\partial}{\partial c}\right)$	$\frac{\partial}{\partial b}-\frac{\partial}{\partial c}$	$-\frac{a}{2c}\frac{\partial}{\partial a}+\frac{b}{2c}\frac{\partial}{\partial b}+\frac{3}{2}\frac{\partial}{\partial c}-\frac{a+b+c}{2c}\frac{\partial}{\partial d}$	$\frac{\partial}{\partial c}$
$\frac{\partial}{\partial d}$	$\frac{\partial}{\partial a}$	$\frac{\partial}{\partial b}$	$\frac{\partial}{\partial c}$	$\frac{\partial}{\partial d}$

Proof. The last column (resp., the last row) of the multiplication table simply reflects the fact that $e = \frac{\partial}{\partial d}$ is the unity vector field. Computation of the product of two first vector fields is easy: just note that

$$\frac{1}{z-1} \cdot \frac{1}{z} = \frac{1}{z-1} - \frac{1}{z}.$$

The products of other pairs of different vector fields are computed similarly. Hence, the only non-immediate computation is that of squares of vector fields.

In order to compute, for example, $\frac{\partial}{\partial a} \circ \frac{\partial}{\partial a}$ at a point $f \in \overline{\mathcal{H}}_{0;1^3}$, we must find a function in $\overline{\mathcal{H}}_{0;1^3}$ having the same values at the critical points of f as the function $\frac{1}{(z-1)^2}$. This can be done in the following way[1]:

- take the difference between $\frac{1}{(z-1)^2}$ and the indefinite tangent vector $\frac{\alpha}{z-1} + \frac{\beta}{z} + \frac{\gamma}{z+1} + \delta$; consider the polynomial which is the numerator of this difference expressed as a fraction of two coprime polynomials;
- divide this polynomial by the numerator of the derivative f' and take the polynomial remainder;
- equate the coefficients of the polynomial remainder to 0 and solve the resulting system of linear equations with respect to the unknowns $\alpha, \beta, \gamma, \delta$.

Indeed, the difference between $\frac{1}{(z-1)^2}$ and the indefinite tangent vector takes zero values at the critical points of f iff this system of linear equations is satisfied. The table shows the results of this computation.

Of course, this procedure works for arbitrary function, not only for $\frac{1}{(z-1)^2}$.

In the case of the space $\overline{\mathcal{H}}_{0;1^3}$, Dubrovin's flat metrics can be constructed from primitive forms having poles at the same points where the functions themselves have poles. Below, we give the results of computations of the metrics for three different primitive forms.

Proposition 4. *In the basis* $\frac{\partial}{\partial a}, \frac{\partial}{\partial b}, \frac{\partial}{\partial c}, \frac{\partial}{\partial d}$, *the metrics at a point* $f = (a,b,c,d) \in \overline{\mathcal{H}}_{0;1^3}$ *is given by the following scalar product tables:*

[1] I am obliged to M. Kazarian for this remark

- *for the primitive form* $\frac{dz}{z^2}$

$(\cdot,\cdot)_{dz/z^2}$	$\frac{\partial}{\partial a}$	$\frac{\partial}{\partial b}$	$\frac{\partial}{\partial c}$	$\frac{\partial}{\partial d}$
$\frac{\partial}{\partial a}$	$\frac{2}{b}$	$\frac{a-b+c}{b^2}$	0	$-\frac{1}{b}$
$\frac{\partial}{\partial b}$	$\frac{a-b+c}{b^2}$	$\frac{2(c-a)}{b^2}$	$-\frac{a-b+c}{b^2}$	$-\frac{a+c}{b^2}$
$\frac{\partial}{\partial c}$	0	$-\frac{a-b+c}{b^2}$	$-\frac{2}{b}$	$-\frac{1}{b}$
$\frac{\partial}{\partial d}$	$-\frac{1}{b}$	$-\frac{a+c}{b^2}$	$-\frac{1}{b}$	0

The determinant of the metric equals $\frac{4}{b^4}$.

- *for the primitive form* $\frac{dz}{z(z-1)}$

$(\cdot,\cdot)_{dz/z(z-1)}$	$\frac{\partial}{\partial a}$	$\frac{\partial}{\partial b}$	$\frac{\partial}{\partial c}$	$\frac{\partial}{\partial d}$
$\frac{\partial}{\partial a}$	$-\frac{2}{a}$	$\frac{1}{a}-\frac{1}{b}$	$\frac{1}{2a}$	$\frac{1}{a}$
$\frac{\partial}{\partial b}$	$\frac{1}{a}-\frac{1}{b}$	$\frac{2}{b}$	$\frac{1}{b}$	$\frac{1}{b}$
$\frac{\partial}{\partial c}$	$\frac{1}{2a}$	$\frac{1}{b}$	0	0
$\frac{\partial}{\partial d}$	$\frac{1}{a}$	$\frac{1}{b}$	0	0

The determinant of the metric equals $\frac{1}{4a^2b^2}$.

- *for the primitive form* $(-1/(z-1)+2/z-1/(z+1))dz$

$(\cdot,\cdot)_{(-1/(z-1)+2/z-1/(z+1))dz}$	$\frac{\partial}{\partial a}$	$\frac{\partial}{\partial b}$	$\frac{\partial}{\partial c}$	$\frac{\partial}{\partial d}$
$\frac{\partial}{\partial a}$	$-\frac{1}{a}$	$\frac{1}{a}-\frac{4}{b}$	$\frac{1}{2a}-\frac{1}{2c}$	$\frac{1}{a}$
$\frac{\partial}{\partial b}$	$\frac{1}{a}-\frac{4}{b}$	0	$\frac{4}{b}-\frac{1}{c}$	$\frac{4}{b}$
$\frac{\partial}{\partial c}$	$\frac{1}{2a}-\frac{1}{2c}$	$\frac{4}{b}-\frac{1}{c}$	$\frac{3}{c}$	$\frac{1}{c}$
$\frac{\partial}{\partial d}$	$\frac{1}{a}$	$\frac{4}{b}$	$\frac{1}{c}$	0

The determinant of the metric equals $\frac{4(a+b+c)^2}{a^2b^2c^2}$.

Note that the third metric, in contrast to the first two ones, vanishes on the hyperplane $a+b+c=0$. The reason for this is that the 1-forms $\frac{dz}{z^2}$ and $\frac{dz}{z(z-1)}$ have no zeroes on the projective line, while the 1-form

$$\left(-\frac{1}{z-1}+\frac{2}{z}-\frac{1}{z+1}\right)dz=\frac{dz}{(z-1)z(z+1)}$$

has zero of order 1 at infinity, and for $a+b+c=0$ infinity is as well a critical point of f.

The proof of the Proposition can be obtained either from a direct computation of the sum of residues of an appropriate 1-form at the poles, or by computing the scalar products of the basis vector fields with the unity vector field and using then the multiplication table in the section above.

4.5. Explicit formulas for $\overline{\mathcal{H}}_{0;1^4}$

In contrast to the space $\overline{\mathcal{H}}_{0;1^3}$, the Hurwitz space $\overline{\mathcal{H}}_{0;1^4}$ is no longer a vector space; instead, it is a rank 5 vector bundle over $\overline{\mathcal{M}}_{0;4} = \mathbb{C}P^1$. If we pick a coordinate z on a smooth rational curve C such that the first three poles acquire the coordinates $1, 0, -1$, then the domain in $\overline{\mathcal{H}}_{0;1^4}$ consisting of functions projected to $\mathcal{M}_{0;4}$ acquires the coordinate presentation

$$\left\{\frac{a}{x-1}+\frac{b}{x}+\frac{c}{x+1}+\frac{d}{x-s}+e\right\},$$

where $z = s$ is the coordinate of the fourth pole. The subspace $\mathcal{H}_{0;1^4}$ consists of points (a, b, c, d, e, s) such that $a \neq 0$, $b \neq 0$, $c \neq 0$, $d \neq 0$, $s \neq 1$, $s \neq 0$, $s \neq -1$. The tangent space to $\mathcal{H}_{0;1^4}$ at a point $f \in \mathcal{H}_{0;1^4}$ defined on a smooth rational curve $(C; -1, 0, 1, s)$ can be identified with the vector space

$$T_f = \left\{\frac{\alpha}{x-1}+\frac{\beta}{x}+\frac{\gamma}{x+1}+\frac{\delta 1}{x-s}+\frac{\delta 2}{(x-s)^2}+\varepsilon\right\},$$

of rational functions on C. Indeed, when represented as a fraction of two coprime polynomials, such a function has a numerator of degree 5, which is smaller than the degree 6 of the numerator of the derivative df. Hence, it vanishes at all the critical points of the function f simultaneously if and only if it is identically 0, and the space T_f is thus naturally identified with the space of functions on the set of critical points of f. Denote the coordinate vector fields corresponding to the coordinates α, β, γ, $\delta 1$, $\delta 2$, ε by ξ_α, ξ_β, ξ_γ, $\xi_{\delta 1}$, $\xi_{\delta 2}$, ξ_ε respectively. All these vector fields but $\xi_{\delta 2}$ naturally correspond to the coordinate vector fields for the chosen presentation of T_f. In turn, the vector field $\partial/\partial s$ is nothing but $d\xi_{\delta 2}$: differentiating a function f with respect to s we obtain $d/(x-s)^2$.

Now we can write out the multiplication table for the coordinate vector fields.

Proposition 5. *The multiplication table in the local coordinates α, β, γ, $\delta 1$, $\delta 2$, ε has the following terms on the diagonal:*

$$\xi_\alpha \circ \xi_\alpha = -\frac{3}{2}\xi_\alpha - \frac{b}{2a}\xi_\beta + \frac{c}{2a}\xi_\gamma - \frac{d(1+2s)}{2a}\xi_{\delta 1} - \frac{ds(1+s)}{2a}\xi_{\delta 2} - \frac{a+b+c+d}{2a}\xi_\varepsilon$$

$$\xi_\beta \circ \xi_\beta = \frac{2a}{b}\xi_\alpha - \frac{2c}{b}\xi_\gamma + \frac{2ds}{b}\xi_{\delta 1} - \frac{d(1-s)(1+s)}{b}\xi_{\delta 2} + \frac{a+b+c+d}{b}\xi_\varepsilon$$

$$\xi_\gamma \circ \xi_\gamma = -\frac{a}{2c}\xi_\alpha + \frac{b}{2c}\xi_\beta + \frac{3}{2}\xi_\gamma + \frac{d(1-2s)}{2c}\xi_{\delta 1} + \frac{ds(1-s)}{2c}\xi_{\delta 2} - \frac{a+b+c+d}{2c}\xi_\varepsilon$$

$$\xi_{\delta 1} \circ \xi_{\delta 1} = \xi_{\delta 2}$$

$$\begin{aligned}\xi_{\delta 2} \circ \xi_{\delta 2} = & -\frac{2a(1-s-4s^2)}{d(1-s)^3 s^2(1+s)^2}\xi_\alpha - \frac{2b(1-2s^2)}{ds^3(1-s^2)^2}\xi_\beta + \frac{2c(1+s-4s^2)}{d(1-s)^2 s^2(1+s)^3}\xi_\gamma \\ & - \lambda\xi_{\delta 1} + \frac{-b+d+(-a+2b-c-3d)s^2-2(a-c)s^3-(a+b+c-6d)s^4}{ds^2(1-s^2)^2}\xi_{\delta 2} \\ & - \frac{(a+b+c+d)(1-3s^2)}{ds^2(1-s^2)^2}\xi_\varepsilon\end{aligned}$$

where

$$\lambda = \frac{-2b+2(3b+d)s^2+2(a-c)s^3+2(3a-3b+3c-5d)s^4+6(a-c)s^5+2(a+b+c+4d)s^6}{ds^3(1-s^2)^3}.$$

In addition, the results of multiplication by the vector field $\xi_{\delta 2}$ are

$$\xi_\alpha \circ \xi_{\delta 2} = \frac{1}{(1-s)^2}\xi_\alpha - \frac{1}{(1-s)^2}\xi_{\delta 1} - \frac{1}{1-s}\xi_{\delta 2}$$

$$\xi_\beta \circ \xi_{\delta 2} = \frac{1}{s^2}\xi_\beta - \frac{1}{s^2}\xi_{\delta 1} + \frac{1}{s}\xi_{\delta 2}$$

$$\xi_\gamma \circ \xi_{\delta 2} = \frac{1}{(1+s)^2}\xi_\gamma - \frac{1}{(1+s)^2}\xi_{\delta 1} + \frac{1}{1+s}\xi_{\delta 2}$$

$$\begin{aligned}\xi_{\delta 1} \circ \xi_{\delta 2} = & \frac{2a}{d(1-s)^2 s(s+s)}\xi_\alpha + \frac{b}{ds^2(1-s^2)}\xi_\beta - \frac{2c}{d(1-s)s(1+s)^2} \\ & - \frac{b+(a-2b+c-3d)s^2+2(a-c)s^3+(a+b+c+3d)s^4}{ds^2(1-s^2)^2}\xi_{\delta 1} \\ & - \frac{1-3s^2}{s(1-s^2)}\xi_{\delta 2} + \frac{(a+b+c+d)}{ds(1-s^2)}\xi_\varepsilon.\end{aligned}$$

The other terms in the multiplication table can be obtained easily as in the proof of Proposition 3.

Proposition 6. *In the basis α, β, γ, $\delta 1$, $\delta 2$, ε, the metric at a point $f = (a, b, c, d, s, e) \in \overline{\mathcal{H}}_{0;1^4}$ for the primitive form $\frac{dz}{z(z-1)}$ is given by the* *following scalar product table*:

$(\cdot,\cdot)_{dz/z(z-1)}$	ξ_α	ξ_β	ξ_γ	$\xi_{\delta 1}$	$\xi_{\delta 2}$	ξ_ε
ξ_α	$-\frac{2}{a}$	$\frac{1}{a}-\frac{1}{b}$	$\frac{1}{2a}$	$\frac{1}{a(1-s)}$	$\frac{d}{a(1-s)^2}$	$\frac{1}{a}$
ξ_β	$\frac{1}{a}-\frac{1}{b}$	$\frac{2}{b}$	$\frac{1}{b}$	$-\frac{1}{bs}$	$\frac{d}{bs^2}$	$\frac{1}{b}$
ξ_γ	$\frac{1}{2a}$	$\frac{1}{b}$	0	0	0	0
$\xi_{\delta 1}$	$\frac{1}{a(1-s)}$	$-\frac{1}{bs}$	0	0	$\frac{1}{(1-s)^2 s^2}$	0
$\xi_{\delta 2}$	$\frac{d}{a(1-s)^2}$	$\frac{d}{bs^2}$	0	$\frac{1}{(1-s)^2 s^2}$	$-\frac{2d(1-2s)}{(1-s)^3 s^3}$	0
ξ_ε	$\frac{1}{a}$	$\frac{1}{b}$	0	0	0	0

The determinant of the metric equals $-\frac{1}{4a^2b^2(1-s)^4s^4}$.

The **proof** of the Proposition is the result of a direct computation.

4.6. Compactified Hurwitz manifolds and extension of Frobenius structures to the boundary: a problem

In all the explicit computations above the only singularities the elements of the Frobenius structures possess at the boundary $\overline{\mathcal{H}} \setminus \mathcal{H}$ are poles. Indeed, in the case of the Hurwitz space $\overline{\mathcal{H}}_{0;1^3}$, the boundary is the union of the three hyperplanes $a = 0, b = 0, c = 0$, and the support of the poles of multiplication and flat metrics is contained in this union. In its own turn, for the coordinate presentation of $\overline{\mathcal{H}}_{0;1^4}$, the boundary is the union of the subvarieties $a = 0, b = 0, c = 0, d = 0, s = 1, s = 0, s = -1$. The formulas show that the support of the poles of multiplication and flat metrics is contained in this union. This is a basis for the following question:

Whether it is possible, for an arbitrary Hurwitz space $\overline{\mathcal{H}}_{g;k_1,\dots,k_n}$, *to chose a primitive form in such a way that the flat metric defined by this primitive form has singularities at most on the boundary, and these singularities are at most poles?*

The existence of such a primitive form would mean the existence of distinguished Frobenius structures on Hurwitz spaces.

References

[1] Arnol'd, V. I. *Topological classification of complex trigonometric polynomials and the combinatorics of graphs with an identical number of vertices and edges.* (Russian) Funktsional. Anal. i Prilozhen. 30 (1996), no. 1, 1–17, 96; translation in Funct. Anal. Appl. 30 (1996), no. 1, 1–14

[2] Dubrovin, Boris *Geometry of 2D topological field theories*, in: Integrable systems and quantum groups (Montecatini Terme, 1993), 120–348, Lecture Notes in Math., **1620**, Springer, Berlin, 1996.

[3] Ekedahl, Torsten; Lando, Sergei; Shapiro, Michael; Vainshtein, Alek *Hurwitz numbers and intersections on moduli spaces of curves*, Invent. Math. 146 (2001), no. 2, 297–327

[4] Lando, S. K. *Ramified coverings of the two-dimensional sphere and intersection theory in spaces of meromorphic functions on algebraic curves*, (Russian) Uspekhi Mat. Nauk 57 (2002), no. 3(345), 29–98; translation in Russian Math. Surveys 57 (2002), no. 3, 463–533

[5] Lando, Sergei K.; Zvonkin, Alexander K. *Graphs on surfaces and their applications.* With an appendix by Don B. Zagier. Encyclopaedia of Mathematical Sciences, 141. Low-Dimensional Topology, II. Springer-Verlag, Berlin, 2004

[6] Saito, Kyoji. *Theory of logarithmic differential forms and logarithmic vector fields*, J. Fac. Sci. Univ. Tokyo Sect. IA Math. 27 (1980), no. 2, 265–291.

[7] Saito, Kyoji. *Period mapping associated to a primitive form*, Publ. Res. Inst. Math. Sci. 19 (1983), no. 3, 1231–1264.

[8] Zvonkin, D., Lando, S. K. *On multiplicities of the Lyashko-Looijenga mapping on strata of the discriminant*, (Russian. Russian summary) Funktsional. Anal. i Prilozhen. 33 (1999), no. 3, 21–34, 96; translation in Funct. Anal. Appl. 33 (1999), no. 3, 178–188 (2000)

National Research University Higher School of Economics,
Skolkovo Institute of Science and Technology
E-mail address: `lando@hse.ru`

Advanced Studies in Pure Mathematics 83, 2019
Primitive Forms and Related Subjects — Kavli IPMU 2014
pp. 237–268

Calabi-Yau geometry, Primitive form and Mirror symmetry

Si Li

Abstract.

This note comes out of the author's lecture presented at the workshop *Primitive forms and related subjects*, Feb 10-14 2014.

Contents

§1. Introduction

Primitive forms are introduced by K. Saito [40, 41] around early 1980's as a generalization of the elliptic period integral theory associated to an isolated singularity. It leads to systematic examples of Frobenius manifold structure on the universal unfoldings of isolated singularities, characterizing the mathematical structure of topological Landau-Ginzburg B-model. Motivated by mirror symmetry, there is a vast interest in understanding the structure of primitive forms nowadays.

One of the recent development is the mathematical theory of Landau-Ginzburg A-models constructed by Fan, Jarvis and Ruan [17], popularized as the FJRW theory (see also [8, 39] for several purely algebraic constructions of Landau-Ginzburg A-model and their relationships with

---

Received February 28, 2015.
Revised February 23, 2018.
2010 *Mathematics Subject Classification.* 14N35, 53D45.
Key words and phrases. Calabi-Yau, Primitive form, Mirror symmetry, Landau-Ginzburg, Singularity, Frobenius manifold, moduli space, Hodge theory, Period map.

FJRW theory). With the huge success toward understanding the mirror symmetry between Calabi-Yau manifolds and Calabi-Yau/Landau-Ginzburg correspondence, it is desirable to see whether mirror symmetry between Landau-Ginzburg models hold directly. A first step toward establishing such Landau-Ginzburg mirror symmetry is to understanding the relation between FJRW theory in the A-model and Saito's primitive form in the B-model. The purpose of this lecture is to present some recent progress in this direction related to the author's works [13, 23, 28, 29].

There are two goals in this lecture. The first is to give a unified presentation for both Calabi-Yau and Landau-Ginzburg B-models via the geometry of polyvector fields. On compact Calabi-Yau manifolds, Barannikov and Kontsevich [5] construct a large class of Frobenius manifold structures on their extended moduli spaces of complex structures, which is closely related to the Kodaira-Spencer gauge theory for polyvector fields introduced in [7]. With deformation theory, Barannikov developed the generalized period map via variation of semi-infintie Hodge structures [3] to describe the mirror of Gromov-Witten invariants. The prototype of semi-finite Hodge structures and generalized period maps can be traced back to Landau-Ginzburg models as Saito's higher residue theory and primitive forms [40, 41]. The first part of the lecture is to explain how the geometry of polyvector fields in Calabi-Yau geometry is related to Landau-Ginzburg models [28].

The second part is to present the recent development of perburbative method for primitive forms [28, 29] and mirror symmetry [23]. For weighted homogenous cases, explicit expressions of primitive forms are only known for ADE and simple elliptic singularities. This becomes one of the main obstacles to test mirror symmetry between Landau-Ginzburg models. However, WDVV equation is a powerful integrable equation, which usually reduces the computation of generating functions to a finite few orders. A recursive algorithm is developed in [28, 29] which allows us to compute any primitive forms up to arbitrary finite order. This solves the computation difficulty in Landau-Ginzburg B-model. As an application, a version of Landau-Ginzburg mirror symmetry conjecture for a large class of singularities is established [23, 29].

Acknowledgement: The author would like to thank the organizers and participants of the workshop on Primitive forms and related subjects, and the hospitality of the Kavli IPMU at University of Tokyo.

§2. Calabi-Yau geometry

2.1. Polyvector fields

Let X be a compact Calabi-Yau manifold of dimension d with holomorphic volume form Ω_X. Let

$$\mathrm{PV}(X) = \bigoplus_{0\leq i,j\leq d} \mathrm{PV}^{i,j}(X), \quad \mathrm{PV}^{i,j}(X) = \mathcal{A}^{0,j}(X, \wedge^i T_X)$$

be the space of polyvector fields on X. Here T_X is the holomorphic tangent bundle, and $\mathcal{A}^{0,j}(X, \wedge^i T_X)$ is the space of smooth $(0,j)$-forms valued in $\wedge^i T_X$. $\mathrm{PV}(X)$ is a differential bi-graded commutative algebra: the differential is

$$\bar{\partial} : \mathrm{PV}^{i,j}(X) \to \mathrm{PV}^{i,j+1}(X),$$

and the algebra structure arises from wedge product. Our degree convention is that elements of $\mathrm{PV}^{i,j}(X)$ are of degree $j-i$. The graded-commutativity says

$$\alpha\beta = (-1)^{|\alpha||\beta|}\beta\alpha$$

where $|\alpha|, |\beta|$ denote the degree of α, β respectively. Ω_X induces an identification between the space of polyvector fields and differential forms

$$\begin{aligned} \mathrm{PV}^{i,j}(X) &\overset{\lrcorner\Omega_X}{\cong} \mathcal{A}^{d-i,j}(X) \\ \alpha &\to \alpha\lrcorner\Omega_X \end{aligned}$$

where $\lrcorner$ is the contraction, and $\mathcal{A}^{i,j}(X)$ denotes smooth differential forms of type (i,j). The holomorphic de Rham differential ∂ on forms defines an operator on $\mathrm{PV}(X)$ via the above isomorphism, which we still denote by

$$\partial : \mathrm{PV}^{i,j}(X) \to \mathrm{PV}^{i-1,j}(X)$$

i.e.

$$(\partial\alpha)\lrcorner\Omega_X \equiv \partial(\alpha\lrcorner\Omega_X), \quad \alpha \in \mathrm{PV}(X).$$

The definition of ∂ doesn't depend on the choice of Ω_X on compact Calabi-Yau manifolds. It induces a bracket on polyvector fields (Tian-Todorov lemma)

$$\{\alpha, \beta\} = \partial(\alpha\beta) - (\partial\alpha)\beta - (-1)^{|\alpha|}\alpha(\partial\beta)$$

which coincides with the Schouten-Nijenhuis bracket (up to a sign).

We can integrate polyvector fields by the *trace map* $\mathrm{Tr} : \mathrm{PV}(X) \to \mathbb{C}$

$$\mathrm{Tr}(\alpha) := \int_X (\alpha \lrcorner \Omega_X) \wedge \Omega_X .$$

Let $\langle -, - \rangle$ be the induced pairing $\mathrm{PV}(X) \otimes \mathrm{PV}(X) \to \mathbb{C}$

$$\alpha \otimes \beta \to \langle \alpha, \beta \rangle \equiv \mathrm{Tr}\,(\alpha\beta) .$$

It is easy to see that $\bar{\partial}$ is (graded) skew symmetric for this pairing and ∂ is (graded) symmetric.

2.2. Symplectic structure

Following Givental's symplectic formalism, let us add a formal variable z representing the "gravitational descendant" and introduce the following spaces

$$S(X) := \mathrm{PV}(X)((z)),\ S_+(X) := \mathrm{PV}(X)[[z]],\ S_-(X) := z^{-1}\,\mathrm{PV}(X)[z^{-1}].$$

There exists a natural symplectic pairing on $S(X)$ given by

$$\omega(f(z)\alpha, g(z)\beta) := \mathrm{Res}_{z=0}\,(f(z)g(-z)dz)\,\mathrm{Tr}(\alpha\beta).$$

It is direct to check that the differential

$$Q = \bar{\partial} + z\partial$$

is (graded) skew-symmetric with respect to the symplectic pairing ω. Let us denote by

$$\mathcal{H} = H^*(S(X), Q), \quad \mathcal{H}_+ = H^*(S_+(X), Q).$$

ω descends to define a pairing on the cohomology $\mathcal{H}$, still denoted by

$$\omega : \mathcal{H} \otimes_{\mathbb{C}} \mathcal{H} \to \mathbb{C}.$$

$\mathcal{H}_+$ becomes an isotropic subspace. Let us define the isomorphism

$$\Gamma_\Omega : \mathrm{PV}(X)((z)) \to \mathcal{A}(X)((z)),\ z^k\alpha \to z^{k+i-1}\alpha \lrcorner \Omega_X,\ \alpha \in \mathrm{PV}^{i,j}(X).$$

It transfers Q to the de Rham differential

$$\Gamma_\Omega \circ Q = d \circ \Gamma.$$

Under the isomorphism Γ_Ω, we have

$$\Gamma_\Omega(S_+(X)) = \prod_{p \in \mathbb{Z}} z^{d-p+1} F^p \mathcal{A}(X),$$

where $F^p\mathcal{A}(X) = \mathcal{A}^{\geq p,*}(X)$. At the cohomology level we have isomorphisms

$$\Gamma_\Omega : \mathcal{H} \overset{\sim}{\to} H^*(X, \mathbb{C})((z))$$

and

$$\Gamma_\Omega : \mathcal{H}_+ \overset{\sim}{\to} \prod_{p \in \mathbb{Z}} z^{d-p+1} F^p H^*(X, \mathbb{C}).$$

In particular, we see that the isotropic embedding $S_+(X) \subset S(X)$ (or $\mathcal{H}_+ \subset \mathcal{H}$) plays the role of Hodge filtration.

The symplectic space $(\mathcal{H}, \omega)$ is basically Givental's loop space formalism in the B-model, where $(S(X), \omega)$ can be viewed as the lifting to the chain level. Givental has formulated the generating function of Gromov-Witten invariants as the geometry of lagrangian cones inside the symplectic space. The B-model aspect is established via Barannikov's generalized period map [2, 3] using deformation theory [5].

In fact, the lagrangian cone construction can be lifted to the chain level in the B-model, which gives rise to a gauge theory of polyvector fields on Calabi-Yau manifolds. This is developed in [13], which we call BCOV theory, as a generalization of the Kodaira-Spencer gravity theory on Calabi-Yau three-folds discovered by [7]. It turns out that the tree level BCOV theory is equivalent to Givental's formalism and Barannikov's generalized period map at the cohomology level [33], which we now describe.

2.3. BCOV theory

Define a formal graded submanifold $\mathcal{L}_X$ of $S(X)$ based at 0

$$\mathcal{L}_X = \{z - ze^{\mu/z} | \mu \in S_+(X)\}.$$

By "formal" we mean it is defined via its functor of points. More precisely, this is a functor from nilpotent Artinian graded algebras R (with maximal ideal $m \subset R$) to sets. If R is an Artinian graded algebra, then the R-points of $S(X)$ is the set of degree 0 elements of $S(X) \otimes m$. We define $\mathcal{L}_X(R)$ to be the set of those $a \in S(X) \otimes m$ which are of degree 0, and which can be expressed (necessarily in a unique way) in the form

$$a = z - ze^{\mu/z}$$

for some $\mu \in S_+(X) \otimes m$. This expression makes sense because the maximal ideal $m \subset R$ is nilpotent.

The fundamental property of $\mathcal{L}_X$ is that it is a formal lagrangian submanifold of $S(X)$ which is preserved by the differential Q [13]. If we formally identify

$$S(X) = T^*(S_+(X))$$

via the natural splitting

$$S(X) = S_+(X) \oplus S_-(X),$$

then there exists a formal functional I^{BCOV} on $S_+(X)$ such that

$$\mathcal{L}_X = \mathrm{Graph}(dI^{BCOV}).$$

We have the following explicit formula [13]

$$I^{BCOV}(\mu) = \mathrm{Tr}\,\langle e^{\mu}\rangle_0\,,$$

where

$$\langle - \rangle_0 : \mathrm{Sym}(S_+(X)) \to \mathrm{PV}(X)$$

is the map via intersection on $\overline{M}_{0,n}$

$$\left\langle z^{k_1}\mu_1, \ldots, z^{k_n}\mu_n \right\rangle_0 = \mu_1 \wedge \cdots \wedge \mu_n \int_{\overline{M}_{0,n}} \psi_1^{k_1} \cdots \psi_n^{k_n} = \binom{n-3}{k_1 \cdots k_n} \mu_1 \wedge \cdots \wedge \mu_n.$$

I^{BCOV} is called the BCOV interaction in [13], which also appears in a finite dimensional toy model in [36]. Note that if we "turn off" gravitational descendants by setting $z = 0$ for the inputs, then I^{BCOV} is reduced to the cubic interaction of Kodaira-Spencer gauge theory [7]. The geometric fact that $\mathcal{L}_X$ is preserved by the differential Q is translated to a classical master equation for I^{BCOV}, which describes the infinitesimal gauge transformation in the BV formalism [13].

2.4. Generating function

The genus g generating function of B-model invariants could be obtained via g-loop Feynman diagrams with vertices being I^{BCOV}. For $g > 0$, this requires suitable renormalizations (see [13]). For $g = 0$, however, renormalization is not needed and we obtain the genus zero generating function as a sum of tree diagrams in BCOV theory. This is basically equivalent to a version of homological perturbation lemma, which amounts to go to Q-cohomology.

The Q-fixed point of $\mathcal{L}_X$ can be described as follows. Observe that

$$Q(ze^{\mu/z} - z) = \left(Q\mu + \frac{1}{2}\{\mu, \mu\}\right) e^{\mu/z}.$$

The Q-fixed point of $\mathcal{L}_X$ are described by solutions of the Maurer-Cartan equation

$$Q\mu + \frac{1}{2}\{\mu, \mu\} = 0,$$

which can be viewed as an extended version of deforming a pair of complex structure on X together with a holomorphic volume form [3].

Let $\mathcal{M}_X$ denote the formal moduli of gauge equivalent solutions

$$\mathcal{M}_X = \left\{ \mu \in S_+(X) | Q\mu + \frac{1}{2}\{\mu, \mu\} = 0 \right\} / \sim,$$

which again is defined via its functor of points on nilpotent Artinian graded algebras. It follows from Calabi-Yau geometry that the moduli $\mathcal{M}_X$ is smooth [3, 5]. In the same fashion, we can view it as a formal lagrangian submanifold

$$\mathcal{M}_X \hookrightarrow \mathcal{H}, \quad \mu \to z - ze^{\mu/z}.$$

To obtain the generating function at the cohomology level, we need a choice of splitting since $S_-(X)$ is not preserved by Q.

Let $\mathcal{L} \subset \mathcal{H}$ a linear isotropic subspace of $\mathcal{H}$ such that

(1) $\mathcal{H} = \mathcal{H}_+ \oplus \mathcal{L}$,
(2) $\mathcal{L}$ is preserved by the operator $z^{-1} : \mathcal{H} \to \mathcal{H}$.

The splitting $\mathcal{L}$ allows us to identify

$$H^*(X) = \mathcal{H}_+/z\mathcal{H}_+ \cong \mathcal{H}_+ \cap z\mathcal{L},$$

hence

$$\mathcal{H}_+(X) \cong H^*(X)[[z]], \quad \mathcal{L} \cong z^{-1}H^*(X)[z], \quad \mathcal{H} \cong H^*(X)((z)).$$

In particular, we can formally identify

$$\mathcal{H} \cong T^*(H^*(X)[[z]])$$

from which we obtain a formal generating function $F_0^{\mathcal{L}}$ on $H^*(X)[[z]]$ such that

$$\mathcal{M}_X \cong \text{Graph}(d\mathbf{F}_0^{\mathcal{L}}).$$

We can make this construction explicitly. Let $\{\phi_\alpha\}$ be a basis of $H^*(X)$, and $\{\phi^\alpha\}$ be a dual basis of $H^*(X)$ such that

$$\int \phi_\alpha \wedge \phi^\beta = \delta_\alpha^\beta.$$

Let $\mathbf{s}$ be a coordinate on $\mathcal{M}_X$ and $\mu(\mathbf{s})$ a universal solution of the Maurer-Cantan equation. We can always expand in $\mathcal{H}$

$$ze^{\mu(\mathbf{s})/z} = z + \sum_{k \geq 0} \sum_\alpha \tau_k^\alpha(\mathbf{s})\phi_\alpha(-1)^k z^k + \sum_{k \geq 0} \sum_\alpha p_{k,\alpha}(\mathbf{s})\phi_\alpha z^{-k-1}.$$

$\{\tau_k^\alpha\}$ can be viewed as the linear coordinates on $H^*(X)[[z]]$. The transformation

$$\mathbf{s} \to \{\tau_k^\alpha(\mathbf{s})\}$$

is invertible, which leads to different choice of the coordinates $\{\tau_k^\alpha\}$ on $\mathcal{M}_X$. In terms of $\{\tau_k^\alpha\}$, we find

$$p_{k,\alpha} = \frac{\partial \mathbf{F}_0^{\mathcal{L}}(\tau)}{\partial \tau_k^\alpha}.$$

If we set

$$\tau_k^\alpha = 0, \quad k > 0,$$

then the expansion

$$e^{\mu(\mathbf{s}(\tau_0^\alpha, \tau_1^\alpha = 0, \cdots))/z} = 1 + \sum_\alpha \tau_0^\alpha \phi_\alpha z^{-1} + O(z^{-2})$$

plays the role of Givental's J-function, or Barannikov's generalized period map. In particular, $\{\tau_0^\alpha\}$ gives the flat coordinates of the underlying Frobenius manifold structure. As we will see, this is the analogue of K. Saito's primitive period map [40] in Landau-Ginzburg models.

The generating function $\mathbf{F}_0^{\mathcal{L}}$ depends on the choice of $\mathcal{L}$. In contrast to the A-model, there is a prior no canonical choice of the splitting $\mathcal{L}$. In fact, the freedom of the splitting is exactly responsible for the famous holomorphic anomaly equation [7] when $\mathcal{L}$ is the complex conjugate splitting. See also [11, 13]. Around the large complex limit, there exists monodromy splitting from which $\mathbf{F}_0^{\mathcal{L}}$ gives the mirror of Gromov-Witten invariants [2, 19, 34].

2.5. Feynman diagrams

$\mathbf{F}_0^{\mathcal{L}}$ and the BCOV interaction I^{BCOV} are related by tree-level Feynman diagrams. In the absence of gravitational descendants, this is observed in [7], leading to the string field theory of B-twisted topological string by Kodaira-Spencer gauge theory. It is further illustrated in [5] as a homological perturbation related to the deformation theory on Calabi-Yau manifolds. We give a brief discussion here and refer to [33] for details.

For simplicity, we take $\mathcal{L}$ to be the complex conjugate splitting and write the generating function as $\mathbf{F}_0^X$. This splitting is related to the Harmonic theory as follows. Let g be a chosen Kähler metric on X. Let $\bar\partial^*$ be the adjoint of $\bar\partial$ on $\mathrm{PV}(X)$, and

$$\Delta = \bar\partial\bar\partial^* + \bar\partial^*\bar\partial$$

be the Laplacian. Let

$$h_u \in \mathrm{PV}(X) \otimes \mathrm{PV}(X), \quad u > 0$$

be the kernel of the heat operator $e^{-u\Delta}$. Formally,

$$e^{-u\Delta}(\alpha)(x) = \mathrm{Tr}(h_u(x, y) \wedge \alpha(y)), \quad \forall \alpha \in \mathrm{PV}(X),$$

where x, y represents two copies of coordinate on X, and the trace map is the integration over y with respect to the Calabi-Yau volume form as before. We define the *BCOV propagator* by the kernel

$$P = \int_0^\infty (\bar{\partial}^* \partial \otimes 1) h_u du,$$

which represents the operator

$$\bar{\partial}^* \partial \frac{1}{\Delta}.$$

Here $\frac{1}{\Delta}$ is Green's operator for the Laplacian Δ. Formally, P is the gauge fixed expression of $\frac{\partial}{\bar{\partial}}$, which is the inverse of the free part of the Kodaira-Spencer gauge action [7]. Let

$$\mathbf{H} = \{\mu \in \mathrm{PV}(X) | \Delta\mu = 0\}$$

be the harmonic elements. Since $\bar{\partial}\mathbf{H} = \partial\mathbf{H} = 0$, Hodge theory implies a natural isomorphism

$$\mathcal{H}_+ = \mathbf{H}[[z]], \quad \mathcal{H} = \mathbf{H}((z)),$$

which can be identified with the complex conjugate splitting [13].

Definition 2.1. The genus zero partition function $\mathbf{F}_0^{BCOV}$ of BCOV theory is the formal function on $\mathbf{H}[[z]]$ defined by

$$\mathbf{F}_0^{BCOV} = \sum_{\Gamma:\mathrm{Tree}} \frac{W_\Gamma(P, I^{BCOV})}{|Aut(\Gamma)|}$$

$W_\Gamma(P, I^{BCOV})$ is the Feynman diagram integrals with I^{BCOV} as the vertices and P as the propagator. $Aut(\Gamma)$ is the size of the automorphism group as graphs. The summation is over all connected tree diagrams with external edges where we put harmonic polyvector fields $\mathbf{H}[[z]]$.

We can also view $\mathbf{F}_0^X$ as a function on $\mathbf{H}[[z]]$. Then $\mathbf{F}_0^X$ has the following perturbative tree diagram expansion

$$\mathbf{F}_0^X = \mathbf{F}_0^{BCOV}.$$

This formula says that $\mathbf{F}_0^X$ arising from the deformation theory of Calabi-Yau manifolds can be identified with the tree-level amplitude of a gauge theory (BCOV theory). The importance of this observation is that it allows us to generalize to construct the higher genus generating function via higher loop Feynman diagrams, though suitable renormalization of the ultraviolet divergence is required. This is the point of view fully developed in [13].

§3. Landau-Ginzburg model

Now we move on to the Landau-Ginzburg B-model. We will focus on an isolated singularity defined by a weighted homogeneous polynomial

$$f : X = \mathbb{C}^n \to \mathbb{C}, \quad f(\lambda^{q_1} x_1, \ldots, \lambda^{q_n} x_n) = \lambda f(x_1, \ldots, x_n).$$

q_i are called the weights of x_i, and the central charge of f is defined by

$$\hat{c}_f = \sum_i (1 - 2q_i).$$

Associated to f, K. Saito has introduced the concept of a primitive form [40], which induces a Frobenius manifold structure (originally called a flat structure) on the local universal deformation space of f. The generalization to arbitrary isolated singularities is later fully established by M. Saito [43]. See also [4, 14, 15, 45] for a certain class of Laurent polynomials. This gives rise to the genus zero correlation functions in the Landau-Ginzburg B-model. For string theoretical point of view, we refer to [35].

In the rest of this section, we will give a brief review of primitive forms. Our presentation will base on the work [28], which exhibits a unified geometry of two models. We will also describe the perburbative formula of primitive forms [28] which is fully developed in [29] for applications to mirror symmetry between Landau-Ginzburg models.

3.1. Residue revisited

We would like to extend the discussion on Calabi-Yau manifolds to Landau-Ginzburg models on the pair $(X = \mathbb{C}^n, f)$. Let us fix the holomorphic volume form

$$\Omega = dx_1 \wedge \cdots \wedge dx_n$$

for convenience. Different choice of Ω will lead to essentially equivalent results in the following discussions. In the Landau-Ginzburg model, we have the twisted operator

$$\bar{\partial}_f := \bar{\partial} + \{f, -\} : \mathrm{PV}(X) \to \mathrm{PV}(X).$$

Its cohomology is given by

$$H^*(\mathrm{PV}(X), \bar{\partial}_f) = H^0(\mathrm{PV}(X), \bar{\partial}_f) \cong \mathrm{Jac}_0(f),$$

where $\mathrm{Jac}_{\mathbf{0}}(f) = \mathbb{C}\{x_i\}/\{\partial_i f\}$ is the Milnor ring of the isolated singularity. Let $\Omega^k_{X,\mathbf{0}}$ be the germ of holomorphic k-forms at $\mathbf{0}$. Let us define

$$\Omega_f := \Omega^n_{X,\mathbf{0}}/df \wedge \Omega^{n-1}_{X,\mathbf{0}}.$$

With our choice of Ω, we can identify

$$\mathrm{Jac}_{\mathbf{0}}(f) \to \Omega_f, \quad [\phi] \to [\phi\Omega].$$

There exists a classical residue pairing defined on Ω_f:

$$\eta_f : \Omega_f \otimes \Omega_f \to \mathbb{C}.$$

Our first task is to understand the residue pairing η_f in terms of trace map on $\mathrm{PV}(X)$. This is done in [28] as follows. Since X is noncompact, the naive trace map via integration doesn't work. Let

$$\mathrm{PV}_c(X) \subset \mathrm{PV}(X)$$

be the subspace of compactly supported polyvector fields. The trace map is defined on $\mathrm{PV}_c(X)$

$$\mathrm{Tr}(\mu) = \int_X (\mu \lrcorner \Omega) \wedge \Omega, \quad \mu \in \mathrm{PV}_c(X).$$

As in the Calabi-Yau case, Tr descends to cohomologies

$$\mathrm{Tr} : H^*(\mathrm{PV}_c(X), \bar{\partial}_f) \to \mathbb{C}.$$

The key observation is that the embedding of complexes

$$(\mathrm{PV}_c(X), \bar{\partial}_f) \hookrightarrow (\mathrm{PV}(X), \bar{\partial}_f)$$

is in fact a quasi-isomorphism, inducing a canonical isomorphism

$$H^*(\mathrm{PV}_c(X), \bar{\partial}_f) \simeq H^*(\mathrm{PV}(X), \bar{\partial}_f).$$

It follows that we have a induced trace map

$$\mathrm{Tr} : \mathrm{Jac}_0(f) \to \mathbb{C}.$$

It is proved in [28] that given $[\alpha], [\beta] \in \mathrm{Jac}_0(f)$,

$$\eta_f([\alpha\Omega], [\beta\Omega]) = \mathrm{Tr}([\alpha\beta]).$$

To get a feeling on how this works, let us look at the simplest example when $X = \mathbb{C}$ is one-dimensional. Let ρ be a cut-off function, which is 1 around the isolated singularity 0 and vanishes outside a big ball. Let us define maps

$$T_\rho : \mathrm{PV}(X) \to \mathrm{PV}_c(X), \quad \mu \to \rho\mu + (\bar{\partial}\rho)\frac{\partial_x}{f'} \wedge \alpha.$$

and

$$R_\rho : \mathrm{PV}(X) \to \mathrm{PV}(X), \quad \mu \to (1-\rho)\frac{\partial_x}{f'} \wedge \alpha.$$

Note that both T_ρ and R_ρ are well-defined since $\bar{\partial}\rho$ and $(1-\rho)$ vanishes around the critical point of f. It is straightforward to check that

$$\bar{\partial}_f R_\rho + R_\rho \bar{\partial}_f = 1 - T_\rho.$$

This in fact proves the quasi-isomorphic embedding $\mathrm{PV}_c(X) \hookrightarrow \mathrm{PV}(X)$ with an explicit homotopic inverse. Let g be a holomorphic function representing an element $[g] \in \mathrm{Jac}_0(f)$. Via the above explicit homotopy, we have

$$\mathrm{Tr}([g]) = \int_{\mathbb{C}} (T_\rho(g) \lrcorner dx) \wedge dx = \int_{\mathbb{C}} \bar{\partial}\rho \frac{g dx}{f'} = \oint \frac{g dx}{f'} = \mathrm{Res}_0 \left(\frac{g dx}{f'} \right).$$

In higher dimensions, the proof is similar though via a more complicated homotopy. We refer to [28] for details.

We remark that $\mathrm{PV}_c(X)$ can be replaced by L^2 space, where Hodge theory and L^2-cohomology could be developed. The construction first appeared in [24], which is also fully developed recently in [16].

3.2. Higher residues

The advantage of revisiting residue via polyvector fields is that it allows us to generalize easily to higher residues discovered in [41]. The construction is parallel to Calabi-Yau models. Let us introduce a descendant variable z and consider the extended complex

$$(\mathrm{PV}(X)[[z]], Q_f = \bar{\partial}_f + z\partial).$$

The cohomology is given by

$$H^*(\mathrm{PV}(X)[[z]], Q_f) \simeq \mathbb{C}\{x_i\}[[z]] / \left(\partial_{x_i} f + z \frac{\partial}{\partial x_i} \right) \mathbb{C}\{x_i\}[[z]].$$

Under Ω, this can be identified with the (formally completed) Brieskorn lattice

$$\mathcal{H}_f^{(0)} := \Omega^n_{X,\mathbf{0}}[[z]] / (df + zd)\Omega^{n-1}_{X,\mathbf{0}}[[z]].$$

Note that

$$\Omega_f = \mathcal{H}_f^{(0)}/z\mathcal{H}_f^{(0)}.$$

There is a natural $\mathbb{Q}$-grading on $\mathcal{H}_f^{(0)}$ defined by assigning the degrees

$$\deg(x_i) = q_i, \quad \deg(dx_i) = q_i, \quad \deg(z) = 1.$$

For a homogeneous element of the form $\varphi = z^k g(x_i) dx_1 \wedge \cdots \wedge dx_n$, we define

$$\deg(\varphi) = \deg(g) + k + \sum_i q_i.$$

In [41], K. Saito constructed the *higher residue pairing*

$$K_f : \mathcal{H}_f^{(0)} \otimes \mathcal{H}_f^{(0)} \to z^n \mathbb{C}[[z]]$$

which satisfies the following properties

(1) K_f is equivariant with respect to the $\mathbb{Q}$-grading, i.e.,

$$\deg(K_f(\alpha, \beta)) = \deg(\alpha) + \deg(\beta)$$

for homogeneous elements $\alpha, \beta \in \mathcal{H}_f^{(0)}$.

(2) $K_f(\alpha,\beta) = (-1)^n \overline{K_f(\beta,\alpha)}$, where the $-$ operator takes $z \to -z$.

(3) $K_f(v(z)\alpha, \beta) = K_f(\alpha, v(-z)\beta) = v(z) K_f(\alpha, \beta)$ for $v(z) \in \mathbb{C}[[z]]$.

(4) The leading z-order of K_f defines a pairing

$$\mathcal{H}_f^{(0)}/z\mathcal{H}_f^{(0)} \otimes \mathcal{H}_f^{(0)}/z\mathcal{H}_f^{(0)} \to \mathbb{C}, \quad \alpha \otimes \beta \mapsto \lim_{z \to 0} z^{-n} K_f(\alpha, \beta)$$

which coincides with the usual residue pairing

$$\eta_f : \Omega_f \otimes \Omega_f \to \mathbb{C}.$$

The last property implies that K_f defines a semi-infinite extension of the residue pairing, which explains the name "higher residue". An alternate way to understand the higher residue pairing is through the trace map in the spirit of our construction for residue. The natural embedding

$$(\mathrm{PV}_c(X)[[z]], Q_f) \hookrightarrow (\mathrm{PV}(X)[[z]], Q_f)$$

is again a quasi-isomorphism. Let us define a pairing

$$\begin{aligned} \tilde{K}_f : &\mathrm{PV}_c(X)[[z]] \times \mathrm{PV}_c(X)[[z]] \to z^n \mathbb{C}[[z]], \\ &\tilde{K}_f(g(z)\alpha, h(z)\beta) = z^n g(z) h(-z) \operatorname{Tr}(\alpha\beta). \end{aligned}$$

It is easy to see that $\tilde{K}_f$ descends to $H^*(\mathrm{PV}_c(X)[[z]], Q_f)$ which is canonically isomorphic to $H^*(\mathrm{PV}(X)[[z]], Q_f)$. Under the identification of $H^*(\mathrm{PV}(X)[[z]], Q_f)$ with $\mathcal{H}_f^{(0)}$, we obtain a pairing

$$\tilde{K}_f : \mathcal{H}_f^{(0)} \otimes \mathcal{H}_f^{(0)} \to z^n \mathbb{C}[[z]].$$

It is proved in [28] that $\tilde{K}_f$ is precisely the higher residue pairing

$$\tilde{K}_f = K_f.$$

3.3. Universal unfolding

In the Calabi-Yau case, the (extended) moduli space of complex structures is controlled by the differential graded Lie algebra $(\mathrm{PV}(X), \bar{\partial}, \{,\})$. In the Landau-Ginzburg case, this is twisted to be $(\mathrm{PV}(X), \bar{\partial}_f, \{,\})$. The universal solutions of the associated Maurer-Cartan equation is greatly simplified, and can be represented as a deformation of $f(\mathbf{x})$ via the universal unfolding:

$$F : \mathbb{C}^n \times \mathbb{C}^\mu \to \mathbb{C}, \quad F = f(\mathbf{x}) + \sum_{\alpha=1}^{\mu} s^\alpha \phi_\alpha(\mathbf{x}),$$

where $\mu = \dim_{\mathbb{C}} \mathrm{Jac}(f)$, and $\{\phi_\alpha(\mathbf{x})\}$ is a basis of $\mathrm{Jac}_{\mathbf{0}}(f)$.

In our case f being weighted homogenous, we can further assume that ϕ_α are all weighted homogeneous with increasing degrees

$$0 = \deg(\phi_1) \leq \deg(\phi_2) \leq \cdots \leq \deg(\phi_\mu) = \hat{c}_f, \quad \text{where } \deg(x_i) = q_i.$$

The Brieskorn lattice and the higher residue pairing can be extended to the family case on the germ $\mathcal{M} = (\mathbb{C}^\mu, 0)$ associated to the unfolding F. We have

$$\mathcal{H}_F^{(0)} := \Omega^n_{X\times\mathcal{M}/\mathcal{M},\mathbf{0}}[[z]]/(dF + zd)\Omega^{n-1}_{X\times\mathcal{M}/\mathcal{M},\mathbf{0}}[[z]]$$

where $\Omega^*_{X\times\mathcal{M}/\mathcal{M}}$ is the sheaf of relative holomorphic differential forms. It can be viewed as a free sheaf of rank μ on $\mathcal{M} \times \hat{\Delta}$, where $\hat{\Delta}$ is the formal disk with parameter z. $\mathcal{H}_F^{(0)}$ is equipped with a flat Gauss-Manin connection on $\mathcal{M} \times \hat{\Delta}$, denoted by ∇^{GM}. The higher residue pairing extends to

$$K_F : \mathcal{H}_F^{(0)} \otimes_{\mathcal{O}_{\mathcal{M}}} \mathcal{H}_F^{(0)} \to z^n \mathcal{O}_{\mathcal{M}}[[z]]$$

satisfying the following properties [41]

(1) $K_F(s_1, s_2) = (-1)^n \overline{K_F(s_2, s_1)}$, where $-$ is the operator $z \to -z$.

(2) $K_F(g(z)s_1, s_2) = K_F(s_1, g(-z)s_2) = g(z)K_F(s_1, s_2)$ for any $g \in \mathcal{O}_{\mathcal{M}}[[z]]$.
(3) $\partial_V K_F(s_1, s_2) = K_F(\nabla_V^{GM} s_1, s_2) + K_F(s_1, \nabla_V^{GM} s_2)$ for any $V \in T_{\mathcal{M}}$.
(4) $z\partial_z K_F(s_1, s_2) = K_F(\nabla_{z\partial_z}^{GM} s_1, s_2) + K_F(s_1, \nabla_{z\partial_z}^{GM} s_2)$.
(5) The induced pairing

$$\mathcal{H}_F^{(0)}/z\mathcal{H}_F^{(0)} \otimes_{\mathcal{O}_{\mathcal{M}}} \mathcal{H}_F^{(0)}/z\mathcal{H}_F^{(0)} \to \mathcal{O}_{\mathcal{M}}$$

coincides with the classical residue pairing.

In a similar fashion, K_F can be constructed via fiberwise cut-off function [28].

3.4. Good basis and Generating function

Let us extend the higher residue pairing to

$$K_f : \mathcal{H}_f \otimes \mathcal{H}_f \to \mathbb{C}((z)).$$

This defines a symplectic pairing ω_f on $\mathcal{H}_f$ by

$$\omega_f(\alpha, \beta) := \mathrm{Res}_{z=0}\, z^{-n} K_f(\alpha, \beta) dz,$$

with $\mathcal{H}_f^{(0)}$ being an isotropic subspace. At this stage, we are in completely analogue situation as in Calabi-Yau models. The next step is to use deformation theory to describe a lagrangian cone inside $\mathcal{H}_f$ from which we obtain the generating function as the genus zero invariants in Landau-Ginzburg B-model.

The first thing we need is a splitting $\mathcal{L}$

$$\mathcal{H}_f = \mathcal{H}_f^{(0)} \oplus \mathcal{L}$$

such that:

(1) $\mathcal{L}$ preserves the $\mathbb{Q}$-grading;
(2) $\mathcal{L}$ is an isotropic subspace;
(3) $z^{-1} : \mathcal{L} \to \mathcal{L}$ maps $\mathcal{L}$ into $\mathcal{L}$.

Here for f being weighted homogeneous, (1) is equivalent to the conventional condition that $\nabla_{z\partial_z}^{GM}$ preserves $\mathcal{L}$. The choice of $\mathcal{L}$ is essentially equivalent to the notion of a good section [40]. Let

$$B = \mathcal{H}_f^{(0)} \cap z\mathcal{L}.$$

We can represent a basis of B by homogeneous polynomials $\{\phi_\alpha\}_{\alpha=1}^{\mu}$ such that

$$B = \mathrm{Span}_{\mathbb{C}}\{[\phi_\alpha \Omega]\}.$$

$\{\phi_\alpha\}$ can be viewed as a "good" representative of $\mathrm{Jac}_0(f)$, defining a lifting

$$\Omega_f \to \mathcal{H}_f^{(0)}.$$

The isotropic property of $\mathcal{H}_f^{(0)}$ and $\mathcal{L}$ implies that the higher residue paring

$$K_f(B, B) \in z^n\mathbb{C}.$$

In fact, $B \subset \mathcal{H}_f^{(0)}$ implies that

$$K_f(B, B) \subset z^n\mathbb{C}[[z]]$$

while $B \subset z\mathcal{L}$ implies that

$$K_f(B, B) \subset z^n\mathbb{C}[z^{-1}].$$

This says that for such choice of representatives $\{\phi_\alpha\}$, the higher residue pairings vanish beyond the leading ordinary residue pairing. Such vanishing property is the key property of so-called "*good basis*" [40].

Given a good basis B, we can identify

$$\mathcal{H}_f^{(0)} = B[[z]], \quad \mathcal{L} = z^{-1}B[z^{-1}], \quad \mathcal{H}_f = B((z)).$$

Now we follow the Calabi-Yau situation to describe the lagrangian cone. A universal solution of the Maurer-Cantan equation

$$Q_f\mu + \frac{1}{2}\{\mu, \mu\} = 0, \quad \mu \in \mathrm{PV}(X)[[z]],$$

can be described by

$$\mu(\mathbf{s}) = \sum_{k\geq 0}\sum_{\alpha} s_k^\alpha \phi_\alpha z^k,$$

parametrized by $\mathbf{s} = \{s_k^\alpha\}$. $\{s_0^\alpha\}$ parametrizes the universal unfolding, while s_k^α for $k > 0$ corresponds to descendant deformations.

Let $\phi^\alpha \in B$ be a dual basis of B such that

$$K_f([\phi_\alpha\Omega], [\phi^\beta\Omega]) = z^n\delta_\alpha^\beta.$$

The lagrangian cone is then specified by

$$z - ze^{\mu/z}$$

from which we can always expand

$$z[e^{\mu(\mathbf{s})/z}\Omega] = z[\Omega] + \sum_{k\geq 0}\sum_{\alpha} \tau_k^\alpha(\mathbf{s})[\phi_\alpha\Omega](-1)^k z^k + \sum_{k\geq 0}\sum_{\alpha} p_{k,\alpha}(\mathbf{s})[\phi^\alpha\Omega]z^{-k-1} \in \mathcal{H}_f[[\mathbf{s}]]$$

where [] refers to the class in $\mathcal{H}_f$. $\{\tau_k^\alpha\}$ can be viewed as the linear coordinates on $B[[z]]$. The transformation

$$\mathbf{s} \to \{\tau_k^\alpha(\mathbf{s})\}$$

defines a coordinate transformation. In terms of $\{\tau_k^\alpha\}$, we find

$$p_{k,\alpha} = \frac{\partial \mathbf{F}_0^{\mathcal{L}}(\tau)}{\partial \tau_k^\alpha},$$

where $\mathbf{F}_0^{\mathcal{L}}$ is the generating function of genus zero invariants in Landau-Ginzburg B-model with respect to the splitting $\mathcal{L}$, or the good basis B.

3.5. Primitive form and perturbation theory

A section $\zeta \in \mathcal{H}_F^{(0)}$ is called a *primitive form* if it satisfies the following conditions:

(1) (Primitivity) The section ζ induces an $\mathcal{O}_{\mathcal{M}}$-module isomorphism

$$z\nabla^{GM}\zeta : T_{\mathcal{M}} \to \mathcal{H}_F^{(0)}/z\mathcal{H}_F^{(0)}; \quad V \mapsto z\nabla_V^{GM}\zeta.$$

(2) (Orthogonality) For any local sections V_1, V_2 of $T_{\mathcal{M}}$,

$$K_F\left(\nabla_{V_1}^{GM}\zeta, \nabla_{V_2}^{GM}\zeta\right) \in z^{n-2}\mathcal{O}_{\mathcal{M}}.$$

(3) (Holonomicity) For any local sections V_1, V_2, V_3 of $T_{\mathcal{M}}$,

$$K_F\left(\nabla_{V_1}^{GM}\nabla_{V_2}^{GM}\zeta, \nabla_{V_3}^{GM}\zeta\right) \in z^{n-3}\mathcal{O}_{\mathcal{M}} \oplus z^{n-2}\mathcal{O}_{\mathcal{M}};$$
$$K_F\left(\nabla_{z\partial_z}^{GM}\nabla_{V_1}^{GM}\zeta, \nabla_{V_2}^{GM}\zeta\right) \in z^{n-3}\mathcal{O}_{\mathcal{M}} \oplus z^{n-2}\mathcal{O}_{\mathcal{M}}.$$

(4) (Homogeneity) There is a constant $r \in \mathbb{C}$ such that

$$\left(\nabla_{z\partial_z}^{\Omega} + \nabla_E^{\Omega}\right)\zeta = r\zeta.$$

where E is the Euler vector field. In our case of weighted homogeneous singularity, we have $r = \sum_i q^i$.

It is proved in [40] that the space of primitive forms of f up to a constant rescaling is isomorphic to the space of good basis. The generalization to arbitrary isolated singularities is established by M. Saito [43, 44].

Closed formula for primitive forms are known only for ADE singularities ($\hat{c}_f < 1$) and simple elliptic singularities ($\hat{c}_f = 1$) [40]. In the following we will describe a perturbative theory of primitive forms

[28] from which we can compute the primitive form recursively order by order.

Givental's J-function is essentially the analogue of primitive form. Starting from the formula for the lagrangian cone

$$z[e^{\mu(\mathbf{s})/z}\Omega] = z[\Omega] + \sum_{k\geq 0}\sum_{\alpha} \tau_k^\alpha(\mathbf{s})[\phi_\alpha\Omega](-1)^k z^k + \sum_{k\geq 0}\sum_{\alpha} p_{k,\alpha}(\mathbf{s})[\phi^\alpha\Omega]z^{-k-1} \in \mathcal{H}_f[[\mathbf{s}]].$$

If we set

$$\tau_k^\alpha = 0, \quad k > 0,$$

then we find the expansion

$$[e^{\mu(\mathbf{s}(\tau_0^\alpha,\tau_1^\alpha=0,\cdots))/z}\Omega] = [\Omega] + \sum_\alpha \tau_0^\alpha[\phi_\alpha\Omega]z^{-1} + O(z^{-2}).$$

It is shown in [28] that the left hand side

$$e^{\mu(\mathbf{s}(\tau_0^\alpha,\tau_1^\alpha=0,\cdots))/z}\Omega$$

gives exactly the expression of primitive form with respect to the good basis $\{\phi_\alpha\}$ under the trivialization map via Gauss-Manin connection

$$\mathcal{H}_F \overset{e^{(F-f)/z}}{\to} \mathcal{H}_f \otimes_{\mathbb{C}} \mathbb{C}[[\tau_0^\alpha]].$$

This is translated into the following algorithm. Let $s = \{s^\alpha\}$ parametrizes the universal unfolding with respect to the good basis ϕ_α

$$F = f + \sum_\alpha s^\alpha\phi_\alpha.$$

Let us represent the primitive ζ as a power series

$$\zeta = \sum_{k\geq 0}\zeta_{(k)} = \sum_{k\geq 0}\sum_\alpha \zeta_{(k)}^\alpha\phi_\alpha\Omega, \quad \zeta_{(k)}^\alpha \in \mathbb{C}[[z]]\otimes_{\mathbb{C}}\mathbb{C}[s]_k$$

where $\mathbb{C}[s]_k$ is the space of homogenous polynomials in $\{s^\alpha\}$ of degree k. Then ζ is characterized by the equation [28]

$$(\dagger) \qquad [e^{(F-f)/z}\zeta] \in [\Omega] + z^{-1}B[z^{-1}][[s]].$$

This equation can be solved recursively in terms of the order k as follows.

Since $e^{(F-f)/z} \equiv 1 \mod (s)$, the leading order of $(\dagger)$ is

$$\zeta_{(0)} \in [\Omega] + z^{-1}B[z^{-1}]$$

which is uniquely solved by $\zeta_{(0)} = \Omega$. Suppose we have solved (†) up to order N, i.e., $\zeta_{(\leq N)} := \sum_{k=0}^{N} \zeta_{(k)}$ such that

$$[e^{(F-f)/t}\zeta_{(\leq N)}] \in [\Omega] + z^{-1}B[z^{-1}][[s]] \mod (s^{N+1}).$$

Let $R_{N+1} \in B((z)) \otimes_{\mathbb{C}} \mathbb{C}[s]_{(N+1)}$ be the $(N+1)$-th order component of $e^{(F-f)/t}\zeta_{(\leq N)}$. Let

$$R_{N+1} = R^{+}_{N+1} + R^{-}_{N+1}$$

where $R^{+}_{N+1} \in B[[z]] \otimes_{\mathbb{C}} \mathbb{C}[s]_{(N+1)}, R^{-}_{N+1} \in z^{-1}B[z^{-1}] \otimes_{\mathbb{C}} \mathbb{C}[s]_{(N+1)}$. Let $\tilde{R}^{+}_{N+1} \in B_F[[z]] \otimes_{\mathbb{C}} \mathbb{C}[s]_{(N+1)}$ correspond to R^{+}_{N+1} under the manifest identification between B and B_F. Then

$$\zeta_{(\leq N+1)} := \zeta_{(\leq N)} - \tilde{R}^{+}_{N+1}$$

gives the unique solution of (†) up to order $N+1$. This algorithm allows us to solve ζ perturbatively to arbitrary order. Moreover, we can find the expansion

$$[e^{(F-f)/z}\zeta] = [\Omega] + \sum_{\alpha} \tau^{\alpha}(s)[\phi_{\alpha}\Omega]z^{-1} + \sum_{\alpha} p_{\alpha}(s)[\phi^{\alpha}\Omega]z^{-2} + O(z^{-3}).$$

Then the coordinate transformation

$$s^{\alpha} \to \tau^{\alpha}(s)$$

gives the flat coordinate τ^{α} of the underlying Frobenius manifold structure. In terms of τ^{α}, we have

$$p_{\alpha}(s(\tau)) = \frac{\partial \mathbf{F}_0(\tau)}{\partial \tau^{\alpha}}$$

where $\mathbf{F}_0(\tau)$ is the potential function satisfying the WDVV equation.

3.6. Examples of primitive forms

Let us illustrate how the perturbative formula of primitive form works in some examples.

3.6.1. *ADE singularity* Let f be a weighted homogenous polynomial of ADE type. This is equivalent to $\hat{c}_f < 1$. There exists a unique choice of good basis, and any representative of homogenous basis $\{\phi_{\alpha}\}$ of $\mathrm{Jac}_0(f)$ are equivalent. Their degrees are bounded by

$$\deg(\phi_{\alpha}) \leq \hat{c}_f < 1.$$

The fact that an arbitrary representative $\{\phi_\alpha\}$ being a good basis is a very special property of ADE singularities and follows from a degree argument as follows [40]. We need to show that

$$K_f(\phi_\alpha\Omega, \phi_\beta\Omega) \in \mathbb{C}z^n, \forall\alpha, \beta.$$

In fact, since K_f preserves the grading, we have

$$\begin{aligned}\deg(K_f(\phi_\alpha\Omega, \phi_\beta\Omega)) &= \deg(\phi_\alpha) + \deg(\phi_\alpha) + 2\deg(\Omega)\\ &= \deg(\phi_\alpha) + \deg(\phi_\alpha) + n - \hat{c}_f.\end{aligned}$$

It follows that

$$n \leq \deg(K_f(\phi_\alpha\Omega, \phi_\beta\Omega)) \leq n + \hat{c}_f.$$

Since $0 < \hat{c}_f < 1$ and $\deg(K_f(\phi_\alpha\Omega, \phi_\beta\Omega))$ is an integer, it follows that $\deg(K_f(\phi_\alpha\Omega, \phi_\beta\Omega)) = n$, i.e.,

$$K_f(\phi_\alpha\Omega, \phi_\beta\Omega) \in \mathbb{C}z^n.$$

Let F be the universal unfolding

$$F = f + \sum_\alpha s^\alpha\phi_\alpha.$$

We extend the weighted homogeneous degree to s^α, z by

$$\deg(s^\alpha) = 1 - \deg(\phi_\alpha), \quad \deg(z) = 1$$

such that F is homogenous of degree 1. Equation (†) respects the degree. By the fact that $\deg(s^\alpha) > 0$ and simple degree counting, we find that

$$[e^{(F-f)/z}\Omega] \in [\Omega] + z^{-1}B[z^{-1}][[s^\alpha]],$$

i.e. $\zeta = \Omega$ is the primitive form. It is again a special property of ADE singularity that the primitive form doesn't depend on the deformation parameter.

3.6.2. *Simple elliptic singularity* Simple elliptic singularities are characterized by $\hat{c}_f = 1$. We consider an example of type $E_6^{(1,1)}$ with

$$f = \frac{1}{3}(x_1^3 + x_2^3 + x_3^3).$$

A monomial basis is given by

$$\{\phi_1, \ldots, \phi_8\} = \{1, x_1, x_2, x_3, x_1x_2, x_2x_3, x_3x_1, x_1x_2x_3\}.$$

Lemma 3.1. *The above basis $\{\phi_i\}$ is a good basis.*

Proof. Similar to the ADE singularities, we have the degree estimate

$$3 = n \leq \deg(K_f(\phi_i\Omega, \phi_j\Omega)) \leq n + \hat{c}_f = 4.$$

The only possibility that the above degree reaches 4 is when $\phi_i = \phi_j = \phi_8$. Therefore we only need to show that

$$K_f(\phi_8\Omega, \phi_8\Omega) = 0.$$

This follows from the symmetry of the higher residue pairing

$$K_f(\phi_8\Omega, \phi_8\Omega) = (-1)^3\overline{K_f(\phi_8\Omega, \phi_8\Omega)} = -K_f(\phi_8\Omega, \phi_8\Omega).$$

Q.E.D.

We consider the universal unfolding

$$F = f + \sigma\phi_8 + \sum_{\alpha=1}^{7} u^\alpha \phi_\alpha$$

where $(u^1, \ldots, u^7, \sigma)$ are the deformation parameters. There exists a one-parameter choice of good basis for simple elliptic singularities. In this example, they are

$$B(c) = \mathrm{Span}_{\mathbb{C}}\{\phi_1, \ldots, \phi_7, \phi_8 + cz\}, \quad c \in \mathbb{C}.$$

$B(c)$ being a good basis follows immediately from the above lemma.

We extend the weighted homogenous degree as before. Then

$$\deg(u^\alpha) > 0, \quad \deg(\sigma) = 0.$$

We can use similar degree argument as the ADE case, with some extra care on the degree zero parameter σ. The first observation is that as elements in $\mathcal{H}_f$,

$$[e^{\sigma\phi_8/z}\Omega] = g(\sigma)[\Omega] + z^{-1}h(\sigma)[\phi_8\Omega],$$

where $g(\sigma)$ and $h(\sigma)$ are respectively given by

$$g(\sigma) = 1 + \sum_{r=1}^{\infty} \frac{(-1)^r \sigma^{3r} \prod_{j=1}^{r}(3j-2)^3}{(3r)!},$$

$$h(\sigma) = \sigma + \sum_{r=1}^{\infty} \frac{(-1)^r \sigma^{3r+1} \prod_{j=1}^{r}(3j-1)^3}{(3r+1)!}.$$

This can be computed explicitly by expanding $e^{\sigma\phi_8/z}$ and using the equivalence relation in $\mathcal{H}_f$. $g(\sigma), h(\sigma)$ are the fundamental solutions to

$$\left((1+\sigma^3)\partial_\sigma^2 + 3\sigma^2\partial_\sigma + \sigma\right)\nu(\sigma) = 0,$$

which is the Picard-Fuchs equation for the period integrals on the elliptic curve inside $\mathbb{P}^2$ defined by the cubic equation $f + \sigma\phi_8 = 0$.

By a similar degree argument, we have

$$\begin{aligned}
[e^{(F-f)/z}\Omega] &\in [e^{\sigma\phi_8/z}\Omega] + z^{-1}B(c)[z^{-1}] \\
&= (g(\sigma) - ch(\sigma))[\Omega] + z^{-1}h(\sigma)[(\phi_8 + cz)\Omega] + z^{-1}B(c)[z^{-1}] \\
&= (g(\sigma) - ch(\sigma))[\Omega] + z^{-1}B(c)[z^{-1}].
\end{aligned}$$

It follows that

$$\left[e^{(F-f)/z}\frac{\Omega}{g(\sigma) - ch(\sigma)}\right] \in \Omega + z^{-1}B(c)[z^{-1}].$$

In particular, the primitive form associated to the good basis $B(c)$ is given by [40]

$$\zeta(c) = \frac{\Omega}{g(\sigma) - ch(\sigma)}.$$

In other words, the normalization is given by a period on the elliptic curve. Different choices of primitive forms correspond to the choices of the period.

3.6.3. *Exceptional unimodular singularity* For singularities with $\hat{c}_f > 1$, there exists parameters of the deformation which have negative degree. The degree argument that we have used in ADE and simple elliptic singularity doesn't work well here since there would exist non-trivial mixing between positive degree and negative degree parameters. This is the essential difficulty to find the expression of primitive forms in such situation. Nevertheless, we can use the perturbative algorithm to find order expansion of the primitive form.

The first nontrivial examples beyond simple elliptic singularities are given by Arnold's exceptional unimodular singularities. There are in total 14 types, and let us consider an example of type E_{12}

$$f = x^3 + y^7.$$

The central charge is given by $\hat{c}_f = \frac{22}{21}$. There is a unique good basis given by

$$\{\phi_1, \dots, \phi_{12}\} := \{1, y, y^2, x, y^3, xy, y^4, xy^2, y^5, xy^3, xy^4, xy^5\}.$$

This follows from a similar degree argument as ADE singularities and we leave the details to the readers. See also Section 4.2.

We represent the universal unfolding by

$$F = x^3 + y^7 + \sum_{i=1}^{12} u_i\phi_i.$$

By direct calculations (with a computer), the primitive form ζ up to order 10 is given by [28]

$$\begin{aligned}\zeta/\Omega = 1 &+ \frac{4}{3\cdot 7^2}u_{11}u_{12}^2 - \frac{64}{3\cdot 7^4}u_{11}^2u_{12}^4 - \frac{76}{3^2\cdot 7^4}u_{10}u_{12}^5 + \frac{937}{3^3\cdot 7^5}u_9u_{12}^6\\ &+ \frac{218072}{3^4\cdot 5\cdot 7^6}u_{11}^3u_{12}^6 + \frac{1272169}{3^4\cdot 5\cdot 7^7}u_{10}u_{11}u_{12}^7 + \frac{28751}{3^4\cdot 7^7}u_8u_{12}^8\\ &- \frac{1212158}{3^4\cdot 7^8}u_9u_{11}u_{12}^8 - \frac{38380}{3^3\cdot 7^8}u_7u_{12}^9 + \left(\frac{1}{7^2}u_{12}^3 - \frac{101}{5\cdot 7^4}u_{11}u_{12}^5\right.\\ &\left.+ \frac{1588303}{3^4\cdot 5\cdot 7^7}u_{11}^2u_{12}^7 + \frac{378083}{3^4\cdot 5\cdot 7^7}u_{10}u_{12}^8 - \frac{108144}{3\cdot 7^8}u_9u_{12}^9\right)x\\ &+ \left(\frac{1447}{3^3\cdot 7^6}u_{12}^7 - \frac{71290}{3^3\cdot 7^8}u_{11}u_{12}^9\right)y - \frac{45434}{3^4\cdot 7^8}u_{12}^{10}xy\\ &- \left(\frac{53}{3^2\cdot 7^4}u_{12}^6 - \frac{46244}{3^3\cdot 7^7}u_{11}u_{12}^8\right)x^2 + \frac{22054}{3^4\cdot 7^7}u_{12}^9x^3 + O(u^{10}).\end{aligned}$$

§4. Mirror symmetry between Landau-Ginzburg models

4.1. Mirror singularities

The main application of the perturbative primitive form is to mirror symmetry. The Calabi-Yau B-model around the large complex limit is mirror to Gromov-Witten theory. For Landau-Ginzburg B-models that we have discussed, they are expected to be mirror to FJRW theory [17]. The computation tools that we have developed will enable us to approach this.

The Landau-Ginzburg mirror pairs originate from an old physical construction of Berglund-Hübsch [6] that was completed by Krawitz [25]. Let

$$W : \mathbb{C}^N \to \mathbb{C}$$

be a weighted homogeneous polynomial with an isolated critical point at the origin. We define its *maximal group of diagonal symmetries* to be

$$G_W = \left\{(\lambda_1, \ldots, \lambda_N) \in (\mathbb{C}^*)^N \,\middle|\, W(\lambda_1 x_1, \ldots, \lambda_N x_N) = W(x_1, \ldots, x_N)\right\}.$$

In the BHK mirror construction, the polynomial W is required to be *invertible* [9, 25], i.e., the number of variables must equal the number of monomials of W and it contains no monomial of the form x_ix_j for $i \neq j$. By rescaling the variables, we can always write W as

$$W = \sum_{i=1}^{N} \prod_{j=1}^{N} x_j^{a_{ij}}.$$

We denote its *exponent matrix* by $E_W = (a_{ij})_{N\times N}$. The mirror polynomial of W is

$$W^T = \sum_{i=1}^{N} \prod_{j=1}^{N} x_j^{a_{ji}},$$

i.e., the exponent matrix E_{W^T} of the mirror polynomial is the transpose of E_W.

All invertible polynomials have been classified by Kreuzer and Skarke.

Theorem 4.1 ([27], Theorem 1)**.** *A polynomial is invertible if and only if it is a disjoint sum of the three following atomic types, where $a \geq 2$ and $a_i \geq 2$:*

- Fermat: x^a.
- Chain: $x_1^{a_1} + x_1x_2^{a_2} + \cdots + x_{N-1}x_N^{a_N}$.
- Loop: $x_1^{a_1}x_N + x_1x_2^{a_2} + \cdots + x_{N-1}x_N^{a_N}$.

The inverse matrix E_W^{-1} plays an important role in the mirror map constructed by Krawitz in [25]. Let us write

$$E_W^{-1} = \begin{pmatrix} q_{11} & \cdots & q_{1N} \\ \vdots & \vdots & \vdots \\ q_{N1} & \cdots & q_{NN} \end{pmatrix},$$

and define

$$\begin{aligned} \rho_j^W &:= \left(\exp(2\pi\sqrt{-1}q_{1j}), \ldots, \exp(2\pi\sqrt{-1}q_{Nj})\right), \\ \rho_j^{W^T} &:= \left(\exp(2\pi\sqrt{-1}q_{j1}), \ldots, \exp(2\pi\sqrt{-1}q_{jN})\right). \end{aligned}$$

According to [25], the group G_W is generated by $\{\rho_j^W\}_{j=1}^N$ and G_{W^T} is generated by $\{\rho_j^{W^T}\}_{j=1}^N$. Recall q_i is the weight of x_i in W. Let q_i^T be the weight of x_i in W^T. We remark that

$$q_i = \sum_{j=1}^{N} q_{ij} \quad \text{and} \quad q_i^T = \sum_{j=1}^{N} q_{ji}.$$

4.2. Good basis of invertible polynomials

To obtain the Frobenius manifold structure on the deformation space of the singularity, we need to identify a good basis. The abstract existence of such a good basis is proved in [40] for the quasi-homogenous cases and in [43] for general isolated singularity. Moreover, such a good basis is generally not unique. However, mirror symmetry favors for a particular one as the mirror of FJRW theory, and we need explicit information about the good basis to obtain the enumerative data. This involves the exact computation of higher residue pairing which is usually hard to perform.

In case of ADE and simple elliptic singularities, the justification of good basis follows from a simple degree counting argument (see Section 3.6). The effectivity of this method fails when the central charge of the polynomial gets large. The way out is that we have to incorporate the full symmetry group G_W to improve the degree counting argument. This is developed in [23] to find explicit good basis for all invertible polynomials. We review this result here.

In the rest of this section, we will use f instead of W to adopt the conventional notation in singularity theory.

Firstly, if $f(\mathbf{x},\mathbf{y}) = f_1(\mathbf{x}) + f_2(\mathbf{y})$ is the disjoint sum of two polynomials f_1, f_2, and $\{\eta_i(\mathbf{x})\}_{i\in I}$ and $\{\varphi_\alpha(\mathbf{y})\}_{\alpha\in A}$ are good basis for f_1 and f_2 respectively, then it is easy to show that $\{\eta_i(\mathbf{x})\varphi_\alpha(\mathbf{y})\}_{(i,\alpha)\in I\times A}$ is a good basis of W. This reduces the search of good basis for invertible polynomials to atomic types by Theorem 4.1. They are found in [23] as follows.

Theorem 4.2 ([23]). *Let f be an invertible polynomial of atomic types. Then the following choice $\{\phi_\alpha\}$ is a good basis of f.*

- *Let $f = x^a$ be a Fermat, then $\{\phi_\alpha\} = \{x^r \mid 0 \le r \le a-2\}$.*
- *Let $f = x_1^{a_1} + x_1x_2^{a_2} + \cdots + x_{N-1}x_N^{a_N}$ be a chain, then*

$$\{\phi_\alpha\} = \left\{\prod_{i=1}^{N} x_i^{r_i}\right\}_{\mathbf{r}}$$

where $\mathbf{r} = (r_1, \ldots, r_N)$ with $r_i \le a_i - 1$ for all i and $\mathbf{r}$ is not of the form $(*, \ldots, *, k, a_{N-2l} - 1, \ldots, 0, a_{N-2} - 1, 0, a_N - 1)$ with $k \ge 1$.

- *Let $f = x_1^{a_1}x_N + x_1x_2^{a_2} + \cdots + x_{N-1}x_N^{a_N}$ be a loop, then*

$$\{\phi_\alpha\} = \left\{\prod_{i=1}^{N} x_i^{r_i} \;\middle|\; 0 \le r_i < a_i\right\}.$$

We will call the above monomials the *standard basis*.

The Fermat type $f = x^a$ is of type A_{a-1}-singularity, and we already know that its standard basis is a good basis. We will focus on the remaining two cases.

Recall G_f is the diagonal symmetry group of f. $H_f^{(0)}$ inherits the structure of G_f-representation. Explicitly, let

$$\{\rho_j^f = (e^{2\pi\sqrt{-1}q_{1j}}, \dots, e^{2\pi\sqrt{-1}q_{Nj}})\}_{j=1}^N$$

be the generator of G_f, where $E_f^{-1} = (q_{ij})_{N\times N}$ is the inverse of the exponent matrix E_f. Then the G_f-action on $H_f^{(0)}$ is generated by

$$\rho_j^f : x_i \to e^{2\pi\sqrt{-1}q_{ij}} x_i, \quad dx_i \to e^{2\pi\sqrt{-1}q_{ij}} dx_i, \quad z \to z.$$

Since G_f preserves f, it also preserves the higher residue pairing K_f. Therefore we have a G_f-invariant pairing

$$K_f : H_f^{(0)} \otimes H_f^{(0)} \to z^n\mathbb{C}[[z]].$$

Let $x_1^{r_1}\cdots x_N^{r_N}$ and $x_1^{r_1'}\cdots x_N^{r_N'}$ be monomials in the standard basis for either the chain or loop type. We need to show that

$$K_f(x_1^{r_1}\cdots x_N^{r_N} d^N x, x_1^{r_1'}\cdots x_N^{r_N'} d^N x) \in z^N\mathbb{C}.$$

Let

$$(m_1, \dots, m_N) = (r_1 + r_1', \dots, r_N + r_N').$$

The ρ_j^f-invariance of K_f implies the *integral conditions*

$$\sum_{i=1}^N (m_i + 2)q_{ij} = k_j \in \mathbb{Z}, \text{ for all } j,$$
$$\text{if } K_f(x_1^{r_1}\cdots x_N^{r_N} d^N x, x_1^{r_1'}\cdots x_N^{r_N'} d^N x) \neq 0.$$

(the extra 2 comes from two copies of $d^N x$). This is equivalent to

$$(\star) \qquad (k_1, k_2, \dots, k_N)\, E_f = (m_1 + 2, m_2 + 2, \dots, m_N + 2).$$

Remark 4.1. The integral conditions is mirror to the Integer Degree Axiom for the degree of orbifold Line bundles in FJRW theory [23].

4.2.1. *chain type* Let $f = x_1^{a_1} + x_1 x_2^{a_2} + \cdots + x_{N-1} x_N^{a_N}$. The exponent matrix has the form

$$E_f = \begin{pmatrix} a_1 & & & & \\ 1 & a_2 & & & \\ & \ddots & \ddots & & \\ & & & a_{N-1} & \\ & & & 1 & a_N \end{pmatrix}.$$

Equation $(\star)$ in this case becomes

$$m_j = k_j a_j + k_{j+1} - 2,\ j = 1, \ldots, N-1; \quad m_N = k_N a_N - 2$$

where $0 \le m_i \le 2a_i - 2$ and the k_i are integers. Tracing all possible values of k_i starting from k_N, the only possibilities are

(1) $(k_1, \ldots, k_N) = (1, 1, \ldots, 1)$.
(2) $(k_1, \ldots, k_N) = (1, \ldots, 1, 0, 2, \ldots, 0, 2)$.
(3) $(k_1, \ldots, k_N) = (1, \ldots, 1, 2, 0, 2, \ldots, 0, 2)$.

In case (3), the N-tuple $(m_1, \ldots, m_N)$ is given by

$$(a_1 - 1, \ldots, a_{N-2l-2} - 1, a_{N-2l-1}, 2a_{N-2l} - 2, \ldots, 0, 2a_{N-2} - 2, 0, 2a_N - 2).$$

This can not appear if both $x_1^{r_1} \cdots x_N^{r_N}$ and $x_1^{r'_1} \cdots x_N^{r'_N}$ are in the standard basis. For cases (1) and (2), we check directly that

$$\deg(x_1^{m_1} \cdots x_N^{m_N}) = \deg(x_1^{r_1} \cdots x_N^{r_N}) + \deg(x_1^{r'_1} \cdots x_N^{r'_N}) = \hat{c}_f.$$

Since K_f preserves the $\mathbf{Q}$-grading, we have

$$\begin{aligned} &\deg K_f(x_1^{r_1} \cdots x_N^{r_N} d^N x, x_1^{r'_1} \cdots x_N^{r'_N} d^N x) \\ &= \deg(x_1^{r_1} \cdots x_N^{r_N}) + \deg(x_1^{r'_1} \cdots x_N^{r'_N}) + 2\sum_i q_i \\ &= N. \end{aligned}$$

It follows that $K_f([x_1^{r_1} \cdots x_N^{r_N}]_f, [x_1^{r'_1} \cdots x_N^{r'_N}]_f)$ lies in $z^N \mathbb{C}$.

4.2.2. *loop type* Let $f = x_1^{a_1} x_N + x_1 x_2^{a_2} + \cdots + x_{N-1} x_N^{a_N}$. The exponent matrix is

$$E_f = \begin{pmatrix} a_1 & & & & 1 \\ 1 & a_2 & & & \\ & \ddots & \ddots & & \\ & & & a_{N-1} & \\ & & & 1 & a_N \end{pmatrix}.$$

With the convention $k_1 \equiv k_{N+1}$, Equation $(\star)$ above implies

$$m_i + 2 = k_i a_i + k_{i+1}, \quad i = 1, \ldots, N.$$

Let $h_i = k_i - 1$ for each i. The above equation becomes

$$m_i + 2 = (h_i + 1)a_i + h_{i+1} + 1.$$

Since $0 \leq m_i \leq 2a_i - 2$, we get

$$(a) \qquad 1 - a_i \leq h_i a_i + h_{i+1} \leq a_i - 1, \quad i = 1, \ldots, N.$$

If there is some $h_{i+1} = 0$, then the above equation implies $h_i = 0$, and recursively,

$$(h_1, \ldots, h_N) = (0, 0, \ldots, 0).$$

Otherwise, we can assume none of the h_i is zero. There are two situations. Either there is one h_i with $|h_i| = 1$ or all $|h_i| \geq 2$. For the first case, we assume some $h_{i+1} = \pm 1$. Since $h_i \neq 0$ by assumption, the inequality (a) implies $h_i = \mp 1$. We can repeat this process and get the following solution when N is an even number:

$$(h_1, \ldots, h_N) = (\pm 1, \mp 1, \ldots, \pm 1, \mp 1).$$

Finally we prove it is impossible to have all $|h_i| \geq 2$. Equation (a) implies

$$(b) \qquad -1 + \frac{1 - h_{i+1}}{a_i} \leq h_i \leq 1 - \frac{1 + h_{i+1}}{a_i}.$$

If all $|h_{i+1}| \geq 2$, this implies

$$(c) \qquad |h_i| < |h_{i+1}|.$$

In fact, if $h_{i+1} \geq 2$, then the RHS of inequality (b) implies $h_i < 1$. By assumption, we know $h_i \leq -2$. However, since

$$-h_{i+1} < -1 + \frac{1 - h_{i+1}}{a_i},$$

inequality (c) follows from the LHS of (b). A similar argument works for $h_{i+1} \leq -2$. We repeat this process and we find

$$|h_i| = |h_{i+N}| < \cdots < |h_{i+1}| < |h_i|,$$

which is absurd. Thus the only possibilities for the k_i's are

(1) $(k_1, \ldots, k_N) = (1, 1, \ldots, 1)$, and

(2) $(k_1, \ldots, k_N) = (1 \pm 1, 1 \mp 1, \ldots, 1 \pm 1, 1 \mp 1)$, if N is even.
In each case, again we have

$$\deg(x_1^{m_1} \cdots x_N^{m_N}) = \deg(x_1^{r_1} \cdots x_N^{r_N}) + \deg(x_1^{r_1'} \cdots x_N^{r_N'}) = \hat{c}_f.$$

By the same degree reason as in the chain case, we know that $K_f(x_1^{r_1} \cdots x_N^{r_N} d^N x, x_1^{r_1'} \cdots x_N^{r_N'} d^N x)$ lies in $z^N \mathbb{C}$.

4.3. A Landau-Ginzburg Mirror theorem

The standard basis for an invertible polynomial determines a primitive form, hence a Frobenius manifold structure on the germ of its universal unfolding. In particular, we obtain the genus zero potential function $\mathbf{F}_0$ of the associated Landau-Ginzburg B-model. The exact formula of $\mathbf{F}_0$ is not known in general. The difficulty goes back to the lack of knowledge on its primitive form, which is constructed from the good basis via an abstract Riemann-Hilbert-Birkhoff problem [40]. Nevertheless, the perturbative algorithm in Section 3.5 allows us to compute $\mathbf{F}_0$ order by order recursively.

The perturbative theory becomes extremely useful when integrability is taken into account. In fact, WDVV equation reduces the computation of $\mathbf{F}_0$ to only a finite order computation. The most general form of such reconstruction type theorem is established in [23].

Theorem 4.3 ([23]). *Let W be an invertible polynomial with no chain variables of weight $1/2$. Then for both FJRW theory of (W, G_W) and primitive form of W^T with respect to the standard good basis, the genus 0 invariants are completely determined by 2-point, 3-point and 4-point functions accompanied with WDVV equation, String equation, Dimension Axiom and Integer Degree Axiom.*

Here the selection rules of Dimension Axiom and Integer Degree Axiom are natural geometric properties (see [23] for details). There is a minor situation for chain types with weight $1/2$ not covered, i.e., $W = x_1^{a_1} x_2 + x_2^{a_2} x_3 + \cdots + x_{N-1}^{a_{N-1}} x_N + x_N^{a_N}$ with $a_N = 2$. This is a technical difficulty of missing information about certain FJRW 3-point functions due to the non-algebraic nature of FJRW theory. This theorem says that $\mathbf{F}_0$ is completely determined by symmetries and its Taylor series up to order 4. In particular, the enumerative data of genus 0 Landau-Ginzburg models can be fully computed.

A straight-forward application of our reconstruction theorem is to prove a general form of the Landau-Ginzburg mirror symmetry conjecture between invertible polynomials. At genus zero, the Landau-Ginzburg mirror symmetry conjecture says that the generating function of orbifold FJRW theory of the pair (W, G_W) is equivalent to the generating function of the Landau-Ginzburg B-model of W^T associated to a primitive form. This is established for ADE singularities [17] and simple elliptic singularities [26, 37]. The method of perturbative primitive form is first applied to exceptional unimodular singularities [29]. Those are the first examples of Landau-Ginzburg mirror symmetry for singularites of central charge greater than 1. In general, we have the following theorem.

Theorem 4.4 (Landau-Ginzburg Mirror Symmetry Theorem [23]). *Let W be an invertible polynomial with no chain variables of weight $1/2$. Then the FJRW theory of (W, G_W) is equivalent to Saito-Givental theory of W^T at all genera.*

The proof is reduced to a finite check of 4-point functions, thanks to Theorem 4.3. We refer to [23] for details.

References

[1] S. Barannikov, *Extended moduli spaces and mirror symmetry in dimensions $n > 3$*, ProQuest LLC, Ann Arbor, MI, 1999, Thesis (Ph.D.)–University of California, Berkeley.

[2] ______, *Non-commutative periods and mirror symmetry in higher dimensions*, Comm. Math. Phys. **228** (2002), no. 2, 281–325.

[3] ______, *Quantum periods. I. Semi-infinite variations of Hodge structures*, Internat. Math. Res. Notices, **23** (2001), 1243–1264.

[4] ______, *Semi-infinite Hodge structures and mirror symmetry for projective spaces*, preprint at arXiv:0010157 [math.AG].

[5] S. Barannikov and M. Kontsevich, *Frobenius manifolds and formality of Lie algebras of polyvector fields*, Internat. Math. Res. Notices, **4** (1998), 201–215.

[6] P. Berglund and T. Hübsch, *A Generalized Construction of Mirror Manifolds*, Nucl. Phys. B 393 (1993) 377–391.

[7] M. Bershadsky, S. Cecotti, H. Ooguri and C. Vafa, *Kodaira-Spencer theory of gravity and exact results for quantum string amplitudes*, Comm. Math. Phys. **165** (1994), no. 2, 311–427. MR1301851 (95f:32029).

[8] H.-L. Chang, J. Li and W. Li, *Witten's top Chern class via cosection localization*, to appear in Invent. Math., preprint at arXiv:1303.7126 [math.AG].

[9] A. Chiodo and Y. Ruan, *A global mirror symmetry framework for the Landau-Ginzburg/Calabi-Yau correspondence*, Ann. Inst. Fourier (Grenoble) vol. 61, no. 7 (2011), 2803–2864.

[10] T. Coates and A. B. Givental, *Quantum Riemann-Roch, Lefschetz and Serre*, Ann. of Math. (2) **165** (2007), no. 1, 15–53.

[11] T. Coates and H. Iritani, *A Fock sheaf for Givental quantization*, preprint at arXiv:1411.7039 [math.AG].

[12] K. Costello, *Renormalization and effective field theory*, Mathematical Surveys and Monographs, vol. 170, American Mathematical Society, Providence, RI, 2011. MR2778558.

[13] K. Costello and S. Li, *Quantum BCOV theory on Calabi-Yau manifolds and the higher genus B-model*, preprint at arXiv:1201.4501 [math.QA].

[14] A. Douai and C. Sabbah, *Gauss-Manin systems, Brieskorn lattices and Frobenius structures. I*, Proceedings of the International Conference in Honor of Frédéric Pham (Nice, 2002), 2003, pp. 1055–1116.

[15] ———, *Gauss-Manin systems, Brieskorn lattices and Frobenius structures. II*, Frobenius manifolds, Aspects Math., E36, Vieweg, Wiesbaden, 2004, pp. 1–18.

[16] H. Fan, *Schrödinger equations, deformation theory and tt^*-geometry*, preprint at arXiv:1107.1290 [math-ph].

[17] H. Fan, T. Jarvis and Y. Ruan, *The Witten equation, mirror symmetry, and quantum singularity theory*, Ann. of Math. (2) **178** (2013), no. 1, 1–106.

[18] A. Givental, *A tutorial on quantum cohomology*, Symplectic geometry and topology (Park City, UT, 1997), IAS/Park City Math. Ser., vol. 7, Amer. Math. Soc., Providence, RI, 1999, pp. 231–264.

[19] A. B. Givental, *A mirror theorem for toric complete intersections*, Topological field theory, primitive forms and related topics (Kyoto, 1996), Progr. Math., vol. 160, Birkhäuser Boston, Boston, MA, 1998, pp. 141–175.

[20] ———, *Gromov-Witten invariants and quantization of quadratic Hamiltonians*, Mosc. Math. J. **1** (2001), no. 4, 551–568, 645. Dedicated to the memory of I. G. Petrovskii on the occasion of his 100th anniversary.

[21] ———, *Symplectic geometry of Frobenius structures*, Frobenius manifolds, Aspects Math., E36, Vieweg, Wiesbaden, 2004, pp. 91–112.

[22] M. Kaneko and D. Zagier, *A generalized Jacobi theta function and quasimodular forms*, The moduli space of curves (Texel Island, 1994), Progr. Math., vol. 129, Birkhäuser Boston, Boston, MA, 1995, pp. 165–172.

[23] W. He, S. Li, R. Webb and Y. Shen, *Landau-Ginzburg mirror symmetry conjecture*, preprint at arXiv:1503.01757 [math.AG].

[24] S. Klimek and A. Lesniewski, *Local Rings of Singularities and $N = 2$ Supersymmetric Quantum mechanics*, Comm. Math. Phys, 136, 327–344 (1991).

[25] M. Krawitz, *FJRW rings and Landau-Ginzburg Mirror Symmetry*, Ph.D. thesis, University of Michigan, 2010.

[26] M. Krawitz and Y. Shen, *Landau-Ginzburg/Calabi-Yau Correspondence of all Genera for Elliptic Orbifold $\mathbb{P}^1$*, preprint at arXiv:1106.6270 [math.AG].

[27] M. Kreuzer and H. Skarke, *On the classification of quasihomogeneous functions*, Comm. Math. Phys. **150** (1992), no. 1, 137–147.

[28] C. Li, S. Li and K. Saito, *Primitive forms via polyvector fields*, arXiv:1311.1659 [math.AG].

[29] C. Li, S. Li, K. Saito and Y. Shen, *Mirror symmetry for exceptional unimodular singularities*, To appear in J. Eur. Math. Soc. Preprint at arXiv:1405.4530 [math.AG].

[30] S. Li, *Feynman graph integrals and almost modular forms*, Comm. in Number Theory and Physics **6** (2012).

[31] ______, *BCOV theory on the elliptic curve and higher genus mirror symmetry*, arXiv:1112.4063 [math.QA].

[32] ______, *Calabi-Yau geometry and higher genus mirror symmetry* (2011). Thesis, Harvard.

[33] ______, *Variation of Hodge structures, Frobenius manifolds and Gauge theory*, arXiv:1303.2782 [math.QA].

[34] B. H. Lian, K. F. Liu and S. T. Yau, *Mirror principle. I*, Asian J. Math. **1** (1997), no. 4, 729–763.

[35] A. Losev, *Hodge strings and elements of K. Saito's theory of primitive form*, Topological field theory, primitive forms and related topics (Kyoto, 1996), Progr. Math., vol. 160, Birkhäuser Boston, Boston, MA, 1998, pp. 305–335.

[36] A. Losev, S. Shadrin and I. Shneiberg, *Tautological relations in Hodge field theory*, Nuclear Phys. B **786** (2007), no. 3, 267–296.

[37] T. Milanov, *Analyticity of the total ancestor potential in singularity theory*, Advances in Math. **255** (2014), 217–241.

[38] A. Okounkov and R. Pandharipande, *Virasoro constraints for target curves*, Invent. Math. **163** (2006), no. 1, 47–108.

[39] A. Polishchuk and A. Vaintrob, *Matrix factorizations and Cohomological Field Theories*, to appear in Journal für die reine und angewandte Mathematik, preprint at arXiv:1105.2903 [math.AG].

[40] K. Saito, *Period mapping associated to a primitive form*, Publ. Res. Inst. Math. Sci. **19** (1983), no. 3, 1231–1264.

[41] ______, *The higher residue pairings $K_F^{(k)}$ for a family of hypersurface singular points*, Singularities, Part 2 (Arcata, Calif., 1981), 1983, pp. 441–463.

[42] K. Saito, *From primitive form to mirror symmetry*, arXiv:1408.4208 [math.AG].

[43] M. Saito, *On the structure of Brieskorn lattice*, Ann. Inst. Fourier (Grenoble) **39** (1989), no. 1, 27–72.

[44] ______, *On the structure of Brieskorn lattices, II*, arXiv:math.AG/1312.6629.

[45] A. Takahashi, *Primitive forms, Topological LG models coupled to Gravity and Mirror Symmetry*, arXiv:math.AG/9802059.

Yau Mathematical Sciences Center, Tsinghua University,
Beijing, China
E-mail address: sili@mail.tsinghua.edu.cn

Advanced Studies in Pure Mathematics 83, 2019
Primitive Forms and Related Subjects — Kavli IPMU 2014
pp. 269–293

TQFT, homological algebra and elements of K. Saito's theory of Primitive form: an attempt of mathematical text written by mathematical physicist

Andrey Losev

Abstract.

The text is devoted to explanation of the concept of Topological Quantum Field Theory (TQFT), its application to homological algebra and to the relation with the theory of good section from K. Saito's theory of Primitive forms. TQFT is explained in Dirac-Segal framework, 1 dimensional examples are explained in detail. As a first application we show how it can be used in explicit construction of reduction of ∞-structure after contraction of a subcomplex. Then we explain Associativity and Commutativity equations using this language. We use these results to construct solutions to Commutativity equations and find a new proof of for the fact that tree level BCOV theory solved Oriented Associativity equations.

§1. Preface for mathematicians: please do not be afraid to try to read this text

Nowadays it is clear that what people are calling modern mathematical physics (some people are calling it mathematical quantum field theory) is a draft of a future chapter of pure mathematics. It is a very preliminary heuristic draft full of heuristic ideas and unproven conjectures. Moreover, it is a bad habit in mathematical physics texts to omit definitions and not to distinguish conjectures and theorems. Therefore, such texts are unreadable by normal mathematicians. These texts are mostly considered as a source of heuristic ideas that should be properly formulated, studied and turned into mathematics. This work of translating heuristics into mathematics is an extra skill that mathematicians

Received May 9, 2016.
Revised January 17, 2018.
2010 *Mathematics Subject Classification.* 18G99 and 81T45.
Key words and phrases. Theory of Primitive form, homological algebra, topological quantum field theory (TQFT).

may not have. Moreover, such translation can be done only if mathematician already spent a lot of time reading this heuristics and thinking about it. Therefore, even mathematicians that have this skill prefer to save their time and wait until someone translates the mathematical physics text for them.

It is quite logical to expect that mathematical physicists would sometime write texts that are readable by mathematicians. Due to the current reputation of mathematical physicists the potential readability of the texts should be mentioned in the title. Here I present an attempt to write such a text.

At the first glance the text contains many "mathematical physics wording", like quantum mechanics, quantum field theory, amplitudes and so on. However, I will try to define everything that I will use keeping for these objects their standard names.

This may be compared with text on homological algebra that is using names like differential, cycles and boundaries. In order to understand such text the reader does not need to know differential geometry. All object are defined in terms of linear algebra, they are just keeping names from the differential geometric realization.

The text is devoted to explanation of the concept of Topological Quantum Field Theory (TQFT), its application to homological algebra and to the relation with the theory of good section from K. Saito's theory of Primitive forms. TQFT is explained in Dirac-Segal framework, 1 dimensional examples are explained in detail. As a first application we show how it can be used in explicit construction of reduction of ∞-structure after contraction of a subcomplex. Then we explain Associativity and Commutativity equations using this language. We use these results to construct solutions to Commutativity equations and find a new proof of for the fact that tree level BCOV theory solved Oriented Associativity equations.

§2. TQFT for mathematicians

2.1. Dirac-Segal Quantum Field Theory

Definition. A D-dimensional quantum field theory according to Segal [1] may be considered as monoidal functor I between monoidal category of decorated smooth oriented D-dimensional cobordisms and monoidal category of vector spaces. By decorated cobordism we mean that the oriented D-dimensional smooth manifold is equipped with some kind of geometrical data, such as metric, complex or almost complex structure etc. When manifold is cut along the $D-1$ dimensional submanifold the geometrical data is induced on the result of the cutting.

Explanation of the definition. The functor was motivated by functional integral that is why it is called I.

The objects are $D-1$ oriented dimensional manifolds Γ and $I(\Gamma)$ are vector spaces, such that if we change orientation of Γ to Γ^{opp} the image is the dual space:

$$I(\Gamma^{opp}) = I(\Gamma)^* \tag{1}$$

Let us call the D dimensional manifolds by Σ, and geometrical data on it by g. The functor sends the pair (Σ, g) to $I(\Sigma, g)$ that for each fixed g belongs to $I(\partial\Sigma)$, so the functor I may be considered as an element of the tensor product of $I(\partial\Sigma)$ and the space of functions on the space $Geom_\Sigma$ of possible geometric data on Σ:

$$I \in I(\partial\Sigma) \otimes Fun(Geom_\Sigma) \tag{2}$$

The functoriality condition means the following. Suppose that Σ is cut by not self-intersecting and not crossing the boundaries Γ into Σ^Γ. The geometrical data on a cut manifold is induced from the original manifold, so we have a map:

$$Cut : Geom(\Sigma) \to Geom(\Sigma^\Gamma) \tag{3}$$

Its inverse image Cut^* is a linear map

$$Cut^* \colon Fun(Geom(\Sigma^\Gamma) \to Fun(Geom(\Sigma)) \tag{4}$$

The boundary of Σ^Γ is a disjoint union of the boundary of Σ and two copies of Γ with opposite orientation, therefore

$$I(\partial\Sigma^\Gamma) = I(\partial\Sigma \cup \Gamma \cup \Gamma^{opp}) = I(\partial\Sigma) \otimes I(\Gamma) \otimes I(\Gamma)^* \tag{5}$$

where we use monoidal nature of the functor that takes disjoint union into a tensor product.

Note, that there is a natural map $Glue_\Gamma$

$$Glue_\Gamma : I(\partial\Sigma^\Gamma) \to I(\partial\Sigma) \tag{6}$$

induced by a canonical contraction between $I(\Gamma)$ and $I(\Gamma)^*$.

The functoriality means that

$$Glue_\Gamma \otimes Cut^*(I(\Sigma^\Gamma)) = I(\Sigma) \tag{7}$$

One dimensional example – Quantum Mechanics.

If this still looks to abstract I will give an example for $D = 1$. Let Σ be an interval, and let geometric data be a metric. Then the space of geometric data is the space of positive numbers $\mathbb{R}_+$. The boundary of an interval is a set of two points. If we denote by $V = I(point)$ for positive orientation of the point, then

$$I(\partial\Sigma) = I(point \cup point^{opp}) = V \otimes V^* \tag{8}$$

and

$$I(\Sigma) \in V \otimes V^* \otimes Fun(\mathbb{R}_+) = Maps(\mathbb{R}_+, End(V)) \tag{9}$$

Then, the equation (7) means that this map is a representation of the semigroup $\mathbb{R}_+$ in $End(V)$.

To see this, let us denote by I_t the evaluation of I at t:

$$I_t = ev_t I(\Sigma)$$

If we cut an interval of length t by a point P into two intervals with the lengths t_1 and t_2, then, due to monoidal functoriality

$$I_t(\Sigma^P) = I_{t_1} \otimes I_{t_2} \tag{10}$$

and the main equation (7) reads

$$I_{t_1+t_2} = I_{t_1} \circ I_{t_2} \tag{11}$$

where $\circ$ stands for composition in $End(V)$. So, it is the semigroup representation condition.

The universal solution to this condition is

$$I_t = \exp(-tH) \tag{12}$$

where $H \in End(V)$ is called Hamiltonian operator in physics.

This formulation of quantum mechanics (1-dimensional QFT) was discovered by Dirac around 1930. That is why I often call Segal formulation of QFT the Dirac-Segal formulation.

2.2. Quantum mechanics on oriented graphs

It is quite interesting that although Dirac-Segal formulation was originally about manifolds, the one-dimensional quantum field theory may be generalized to include graphs. This generalization can be made by applying the cutting axiom (7) to oriented graphs equipped with lenghts of their edges.

Motivation for the definition. Vertices of the graph may be decomposed into the 1-valent that we will call external and the rest that we will call internal. Cutting out neighborhoods of internal vertices we are cutting the graph into intervals and "p-corollas" (by corolla we mean a connected tree with p 1-valent vertices and p edges that link these vertcies to a single p-valent vertices). Depending on the orientation of the graph p-corollas should be represented as

$$Cor_p \in V^{*\otimes(p-q)} \otimes V^{\otimes q} \otimes Fun(\mathbb{R}^p_+) \tag{13}$$

Taking the limit when lengths of all edges of the corolla are tending to zero we get PROR - multi linear operation with $p-q$ inputs and q outputs:

$$m^q_{p-q} = \lim_{t_1,\dots,t_p \to 0} Cor_p \in V^{*\otimes(p-q)} \otimes V^{\otimes q} \tag{14}$$

Definition. The generalization of monoidal functor associates to the oriented graph γ a tensor-valued function on the space of lengths of edges that we call I_γ:

$$I_\gamma = < \otimes^{n_e}_{i=1} I(t_i), \otimes^{n_c}_{\alpha=1} m_\alpha >_\gamma \in Fun(\mathbb{R}^{n_e}_+), \tag{15}$$

where n_e is the number of edges of the graph, n_c - the number of internal vertices(corollas), $<,>_\gamma$ is the pairing determined by the incident relations of the graph.

Example 2.2.1.

For the Y-shaped graph with two inputs and one output the $I_Y \in V^{*\otimes 2} \otimes V$, and its value on $v_1 \otimes v_2$ is equal to:

$$I_Y(v_1, v_2) = \exp(-t_3 H)\left(m^1_2(\exp(-t_1 H)v_1, \exp(-t_2 H)v_2)\right) \tag{16}$$

Example 2.2.2. For a graph that is an interval consisting of 3 intervals of lengths t_1, t_2, t_3 the corollas are linear operators ϕ and operator corresponding to such graph is

$$I_{3int} = \exp(-t_3 H)\phi \exp(-t_2 H)\phi \exp(-t_1 H) \tag{17}$$

Definition of preamplitude. Suppose that H is semi positive defined. In this case it is possible to consider so-called preamplitudes PA that are limits of I_γ when lengths of all external edges (edges that connect external vertices to the rest of the graph) tend to $+\infty$. In this case operators corresponding to these edges turn into projectors on the kernel of H that we will denote by Π. Thus, preamplitude corresponding to the Y graph equals

$$PA_Y = \Pi\left(m^1_2(\Pi, \Pi)\right) \in KerH \otimes KerH \to KerH \tag{18}$$

and preamplitude corresponding to the 3-interval graph is a linear operator on $KerH$ that equals to

$$PA_{3int} = \Pi\phi\exp(-t_2 H)\phi\Pi \tag{19}$$

Comment. Physicists are studying amplitudes. I think it is more appropriate to introduce preamplitudes such that amplitudes are obtained from them by integration over the space of geometrical data, see below. In particular, in string theory the measures on the moduli space are the most known examples of preamplitudes.

2.3. Topological quantum theory in Segal formulation

Definition of TQFT. The notion of topological quantum field theory (TQFT) can be obtained by the following "supersymmetrization" of Dirac-Segal formulation of quantum field theory. The first "supersymmetrization" is the replacement the vector spaces by complexes. We will denote differentials in these complexes by letter Q that is traditional in mathematical physics literature. The second "supersymmetrization" is the replacement of the space $Geom$ of geometrical data by a superspace $\Pi(T)Geom$, that stands for a tangent bundle to the space $Geom$ with inversed parity of the fibers. The space of functions on $\Pi(T)Geom$ is just the supercommutative DeRham superalgebra $\Omega^*(Geom)$. Thus, the functor of TQFT belongs to differential form valued element of the tensor algebra of complexes:

$$I(\Sigma) \in V(\partial\Sigma) \otimes \Omega^*(Geom(\Sigma)) \tag{20}$$

Note, that supermanifold is not a set of points, so the above definition cannot be stricly speaking considered as a specialization of the definition of QFT but rather its generalization. The functoriality axiom (7) has exactly the same form.

Closeness Axom of TQFT. Topological QFT have an additional axiom, the axiom of total closeness of the functor. Namely,

$$(d + Q)I(\Sigma) = 0 \tag{21}$$

where d is a de Rham differential acting on $\Omega^*(Geom)$.

Comment. The closeness axiom means that the differential form $I(\Sigma)$ becomes closed after restriction to the closed subspace in $V(\partial)$. If we further specialize to degree zero forms in $\Omega^*(Geom)$ it would mean that such functions are constant on connected components, i.e. they do not depend on geometrical data, thus, they are topological invariants. This property gave the name "topological" to such QFT, and replacement functions by differential forms and condition of being locally constant by condition of being just closed is standard in derived mathematics.

2.4. Universal TQFT in dimension 1

Similarly to QFT is dimension 1, it is easy to find the universal TQFT in dimension 1. Namely, this TQFT is governed by a complex V with differential Q and additional differential G, and

$$\begin{aligned} I(t,dt) &= \exp(\{d+Q,-tG\}) \\ &= \exp(-t\{Q,G\}-dtG) = \exp(-t\{Q,G\})(1-dtG) \end{aligned} \tag{22}$$

From the first equality above it is clear that such I satisfies the closeness axiom.

It follows from the (22) that after restriction to function we have a 1-dimensional quantum field theory with a very special Hamiltonian

$$H = \{Q,G\} \tag{23}$$

Such Hamiltonians are known as supersymmetric Hamiltonians. Such Hamiltonians were known in differential geometry.

2.5. Examples of supersymmetric Hamiltonians

Example 2.5.1. The first example is Laplacian Δ acting on the complex $\Omega^*(X)$ of differential forms on Riemannian manifold X with differential Q being the de Rham differential on X:

$$\Delta = \{d,d^*\} \tag{24}$$

where

$$G = d^* = - * d* \tag{25}$$

is the Hodge conjugated operator.

Example 2.5.2. The second example of such Hamiltonian acting on the same complex $\Omega^*(X)$ is the Lie derivative Lie_v along the vector field v on X:

$$Lie_v = \{d,\iota_v\} \tag{26}$$

where

$$G = \iota_v \tag{27}$$

is the operator of contraction of differential form with the vector field v.

Example 2.5.3. The third example comes from the homological algebra. Consider a complex V together with its decomposition into a direct sum of contractable subcomplex V_c and residual subcomplex V_r:

$$V = V_r \oplus V_c \tag{28}$$

so that we have inclusion and projection operators

$$i : V_r \to V, \quad \pi : V \to V_r \tag{29}$$

and also an odd homotopy operator (that inverts differential on contractable subcomplex V_c)

$$h : V \to V, \ \ h \circ h = h \circ i = \pi \circ h = 0, \ \ \{Q, h\} = Proj_{V_c} = id - i \circ \pi, \tag{30}$$

here $Proj_{V_c}$ stands for the projection operator on a contractable subcomplex.

This data means that we have a topological quantum mechanics with

$$G = h, \quad H = Proj_{V_c} \tag{31}$$

The I of this example has another important property that we will use below, namely, it has a limit when t turns to $+\infty$:

$$\lim_{t \to +\infty} \exp(-tH) = \pi \circ i = Proj_{V_r}, \tag{32}$$

where $Proj_{V_r}$ is the projector on the space V_r.

2.6. Local observables in QFT

By Segal's definition the main object I in QFT corresponds to manifolds with boundaries. Now we define the notion of observables in QFT. Consider the following construction: take a submanifold C in Σ and consider its small tubular neighborhood $Tub\ C_\epsilon$, i.e. the space of points in Σ such that their distance to C is not greater than ϵ. Let $\Gamma_{C,\epsilon}$ be the boundary of $Tub\ C_\epsilon$. Consider the space $\Sigma_{Tub\ C_\epsilon}$ obtained by cutting the tubular neighborhood $Tub\ C_\epsilon$ out of Σ:

$$\Sigma_{Tub\ C_\epsilon} = \Sigma \backslash Tub\ C_\epsilon \tag{33}$$

The boundary of $\Sigma_{Tub\ C_\epsilon}$ is a disjoint union of the boundary of Σ and $\Gamma_{C,\epsilon}$. Therefore

$$I_{\Sigma_{Tub\ C_\epsilon}} \in V_{\partial\Sigma} \otimes V_{\Gamma_{C,\epsilon}} \tag{34}$$

Definition of observables. We will call the family of vectors $v_\epsilon \in V^*_{\Gamma_{C,\epsilon}}$ good if there is a limit

$$\lim_{\epsilon\to 0} < v_\epsilon, I_{\Sigma_{Tub\ C_\epsilon}} >$$

The good family of vectors is called null if this limit equals to zero. We define the space of observables Obs_C associated to C as the coset of good families over the space of null families. The limit mentioned above is defined on the space of observables, and we will call the value of this limit the correlator of an observable associated with the submanifold C:

$$I(O_C) = \lim_{\epsilon\to 0} < v_\epsilon, I_{\Sigma_{Tub\ C_\epsilon}} > \tag{35}$$

If C is point we will call such observables local observables.

Example. Consider the one-dimensional example. The only observables are local observables. The tubular neighborhood of a point on an interval is also an interval, so local observables are just linear operators:

$$Obs_P = End(V)$$

and correlator of an observable in topological quantum mechanics equals to

$$I(O_P)(t_1, t_2) = \exp(-t_1 H - dt_1 G) O \exp(-t_2 H - dt_2 G) \tag{36}$$

where t_1 and t_2 are the distances between the point P and the left and right ends of the interval respectively. We see that in 1-dimensional theory the observable corresponds to the 2-valent internal vertex.

Remark. Local observables as tangent vectors to space of theories. It is important to note that local observables correspond to tangent space to the space deformations of QFT that do not change the vector spaces associated to boundaries. It is a general statement, but we may illustrate it in topological quantum mechanics. Actually, if we change Q by δQ, the I corresponding to the interval of length t is changing according to:

$$\delta I = \int_{t_1+t_2=t, t_i\geq 0} \exp(-t_1 H - dt_1 G)\delta Q \exp(-t_2 H - dt_2 G) \tag{37}$$

§3. Topological quantum mechanics on trees and constructive induction of operations after contraction of the subcomplex

Here I will explain how topological quantum mechanics (another name of 1-dimensional TQFT) of Example 2.5.3 on a tree may be used to give a constructive proof of the Kadeishvili-type theorems [2] about induction of ∞-structures on residual subcomplexes (like subcomplex V_r in (28)).

3.1. Amplitudes and Kadeishvili theorem.

Consider preamplitudes in topological quantum mechanics on rooted oriented trees. To define such theory we need a triple consisting of Q, G, and a set of operations $m_1, \ldots, m_k$, where Q and G will be as in Example 2.5.3 and $m_n \in V^{*\otimes n} \otimes V$.

In the case from example 2.5.3 the projector Π associated to leaves may be considered as operator i of inclusion of V_r into V, and projector Π associated to the root may be considered as a projection π from V to V_r. Let n_l be the number of leaves of the rooted tree and n_c the number of internal vertexes (corollas). The preamplitude associated to tree γ is a n_l-operation on V_r taking value in differential forms on the lengths of edges. It has the following form:

$$PA_\gamma = < \otimes_{i=1}^{n_e} I(t_i), \otimes_{\alpha=1}^{n_c} m_\alpha \otimes i^{\otimes n_l} \otimes \pi >_\gamma \in \Omega^*(\mathbb{R}_+^{n_e}) \otimes V_r^{*\otimes n_l} \otimes V_r, \tag{38}$$

where contraction of tensors are made according to the tree.

For example, for the Y tree the preamplitude is just the restriction of the binary operation on V_r, while for the 3-interval tree (that tree has only m_1 operation ϕ), the preamplitude equals

$$PA_{3int} = \pi\phi \exp(-tProj_{V_c} - dth)\phi i \tag{39}$$

Definition of a tree amplitude. Let as define a tree amplitude $A_{n_l} \in V_r^{*\otimes n_l} \otimes V_r$ as a sum over all rooted trees with n_l leaves of the integrals of preamplitudes over spaces of lengths of edges (and we will divide each integral over the number n_γ of elements of the symmetry group of the tree):

$$A_{n_l} = \sum_\gamma 1/n_\gamma \int_{\mathbb{R}_+^{n_e}} PA \tag{40}$$

Such integral can be easily calculated, since

$$\int_{\mathbb{R}_+} \exp(-tProj_{V_c} - dth) = -h, \tag{41}$$

and we are coming to the following formula for the amplitude:

$$A_{n_l} = \sum_\gamma (-1)^{n_e}/n_\gamma < h^{\otimes n_e}, \otimes_{\alpha=1}^{n_c} m_\alpha \otimes i^{\otimes n_l} \otimes \pi >_\gamma, \tag{42}$$

One leaf example. In particular, if the only operation is m_1 (that we still denote by π) the amplitude for one leaf equals to

$$A_1 = \pi\phi i - \pi\phi h\phi i + \pi\phi h\phi h\phi i - \dots \tag{43}$$

Kadeishvili Theorem.

Suppose that operations $Q + m_1, m_2, m_3 \dots$ form an L_∞ algebra, then operations constructed from Q and amplitudes $Q + A_1, A_2, A_3 \dots$ also form an L_∞ algebra. This algebra is called the algebra obtained by the contraction of the subcomplex.

3.2. Kadeishvili theorem in one leaf case.

Before we proceed to discussion of the proof it is instructive to look at the particular case, namely, the case where the only nonzero operation is $m_1 = \phi$. Then the theorem actually states that if ϕ is a solution to Maurer-Cartan equation

$$\{Q, \phi\} + \phi^2 = 0 \tag{44}$$

then

$$\{Q, A_1\} + A_1^2 = 0 \tag{45}$$

Remark. We may specialize even further, and consider Q-closed ϕ and take V_r to be cohomology of Q, so the input of the construction is a bicomplex. Terms in (43) look like differentials in spectral sequence construction. However, there is an important difference. In spectral sequence construction differentials act on different spaces, each of these spaces are cohomology of the previous differential, while all summands of (43) act on the same space, namely, cohomology of Q (the first differential of the spectral sequence). Moreover, differentials in spectral sequence construction are defined canonically while operators in (43) depend on the choice of the direct decomposition of the space V into cohomology and contractable complex, and also on the choice of homotopy. So, differential A_1 seem to contain more information than differentials of spectral sequence, however, one can show that all information that is invariant under the choice of decomposition and homotopy is contained in spectral sequence differentials. On the other hand, construction (43) provides an explicit answer to the question: how to construct a differential on the cohomology of Q, such that cohomology of the constructed differential would be equal to cohomology of the total differential (if the sum in (43) contains finite number of terms). We will later see the similar phenomena in discussion of Massey operations.

Now let us discuss how to prove a theorem.

3.3. Outline of the direct proof of Kadeishvili theorem

The subtle issue in the one leaf case is that ϕ has grading zero, so the sum (43) may diverge. Thus, we have to consider it as a formal one, namely, we take

$$\phi = \sum_{k=1}^{\infty} \phi_k \epsilon^k \tag{46}$$

and consider ϵ as formal parameter. From Maurer-Cartan equation we get

$$\{Q, \phi_1\} = 0, \quad \{Q, \phi_2\} = -\phi_1^2, \quad \ldots \tag{47}$$

In the first order in ϵ statement of the theorem is Q-closeness of $\pi\phi_1 i$ that follows from Q-closeness of ϕ_1. In the second order the substitution of expression (43) into (45) gives

$$\begin{aligned} &\{Q, \pi\phi_2 i - \pi\phi_1 h\phi_1 i\} + \pi\phi_1 i\pi\phi_1 i = \\ &\pi\{Q, \phi_2\}i + \pi\phi_1\{Q, h\}\phi_1 i + \pi\phi_1(1 - \{Q, h\})\phi_1 i = 0 \end{aligned} \tag{48}$$

where we used (47)

One may show that similar cancellations happen at higher orders.

The multileaf case can be treated similarly, by direct computation and combinatorics of trees. The composition of amplitudes would produce $i\pi$ term that can be replaced by $(1 - \{Q, h\})$. The contribution coming from 1 being combined with the action of Q on the operation m_n would lead to the Maurer-Cartan expressions, while $\{Q, h\}$ terms would be cancelled by the action of Q on homotopy. This may be considered as a combinatorics trick. It is reasonable to look for more conceptual explanation, that we will give in the next subsection. It is remarkable that the more conceptual explanation has generalizations in dimensions higher than 1.

3.4. Proof of the induction theorem based on the properties of the preamplitude

In order to explain main ideas we will first give the proof in the simplest case when V_r is the space of cohomology of Q and when both terms in Maurer-Carttan equation are equal to zero separately, i.e. when all operations m_k are Q-closed and when these operation form the infinity algebra. Later we will comment on how this proof may be modified to include the general case.

The proof that we are going to present here is based on two properties: closeness and factorization. Closeness means that each preamplitude is closed as a form on the space of lengths of edges. It follows from

$d + Q$ closeness of the evolution operator $I(t, dt)$ associated to intervals and from Q-closeness of operations m_i, projection π and inclusion i.

Factorization means that when length of one of the edges goes to infinity, the preamplitude factorizes into composition of preamplitudes corresponding to trees obtained by cutting the original tree along such an edge. This follows from

$$\lim_{t\to+\infty} \exp(-tProjV_c)(1 - dth) = ProjV_r = i \circ \pi \tag{49}$$

The idea of the proof is to compactify the space of lengths of the tree to a n_e hypercube, to take a $n_e - 1$-form component of the preamplitude and to integrate it along the boundary of the hypercube. Since the preamplitude is closed its $n_e - 1$-form component is also closed. Thus, the integral will be equal to zero. Contributions from the faces corresponding to infinite lengths trees would form compositions of amplitudes. Contributions from zero length faces would form an amplitude corresponding to the tree that has a composition of operations in one of its internal vertices. When we sum over all trees contribution of such vertices would cancel due to infinity algebra conditions on the operations. For the one leaf trees this derivation was obtained by Lysov [3].

It is important to mention that the space of lengths for the one leaf case may be considered as a moduli space RM_n: for points on a real line moduli common shifts:

$$RM_n = \mathbb{R}^n/\mathbb{R} \tag{50}$$

In the next section we will study the complex version of such moduli space.

Remark 1. This idea was used by Kontsevitch [4] in his proof of ∞-morphism theorem in the work on deformation quantization. He used the moduli space of complex structures on a disk with marked points in the bulk of the disk and on the boundary of the disk. The two-dimensional topological theory he constructed produced closed differential form on the moduli space. The boundary contribution turn out to be exactly the ∞-morphism statement.

Remark 2. If V_r equals to cohomology of Q then the amplitudes for the multileaf case may be considered as generalizations of Massey operations in the same way in which A_1 is the generalization of the spectral sequence differentials. While Massey operations are conventionally understood as a partly defined canonical operations (when some products in cohomology are zero) the amplitudes are always defined but depend on the choice of decomposition of the complex and the choice of homotopy. The amplitudes form an L_∞ structure, and changes of auxillary data just change this structure into an equivalent one.

§4. Configuration spaces in two dimensional theories

In the previous section we observed that it is useful to consider preamplitudes in one dimensional quantum field theories. These preamplitudes are closed forms on moduli spaces (that sometimes are configuration spaces). Integrals of preamplitudes over moduli spaces turn out to satisfy interesting quadratic equations. In one-dimensional case amplitudes turn out to be induced operations on cohomology, while quadratic equations become infinity structure equations.

Now we would like to play similar game for 2-dimensional conformal field theories.

4.1. WDVV equations

Consider Deligne-Mumford compactification $\bar{M}_{0,n+1}$ of the moduli space of $\mathbb{CP}^1$ with $n+1$ marked points. Let us take n points with one orientation (we will call these points input points) and one point with the opposite orientation (we will call this point an output point). Due to general properties of TQFT explained in the previous sections, the preamplitude in topological theory associates vector spaces W to input points, vector space W^* to an output point (where W is the space of cohomology of Q in the space of local observables). QFT associates to $\mathbb{CP}^1$ with n marked points a preamplitude

$$PA_n \in \Omega^*(\bar{M}_{0,n+1}) \otimes Hom(W^{\otimes n}, W) \tag{51}$$

This preamplitude satisfies two main properties:
1. Symmetry under permutations of inputs
2. Closeness, i.e.

$$dPA_n = 0 \tag{52}$$

where d is De Rham differential on $\bar{M}_{0,n+1}$.
3. Factorization. Consider the compactification divisor D_{n_1,n_2} of the moduli space $\bar{M}_{0,n+1}$ where projective space degenerates into two projective spaces with n_1+1 on the component that does not contain an output point and with n_2+1 points on a component that has an output point. Being restricted to such a divisor the preamplitude factorizes:

$$PA_n|_{D_{n_1,n_2}} = PA_{n_2} \circ PA_{n_1} \tag{53}$$

Therefore we can get quadratic relations on amplitudes by integrating preamplitude against boundary divisors that equal to zero in homology of moduli space.

The simplest case is provided by $\bar{M}_{0,4}$ where we will label input points as 1, 2 and 3. This moduli space is just CP^1. Compactification divisor represents surfaces that are two projective spaces intersecting by a point with two input points belonging to first $\mathbb{CP}^1$ and the third input point and the output point belonging to the second $\mathbb{CP}^1$. Since the moduli space is connected the difference of divisors (that we will call D_a) equals to zero in homology of moduli space. We call it D_a because evaluation of amplitudes on this divisor leads to associativity of two to one amplitudes.

This relation leads to relations in other moduli spaces (known as Keel's relations [5]). Consider a forgetful map

$$f : \bar{M}_{0,4+k} \to \bar{M}_{0,4} \tag{54}$$

The divisors $f^*(D_a)$ are compactification divisors. They correspond to surfaces that are two copies of $\mathbb{CP}^1$ intersecting at a point with input points spread among components (such that there are at least two input points on a component that does not contain an output point, and at least one input point on the component that has an output point).

Consider the total amplitude TA that is a sum over all amplitudes and, therefore, the map

$$TA \in \oplus_{n=2}^{+\infty} S^n W \to W, \tag{55}$$

as a formal vector field v on W.

In terms of the vector field v the Keel's relations take the following form. Let T^a, $a = 1, \dots, \mu$ be linear coordinates on W. Consider second partial derivatives f^a_{bc} :

$$f^a_{bc} = \frac{\partial^2 v^a}{\partial T^b \partial T^c} \tag{56}$$

The Keel's relation state that f^a_{bc} form structure constants of associative commutative algebra - the Oriented Associativity equation:

$$\sum_{a=1}^{\mu} f^a_{bc}\, f^e_{ad} = \sum_{a=1}^{\mu} f^a_{cd}\, f^e_{ba} \tag{57}$$

Suppose, that the space W has a non degenerate scalar product η, such that after identification of W and W^* the n to one amplitudes become symmetric in all $n+1$ arguments. In this case the vector field has a potential form, namely

$$v^a = \sum_{b=1}^{\mu} \eta^{ab} \frac{\partial F}{\partial T^b} \tag{58}$$

The Oriented Associativity equations for potential vector field are known as WDVV or simply Associativity equations.

4.2. Losev-Manin moduli spaces, simplest space and Commutativity equations

In [6] Losev and Manin introduced a new moduli space for Riemann surfaces with marked points.

Definition of Losev-Manin moduli spaces. The marked points are colored into to colors - black and white. When a handle degenerates or a surface is decoupled into two surfaces glued by a point the degeneration is described by white points (one input and one output). White points behave exactly like marked points in Deligne-Mumford compactification of the moduli space. Black points are distinguished from the white points because they can collide with other black points. However, black points cannot collide with white points.

In construction of the amplitudes vector spaces associated to black points are different from that associated to white points - I will denote these spaces by V.

The simplest example of Losev-Manin moduli space is the moduli space L_k of projective space with two white points (one input and one output) and k black points (for stability k should be positive).

The space L_k may be described as a compactification of a coset $(\mathbb{C}^*)^k/\mathbb{C}^*$, where $\mathbb{C}^*$ is acting diagonally by multiplication. Really, input white point may be placed at zero, output point - at infinity, $(\mathbb{C}^*)^k$ is the space of positions of black points and $\mathbb{C}^*$ is the action of automorphism of the $\mathbb{CP}^1$ with two marked white points.

The non-compact directions in complex codimension 1 are formed when a group of k_1 points tends to zero (or, equivalently, a group of $k-k_1$ points tends to infinity). Such limits are compactified by two $\mathbb{CP}^1$ such that infinity of the first $\mathbb{CP}^1$ is identified with the zero of the second $\mathbb{CP}^1$. First $\mathbb{CP}^1$ carries k_1 black points while second $\mathbb{CP}^1$ carries $k-k_1$ black points. We will call such divisors D_{k_1,k_2}.

The moduli space restricted to divisor D_{k_1,k_2} is $L_{k_1} \times L_{k_2}$.

The preamplitudes are closed differential forms on L_k with values in $S^kV^* \otimes End(W)$, and similarly to the Deligne-Mumford case we should have a factorization condition:

$$PA_k|_{D_{k_1,k_2}} = PA_{k_2} \circ PA_{k_1} \tag{59}$$

where composition is a composition in $End(W)$.

Like in the case of Keel's relations there is a fundamental compactification divisor that equals to zero in homology. Consider L_2, it may be parametrized by ratio of coordinates of two black points: $w = z_1/z_2$. Since L_2 is connected the divisor

$$D_c = \{w = 0\} - \{w = \infty\}$$

is zero in homology of L_2. This mean that if we consider PA_1 as a map from V to $End(W)$ then

$$[PA_1(u_1), PA_1(u_2)] = 0 \tag{60}$$

Due to this commutativity we call divisor D_c commutative divisor.

Like in the Deligne-Mumford case there are forgetful maps

$$f : L_{k+2} \to L_2 \tag{61}$$

Therefore, preimages of commutative divisor $f^*(D_c)$ also equal to zero in homology. It turns out that these equations may be written as follows. Let us consider A_n as a n-th coefficient of Teylor seria of the $End(W)$ valued function A_{tot} on V. Let us denote as d^v the De Rham operator on V. Then equations, that we would call Commutativity equations, look as follows:

$$d^v A_{tot} \circ d^v A_{tot} = 0, \tag{62}$$

where composition implies simultaneous composition as elements of $End(W)$ and external multiplication as 1-forms on V. Quite remarkably, this equation first appeared in the Theory of Primitive form of K. Saito.

4.3. Construction of solutions to Commutativity equation from bicomplexes with strong Hodge property

In this subsection I will explicitly construct preamplitudes on spaces L_k in terms of linear algebra data. I will give explicite construction in simplest case and briefly explain how it may be generalized to more general case. Moreover, I will concentrate on linear data with strong Hodge condition, for generalization to Hodge condition in terms of Khoroshkin, Markarian and Shadrin, see their paper [7].

Definition of strong Hodge property. Consider Z_2 graded bicomplex V with differentials Q and G_-. Let W be cohomology of Q, G-homotopy, and let i_W and π_W be inclusion of cohomology and projection to cohomology respectively. By strong Hodge property I mean the following conditions:

$$G_- i_W = \pi_W G_- = 0 \quad \{G, G_-\} = 0 \tag{63}$$

In order to construct solution to Commutativity equation we need a commutative family.

Definition of simplified commutative family. A simplified commutative family is an even formal map U from V to commutative subalgebra of $End(W)$. We demand that this map has the simplified Maurer-Cartan property:

$$[Q, U] = 0; \quad [[G_-, U], U] = 0 \tag{64}$$

Construction of preamplitude for L_k spaces.
We construct a differential form on L_k as follows. Let us parametrize points on $\mathbb{C}^k$ by their coordinates

$$z_a = \exp(t_a + i\phi_a)$$

and denote

$$t_{aa-1} = t_a - t_{a-1}, \quad \phi_{aa-1} = \phi_a - \phi_{a-1}$$

Then the preamplitude PA_k has the following form:

$$\begin{aligned} PA_k = \pi_W U \exp(-t_{kk-1}\{G, Q\} - dt_{kk-1}G + id\phi_{kk-1}G_-)U \\ \dots U \exp(-t_{21}\{G, Q\} - dt_{21}G + id\phi_{21}G_-)U i_W \end{aligned} \tag{65}$$

Now consider $d^V PA_k$. It follows from the properties described above that $d^V PA_k$ is closed differential form on the space L_k. Actually, we need to check that this form has no jumps when t_{aa-1} reaches zero, and that it can be continued to compactification. The second property follows from the fact that G_- annihilates cohomology W. The first property is more tricky, it follows from commutativity of the algebra and from

$$\{d^V U, [G_-, U]\} = 0 \tag{66}$$

that can be derived from the simplified Maurer-Cartan conditions (together with the commutativity of the target algebra). Factorization property of PA_k is obvious, therefore by integrating over spaces L_k and summing over k we obtain that

$$A = d^V \left(\sum_{k=0}^{+\infty} \pi_W (U(-GG_- U)^k) i_W \right) \tag{67}$$

squares to zero, i.e. A solves Commutativity equation. In the proof presented above it is clear that G_- is an operator of contraction with vector field of the circle action, and K. Saito differential (see last section):

$$Q_S = Q + zG_- \tag{68}$$

is an equivariant differential with respect to the circle action.

Purely homological proof.

It is also instructive to give another proof of the formula (67), that belongs purely to standard homological algebra. Consider the bicomplex $V \otimes \Omega^*(V)$, where the first differential equals to Q and the second differential ϕ is given by

$$\phi = d^V U + [G_-, U] \tag{69}$$

The ϕ squares to zero due to (66).

Now we induce the action of ϕ on cohomology of Q, i.e. on W according to (43). We get A_1 that contains differential forms on V of different degrees. Since G_- annihilates W the degree zero form is absent. The relation $A_1^2 = 0$ starts with equations on one forms, that we will denote a simply A. From explicit form of ϕ we deduce that A actually has the form as in (67), that completes the proof.

Despite the simplicity of this proof, it hides the two-dimensional origin of the formulas.

Both proofs may be generalized to the non simplified Maurer-Cartan case, i.e. that is

$$[Q, U] + [[G_-, U], U] = 0 \tag{70}$$

4.4. Tree level BCOV theory as a solution to Oriented Associativity equations

In the previous section we constructed solution to Commutativity equation in terms of linear algebra. Recall, that moduli space $\bar{M}_{0,k}$ may be obtained from spaces L_{k-2} by blowing up diagonal in the configuration space of black points. Therefore, one may conjecture that construction similar to construction of solution to Commutativity equation may be obtained by replacing line with points by a trivalent trees.

Another argument that supports this idea is the concept of tropicalization of Riemann surfaces, that replaces general surfaces by trivalent graphs. One may think that cohomology of the moduli space of tropical surfaces are equal to cohomology of $\bar{M}_{0,k}$.

Based on this heuristic arguments we will construct the solution to Associativity equations in terms of algebraic data. Such construction was known as BCOV theory, however, it was only in the work of Losev and Shadrin [8] when it was proven that this construction actually gives solution to Associativity equations. Proof in that paper was combinatorics and quite involved. Here we will outline the simple proof based on the construction of solution to Commutativity equation.

We will explain here oriented version, however, all arguments may be generalized to non oriented one.

Construction of the solution to Oriented Associativity equation.

Suppose we are given the following data:
1. A supercommutative associative differential Z_2-graded algebra B with multiplication m and differential Q.
2. A second order odd differential operator G_- that squares to zero, anticommutes with Q and annihilates Q-cohomology.
3. A homotopy G that contracts an algebra to its cohomology W and that anticommutes with G_-.

We construct the following operation A_γ for a rooted oriented tree γ with n_l leaves:

$$A_\gamma : W^{\otimes n_l} \to W, \tag{71}$$

just like in the theory of induced operations, with the only change - edges correspond not just to homotopy but to a product GG_- like in construction of solution to Commutativity equations.

Losev-Shadrin theorem:

$$A = \sum_\gamma A_\gamma / n_\gamma, \tag{72}$$

where the n_γ is the order of the groop of symmetry of the graph, is a formal vector field on V that satisfies Oriented Associativity equation (57).

Outline of the new proof that uses Commutativity equation:

1. Pick up a leaf of graph and connect it to the root. We get a line with various trees attached to it at points, that are acting on the line

as linear operators of $End(B)$ (roots of trees are inserted at one of the entrances of multiplication operator). Sum over all trees and consider it as an element X of $End(W)$.
2. Check that these operators (after summing up trees) satisfy Maurer-Cartan equation.
3. Therefore, the element X satisfies Commutativity equation and also is symmetric under interchange of two W - one that corresponds to distinguished leaf and a typical leaf of an original tree. Thus, it is a solution to Oriented Associativity equation.

Remark. The expression (72) was defined as an amplitude in the BCOV theory [9]. However, authors did not pay attention to the fact that it actually solves Associativity equations.

§5. Good sections and Commutativity equation in K. Saito theory of Primitive form

Consider W - the quasihomogenious polynomial of n variables $X_1, \ldots, X_N$. Consider the ideal I generated by partial derivatives $\frac{\partial W}{\partial X_i}$ and the Milnor ring:

$$R = \mathbb{C}[X_1, \ldots, X_N]/I \tag{73}$$

Assume that the Milnor ring is finite dimensional and its dimension as a vector space over complex number is μ. Consider a basis in this ring and pick up polynomials Φ_k that represent classes of I. Then we can form a family

$$W_t = W + \sum_{k=1}^{\mu} t_k \Phi_k \tag{74}$$

This family is called a versal deformation of the singularity. In what follows we will consider it only over a formal disk in t variables, i.e. over the $Spec\, \mathbb{C}[[t_1, \ldots, t_\mu]]$.

For example, for $N = 1$, $W = X^n$ (we omit the subscript 1 in the one-dimensional case), the $\mu = n - 1$. If we take the lowest order representatives of the elements of the Milnor ring we will get

$$W_t = X^n + \sum_{k=1}^{n-1} t_k X^{k-1} \tag{75}$$

Consider the de Rham complex Ω of polynomial differential forms on $\mathbb{C}^N$ with differential d. Define

$$\Omega_t = \Omega \otimes \mathbb{C}[[t_1, \ldots, t_\mu]] \tag{76}$$

We consider Ω_t as a complex with the same differential d - that is just de Rham differential in X direction.

All above was kind of standard and well-known. However, based on the study of periods of the hypersurafces

$$W_t = 0$$

and later on the study of exponential integrals of the form

$$\int \exp(W/z)\omega$$

K. Saito [10] extended the complex to

$$\Omega_{t,z} = \Omega_t \otimes \mathbb{C}[[z]] \tag{77}$$

and introduced the differential

$$d^S_{t,z} = zd + dW_t \tag{78}$$

Let $H_{t,z}$ denote the cohomology of $d^S_{t,z}$ in $\Omega_{t,z}$.

Multiplication by Φ_k commutes with $d^S_{t,0}$, therefore, it acts on cohomology $H_{t,0}$ by a linear operator that we will denote by C_k:

$$C_i : H_{t,0} \to H_{t,0} \tag{79}$$

This operator will be very important in what follows.

The cohomology $H_{t,z}$ form a vector bundle over $Spec\, \mathbb{C}[[t_1, \ldots, t_\mu]]$, but this bundle does not have a natural flat connection.

However, if we localize at $z = 0$, i.e. consider

$$\hat{\Omega}_{t,z} = \Omega_t \otimes \mathbb{C} << z >> \tag{80}$$

and corresponding cohomology $\hat{H}_{t,z}$, there is a canonical flat connection called Gauss-Manin connection by K. Saito. This connection is given by its covariantly flat sections

$$S^{GM} = \left[\exp\left(-\frac{1}{z} \sum_{k=1}^{\mu} t_k \Phi_k \right) \omega \right]_{d^S_{t,z}} \tag{81}$$

where ω is t-independent differential form, and $[,]_{d^s}$ stand for a class in corresponding cohomology. The name Gauss-Manin means that in the interpretation of connection on periods of the hypersurface this is actually a Gauss-Manin connection.

One flat connection over the contractible base contains no interesting information. However, K. Saito invented another connection that I will call K. Saito's connection. Its definition is a bit involved and goes in several steps.

Step 1. Consider projection π - evaluation at $z = 0$:

$$\pi : H_{t,z} \to H_{t,0} \tag{82}$$

Let us take S - a section that inverts projection:

$$S : H_{t,0} \to H_{t,z}; \quad \pi \circ S = id \tag{83}$$

Step 2. Let $h \in H_{t,z}$, let ω be a $d^S_{t,0}$ closed representative of this class, $[\omega] = h$. We will define the connection along the tangent vector ϵ_k by the result of a parallel transport along ϵ T^S_ϵ, that acts on h as follows:

$$T^S_\epsilon(h) = \left[\omega + z^{-1}\left(-\sum_{k=1}^{\mu} \epsilon_k \Phi_k + S \circ \pi \left(\sum_{k=1}^{\mu} \epsilon^k \Phi_k\right)\right)\omega\right]_{d^S_{t+\epsilon,z}} \tag{84}$$

The following comments are needed. First, the term in the brackets actually is $d^S_{t+\epsilon,z}$ closed (as can be easily seen by comparison with the GM connection).

Second, both terms in the bracket are closed with respect to $d^S_{t,0}$, and their classes in $d^S_{t,0}$ cohomology are equal, thus, the class of $d^S_{t+\epsilon,z}$ cohomology of bracket is vanishes at $z = 0$, so the results of such transport actually belongs to $H_{t+\epsilon,z}$, and not just to $\hat{H}_{t,z}$ (note, that the parallel transport in GM connection belongs to $\hat{H}_{t+\epsilon,z}$).

Third, if we localize K. Saito connection at $z = 0$ (i.e. consider it as connection on $\hat{H}_{t,z}$) we can compare it with the GM connection and we will get the following remarkable relation

$$\nabla^S = \nabla^{GM} + z^{-1} \sum_{k=1}^{\mu} dt_k C_k \tag{85}$$

Fourth, for a general section the K. Saito connection is not integrable. However, if the section is parallel along the K. Saito connection the K. Saito connection is integrable. Thus, K. Saito defined a good section, such that if we find it at zero it could be transported to the full formal disk $Spec\, \mathbb{C}[[t_1, \ldots, t_\mu]]$.

Moreover, given a good section one can trivialize GM connection and in such trivialization K. Saito connection reads

$$(\nabla^S)_{GM\, frame} = d + z^{-1} A \tag{86}$$

From the flatness of K. Saito connection it follow that:

$$A^2 = 0 \tag{87}$$

$$dA = 0 \tag{88}$$

Note, that system (87) and (88) correspond exactly to Commutativity equations! That is why the Commutativity equations were actually discovered by K. Saito.

It looks like a miracle. In [11] this was explained as follows.

There is a analogue of complex Hodge theory where the role of complex structure is played by function W_t. The differential Q_t depends on W_t while differential G_- does not. Hodge property identifies cohomology of Q and G_- through harmonic forms that are simultaneously annihilated by Q and G_-. K. Saito connection is such a connection in cohomology of Q_t that corresponding class in G_- cohomology is not changing. That is why it is integrable. The good section is a class of holomorphic part of germ of harmonic form at singularity, and Hodge connection restricted to classes of $Q_t + zG_-$ coincides with K. Saito's connection.

The Hodge theory allows to construct the solution to Commutativity equation, that is why K. Saito theory of good section does the same.

References

[1] G. Segal, The definition of conformal field theory. Differential Geometric Methods in Theoretical Physics. Nato Adv. Sci. Inst. Ser. C Math. Phys. Sci, vol. 250, pp. 165–171. Kluwer Academic Publishers, Dordrecht (1988).

[2] T. Kadeishvili, A_∞-algebra Structure in Cohomology and Rational Homotopy Groups, Proc. Conf. in Oberwolfach 01.09-06.09, 1985.

[3] V. Lysov, Anticommutativity Equation in Topological Quantum Mechanics, JETP Lett. 76, 724–727, 2002.

[4] M. Kontsevich, Deformation quantization of Poisson manifolds, I, Lett. Math. Phys. 66:157–216, 2003.

[5] S. Keel, Intersection theory of moduli spaces of stable n-pointed curves of genus zero. Trans. AMS, 330, 545–574, 1992.

[6] A. Losev, Yu. Manin, New moduli spaces of pointed curves and pencils of flat connections, eprint arXiv:math/0001003.

[7] A. Khoroshkin, N. Markarian, S. Shadrin, Hypercommutative Operad as a Homotopy Quotient of BV, Comm. Math. Phys., Volume 322, Issue 3, pp. 697–729.

[8] A. Losev, S. Shadrin, From Zwiebach invariants to Getzler relation, Comm. Math. Phys., Volume 271, Issue 3, 649–679, 2007.

[9] M. Bershadsky, S. Cecotti, H. Ooguri, C. Vafa, Kodaira-Spencer Theory of Gravity and Exact Results for Quantum String Amplitudes, Comm. Math. Phys., Volume 165, Issue 2, pp. 311–428,1994.

[10] K. Saito, Period mapping associated to a primitive form, Publ. RIMS, Kyoto Univ. 19 (3), 1231–1264, 1983.

[11] A. Losev, Hodge Strings and Elements of K. Saito's Theory of Primitive Form Topological Field Theory, Primitive Forms and Related Topics. Edited by Masaki Kashiwara, Atsushi Matsuo, Kyoji Saito and Ikuo Satake. Published by Birkhäuser, Boston, Massachusetts, USA, p. 305, 1998.

Wu Wen-Tsun Key Lab of Mathematics, Chinese Academy of Sciences
Scientific Research Institute of System Development;
National Research University Higher School of Economics, Russian Federation; and Institute for Theoretical and Experimental Physics.
E-mail address: aslosev2@gmail.com

Advanced Studies in Pure Mathematics 83, 2019
Primitive Forms and Related Subjects — Kavli IPMU 2014
pp. 295–326

The phase factors in singularity theory

Todor Milanov

Abstract.

The paper [2] proposed a construction of a twisted representation of the lattice vertex algebra corresponding to the Milnor lattice of a simple singularity. The main difficulty in extending the above construction to an arbitrary isolated singularity is in the so called *phase factors* – the scalar functions produced by composing two vertex operators. They are certain family of multivalued analytic functions on the space of miniversal deformations. The first result in this paper is an explicit formula for the unperturbed phase factors in terms of the classical monodromy operator and the polylogorithm functions. Our second result is that with respect to the deformation parameters the phase factors are analytic functions on the monodromy covering space.

Contents

§1. Introduction

Lattice Vertex Operator Algebras (VOAs) were introduced by R. Borcherds in his work on the Moonshine conjecture (see [3, 4]). After

Received February 26, 2015.
2010 *Mathematics Subject Classification.* 14D05, 14N35, 17B69.
Key words and phrases. period integrals, Frobenius structure, Gromov–Witten invariants, vertex operators.

Borcherds' work, lattice VOAs were used as a tool to generalize the notion of a Kac-Moody Lie algebra. In particular, it is very natural to apply Borcherds' construction to singularity theory, because we have all the necessary ingredients: a Milnor lattice equipped with the intersection pairing. What is even more exciting is that there is a natural candidate for a root system, namely the set of vanishing cycles. It is very tempting to investigate to what extend the classical theory of simple Lie algebras extends to singularity theory. We refer to [18, 21] for further details on Lie algebras in singularity theory. The applications of this class of Lie algebras is still not quite clear. In particular, fundamental questions, such as finding the root multiplicities and classifying the irreducible representations are wide open. Nevertheless, Gromov–Witten theory provides a new motivation to develop further Borcherds' ideas in the settings of singularity theory. Especially, Givental's higher genus reconstruction formalism [8, 10, 22] and the integrable hierarchies introduced by Dubrovin and Zhang [6] suggest to investigate whether the Lie algebras in singularity theory are a source of integrability in the same way as the ADE affine Lie algebras are a source of integrability for the Kac–Wakimoto hierarchies [14].

The relevance of the period integrals associated with a primitive form in the sense of K. Saito (see [17]) for describing integrable hierarchies was first observed by Givental in [10] and developed further in [7]. Partially motivated by these works, it was suggested in [2] that the period integrals associated with a given singularity can be used to construct a σ-twisted representation of the lattice VOA associated with the Milnor lattice, where σ is the classical monodromy. The main issue in confirming the prediction of [2] arises when we compose two vertex operators. We get a set of phase factors that we can not study only in the category of formal Laurent series, due to the fact that exponentiating a formal Laurent series does not make sense in general. The goal of this paper is to establish the analytic properties of the phase factors that are necessary in order to extend the construction of [2] in general.

Our first result is that the vertex operators are mutually local in the sense of VOA. The proof is based on an explicit computation of the unperturbed phase factors. Our main tool is a formula due to Hertling that expresses Saito's higher residue pairing in terms of the *Seifert form*. The final answer is given in terms of the polylogorithm functions. Furthermore, it is not hard to see that the phase factors are convergent and we can express them as integrals along the path of the so called *phase form* (see (2.12)) – quadratic form on the vanishing homology with values 1-forms on the space of deformation parameters. Using the locality and the fact that the Gauss–Manin connection has regular singularities, we prove that the phase factors are analytic functions on the monodromy covering space. This fact was very important in both [2] and [10] for constructing W-constraints and Hirota quadratic equations of the total descendant potential. The compatibility of the phase factors with the monodromy representation is a very non-trivial and somewhat surprising result. It is equivalent to the following question of Givental [10]: given a path C and a pair of integral cycles invariant along C, then is it true that the integral along C of the corresponding phase form is an integer multiple of $2\pi\sqrt{-1}$? Givental answered the above question positively in the case of simple singularities. Unfortunately, his argument is hard to generalize, because it relies on understanding the fundamental group of the complement of the discriminant.

§2. Statement of the main results

2.1. Singularity theory.

Let $W : (\mathbb{C}^{n+1}, 0) \to (\mathbb{C}, 0)$ be the germ of a holomorphic function with an isolated singularity at $x = 0$. We denote by $x = (x_0, x_1, \ldots, x_n)$ the standard coordinate system on $\mathbb{C}^{n+1}$ and by

$$H := \mathcal{O}_{\mathbb{C}^{n+1},0}/(W_{x_0}, W_{x_1}, \ldots, W_{x_n}), \quad W_{x_i} = \partial W/\partial x_i,$$

the Jacobi algebra of W. Let $\{\phi_j(x)\}_{j=1}^N$ be a set of holomorphic germs that represent a basis of H. We may assume that $\phi_1 = 1$. We are interested in the following deformation of W

$$F(x,t) = W(x) + \sum_{j=1}^{N} t_j \phi_j(x), \quad t = (t_1, \ldots, t_N) \in \mathbb{C}^N.$$

Let us cut a Stein domain $X \subset \mathbb{C}^{n+1} \times \mathbb{C}^N$ around $(0,0)$, s.t., the map

$$\varphi : X \to B \times \mathbb{C}_\epsilon, \quad (x,t) \mapsto (t, F(x,t))$$

is well defined and surjective, where $B \subset \mathbb{C}^N$ is a sufficiently small ball around 0 and $\mathbb{C}_\epsilon = \{\lambda \in \mathbb{C} : |\lambda| < \epsilon\}$, and the fibers $X_{t,\lambda} := \varphi^{-1}(t,\lambda)$ satisfy an appropriate transversality condition (see [1]). Put $S := B \times \mathbb{C}_\epsilon$ to avoid cumbersome notation.

Let C be the critical locus $\{F_{x_0} = \cdots = F_{x_n} = 0\} \subset X$ of F relative to B. The map

$$\mathcal{T}_B \to \mathrm{pr}_* \mathcal{O}_C, \quad \partial/\partial t_i \mapsto \partial F/\partial t_i \ (\mathrm{mod}\ F_{x_0}, \ldots, F_{x_n})$$

is called the *Kodaira–Spencer* map. Decreasing X, B, and ϵ if necessary we may arrange that the Kodaira–Spencer map is an isomorphism. In particular, each tangent space T_tB has an associative algebra structure. Let us fix a *primitive* form $\omega \in \Omega^{n+1}_{X/B}(X)$, so that B becomes a Frobenius manifold (see [11, 19]). The flat Frobenius pairing is given by the following residue pairing:

(2.1)
$$(\phi_1(x), \phi_2(x))_t := \frac{1}{(2\pi\sqrt{-1})^{n+1}} \int_\Gamma \frac{\phi_1(y)\phi_2(y)}{F_{y_0}(t,y)\cdots F_{y_n}(t,x)} dy_0 \dots dy_N,$$

where the cycle Γ is a disjoint union of sufficiently small tori around the critical points of F defined by equations of the type $|F_{x_0}| = \cdots = |F_{x_n}| = \epsilon$ and $y = (y_0, \dots, y_n)$ is a local coordinate system near each critical point, s.t., $\omega = dy_0 \wedge \cdots \wedge dy_n$. We have the following identifications:

$$T^*B \cong TB \cong B \times T_0B \cong B \times H, \tag{2.2}$$

where the first isomorphism is given by the residue pairing, the second by the Levi–Civita connection of the flat residue pairing, and the last one is the Kodaira–Spencer isomorphism. Let us denote by ∂_i the flat vector fields corresponding to the basis $\{\phi_i\}_{i=1}^N \subset H$.

2.2. Period vectors

Removing the singular fibers from X we obtain a smooth fibration $X' \to S'$ known as the Milnor fibration, where S' is the subset parametrizing non-singular fibers. Its complement is an irreducible analytic hypersurface known as the *discriminant*. The homology $H_n(X_{t,\lambda};\mathbb{C})$ and the cohomology groups $H^n(X_{t,\lambda};\mathbb{C})$ form vector bundles on S' equipped with a flat Gauss–Manin connection. Let us fix a reference point $(0, \epsilon_0) \in S'$ and denote by

$$\mathfrak{h} := H^n(X_{0,\epsilon_0};\mathbb{C}), \quad \mathfrak{h}^* := H_n(X_{0,\epsilon_0};\mathbb{C})$$

the reference fibers. The vector space $\mathfrak{h}^*$ contains the so called *Milnor lattice* $Q := H_n(X_{0,\epsilon_0};\mathbb{Z})$ and if we assume that n is even then the intersection pairing gives a symmetric bi-linear pairing on Q (which however might be degenerate). We normalize the intersection pairing by the sign $(-1)^{n/2}$ in order to obtain a pairing $(\,|\,)$, s.t., $(\alpha|\alpha) = 2$ for every *vanishing cycle* $\alpha \in Q$. We refer again to [1] for some more details on the vanishing homology.

Using the parallel transport with respect to the Gauss–Manin connection we get the so called monodromy representation

$$\pi_1(S') \to \mathrm{O}(\mathfrak{h}^*), \tag{2.3}$$

where the RHS denotes the group of linear transformations that preserve the intersection pairing. The image of the monodromy representation is a reflection group W generated by the reflections

$$s_\alpha(x) = x - (\alpha|x)\alpha,$$

where α is a vanishing cycle and s_α is the monodromy transformation representing a simple loop corresponding to a path from the reference point to a generic point on the discriminant along which α vanishes.

Let us introduce the notation d_x, where $x = (x_1, \dots, x_m)$ is a coordinate system on some manifold, for the de Rham differential in the coordinates x. This notation is especially useful when we have to apply d_x to functions that might depend on other variables as well. The main object of our interest are the following period integrals (see [10])

$$I^{(k)}_\alpha(t,\lambda) = -d_t\,(2\pi)^{-n/2}\,\partial_\lambda^{k+n/2}\int_{\alpha_{t,\lambda}} d_x^{-1}\omega\ \in T_t^*B \cong H, \tag{2.4}$$

where recall that n is even, $\alpha \in \mathfrak{h}^*$ is a cycle from the vanishing homology, $\alpha_{t,\lambda} \in H_n(X_{t,\lambda}, \mathbb{C})$ is the parallel transport of α along a reference path, and $d_x^{-1}\omega$ is a holomorphic n-form $\eta \in \Omega^n_{X/B}$ defined in a neighborhood of the fiber $X_{t,\lambda}$, s.t., $d_x\eta = \omega$. The periods are multivalued analytic functions in $(t,\lambda) \in S$ with poles along the discriminant. In other words they are analytic functions $\widetilde{S}' \to H$, where $\widetilde{S}'$ is the monodromy covering space of S'.

The periods satisfy the following system of differential equations

$$\partial_i I^{(k)}_\alpha(t,\lambda) = -\phi_i \bullet_t I^{(k+1)}_\alpha(t,\lambda) \tag{2.5}$$

$$\partial_\lambda I^{(k)}_\alpha(t,\lambda) = I^{(k+1)}_\alpha(t,\lambda) \tag{2.6}$$

$$(\lambda - E\bullet_t)\partial_\lambda I^{(k)}_\alpha(t,\lambda) = \Big(\theta - k - \frac{1}{2}\Big) I^{(k)}_\alpha(t,\lambda), \tag{2.7}$$

where the notation in (2.7) is as follows. The vector field $E \in \mathcal{T}_B$ is the *Euler* vector field, which by definition corresponds to F via the Kodaira-spencer isomorphism. Changing the basis $\{\phi_i\}_{i=1}^N \subset H$ if necessary we may arrange that the Euler vector field takes the form

$$E = \sum_{i=1}^N (1-d_i)\tau_i\partial_i + \rho, \tag{2.8}$$

where ρ is a flat vector field and the degree spectrum $0 = d_1 < d_2 \leq \cdots \leq d_{N-1} < d_N =: D$ satisfies $d_i + d_{N+1-i} = D$. The number D is the *conformal dimension* of the Frobenius structure. The linear operator θ is defined by

$$\theta : H \to H, \quad \phi_i \mapsto (D/2 - d_i)\phi_i.$$

It is sometimes called the *Hodge-grading operator*.

2.3. Extending the domain of the period vectors

Using the differential equations (2.5)–(2.7) we would like to extend the domain S in such a way that the period vectors have a translation symmetry and that there exists a point $u^{(0)} \in B$, such that the singularities of the period vectors $I_\alpha^{(m)}(u^{(0)}, \lambda)$ are at points $\lambda = u_i$ $(1 \le i \le N)$ that are sufficiently close to the vertices of a regular N-gon with center 0.

The differential equation (2.7) allows us to extend the definition of $I_\alpha^{(m)}(t, \lambda)$ for all $(t, \lambda) \in B \times \mathbb{C}$. Let us denote by $t = (t_1, \ldots, t_N)$ the standard coordinate system on $B \subset \mathbb{C}^N$ induced from the linear coordinates on $\mathbb{C}^N$. We denote by $t - \lambda\mathbf{1} = (t_1 - \lambda, t_2, \ldots, t_N)$, i.e., this is the time $-\lambda$-flow of t with respect to the flat identity. Note that $X_{t,\lambda} = X_{t-\lambda\mathbf{1},0}$, so the period vectors have the following translation symmetry

$$I_\alpha^{(k)}(t, \lambda) = I_\alpha^{(k)}(t - \lambda\mathbf{1}, 0), \tag{2.9}$$

where $|\lambda| \ll 1$. Note that the RHS is a multivalued function on $B' := B \times \{0\} \cap S'$, while the LHS is a multivalued function on S'. We choose $-\epsilon_0\mathbf{1}$ as a reference point in B' and the reference path for the RHS is obtained from the reference path for the LHS via the translation map

$$\tau : S' \to B', \quad (t, \lambda) \mapsto t - \lambda\mathbf{1}.$$

Using (2.9), we extend the definition of the periods also for all t that belong to a domain in $\mathbb{C}^N$ obtained from B by the flow of the flat vector field ∂_1. Slightly abusing the notation we denote by B the extended domain and redefine S to be $B \times \mathbb{C}$. Note that the Frobenius structure also extends on B and now the translation symmetry (2.9) makes sense for all $(t, \lambda) \in S$.

Let us fix a point $t_0 \in B$, s.t., the critical values of $F(x, t_0)$ are pairwise distinct. There exists an open neighborhood $\mathcal{U} \subset B$ of t_0 such that $\overline{\mathcal{U}} \cap \partial B \neq \emptyset$, the critical values $\{u_i(t)\}_{i=1}^N$ $(t \in \overline{\mathcal{U}})$ are pairwise distinct and form a *canonical* coordinate system on $\mathcal{U}$. Put $1/\Delta_i := (\partial/\partial u_i, \partial/\partial u_i)$ $(1 \le i \le N)$, then the map

$$\Psi_t : \mathbb{C}^N \to T_t U, \quad e_i \mapsto \sqrt{\Delta_i}\partial/\partial u_i$$

gives a trivialization of the tangent bundle $T\mathcal{U}$ in which the Frobenius multiplication and the residue pairing have a diagonal form

$$e_i \bullet e_j = \delta_{i,j}\sqrt{\Delta_j}e_j, \quad (e_i, e_j) = \delta_{i,j}.$$

If we write the differential equations (2.5)–(2.7) in canonical coordinates, then we obtain a system of linear ODEs whose coefficients depend polynomially on the entries of the matrices

$$V := -\Psi^{-1}\theta\Psi, \quad (\lambda - U)^{-1}, \quad \mathrm{ad}_U^{-1}(V),$$

where U is the diagonal matrix $\mathrm{diag}(u_1, \dots, u_N)$. Here we are using that the matrix V is skew-symmetric and we define $\mathrm{ad}_U^{-1}(V)$ to be the unique matrix X with zero diagonal entries, s.t., $[U, X] = V$. The dependence of the matrix V on the canonical coordinates $\{u_i\}_{i=1}^N$ is quite remarkable. It was proved by Dubrovin that $V = V(u)$, $u = (u_1, \dots, u_N)$, is a solution to an integrable system (see equation (3.74) in [5]) and that V extends analytically along any generic path in the domain $\mathfrak{D} = \{u \in \mathbb{C}^N \mid u_i \neq u_j, \forall i \neq j\}$ (Corollary 3.4 in [5], see also Theorem 1.1 in [16]). Using the canonical coordinates, we embed $\overline{\mathcal{U}} \subset \mathfrak{D}$. There exist a point $u^{(0)} \in \mathbb{C}^N$ whose coordinates are sufficiently close to the vertices of a regular polygon with center at 0 and a path in $\mathfrak{D}$ that connects $\overline{\mathcal{U}} \cap \partial B$ and $u^{(0)}$, s.t., $V(u)$ extends analytically along the path. Therefore $V(u)$ is a function analytic in a slightly larger domain $\widetilde{\mathcal{U}}$. In particular, we can extend the system (2.5)–(2.7) on $\widetilde{\mathcal{U}} \times \mathbb{C}$, which gives us the required extension of the period vectors. We redenote by B the extended domain obtained by gluing B and $\widetilde{\mathcal{U}}$ along $\mathcal{U}$.

2.4. The phase factors

Our main goal is to investigate the analytic properties of the following infinite series:

(2.10)

$$\Omega_{\alpha,\beta}(t, \lambda, \mu) = \sum_{n=0}^{\infty} (-1)^{n+1} \big(I_\alpha^{(n)}(t, \lambda), I_\beta^{(-n-1)}(t, \mu)\big) \in \mathbb{C}((\lambda^{-1/|\sigma|})),$$

where σ is the classical monodromy operator and $|\sigma|$ is the order of its semi-simple part. To begin with, it is not hard to determine the radius of convergence. Namely, for fixed $(t, \mu) \in S'$, the series (2.10) is convergent for

$$|\lambda| > \max_i\{|\mu|, |u_i(t)|\},$$

where $u_i(t)$ is the set of critical values of $F(t, x)$. The main motivation to study the above series comes from the so called *phase factors* $B_{\alpha,\beta}(t, \lambda, \mu) := e^{\Omega_{\alpha,\beta}(t,\lambda,\mu)}$, which are produced naturally when we compose two vertex operators (see Section 5).

Recall the variation operator isomorphism

$$\mathrm{Var} : \mathfrak{h} := H^n(X_{0,\epsilon_0}; \mathbb{Z}) \to \mathfrak{h}^* := H_n(X_{0,\epsilon_0}; \mathbb{Z})$$

defined by composing the Lefschetz duality $H^n(X_{0,\epsilon_0};\mathbb{Z}) \cong H_n(X_{0,\epsilon_0}, \partial X_{0,\epsilon_0};\mathbb{Z})$ and the operator $h_* - 1$, where $h : (X_{0,\epsilon_0}, \partial X_{0,\epsilon_0}) \to (X_{0,\epsilon_0}, \partial X_{0,\epsilon_0})$ is the geometric monodromy. Let us recall also the *Seifert forms* (see [1]) in respectively cohomology and homology:

$$\mathrm{SF}(A,B) := (-1)^{n/2+1}\langle A, \mathrm{Var}(B)\rangle, \quad A, B \in \mathfrak{h},$$
$$\mathrm{SF}(\alpha,\beta) := (-1)^{n/2+1}\langle \mathrm{Var}^{-1}(\alpha), \beta\rangle, \quad \alpha, \beta \in \mathfrak{h}^*,$$

where recall that we required n to be even. Both forms are non-degenerate, integer valued, and σ-invariant. The sign here is chosen in such a way that the symmetrization of the Seifert form yields the sign-normalized intersection form introduced above

$$\mathrm{SF}(\alpha,\beta) + \mathrm{SF}(\beta,\alpha) = (\alpha|\beta).$$

Using the variation isomorphism we introduce intersection form in cohomology $(A|B) := (\mathrm{Var}(A)|\,\mathrm{Var}(B))$. Note that the cohomological intersection form is also the symmetrization of the cohomological Seifert form.

Finally, let us introduce the following polylogorithm operator series depending on a linear operator σ whose eigenvalues are roots of unity:

$$\mathrm{Li}_\sigma(x) = \sum_{k=1}^{\infty} \frac{x^{k+\mathcal{N}}}{k+\mathcal{N}},$$

where $\mathcal{N} := -\frac{1}{2\pi\sqrt{-1}}\log\sigma$, s.t., the eigenvalues of $\mathcal{N}$ belong to the set $(-1,0]\cap\mathbb{Q}$. Note that the series is convergent for $|x|<1$. Moreover, it can be expressed in terms of the standard polylogorithm functions

$$\mathrm{Li}_\sigma(x) = x^{\mathcal{N}_n} \sum_{p=1}^{\infty}\sum_{r=1}^{|\sigma|} (-\mathcal{N}_n|\sigma|)^{p-1}\,\mathrm{Li}_p(\eta^r x^{1/|\sigma|})\sigma_s^r, \tag{2.11}$$

where we decomposed $\mathcal{N} = \mathcal{N}_s + \mathcal{N}_n$ into a diagonal and nilpotent operator, $\sigma_s := e^{-2\pi\sqrt{-1}\mathcal{N}_s}$, $|\sigma|$ is the order of σ_s, and $\eta = e^{2\pi\sqrt{-1}/|\sigma|}$. In particular, the series $\mathrm{Li}_\sigma(x)$ extends analytically along any path avoiding $x=0$ and $x=1$.

Theorem 2.1. *The following formula holds*

$$\Omega_{\alpha,\beta}(0,\lambda,\mu) = -(\mathrm{Li}_\sigma(\mu/\lambda)\alpha|\beta) + P_{\alpha,\beta}(\lambda,\mu),$$

where $P_{\alpha,\beta}(\lambda,\mu)$ is a polynomial in $\lambda^{\pm 1/|\sigma|}$, $\mu^{\pm 1/|\sigma|}$, $\log\lambda$, and $\log\mu$ satisfying

$$P_{\alpha,\beta}(\lambda,\mu) - P_{\beta,\alpha}(\mu,\lambda) = \mathrm{SF}\Big(\frac{e^{-2\pi\sqrt{-1}\mathcal{N}}-1}{\mathcal{N}}(\mu/\lambda)^{\mathcal{N}}\alpha_1, \beta\Big),$$

where α_1 is the projection of α on the generalized eigen subspace $\mathfrak{h}_1^* :=$ $\mathrm{Ker}(\mathcal{N}_s)$.

As a corollary of Theorem 2.1 we obtain the following result (for the proof see Lemma 4.1).

Corollary 2.2. *Let $(\lambda, \mu) \in \mathbb{C}^2$ be a fixed point, s.t., $|\lambda| > |\mu| > 0$ and $|\lambda - \mu| \ll 1$. Let $C \subset \mathbb{C}^2$ be a path from (λ, μ) to (μ, λ) avoiding the diagonal and contained in a sufficiently small neighborhood of the line segment $[(\lambda, \mu), (\mu, \lambda)]$. The following symmetry holds $B_{\alpha,\beta}(0, \lambda, \mu) = B_{\beta,\alpha}(0, \mu, \lambda)$, where the second phase factor is obtained from $B_{\beta,\alpha}(0, \lambda, \mu)$ via analytic continuation along C.*

Corollary 2.2 is important for the applications to VOA representations, because it essentially means that the vertex operators are mutually local. In particular, we can construct a representation of the lattice VOA corresponding to the Milnor lattice. In Section 5 we give more details on how to construct such a representation, although a more systematic investigation will be presented elsewhere.

2.5. The phase form and analytic extension

The phase 1-form is by definition the following 1-form on B' depending on a parameter ξ:

$$\mathcal{W}_{\alpha,\beta}(t, \xi) = I^{(0)}_\alpha(t, 0) \bullet_t I^{(0)}_\beta(t, \xi), \quad \alpha, \beta \in \mathfrak{h}^*, \tag{2.12}$$

where the RHS is interpreted as a cotangent vector in $T_t^* B$ via the identifications (2.2) and the parameter ξ is assumed to be sufficiently small. In other words, we expand the RHS into a Taylor series at $\xi = 0$

$$\sum_{m=0}^{\infty} \mathcal{W}^{(m)}_{\alpha,\beta}(t) \frac{\xi^m}{m!}, \quad \mathcal{W}^{(m)}_{\alpha,\beta}(t) := I^{(0)}_\alpha(t, 0) \bullet I^{(m)}_\beta(t, 0) \in T_t^* B,$$

where for each $t \in B'$ the radius of convergence of the above series is non-zero.

The phase form can be used to describe the dependence of the phase factors in the deformation parameters

$$\Omega_{\alpha,\beta}(t, \lambda, \mu) = \Omega_{\alpha,\beta}(0, \lambda, \mu) + \int_{-\lambda\mathbf{1}}^{t-\lambda\mathbf{1}} \mathcal{W}_{\alpha,\beta}(t', \mu - \lambda), \tag{2.13}$$

where (t, λ, μ) is in the domain of convergence of (2.10), $|\mu - \lambda| \ll 1$, and the integration path should be chosen appropriately. The main result of this paper can be stated as follows.

Theorem 2.3. *Let (t,λ,μ) be a point in the domain of convergence of (2.10), s.t., $|\mu-\lambda|\ll 1$ and $C\subset B'$ be a closed loop based at $t-\lambda\mathbf{1}$, then*

$$\oint_{t'\in C}\mathcal{W}_{\alpha,\beta}(t',\mu-\lambda)-\Omega_{w(\alpha),w(\beta)}(t,\lambda,\mu)+\Omega_{\alpha,\beta}(t,\lambda,\mu)\ \in\ 2\pi\sqrt{-1}\,\mathbb{Z},$$

where w is the monodromy transformation along C.

The integral (2.13) provides analytic extension along any path, so the phase factors can be interpreted as Laurent series in $(\mu-\lambda)$. Theorem 2.3 implies that the coefficients of the Laurent series expansion are analytic on the monodromy covering space, i.e, if a closed loop C is in the kernel of the monodromy representation, then the coefficients are invariant with respect to the analytic continuation along C. Finally, when the cycles α and β are invariant along C, we get an affirmative answer to Givental's question from [10].

Corollary 2.4. *Let C be a closed loop in B' and $\alpha,\beta\in Q$ be cycles invariant with respect to the monodromy transformation along C, then $\oint_{t\in C}\mathcal{W}_{\alpha,\beta}(t,\xi)\in 2\pi\sqrt{-1}\,\mathbb{Z}$.*

§3. The phase factors at $t=0$

Let $\mathcal{H}=H(\!(z^{-1})\!)$ be Givental's symplectic loop space, where the symplectic form is

$$\Omega(f(z),g(z))=\mathrm{Res}_{z=0}(f(-z),g(z))dz,\quad f,g\in\mathcal{H}.$$

There is a natural polarization $\mathcal{H}=\mathcal{H}_+\oplus\mathcal{H}_-$ with $\mathcal{H}_+:=H[z]$ and $\mathcal{H}_-:=H[\![z^{-1}]\!]z^{-1}$ which allows us to identify $\mathcal{H}\cong T^*\mathcal{H}_+$. Let us introduce the generating series

$$\mathbf{f}_a(t,\lambda;z)=\sum_{n\in\mathbb{Z}}I_a^{(n)}(t,\lambda)\,(-z)^n,\quad a\in\mathfrak{h}^*$$

and note that $\Omega_{\alpha,\beta}(t,\lambda,\mu)=\Omega(\mathbf{f}_\alpha(t,\lambda;z)_+,\mathbf{f}_\beta(t,\mu;z))$, where the index $+$ denotes the projection on $\mathcal{H}_+$ along $\mathcal{H}_-$.

3.1. The Virasoro grading operator

The symplectic vector space $\mathcal{H}$ has the following Virasoro grading operator:

$$\ell_0(z)=z\partial_z+\frac{1}{2}-\theta+\rho/z,$$

where $\rho \in \mathrm{End}(H)$ is the operator of multiplication by W. Note that ρ/z is a nilpotent operator, commuting with $\ell_0(z)$, while $\ell_0(z) - \rho/z$ is diagonalizable. Let us decompose, the symplectic vector space

$$\mathcal{H} = \mathcal{H}_{<0} \oplus \mathcal{H}_0 \oplus \mathcal{H}_{>0}, \tag{3.1}$$

according to the sign of the eigenvalues of $\ell_0(z)$, i.e., $\mathcal{H}_{<0}$ is the sum of all generalized eigen subspaces of ℓ_0 with eigenvalue < 0, $\mathcal{H}_0$ – with eigenvalue 0, and $\mathcal{H}_{>0}$ – with eigenvalue > 0.

We are going to compute the symplectic pairing

$$\Omega(\mathbf{f}_\alpha(0,\lambda;z)_{>0}, \mathbf{f}_\beta(0,\mu;z)) = \Omega(\mathbf{f}_\alpha(0,\lambda;z), \mathbf{f}_\beta(0,\mu;z)_{<0}), \quad \alpha,\beta \in \mathfrak{h}^*,$$

where the index > 0 (resp. < 0) corresponds to projection with respect to the spectral decomposition of the operator $\ell_0(z)$ onto the subspace spanned by all generalized eigen subspaces with eigenvalue > 0 (resp. < 0). It is not hard to check that the difference between the phase factor $\Omega_{\alpha,\beta}(0,\lambda,\mu)$ and the above symplectic pairing is a function $P_{\alpha\beta}$ that has the form described in Theorem 2.1. So the main difficulty is in finding an explicit formula for the above symplectic pairing.

3.2. Fundamental solution

The differential equation (2.7) can be solved explicitly when $t = 0$. Assume that $k = -m$ with $m \gg 0$, then the fundamental solution is given by the following operator-valued function

$$\Phi_m(\lambda) = e^{\rho\partial_\lambda\partial_m}\Big(\frac{\lambda^{\theta+m-1/2}}{\Gamma(\theta+m+1/2)}\Big),$$

where the RHS is defined first for m a complex number, so that ∂_m makes sense and the Gamma function is defined through its Taylor's series expansion at $\theta = 0$, so it makes sense to substitute a linear operator for θ. Note that if we assume that $1/\Gamma(m) = 0$ when m is a negative integer or 0, then the above formula makes sense for all $m \in \mathbb{Z}$.

Using the commutation relations

$$[\rho,\theta] = \rho, \quad \rho\partial_\lambda\Phi_m(\lambda) = \Phi_m(\lambda)\rho, \tag{3.2}$$

it is easy to prove that the analytic continuation around $\lambda = 0$ transforms $\Phi_m(\lambda)$ into

$$\Phi_m(\lambda)\, e^{2\pi\sqrt{-1}\rho}\, e^{2\pi\sqrt{-1}(\theta-1/2)}.$$

3.3. Geometric sections

If $\omega \in \Omega^{n+1}_{X/B}(X)$ is a holomorphic form then $s(\omega) := \nabla_{\partial_\lambda} \int d^{-1}\omega$, where ∇ is the Gauss–Manin connection, defines a holomorphic section of the vanishing cohomology bundle known as *geometric section*. Let us denote by $s(\omega, \lambda)$ the value of the section at the point $(0, \lambda) \in S'$.

Put $\ell = n/2$ (we assume that n is even). Note that by definition the period vector

$$I^{(-\ell)}(0, \lambda) = (2\pi)^{-\ell} \sum_{i=1}^{N} \phi^i \otimes s(\omega_i, \lambda),$$

where $\{\phi^i\} \subset H$ is a basis of H dual to $\{\phi_i\}$ with respect to the residue pairing and $\omega_i \in \Omega^{n+1}_{X/B}(X)$ is defined by

$$-\nabla_{\partial_i} \int d^{-1}\omega = \nabla_{\partial_\lambda} \int d^{-1}\omega_i.$$

On the other hand, since $\Phi_\ell(\lambda)$ is a fundamental solution for the differential equation (2.7) with $k = -\ell$, we get that

$$I^{(-\ell)}(0, \lambda) = \sum_{i=1}^{N} \Phi_\ell(\lambda)\phi^i \otimes A_i,$$

where $\{A_i\}_{i=1}^N \subset \mathfrak{h}$ is a basis with some very special properties that guarantee that ω is a primitive form (see [11] for more details). Let us introduce the linear map

$$^{\#} : \mathrm{End}(H) \to \mathrm{End}(\mathfrak{h}), \quad \text{s.t.,} \quad \sum_{i=1}^{N} T\phi^i \otimes A_i = \sum_{i=1}^{N} \phi^i \otimes T^{\#} A_i.$$

Lemma 3.1. *The map* $^{\#}$ *is anti-homomorphism of algebras, i.e.,* $(T_1T_2)^{\#} = T_2^{\#} T_1^{\#}$.

Proof. We just have to use that $T^{\#} A_i = \sum_{j=1}^{N} (T\phi^j, \phi_i) A_j$. Q.E.D.
The numbers

$$s_i := d_i + \ell - \frac{1}{2} - \frac{D}{2}, \quad 1 \le i \le N,$$

are known as the *Steenbrink spectrum* of the singularity. The key result that reveals the Hodge-theoretic origin of the period integrals is the following Lemma.

Proposition 3.2. *The following formula holds*

$$s(\omega_i,\lambda) = (2\pi)^\ell\, \lambda^{s_i+\rho^\#}\,\Gamma(s_i+\rho^\#+1)^{-1}A_i,$$

where the value of the Gamma function is defined through its Taylor's series at $\rho^\# = 0$.

Proof. We have to compute $\Phi_\ell(\lambda)^\# A_i$. Using the commutation relations (3.2) we get

$$\begin{aligned}\Phi_\ell(\lambda) &= \sum_{k=0}^\infty \frac{1}{k!}\rho^k\partial_\lambda^k\partial_\ell^k\left(\lambda^{\theta+\ell-1/2}\Gamma(\theta+\ell+1/2)^{-1}\right)\\ &= \sum_{k=0}^\infty \frac{1}{k!}\partial_\ell^k\left(\lambda^{\theta+\ell-1/2}\Gamma(\theta+\ell+1/2)^{-1}\right)\rho^k.\end{aligned}$$

Note that

$$\theta^\# A_i = \sum_{j=1}^N (\theta\phi^j,\phi_i)A_j = (d_i - D/2)A_i.$$

To finish the proof we just need to recall Lemma 3.1 and use Taylor's formula. Q.E.D.

Since the geometric sections are single valued, Proposition 3.2 allows us to conclude that the classical monodromy

$$\sigma = e^{-2\pi\sqrt{-1}\rho^\#}\, e^{-2\pi\sqrt{-1}(\theta^\#-1/2)}.$$

Let us write $s_i = p_i + \alpha_i$, where $-1 < \alpha_i \le 0$, then our choice of logarithm $\mathcal{N} = -\frac{1}{2\pi\sqrt{-1}}\log\sigma$ acts on the basis $\{A_i\}\subset\mathfrak{h}$ as follows

$$\mathcal{N}(A_i) = (\alpha_i+\rho^\#)A_i,\quad 1\le i\le N.$$

Let us point out that since

$$\nabla_{\partial_\lambda}^{p_i} s(\omega_i,\lambda) = \lambda^{\alpha_i+\rho^\#}\Gamma(\alpha_i+\rho^\#+1)^{-1}A_i$$

the vectors $\Gamma(\alpha_i+\rho^\#+1)^{-1}A_i \in F_{p_i}\mathfrak{h}$, where $\{F_p\mathfrak{h}\}_{p=0}^n$ is the Steenbrink's Hodge filtration.

3.4. The higher residue pairing

Let us denote by $\mathbb{H}\to B$ the vector bundle whose fiber over $t\in B$ is

$$\mathbb{H}_t = \Omega^{n+1}_{X_t}[\![z]\!]/(zd_x + d_xF(t,x)\wedge)\Omega^n_{X_t}[\![z]\!],$$

where $X_t = X\cap\mathbb{C}^{n+1}\times\{t\}$. By definition, the domain X is chosen so small that $\mathbb{H}_t$ is a $\mathbb{C}[\![z]\!]$-free module of rank N. We can think of sections

ω of $\mathbb{H}$ as formal oscillatory integrals $\int e^{F/z}\omega$, which allows us to see that $\mathbb{H}$ is equipped with a flat connection, called *Gauss–Manin connection*, corresponding to differentiating the integral formally with respect to the deformation parameters.

Recall, also K. Saito's higher residue pairing

$$K_t : \mathbb{H}_t \otimes \mathbb{H}_t \to \mathbb{C}[\![z]\!]z^{n+1}.$$

It is uniquely determined up to a constant by the following properties

(K1) If $\omega_1, \omega_2 \in \mathbb{H}_t$, then $K_t(\omega_1, \omega_2) = (-1)^{n+1}K_t(\omega_2, \omega_1)^*$, where $*$ is the involution $z \mapsto -z$.

(K2) If $p(z) \in \mathbb{C}[z]$, then

$$p(z)K_t(\omega_1, \omega_2) = K_t(p(z)\omega_1, \omega_2) = K_t(\omega_1, p(-z)\omega_2).$$

(K3) The pairing $K_t^{(p)}(\omega_1, \omega_2)$ defined by the coefficient in front of z^{n+1+p} in $K_t(\omega_1, \omega_2)$ depends analytically on t and the Leibnitz rule holds

$$\xi K_t(\omega_1, \omega_2) = K_t(\nabla_\xi \omega_1, \omega_2) + K_t(\omega_1, \nabla_\xi \omega_2),$$

where $\xi = \partial/\partial t_i$ or $z\partial_z$ and ∇ is the Gauss–Manin connection on $\mathbb{H}$.

(K4) If $\omega_i = \phi_i(x)dx_0 \cdots dx_n$ $(i = 1, 2)$, then $K_t^{(0)}(\omega_1, \omega_2) = (\phi_1, \phi_2)$, where the residue pairing (see (2.1)) is with respect to the volume form $dx_0 \cdots dx_n$.

Given a holomorphic form $\omega \in \Omega^{n+1}_{X_0}(X_0)$, let us denote by

$$\widehat{s}(\omega, z) = (-2\pi z)^{-\ell-1/2} \int_0^\infty e^{\lambda/z} s(\omega, \lambda)d\lambda,$$

where the integration path is $\lambda = -tz$, $t \in [0, +\infty)$, the Laplace transform of the corresponding geometric section. We will make use also of the automorphism $\widehat{s}(\omega, z)^* := \widehat{s}(\omega, e^{\pi\sqrt{-1}}z)$. Note that since we might have $\log z$ dependence, * is no longer an involution.

Let us introduce also the so called *elementary sections*. They are defined as follows. Given a vector $A \in \mathfrak{h}$, we can construct a section of the vanishing cohomology bundle over $S' \cap \{t = 0\}$ as follows

$$s(A, \lambda) := \sum_{-1<\alpha\leq 0} \lambda^{\alpha+\rho^\#} A_\alpha,$$

where A_α is the projection of A on the generalized eigen space of σ corresponding to the eigenvalue $e^{-2\pi\sqrt{-1}\alpha}$. Put

$$\widehat{s}(A, z) = (-2\pi z)^{-\ell-1/2} \int_0^\infty e^{\lambda/z} s(A, \lambda)d\lambda,$$

where the integration path is the same as above. The next result is a reformulation of Hertling's formula for Saito's higher residue pairing in terms of the Seifert form.

Lemma 3.3. *The pairing*

$$K_W(\omega_1,\omega_2) := \mathrm{SF}(\widehat{s}(\omega_1,z)^*,\widehat{s}(\omega_2,z))^*\, z^{n+1}$$

coincides with Saito's higher residue pairing $K_0(\omega_1,\omega_2)$.

Proof. We will prove that the Lemma is equivalent to Hertling's formula for Saito's higher residue pairings. In order to compare our formula to Hertling's one, we have to introduce the vector bundle $\mathbb{H}''\to B$ whose fiber over a point $t\in B$ is $\mathbb{H}''_t := \Omega^{n+1}_{X_t}/d_xF(t,x)\wedge d_x\Omega^{n-1}_{X_t}$. If $\omega\in\Omega^{n+1}_{X_t}$ is a holomorphic form, then we denote by $s(\omega)\in\mathbb{H}''_t$ the projection of ω. It is known that $\mathbb{H}''_t$ is a $\mathbb{C}[\partial_\lambda^{-1}]$-module and that if we identify $\mathbb{C}[\![z]\!]\cong\mathbb{C}[\![\partial_\lambda^{-1}]\!]$ via $z\mapsto -\partial_\lambda^{-1}$, then the natural map

$$\mathbb{H}_t\to\mathbb{H}''_t\otimes_{\mathbb{C}[\partial_\lambda^{-1}]}\mathbb{C}[\![\partial_\lambda^{-1}]\!],\quad \omega\mapsto s(\omega),$$

is an isomorphism of $\mathbb{C}[\![z]\!]$-modules. Let us define

$$\widetilde{K}_t(s(\omega_1),s(\omega_2)) := K_t(\omega_2,\omega_1) = (-1)^{n+1}K_t(\omega_1,\omega_2)^*.$$

It is easy to check that $\widetilde{K}$ satisfy properties (K1)–(K4) of the higher residue pairing except that we have to replace z by ∂_λ^{-1}. Since the higher residue pairing is uniquely determined by its properties, $\widetilde{K}$ coincides with the higher residue pairing used in [11]. We have to prove that

$$\widetilde{K}_t(s(\omega_1),s(\omega_2)) = \mathrm{SF}(\widehat{s}(\omega_1,z)^*,\widehat{s}(\omega_2,z))z^{n+1}.$$

Note that using the above formula we can uniquely define a pairing on the space of all elementary sections, s.t., property (K2) holds. We just need to check that the RHS agrees with Hertling's formula for a pair of elementary sections (see formulas (10.81)–(10.83) in [11]).

Let us assume that B_i $(i=1,2)$ are eigenvectors with eigenvalues $e^{-2\pi\sqrt{-1}\beta_i}$, $-1<\beta_i\le 0$. Then

$$\widehat{s}(B_i,z) = (2\pi)^{-\ell-1/2}(-z)^{\beta_i+\rho^\#-\ell+1/2}\Gamma(\beta_i+\rho^\#+1)B_i,\quad i=1,2.$$

Using that the Seifert form is monodromy invariant and infinitesimally $\rho^\#$-invariant, i.e., $\mathrm{SF}(\rho^\# A,B)+\mathrm{SF}(A,\rho^\# B)=0$ we get that

$$\mathrm{SF}(\widehat{s}(B_1,z)^*,\widehat{s}(B_2,z))\, z^{n+1} \tag{3.3}$$

is given by the following formula

$$\frac{z^{\beta_1+\beta_2+2}}{(2\pi\sqrt{-1})^{n+1}}\,\mathrm{SF}\,\big(B_1, e^{\pi\sqrt{-1}(\beta_2+\rho^{\#}+1)}\Gamma(\beta_1-\rho^{\#}+1)\Gamma(\beta_2+\rho^{\#}+1)B_2\big).$$

Since the Seifert form is monodromy invariant, the above pairing vanishes unless $\beta_1+\beta_2\in\mathbb{Z}$, i.e., $\beta_1+\beta_2=-1$, or $\beta_1=\beta_2=0$. Recall, the operator $\mathcal{N}=-\frac{1}{2\pi\sqrt{-1}}\log\sigma$ and note that $\mathcal{N}B_i=-(\beta_i+\rho^{\#})B_i$.

$$\begin{aligned} &e^{\pi\sqrt{-1}(\beta_2+\rho^{\#}+1)}\Gamma(\beta_1-\rho^{\#}+1)\Gamma(\beta_2+\rho^{\#}+1)B_2\\ &\quad=\begin{cases}-2\pi\sqrt{-1}(\sigma-1)^{-1}B_2 & \text{if } \beta_1+\beta_2=-1,\\ 2\pi\sqrt{-1}\mathcal{N}(\sigma-1)^{-1}B_2 & \text{if } \beta_1=\beta_2=0.\end{cases}\end{aligned}$$

Hence the pairing (3.3) takes the form

$$\frac{1}{(2\pi\sqrt{-1})^{n}}\,(-1)^{\ell}\langle B_1,\mathrm{Var}\,(\sigma-1)^{-1}B_2\rangle\, z,\quad \text{for } \beta_1+\beta_2=-1,$$
$$\frac{-1}{(2\pi\sqrt{-1})^{n+1}}\,(-1)^{\ell}\langle B_1,\mathrm{Var}\;2\pi\sqrt{-1}\mathcal{N}(\sigma-1)^{-1}B_2\rangle\, z^2,\ \text{for } \beta_1=\beta_2=0,$$

and it vanishes in all other cases. Q.E.D.

The identity in Lemma 3.3 is really remarkable. It is a relation between two completely different quantities. The LHS is defined via residues of differential forms, while the RHS is purely topological. Let us outline a different way to prove Lemma 3.3, which in particular generalizes the identity for an arbitrary deformation. Given $\omega\in\Omega^{n+1}_{X/B}(X)$ and a critical value $u=u(t)$ of $F(t,x)$ we define the following formal asymptotic series

$$\widehat{s}_u(\omega,z):=(-2\pi z)^{-\ell-1/2}\int_{u(t)}^{\infty}e^{\lambda/z}s(\omega)d\lambda.$$

Note that this is the stationary phase asymptotic as $z\to 0$ of an appropriate oscillatory integral. Let us define

$$K_F(\omega_1,\omega_2)=\sum_u \mathrm{SF}(\widehat{s}_u(\omega_1,z)^*,\widehat{s}_u(\omega_1,z))^*\,z^{n+1},$$

where the sum is over all critical values of F. When $F=W$, this formula reduces to Lemma 3.3. One has to check that the above formula satisfies all properties (K1)–(K4) of the higher residue pairing. The verification is straightforward except for property (K4). In the latter case we take a generic deformation and then we just have to see that the computation on the RHS reduces to proving Lemma 3.3 for A_1-singularity, which is straightforward. It remains only to recall the result of M. Saito [20] that the higher residue pairing is uniquely determined by its properties.

3.5. Proof of Theorem 2.1

Using the explicit formula for the geometric sections in Proposition 3.2 we get that

$$\widehat{s}(\omega_i, z) = (2\pi)^{-1/2}(-z)^{p_i+\mathcal{N}-\ell+1/2}A_i,$$

where $p_i = \lfloor s_i \rfloor$ is the floor of the Steenbrink number s_i. It is convenient, to introduce the operator $p \in \mathrm{End}(\mathfrak{h})$, s.t., $p(A_i) = p_i A_i$. Recalling Lemma 3.3 we get

(3.4)

$$K_W(\omega_i, \omega_j) = (\phi_i, \phi_j)z^{n+1} = \frac{1}{2\pi\sqrt{-1}}\langle A_i, e^{\pi\sqrt{-1}\mathcal{N}}e^{\pi\sqrt{-1}p}A_j\rangle\, z^{s_i+s_j-2\ell+1+n+1},$$

where $\langle\, ,\, \rangle$ is the Seifert form (without the sign normalization) and the 1st identity holds, because ω is a primitive form. Let us define the *residue pairing* $(\, ,\,)$ on $\mathfrak{h}$, s.t., the $(\phi_i, \phi_j) = (A_i, A_j)$, i.e.,

$$(A, B) = \frac{1}{2\pi\sqrt{-1}}\langle A, e^{\pi\sqrt{-1}\mathcal{N}}e^{\pi\sqrt{-1}p}B\rangle, \quad A, B \in \mathfrak{h}.$$

Finally, if $R \in \mathrm{End}(\mathfrak{h})$, then we denote by R^T the transpose with respect to the residue pairing. Let us point out the following relations

$$\mathcal{N} = \mathcal{N}_s + \rho^{\#}, \quad p + \mathcal{N}_s = (\theta + \ell - 1/2)^{\#}, \quad [\mathcal{N}, \rho^{\#}] = 0, \quad [p, \rho^{\#}] = \rho^{\#},$$

where $\mathcal{N}_s$ is the semi-simple part of $\mathcal{N}$. Comparing the powers of z in (3.4) we get

$$p + \mathcal{N}_s + p^T + \mathcal{N}_s^T = 2\ell - 1.$$

By definition

$$\begin{aligned}\mathbf{f}_\alpha(0, \lambda; z) &= \sum_{m\in\mathbb{Z}} \partial_\lambda^{m+\ell} I_\alpha^{(-\ell)}(0, \lambda)\,(-z)^m \\ &= \sum_{i=1}^{N}\sum_{m\in\mathbb{Z}}(-z)^m\partial_\lambda^{m+\ell-p_i}\langle\lambda^{\mathcal{N}}\Gamma(\mathcal{N}+1)^{-1}A_i, \alpha\rangle\phi^i.\end{aligned}$$

Let us introduce also the transpose R^{SF} with respect to the Seifert form

$$\langle RA, B\rangle =: \langle A, R^{SF}B\rangle, \quad \forall A, B \in \mathfrak{h}.$$

Note that since the Seifert form is not symmetric, in general $\langle A, RB\rangle \neq \langle R^{SF}A, B\rangle$. Since

$$\Big(z\partial_z + \frac{1}{2} - \theta\Big)(-z)^m\phi^i = (m+\ell-s_i)(-z)^m\phi^i,$$

after shifting $m \mapsto m-\ell+p_i$ we get that the projection

(3.5)

$$\mathbf{f}_\beta(0,\mu;z)_{<0} = \sum_{j=1}^{N}\sum_{m=-\infty}^{-1}\Big\langle A_j, (-z)^{m-\ell+p^{SF}}\mu^{\mathcal{N}^{SF}-m}\Gamma(\mathcal{N}^{SF}-m+1)^{-1}\beta\Big\rangle\,\phi^j.$$

Similarly,

$$\mathbf{f}_\alpha(0,\lambda;z) = \sum_{i=1}^{N}\sum_{k\in\mathbb{Z}}\Big\langle A_i, (-z)^{k-\ell+p^{SF}}\lambda^{\mathcal{N}^{SF}-k}\Gamma(\mathcal{N}^{SF}-k+1)^{-1}\alpha\Big\rangle\,\phi^i.$$

Let us denote by $\{A^i\}_{i=1}^N$ the basis of $\mathfrak{h}$ dual to $\{A_i\}_{i=1}^N$ with respect to the residue pairing. We need to compute a pairing of the following type:

$$\sum_{i,j=1}^{N}\langle A_i, x\rangle(\phi^i,\phi^j)\langle A_j, y\rangle = (2\pi\sqrt{-1})^2\sum_{i,j=1}^{N}(A_i, M^{-1}x)(\phi^i,\phi^j)(A_j, M^{-1}y),$$

where $M = e^{\pi\sqrt{-1}\mathcal{N}}e^{\pi\sqrt{-1}p}$ and the equality follows from the definition of the residue pairing. Using that $(\phi^i,\phi^j) = (A^i, A^j)$ and the standard properties of dual bases we get

$$(2\pi\sqrt{-1})^2((M^{-1})x,(M^{-1})y) = 2\pi\sqrt{-1}\,\langle M^{-1}x, y\rangle.$$

The symplectic pairing

(3.6)

$$\Omega(\mathbf{f}_\alpha(0,\lambda;z)_{>0},\mathbf{f}_\beta(0,\mu;z)) = \Omega(\mathbf{f}_\alpha(0,\lambda;z),\mathbf{f}_\beta(0,\mu;z)_{<0}) = 2\pi\sqrt{-1}\sum_{m=-\infty}^{-1}\sum_{k\in\mathbb{Z}}\mathrm{Res}_{z=0}$$

$$\Big\langle M^{-1}z^{k-\ell+p^{SF}}\lambda^{\mathcal{N}^{SF}-k}\Gamma(\mathcal{N}^{SF}-k+1)^{-1}\alpha, (-z)^{m-\ell+p^{SF}}\mu^{\mathcal{N}^{SF}-m}\Gamma(\mathcal{N}^{SF}-m+1)^{-1}\beta\Big\rangle.$$

Using that by definition $R^{SF} = MR^TM^{-1}$ we get the following formulas

$$\mathcal{N}_s^T = \mathcal{N}_s^{SF},\quad (\rho^{\#})^{SF} = -\rho^{\#},\quad (\rho^{\#})^T = \rho^{\#},\quad p^{SF} = p^T + \pi\sqrt{-1}\rho^{\#}.$$

Let us point out that the two conjugations do not commute in general and also that $(R^{SF})^{SF} \neq R$ in general. Since $M^{-1}p^{SF} = p^TM^{-1}$ and

$M^{-1}\mathcal{N}^{SF}=\mathcal{N}^TM^{-1}$, the summand in the sum (3.6) can be written in the following form

$$\Big\langle z^{m+k-2\ell+p+p^T}e^{\pi\sqrt{-1}(m-\ell+p)}\lambda^{\mathcal{N}^T-k}\Gamma(\mathcal{N}^T-k+1)^{-1}M^{-1}\alpha,\\ \mu^{\mathcal{N}^{SF}-m}\Gamma(\mathcal{N}^{SF}-m+1)^{-1}\beta\Big\rangle.$$

Using that $e^{\pi\sqrt{-1}p}\mathcal{N}^T=\mathcal{N}^{SF}e^{\pi\sqrt{-1}p}$ and $e^{\pi\sqrt{-1}p}M^{-1}=e^{-\pi\sqrt{-1}\mathcal{N}}$, we get

$$\Big\langle z^{m+k-2\ell+p+p^T}\lambda^{\mathcal{N}^{SF}-k}\Gamma(\mathcal{N}^{SF}-k+1)^{-1}e^{-\pi\sqrt{-1}(\mathcal{N}-m+\ell)}\alpha,\\ \mu^{\mathcal{N}^{SF}-m}\Gamma(\mathcal{N}^{SF}-m+1)^{-1}\beta\Big\rangle.$$

The terms that contribute to the residue must satisfy

$$m+k+p+p^T=2\ell-1=p+p^T+\mathcal{N}_s+\mathcal{N}_s^T=p+p^T+\mathcal{N}+\mathcal{N}^{SF},$$

so we may substitute $p+p^T=2\ell-1-m-k$ and $\mathcal{N}^{SF}-k=-\mathcal{N}+m$

$$(-1)^\ell z^{-1}\left\langle(\mu/\lambda)^{\mathcal{N}-m}\Gamma(\mathcal{N}-m+1)^{-1}\Gamma(-\mathcal{N}+m+1)^{-1}e^{\pi\sqrt{-1}(-\mathcal{N}+m)}\alpha,\beta\right\rangle.$$

Using the well known identity

$$\frac{e^{\pi\sqrt{-1}x}}{\Gamma(1-x)\Gamma(1+x)}=\frac{e^{2\pi\sqrt{-1}x}-1}{2\pi\sqrt{-1}x}$$

with $x=-\mathcal{N}+m$ we get

$$\frac{(-1)^{\ell+1}}{2\pi\sqrt{-1}}z^{-1}\left\langle\frac{(\mu/\lambda)^{\mathcal{N}-m}}{\mathcal{N}-m}(e^{-2\pi\sqrt{-1}\mathcal{N}}-1)\alpha,\beta\right\rangle.$$

It remains only to recall that the intersection form can be expressed in terms of the Seifert form

$$(\alpha|\beta)=(-1)^\ell\langle(\sigma-1)\alpha,\beta\rangle,\quad \alpha,\beta\in\mathfrak{h}^*.$$

We get

$$\Omega(\mathbf{f}_\alpha(0,\lambda;z)_{>0},\mathbf{f}_\beta(0,\mu;z))=-(\mathrm{Li}_\sigma(\mu/\lambda)\alpha|\beta).$$

Put

$$P_{\alpha,\beta}(\lambda,\mu)=\Omega(\mathbf{f}_\alpha(0,\lambda;z)_+,\mathbf{f}_\beta(0,\mu;z))-\Omega(\mathbf{f}_\alpha(0,\lambda;z)_{>0},\mathbf{f}_\beta(0,\mu;z)).$$

Using that Ω is a symplectic pairing and that $\mathcal{H}_{>0}$ is symplectic orthogonal to $\mathcal{H}_{\geq 0}$ we get

$$P_{\beta,\alpha}(\mu,\lambda) = \Omega(\mathbf{f}_\beta(0,\mu;z), \mathbf{f}_\alpha(0,\lambda;z)_-) - \Omega(\mathbf{f}_\beta(0,\mu;z), \mathbf{f}_\alpha(0,\lambda;z)_{<0}).$$

Hence

$$P_{\alpha,\beta}(\lambda,\mu) - P_{\beta,\alpha}(\mu,\lambda) = \Omega(\mathbf{f}_\alpha(0,\lambda;z)_0, \mathbf{f}_\beta(0,\mu;z)_0).$$

The above symplectic pairing can be computed in the same way as above. The only difference is that the summation over m in (3.5) collapses to $m = 0$ and the summation over j reduces only to j, s.t., $s_j = p_j$, i.e., $\mathcal{N}_s A_j = 0$. This implies that in (3.6) we have to put $m = 0$ and replace α (and β) by α_1 (and β_1). The rest of the computation is the same. We get

$$\Omega(\mathbf{f}_\alpha(0,\lambda;z)_0, \mathbf{f}_\beta(0,\mu;z)_0) = (-1)^{\ell+1} \left\langle \frac{(\mu/\lambda)^{\mathcal{N}}}{\mathcal{N}} (e^{-2\pi\sqrt{-1}\mathcal{N}} - 1)\alpha_1, \beta_1 \right\rangle.$$

Finally, the polynomiality statement about $P_{\alpha,\beta}(\lambda,\mu)$ follows from the fact that the vector space $\mathcal{H}_+ \cap \mathcal{H}_{<0}$ is finite dimensional. Q.E.D.

§4. Analytic continuation

We will be interested in the phase factors as multivalued analytic functions. Our starting point is the infinite series (2.10), which by definition is interpreted as a formal Laurent series of the type

$$\sum_{r=0}^{|\sigma|-1} \sum_{\ell=0}^{d-1} \lambda^{m+r/|\sigma|} (\log\lambda)^\ell \sum_{k=0}^{\infty} \Omega_{\alpha,\beta}^{r,\ell,k}(t,\mu)\lambda^{-k}, \tag{4.1}$$

where $d \geq 1$ is the smallest integer number, s.t., $\mathcal{N}_n^d = 0$. If $R > 0$ is the smallest real number, s.t., for every r and ℓ the infinite series in k is convergent for all $|\lambda| > R$, then R^{-1} is called the *radius of convergence*. The radius of convergence does not change if we differentiate the series with respect to λ. On the other hand, the derivative of (2.10) with respect to λ can be computed in terms of the period integrals (see [15] Proposition 2.3)

(4.2)

$$\partial_\lambda \Omega(\mathbf{f}_\alpha(t,\lambda;z)_+, \mathbf{f}_\beta(t,\mu;z)) = \frac{1}{\lambda-\mu}\Big(I_\alpha^{(0)}(t,\lambda), (\theta+1/2)I_\beta^{(-1)}(t,\mu)\Big).$$

Since the period $I_\alpha^{(0)}(t,\lambda)$ is a solution to a Fuchsian differential equation in λ, whose singularities are at $\lambda = \infty$ and the critical values $u_i(t)$ of $F(x,t)$, we get that the series (4.1) is convergent for

$$|\lambda| > \max\{|\mu|, r(t)\},$$

where $r(t) = \max_i\{|u_i(t)|\}$.

4.1. The domains at infinity

Put $S_\infty = \{(t,\lambda) \in S \,|\, |\lambda| > r(t)\}$. Let us fix a real constant $c \in (0,1)$, say $c = |1 - e^{2\pi\sqrt{-1}/|\sigma|}|/3$, s.t., the distance between any two distinct $|\sigma|$-roots of 1 is bigger than c. Let $\epsilon : S_\infty \to \mathbb{R}_{>0}$ be the function

$$\epsilon(t,\lambda) = \sup\nolimits_C \Big(\min\nolimits_{(t',\lambda')\in C}(|\lambda'| - r(t'))\Big),$$

where the sup is over all paths $C \subset S_\infty$ from $(0,\lambda)$ to (t,λ). We introduce the domain

$$D_\infty := \Big\{(t,\lambda,\mu) \in S_\infty \times \mathbb{C} \mid |\lambda - \mu| < c \min\{\epsilon(t,\lambda), \epsilon(t,\mu)\}\Big\}$$

and its subdomain $D_\infty^+ := \{|\lambda| > |\mu|\} \subset D_\infty$.

Let us denote by $\widetilde{S}_\infty$ the universal covering of S_∞. The domains D_∞^+ and D_∞ are (trivial) smooth disk fibrations over S_∞. The phase factors $\Omega_{\alpha,\beta}$ are holomorphic functions on the pullback $\widetilde{D}_\infty^+$ of D_∞^+ via the covering map $\widetilde{S}_\infty \to S_\infty$. In more explicit terms, in order to define the phase factor $\Omega_{\alpha,\beta}(t,\lambda,\mu)$ at some point $(t,\lambda,\mu) \in D_\infty^+$ we have to select a reference path in S_∞ from $(0,\epsilon_0)$ to (t,λ). This choice determines the value of the period vectors $I_\alpha^{(n)}(t,\lambda)$ and using the line segment $[(t,\lambda),(t,\mu)]$ we can specify also the values of $I_\beta^{(-n-1)}(t,\mu)$, so the summands in (2.10) are uniquely defined. Note that if $(t,\lambda,\mu) \in D_\infty$, then $|\lambda - \mu| < |\lambda| - r(t)$, so using the triangle inequality it is easy to verify that the line segment $[(t,\lambda),(t,\mu)]$ does not intersect the discriminant.

4.2. Symmetry of the phase factors at $t = 0$

According to Theorem 2.1, the phase factor $\Omega_{\alpha,\beta}(0,\lambda,\mu)$ extends analytically from D_∞^+ to $D_\infty - \{\lambda = \mu\}$. Note that the polylogorithms that enter in our formula have singularities for (λ,μ), s.t., $\mu = 0$ or $\lambda^{|\sigma|} = \mu^{|\sigma|}$. However, if $(0,\lambda,\mu) \in D_\infty$, then $\mu \neq 0$ and thanks to our choice of the constant c the equality $\lambda^{|\sigma|} = \mu^{|\sigma|}$ implies that $\lambda = \mu$. Therefore, our explicit formula provides the analytic extension along any path in $D_\infty - \{\lambda = \mu\}$. The analytic continuation has the following crucial symmetry. Fix $(0,\lambda,\mu) \in D_\infty^+$ and a path $C \subset D_\infty - \{\lambda = \mu\}$ connecting $(0,\lambda,\mu)$ with $(0,\mu,\lambda)$. The coordinate projections of C along the last two coordinates determine paths C_1 from $(0,\lambda)$ to $(0,\mu)$ and C_2 from $(0,\mu)$ to $(0,\lambda)$ in S_∞.

Lemma 4.1. *Let $C \subset D_\infty - \{\lambda = \mu\}$ be a path from $(0,\lambda,\mu) \in D_\infty^+$ to $(0,\mu,\lambda)$, s.t., the projections C_1 and C_2 of C are homotopic (in S_∞)*

respectively to the line segments $[(0,\lambda),(0,\mu)]$ *and* $[(0,\mu),(0,\lambda)]$, *then*

$$\Omega_{\alpha,\beta}(0,\lambda,\mu)-\Omega_{\beta,\alpha}(0,\mu,\lambda)=-2\pi\sqrt{-1}\Big(\,\mathrm{SF}(\alpha,\beta)+k(\alpha|\beta)\Big),\quad k\in\mathbb{Z},$$

where the 2nd phase factor is obtained from $\Omega_{\beta,\alpha}(0,\lambda,\mu)$ *via analytic continuation along the path* C *and the integer* k *depends on the choice of* C.

Proof. Let us first recall the so called Jonquière's inversion formula [12] (see also Appendix A), which provides a description of the analytic continuation of the polylogorithm functions in terms of Bernoulli polynomials

(4.3)

$$\mathrm{Li}_p(1/x)=(-1)^{p+1}\,\mathrm{Li}_p(x)+(-1)^{p+1}\frac{(2\pi\sqrt{-1})^p}{p!}\,B_p\Big(\frac{1}{2\pi\sqrt{-1}}\log x\Big),\quad 0<|x|<1,$$

where $B_p(x)$ $(p\geq 0)$ are the Bernoulli polynomials defined by

$$\frac{te^{xt}}{e^t-1}=\sum_{p=0}^{\infty}B_p(x)\frac{t^p}{p!}. \tag{4.4}$$

The value of $\mathrm{Li}_p(1/x)$ is specified via analytic continuation along a path $C'\subset\mathbb{C}-\{1\}$ from x to $1/x$, which *does not* wind around $x=1$. The choice of the branch of $\log x$ in (4.3) is such that the formula holds for $p=1$. Let us assume first that C' intersects the real axis once and that $\mathrm{Im}(x)\neq 0$. In order to determine the branch of $\log x$, we have to consider 4 cases depending on whether $\mathrm{Im}(x)>0$ or <0 and whether C' intersects the real interval $(1,\infty)$ or not. However, after analyzing the 4 cases we find that there are two possibilities only. Namely, if we walk along C' from x to $1/x$, then when crossing the real axis, 1 is either on our left or on our right. In the first case $\log x:=\mathrm{Log}\,x$, otherwise $\log x:=\mathrm{Log}\,x+2\pi\sqrt{-1}$, where

$$\mathrm{Log}\,x:=\ln|x|+\sqrt{-1}\ \mathrm{Arg}(x),\qquad -\pi<\mathrm{Arg}(x)\leq\pi$$

is the principal branch of the logarithm. For general C', the choice of the branch can be deduced easily from the above two cases.

Put $x=\mu/\lambda$. Recalling formula (2.11) and using (4.3) we get

$$\mathrm{Li}_\sigma(1/x)=\mathrm{Li}_\sigma(x)^T+\sum_{r=1}^{|\sigma|}\sum_{p=1}^{\infty}\frac{x^{-\mathcal{N}_n}}{\mathcal{N}_n|\sigma|}\,\frac{(2\pi\sqrt{-1}\mathcal{N}_n|\sigma|)^p}{p!}\,B_p\Big(\frac{1}{2\pi\sqrt{-1}}\log x_r\Big)\sigma_s^{-r},$$

where $x_r=\eta^r x^{1/|\sigma|}$ and the paths C'_r from x_r to x_r^{-1} can be described as follows. Let $C':t\mapsto C_2(t)/C_1(t)$ be the path from x to $1/x$ and C'_0

be the induced path from $x^{1/|\sigma|} := e^{\mathrm{Log}(x)/|\sigma|}$ to $x^{-1/|\sigma|} := e^{-\mathrm{Log}(x)|/\sigma|}$. Then C_r' is a composition of a path inside the unit disk from x_r to x_{-r} and $\eta^{-r}C_0'$ (clockwise rotation of C_0' with angle $2\pi r/|\sigma|$).

The infinite sum over p can be computed via (4.4)

$$\mathrm{Li}_\sigma(1/x) = \mathrm{Li}_\sigma(x)^T + \sum_{r=1}^{|\sigma|} \frac{x^{-\mathcal{N}_n}}{\mathcal{N}_n|\sigma|}\Big(2\pi\sqrt{-1}\mathcal{N}_n|\sigma|\frac{e^{\mathcal{N}_n|\sigma \log x_r}}{e^{2\pi\sqrt{-1}\mathcal{N}_n|\sigma|}-1} - 1\Big)\sigma_\varepsilon^{-r}.$$

Since $(0,\lambda,\mu) \in D_\infty^+$ we have $|x-1| < c < 1$ and $|x| < 1$. Moreover, since the branch of the period vectors depending on μ is determined from the branch of the period vectors depending on λ via the straight segment from $[\lambda,\mu]$, we get that the multi-valued functions

$$x^{-\mathcal{N}_n} = e^{-\mathcal{N}_n \,\mathrm{Log}\, x}, \quad x^{1/|\sigma|} = e^{\frac{1}{|\sigma|}\,\mathrm{Log}\, x}$$

are defined via the principal branch of the logarithm. Hence

$$x^{-\mathcal{N}_n}e^{\mathcal{N}_n|\sigma|\log x_r} = \exp\Big(\mathcal{N}_n|\sigma|\sqrt{-1}\Big(\mathrm{Arg}(\eta^r x^{1/|\sigma|}) - \mathrm{Arg}(x^{1/|\sigma|}) + 2\pi\chi_r\Big)\Big),$$

where $\chi_r = 1$ or 0, depending whether 1 is on the left or on the right of C_r'. Since C_0' is in a small neighborhood of 1 we have

$$\mathrm{Arg}(\eta^r x^{1/|\sigma|}) - \mathrm{Arg}(x^{1/|\sigma|}) + 2\pi\chi_r = 2\pi r/|\sigma|, \quad 1 \le r \le |\sigma|-1.$$

The sum over r becomes

$$\frac{2\pi\sqrt{-1}}{\sigma^{-|\sigma|}-1}\Big(\sum_{r=1}^{|\sigma|-1}\sigma^{-r} + \sigma^{-|\sigma|\chi_0}\Big) + \frac{x^{-\mathcal{N}_n}}{-\mathcal{N}_n}\Big(\frac{1}{|\sigma|}\sum_{r=1}^{|\sigma|}\sigma_s^r\Big).$$

Finally, we get

$$\mathrm{Li}_\sigma(1/x) = \mathrm{Li}_\sigma(x)^T - 2\pi\sqrt{-1}\,\frac{\sigma^{1-\chi_0}}{\sigma-1} + \frac{x^{-\mathcal{N}_n}}{-\mathcal{N}_n}\Big(\frac{1}{|\sigma|}\sum_{r=1}^{|\sigma|}\sigma_s^r\Big). \tag{4.5}$$

Using the formula for the phase factors in Theorem 2.1 we compute

$$\begin{aligned}&\Omega_{\alpha,\beta}(0,\lambda,\mu) - \Omega_{\beta,\alpha}(0,\mu,\lambda)\\ &\quad = \Big((\mathrm{Li}_\sigma(1/x) - \mathrm{Li}_\sigma(x)^T)\beta|\alpha\Big) + P_{\alpha,\beta}(\lambda,\mu) - P_{\beta,\alpha}(\mu,\lambda).\end{aligned}$$

Recalling again Theorem 2.1 and using (4.5) we get

$$\Omega_{\alpha,\beta}(0,\lambda,\mu) - \Omega_{\beta,\alpha}(0,\mu,\lambda) = 2\pi\sqrt{-1}\,\mathrm{SF}(\sigma^{1-\chi_0}\beta,\alpha).$$

If $\chi_0 = 0$, then $\mathrm{SF}(\sigma\beta,\alpha) = -\mathrm{SF}(\alpha,\beta)$ otherwise, if $\chi_0 = 1$, then $\mathrm{SF}(\beta,\alpha) = -\mathrm{SF}(\alpha,\beta) + (\alpha|\beta)$. Q.E.D.

4.3. Analytic extension of the phase factors

Recall that the series (2.10) is convergent in the domain

$$D^+ := \{(t,\lambda,\mu) \mid |\lambda-\mu| < |\mu| - r(t) < |\lambda| - r(t)\}.$$

It is not hard to see that the RHS of (4.2) has a singularity at $\lambda=\mu$ of the form $(\alpha|\beta)(\lambda-\mu)^{-1}$. Hence the Laurent series expansion at $\lambda=\infty$ of the series

$$\widetilde{\Omega}_{\alpha,\beta}(t,\lambda,\mu) := \Omega_{\alpha,\beta}(t,\lambda,\mu) - (\alpha|\beta)\log(\lambda-\mu)$$

is convergent in the domain

$$D = \Big\{(t,\lambda,\mu) \mid |\lambda-\mu| < \min\{|\mu|-r(t), |\lambda|-r(t)\}\Big\}.$$

In particular, we get that $\Omega_{\alpha,\beta}(t,\lambda,\mu)$ extends analytically along any path in the domain $D_\infty - \{\lambda=\mu\}$ and that $e^{\Omega_{\alpha,\beta}(t,\lambda,\mu)}$ has a pole of order $-(\alpha|\beta)$ at $\lambda=\mu$.

Lemma 4.2. *If $(t,\lambda,\mu)\in D^+_\infty$ and $C\subset D_\infty - \{\lambda=\mu\}$ is a path from (t,λ,μ) to (t,μ,λ), s.t., the projections of C along the last two coordinates are homotopic to the line segments $[(t,\lambda),(t,\mu)]$ and $[(t,\mu),(t,\lambda)]$, then*

$$\Omega_{\alpha,\beta}(t,\lambda,\mu) - \Omega_{\beta,\alpha}(t,\mu,\lambda) = -2\pi\sqrt{-1}\Big(\mathrm{SF}(\alpha,\beta)+k(\alpha|\beta)\Big), \quad k\in\mathbb{Z},$$

where the 2nd phase factor on the LHS is obtained from $\Omega_{\beta,\alpha}(t,\lambda,\mu)$ via analytic continuation along the path C.

Proof. The main idea is to obtain an integral representation of the infinite series $\Omega_{\alpha,\beta}(t,\lambda,\mu)$. If $(t,\lambda,\mu)\in D_\infty$, then by definition we can find a path $C\subset S_\infty$ from $(0,\lambda)$ to (t,λ), s.t., $|\mu-\lambda|<|\lambda'|-r(t')$ for all $(t',\lambda')\in C$. Put $\mu' := \lambda'+\mu-\lambda$. Using the triangle inequality we get $|\mu'|\geq|\lambda'|-|\mu-\lambda|>r(t')$. Therefore, $(t',\lambda',\mu')\in D$ and it makes sense to think of $\widetilde{\Omega}_{\alpha,\beta}(t',\lambda',\mu')$ as a holomorphic function on the curve C. Using the differential equations (2.5)–(2.7) we get

$$d\widetilde{\Omega}_{\alpha,\beta}(t',\lambda',\mu') = \tau^*\mathcal{W}_{\alpha,\beta}\big|_C\,,$$

where $\tau: S\to B$, $(t,\lambda)\mapsto t-\lambda\mathbf{1}$. Integrating the above identity along C we get

$$\Omega_{\alpha,\beta}(t,\lambda,\mu) = \Omega_{\alpha,\beta}(0,\lambda,\mu) + \int_{-\lambda\mathbf{1}}^{t-\lambda\mathbf{1}} \mathcal{W}_{\alpha,\beta}(t'-\lambda'\mathbf{1},\mu-\lambda). \tag{4.6}$$

The translation $C + (0, \mu - \lambda)$ is a path $\widetilde{C}$ in S_∞ from $(0, \mu)$ to (t, μ). Note that $\widetilde{C}$ is obtained from C also by the map $(t', \lambda') \mapsto (t', \mu')$. Using that by definition

$$\mathcal{W}_{\alpha,\beta}(t' - \lambda' \mathbf{1}, \mu - \lambda) = I_\alpha^{(0)}(t', \lambda') \bullet_{t'} I_\beta^{(0)}(t', \mu'),$$

and translating the integration variables $(t', \lambda') \mapsto (t', \mu')$ we get

$$\int_{-\lambda\mathbf{1}}^{t-\lambda\mathbf{1}} \mathcal{W}_{\alpha,\beta}(t' - \lambda'\mathbf{1}, \mu - \lambda) = \int_{-\mu\mathbf{1}}^{t-\mu\mathbf{1}} \mathcal{W}_{\beta,\alpha}(t' - \mu'\mathbf{1}, \lambda - \mu),$$

i.e., the integral in (4.6) is symmetric with respect to exchanging the pairs (α, λ) and (β, μ). To finish the proof it remains only to recall Lemma 4.1. Q.E.D.

4.4. Proof of Theorem 2.3

Let us choose $t \in B$, s.t., the canonical coordinates $u_i := u_i(t)$ are sufficiently close to the vertices of a regular N-gon with center at 0. Let us fix a reference point (t, λ) in $\{t\} \times \mathbb{C} \subset S_\infty$ and take a closed loop C obtained by approaching one of the points u_i along a path in S_∞ and going around u_i. Since the fundamental group of S' is generated by loops of the above type, it suffices to prove that

(4.7)
$$\oint_C \tau^* \mathcal{W}_{\alpha,\beta}(\,\cdot\,, \mu - \lambda) - \Omega_{w(\alpha),w(\beta)}(t, \lambda, \mu) + \Omega_{\alpha,\beta}(t, \lambda, \mu) \in 2\pi\sqrt{-1}\,\mathbb{Z},$$

where w is the monodromy transformation along C and μ is sufficiently close to λ. The integral in (4.7) can be written also as follows: if $(t, x) \in C$, then $\tau(t, x) = t - x\mathbf{1} \in \tau(C)$ and we have

$$\oint_C \tau^* \mathcal{W}_{\alpha,\beta}(\,\cdot\,, \mu - \lambda) = \oint_{\tau(C)} \mathcal{W}_{\alpha,\beta}(t - x\mathbf{1}, \mu - \lambda).$$

In our argument we always keep t fixed, all paths in S that we use will be identified with paths in $\mathbb{C} \cong \{t\} \times \mathbb{C} \subset S$, and we drop the argument of the phase form.

Lemma 4.3. *If φ is the cycle vanishing over u_i, then $\oint_C \tau^* \mathcal{W}_{\varphi,\varphi} = -4\pi\sqrt{-1}$.*

Proof. The differential equations of the periods imply that the integral is independent of λ and μ. Therefore, we may set $\mu = \lambda$. The computation in this case can be found in [10]. Q.E.D.

Lemma 4.4. *Assume that φ is the cycle vanishing over u_i and α is a cycle invariant along C, then*

$$\oint_C \tau^*\mathcal{W}_{\alpha,\varphi} = -2\Omega_{\alpha,\varphi}(t,\lambda,\mu).$$

Proof. Let us choose $\lambda' \in C$ sufficiently close to u_i with $|\lambda'| > r(t)$ and put $\mu' = \lambda' + \mu - \lambda$. The infinite series (2.10) at $(t,\lambda,\mu) = (t,\lambda',\mu')$ can be interpreted in two different ways: as a convergent Laurent series in $(\lambda')^{-1}$ or as a convergent Laurent series in $(\mu' - u_i)^{1/2}$. The 2nd interpretation allows us to extend the equality $d\Omega_{\alpha,\varphi}(t,\lambda',\mu') = \tau^*\mathcal{W}_{\alpha,\varphi}$ as λ' varies along the loop around u_i. The Lemma follows by the Stoke's theorem. Q.E.D.

If α and β are arbitrary cycles, then we can decompose them as

$$\alpha = \alpha' + (\alpha|\varphi)\varphi/2, \quad \beta = \beta' + (\beta|\varphi)\varphi/2,$$

where α' and β' are invariant along C. We have to compute

$$\oint_C \tau^*\mathcal{W}_{\alpha,\beta} = \frac{(\beta|\varphi)}{2}\oint_C \tau^*\mathcal{W}_{\alpha',\varphi} + \frac{(\alpha|\varphi)}{2}\oint_C \tau^*\mathcal{W}_{\varphi,\beta'} + \frac{(\alpha|\varphi)(\beta|\varphi)}{4}\oint_C \tau^*\mathcal{W}_{\varphi,\varphi}$$

where we used that $\oint_C \tau^*\mathcal{W}_{\alpha',\beta'} = 0$. The last and the first integrals can be computed by respectively Lemma 4.3 and Lemma 4.4. For the middle one, note that changing the variables $(t,x) \to (t, x+\mu-\lambda)$ gives an integral along a loop $\widetilde{C}$ based at (t,μ) which can be computed again by Lemma 4.4

$$\oint_C \tau^*\mathcal{W}_{\varphi,\beta'}(\cdot,\mu-\lambda) = \oint_{\widetilde{C}} \tau^*\mathcal{W}_{\beta',\varphi}(\cdot,\lambda-\mu) = -2\Omega_{\beta',\varphi}(t,\mu,\lambda).$$

In other words, we have

$$\oint_C \tau^*\mathcal{W}_{\alpha,\beta} = -(\beta|\varphi)\Omega_{\alpha',\varphi}(t,\lambda,\mu) - (\alpha|\varphi)\Omega_{\beta',\varphi}(t,\mu,\lambda) - \pi\sqrt{-1}(\alpha|\varphi)(\beta|\varphi).$$

Therefore, (4.7) takes the form

$$(\alpha|\varphi)\Big(\Omega_{\varphi,\beta'}(t,\lambda,\mu) - \Omega_{\beta',\varphi}(t,\mu,\lambda)\Big) - \pi\sqrt{-1}(\alpha|\varphi)(\beta|\varphi).$$

Recalling Lemma 4.2 we get

$$\begin{aligned}&-2\pi\sqrt{-1}(\alpha|\varphi)\,\mathrm{SF}(\varphi,\beta') - \pi\sqrt{-1}(\alpha|\varphi)(\beta|\varphi)\\ &\quad = -2\pi\sqrt{-1}(\alpha|\varphi)\,\mathrm{SF}(\varphi,\beta) + \pi\sqrt{-1}(\alpha|\varphi)(\beta|\varphi)\Big(\mathrm{SF}(\varphi,\varphi) - 1\Big).\end{aligned}$$

Since $\mathrm{SF}(\varphi,\varphi) = 1$, the above number is an integer multiple of $2\pi\sqrt{-1}$. Q.E.D.

§5. VOA representations

In this section we would like to explain the origin of the phase factors and the importance of Theorem 2.3 from the point of view of the representation theory of lattice VOAs.

5.1. The tame Fock space

Let us denote by

$$\mathfrak{W} := \Big(\bigoplus_{m=0}^{\infty} \mathcal{H}^{\otimes m} \otimes \mathbb{C}((\hbar)) \Big) / I$$

the Weyl algebra of $\mathcal{H}$, where I is the two-sided ideal generated by

$$v \otimes w - w \otimes v - \Omega(v, w)\hbar, \quad v, w \in \mathcal{H}.$$

Furthermore we introduce the *Fock space* $\mathbb{V} = \mathfrak{W}/\mathfrak{W}\mathcal{H}_+$. For any $\mathbb{C}$-algebra A, let us denote by $\widehat{\mathbb{V}}_A$ the completion of $\mathbb{V} \otimes A$ corresponding to the filtration $\{\mathcal{H}_-^m \mathbb{V} \otimes A\}_{m=0}^{\infty}$. When $A = \mathbb{C}$ then the index A will be omitted.

As a vector space the Weyl algebra $\mathfrak{W}$ is isomorphic to

$$\bigoplus_{m',m''=0}^{\infty} \mathcal{H}_-^{\otimes m'} \otimes \mathcal{H}_+^{\otimes m''} \otimes \mathbb{C}((\hbar)).$$

We introduce the *tame Weyl algebra*[1]

$$\mathfrak{W}_{\rm tame} \subset \widehat{\mathfrak{W}} := \prod_{m',m''=0}^{\infty} \Big(\underbrace{\mathcal{H}_- \widehat{\otimes} \cdots \widehat{\otimes} \mathcal{H}_-}_{m'} \widehat{\otimes} \underbrace{\mathcal{H}_+ \widehat{\otimes} \cdots \widehat{\otimes} \mathcal{H}_+}_{m''} \widehat{\otimes} \mathbb{C}((\hbar)) \Big),$$

where $\widehat{\otimes}$ is the completion of the tensor product induced by the filtrations $\{z^{-m}\mathcal{H}_-\}_{m=0}^{\infty}$, $\{z^m \mathcal{H}_+\}_{m=0}^{\infty}$, and $\{\hbar^m \mathbb{C}((\hbar))\}_{m=0}^{\infty}$ of respectively $\mathcal{H}_-$, $\mathcal{H}_+$, and $\mathbb{C}((\hbar))$. Every element of $\widehat{\mathfrak{W}}$ is an infinite series of monomials of the type $\hbar^{g-1} v \otimes w$, where $g \in \mathbb{Z}$, v and w have the form respectively $v_1 z^{-k_1-1} \otimes \cdots \otimes v_{m'} z^{-k_{m'}-1}$ and $w = w_1 z^{\ell_1} \otimes \cdots \otimes w_{m''} z^{\ell_{m''}}$. The tame Weyl algebra $\mathfrak{W}_{\rm tame}$ consists of those series for which the non-zero monomials satisfy the inequality

$$k_1 + \cdots + k_{m'} - m' \leq 3(g - 1 + m''/2).$$

Let us point out that if $m'' = 0$, then this is the notion of a tame function introduced in [10]. One can check that $\mathfrak{W}_{\rm tame}$ is in fact an algebra

[1]This notion was invented by H. Iritani.

with the product induced from $\mathfrak{W}$. Finally, let us introduce also the *tame* Fock space $\mathbb{V}_{\rm tame} := \mathfrak{W}_{\rm tame}/\mathfrak{W}_{\rm tame}\mathcal{H}_+$.

5.2. Heisenberg fields and Vertex operators

Let us denote by $\mathcal{O}$ the algebra of analytic functions on the monodromy covering space $\widetilde{S}'$ with at most polynomial growth at $\lambda = \infty$. By definition (and since the Gauss–Manin connection has regular singularities) all period vectors $I^{(m)}_\alpha \in \mathcal{O}\otimes H$. Let us define the following set of linear operators $\mathbb{V}_{\rm tame} \to \widehat{\mathbb{V}}_{\mathcal{O}}$:

$$\phi_a(t,\lambda) = \hbar^{-1/2}\partial_\lambda \mathbf{f}_a(t,\lambda;z), \quad a\in\mathfrak{h}^* \tag{5.1}$$

and

$$\Gamma^\alpha(t,\lambda) = e^{\hbar^{-1/2}\mathbf{f}_\alpha(t,\lambda;z)_-}e^{\hbar^{-1/2}\mathbf{f}_\alpha(t,\lambda;z)_+}, \quad \alpha\in Q, \tag{5.2}$$

where $\pm$ denotes the projection on $\mathcal{H}_\pm$. If we want to define a VOA-representation using the above operators we need to define a composition of vertex operators. Let us first compose two vertex operators ignoring the possible divergence issues. We get

$$\Gamma^\alpha(t,\lambda)\Gamma^\beta(t,\mu) = e^{\Omega_{\alpha,\beta}(t,\lambda,\mu)} :\Gamma^\alpha(t,\lambda)\Gamma^\beta(t,\mu):, \tag{5.3}$$

where $: :$ is the normally ordered product for which the action of all elements of $\mathcal{H}_+$ precedes the action of $\mathcal{H}_-$. The problem is that in general we can not exponentiate elements of the ring $\mathbb{C}((\mu^{-1/|\sigma|}))((\lambda^{-1/|\sigma|}))$. To offset this difficulty we recall that the series (2.10) is convergent for all $(t,\lambda,\mu)\in D_\infty - \{\lambda=\mu\}$. Moreover, using the integral presentation (2.13) we can extend analytically $\Omega_{\alpha,\beta}(t,\lambda,\mu)$ with respect to $(t,\mu)\in S'$ provided λ is sufficiently close to μ. Therefore, the *phase factor*

$$B_{\alpha,\beta}(t,\lambda,\mu) := \exp\left(\Omega_{\alpha,\beta}(t,\lambda,\mu)\right) \tag{5.4}$$

has a convergent Laurent series expansion in $(\mu-\lambda)$ with coefficients multi-valued analytic functions on S', i.e., the coefficients are analytic on the universal cover of S'. According to Theorem 2.3 the coefficients are analytic on the *monodromy* covering space $\widetilde{S}'$. In other words we have the following proposition.

Proposition 5.1. *The phase factor $B_{\alpha,\beta}(t,\lambda,\mu)$ has a Laurent series expansion in $\mu-\lambda$, whose coefficients are elements of $\mathcal{O}$.*

5.3. Applications to VOA

Let us denote by

$$V_Q := \mathrm{Sym}(\mathfrak{h}^*[s^{-1}]s^{-1}) \otimes \mathbb{C}[Q]$$

the vector space underlying the lattice vertex algebra corresponding to the Milnor lattice Q. Note that we did not put a VOA structure on V_Q yet. Following [2], we can introduce a family of products $a_{(k)}b$, $a, b \in V_Q$, $k \in \mathbb{Z}$ and a linear map

$$X : V_Q \to \mathrm{Hom}_{\mathbb{C}}(\mathbb{V}_{\mathrm{tame}}, \widehat{\mathbb{V}}_{\mathcal{O}}),$$

defined recursively by the initial conditions

$$X(as^{-1} \otimes 1) := \phi_a, \quad X(1 \otimes e^{\alpha}) := \Gamma^{\alpha}, \tag{5.5}$$

and the *operator product expansion* (OPE) formula

$$X_t(a_{(M-k-1)}b, \lambda)\mathcal{A} = \frac{1}{k!}\partial_\mu^k \left((\mu - \lambda)^M X_t(a, \mu) X_t(b, \lambda)\mathcal{A}\right)\Big|_{\mu=\lambda}, \tag{5.6}$$

where $a, b \in V_Q$ and $\mathcal{A} \in \mathbb{V}_{\mathrm{tame}}$ are arbitrary, the number M is sufficiently large, $k \geq 0$, and we denote by $X_t(a, \lambda)$ the value of the operator $X(a)$ at a point $(t, \lambda) \in S'$. One has to check that the above definition is correct and that all the fields $X_t(a, \lambda)$, $a \in V_Q$, are mutually local. Let us point out that the products $a_{(k)}b$ will depend on the choice of a point on $\widetilde{S}'$, i.e., a point (t, λ) with a reference path. Our expectation is that V_Q has a unique structure of a VOA, s.t., the products defined via the OPE formula coincides with the Borcherds' products. Moreover, the VOA structure on V_Q is isomorphic to the lattice VOA structure (see [13] for some background). The details will be presented elsewhere.

§Appendix A. Jonquière's inversion formula

We would like to give a proof of formula (4.3), because as it was pointed out by the Wikipedia article about polylogorithms, some authors are not very careful about the choice of the branch of $\log x$, while for our purposes choosing the correct branch is essential.

Let us assume that C' intersects the real axis once and that $\mathrm{Im}(x) \neq 0$. We can always achieve this by replacing C' with a homotopy equivalent path and by deforming slightly x. Put

$$f_p(x) := \mathrm{Li}_p(x) + (-1)^p \mathrm{Li}_p(1/x).$$

Since $x\partial_x f_p(x) = f_{p-1}(x)$, we get that $f_p(x)$ is a polynomial in $\log x$. Let us write it as

$$f_p(x) =: -\frac{(2\pi\sqrt{-1})^p}{p!} h_p\Big(\frac{\log x}{2\pi\sqrt{-1}}\Big).$$

In order to determine the polynomials h_p, let us see what happens when we analytically continue x along a closed loop inside the unit disk going counterclockwise around 0. There are 4 cases to be considered, depending on whether C' intersects $(1,\infty)$ or not and whether $\mathrm{Im}(x) > 0$ or < 0. Let us analyze the case: C' does not intersect $(1,\infty)$ and $\mathrm{Im}(x) < 0$. The remaining cases are similar. Note that in this case $\log x = \mathrm{Log}\, x + 2\pi\sqrt{-1}$. We have the following integral representation

$$\mathrm{Li}_p(x^{-1}) = \frac{x^{-1}}{\Gamma(p)} \int_1^{\infty} \frac{(\mathrm{Log}\ z)^{p-1}}{z - x^{-1}} \frac{dz}{z}.$$

When x is varying around 0, the point x^{-1} will vary clockwise along a closed loop C around the unit disk. Let y be a point on C which we reach slightly after we cross the interval $(1,\infty)$. Using the Cauchy residue theorem, we get

$$\mathrm{Li}_p(y) = \frac{y}{\Gamma(p)} \int_1^{\infty} \frac{(\mathrm{Log}\ z)^{p-1}}{z - y} \frac{dz}{z} + \frac{2\pi\sqrt{-1}}{\Gamma(p)} (\mathrm{Log}\ y)^{p-1}.$$

Extending analytically the above identity as y varies along the remaining part of C, we get the following formula

$$\begin{aligned} \mathrm{a.c.} f_p(x) - f_p(x) &= (-1)^p \Big(\mathrm{a.c.}\, \mathrm{Li}_p(x^{-1}) - \mathrm{Li}_p(x^{-1})\Big) \\ &= -\frac{2\pi\sqrt{-1}}{\Gamma(p)} \Big(-\mathrm{Log}\ x^{-1} + 2\pi\sqrt{-1}\Big)^{p-1}. \end{aligned}$$

Expressing f_p in terms of h_p we arrive at the following difference equation

$$h_p(L+1) - h_p(L) = pL^{p-1}, \quad L := \log x/2\pi\sqrt{-1}, \quad p \geq 1.$$

On the other hand, since $x\partial_x f_p(x) = f_{p-1}(x)$ we also have $\partial_L h_p(L) = p h_{p-1}(L)$. The solution to these relations is unique and it is given by the Bernoulli polynomials.

§ Acknowledgements

I am thankful to K. Hori, C. Li, Si Li, and K. Saito for organizing the workshop "Primitive Forms and Related Subjects" (Kavli IPMU, Feb/2014) and creating an inspiring environment. I am especially thankful to S. Galkin and H. Iritani for pointing out that Dubrovin's classification of semi-simple Frobenius manifolds can be used to extend the domain of the period vectors. This work is supported by JSPS Grant-In-Aid 26800003 and by the World Premier International Research Center Initiative (WPI Initiative), MEXT, Japan.

References

[1] V. I. Arnold, S. M. Gusein-Zade, A. N. Varchenko. *Singularities of differentiable maps. Vol.* II. *Monodromy and asymptotics of integrals.* Monographs in Mathematics, 83. Birkhäuser Boston, Inc., Boston, MA, 1988.

[2] B. Bakalov and T. Milanov. *W-constraints for the total descendant potential of a simple singularity.* Compsitio Math. 149 (2013), 840–888.

[3] R. Borchards. *Vertex algebras, Kac–Moody algebras, and the Monster.* Proc. Natl. Acad. Sci. USA, Vol. 83 (1986), 3068–3071.

[4] R. Borchards. *Monstrous moonshine and monstrous Lie superalgebras.* Invent. Math., Vol. 109 (1992), 405–444.

[5] B. Dubrovin. *Geometry of 2D topological field theories.* In: "Integrable systems and quantum groups" (Montecatini Terme, 1993), 120–348, Lecture Notes in Math., 1620, Springer, Berlin, 1996.

[6] B. Dubrovin, Y. Zhang. *Normal forms of hierarchies of integrable PDEs, Frobenius manifolds and Gromov-Witten invariants.* a new 2005 version of arXiv:math/0108160v1, 295 pp.

[7] E. Frenkel, A. Givental and T. Milanov, *Soliton equations, vertex operators, and simple singularities.* Funct. Anal. Other Math., DOI 10.1007/s11853–010–0035–6

[8] A. Givental. *Semisimple Frobenius structures at higher genus.* IMRN (2001), no. 23, 1265–1286.

[9] A. Givental. *Gromov–Witten invariants and quantization of quadratic Hamiltonians.* Mosc. Math. J. 1 (2001), 551–568.

[10] A. Givental. *A_{n-1} singularities and nKdV Hierarchies.* Mosc. Math. J. 3 (2003), no. 2, 475–505.

[11] C. Hertling. *Frobenius Manifolds and Moduli Spaces for Singularities.* Cambridge Tracts in Mathematics, 151. Cambridge University Press, Cambridge, 2002. x+270 pp.

[12] A. Jonquière. *Notes sur la série (polylogarithm).* Bull. Soc. Math. France 17 (1889), 142–152.

[13] V. G. Kac, *Vertex algebras for beginners*. University Lecture Series, vol. 10, Amer. Math. Soc., Providence, RI, 1996; 2nd ed., 1998.

[14] V.G. Kac and M. Wakimoto, *Exceptional hierarchies of soliton equations*. In: "Theta functions—Bowdoin 1987" (Brunswick, ME, 1987), 191–237, Proc. Sympos. Pure Math., 49, Part 1, Amer. Math. Soc., Providence, RI, 1989. Phys. Lett. B 257 (1991), 329–334.

[15] T. Milanov. *The Eynard–Orantin recursion for ADE singularities*. Commun. Number Theory Phys. 9 (2015), no. 4, 707–739.

[16] T. Milanov. *Lectures on Painleve property for semi-simple Frobenius manifolds*. arXiv:1702.06630

[17] K. Saito. *On periods of primitive integrals I*. Preprint RIMS (1982).

[18] K. Saito and D. Yoshii. *Extended affine root systems IV* (*Simply-laced elliptic Lie algebras*). Publ. RIMS, Kyoto Univ., Vol. 36 (2000), 385–421.

[19] K. Saito and A. Takahashi. *From primitive forms to Frobenius manifolds*. Proc. Sympos. Pure Math. 78 (2008), 31–48.

[20] M. Saito, *On the structure of Brieskorn lattice*. Ann. Inst. Fourier **39** (1989), 27–72.

[21] P. Slodowy. *Beyond Kac–Moody algebras and inside*. Proc. Can. Math. Soc., Vol. 5 (1986), 361–370.

[22] C. Teleman. *The structure of 2D semi-simple field theories*. Invent. Math. 188 (2012), no. 3, 525–588.

Kavli IPMU (WPI)
The University of Tokyo
Kashiwa
Chiba 277-8583
Japan
E-mail address: todor.milanov@ipmu.jp

Advanced Studies in Pure Mathematics 83, 2019
Primitive Forms and Related Subjects — Kavli IPMU 2014
pp. 327–355

On the homological algebra of relative symplectic geometry

Daniel Pomerleano

Abstract.

This note discusses certain deformation theoretic aspects of the symplectic topology of pairs (Y, D), where Y is a smooth projective variety and D is an ample simple normal crossings divisor. Motivated by homological mirror symmetry, we formulate a precise prediction concerning the symplectic cohomology of the affine variety $Y \setminus D$ for a wide class of examples. We show how our answer unites various mirror symmetry predictions in the literature.

§1. Introduction

Mirror symmetry at its most naive level is about the comparison of genus-zero cohomological field theories arising from singularity theory with those arising in symplectic topology. On the algebro-geometric side, it is natural to consider a "globalization" of the setup of singularity theory. Let R denote a complete local ring (R, m) such that $R/m = \mathbb{C}$. A *Landau-Ginzburg model* (X_R, f) is a pair, where X_R is an R-variety whose affinization morphism

$$\pi : X_R \to Spec(\Gamma(\mathcal{O}_{X_R}))$$

is proper and $f \in \mathcal{A}_R := \Gamma(\mathcal{O}_{X_R})$. The mirror partner to such an LG-model is a *pair* (Y, D), where Y is a compact projective variety and D is a simple normal crossings divisor which supports an ample (or at least NEF) divisor [Aur].

The situation is easiest to describe when D is an anti-canonical divisor, when, roughly speaking, we expect the variety X to be mirror

Received March 27, 2015.
Revised November 10, 2017.
2010 *Mathematics Subject Classification.* 53D37, 53D40.
Key words and phrases. Mirror symmetry, symplectic aspects, homological mirror symmetry, Floer homology and cohomology, symplectic aspects.

to the non-compact Calabi-Yau variety $Y^0 = Y \setminus D$. The function f then corresponds to a Hochschild cohomology class in $HH^0(X_R) \cong \mathcal{A}_R$, which can be viewed as mirror to the deformation of a version of the Fukaya category of Y^0, $Fuk(Y^0)$, given by compactification by D. The form of the super potential f is known to be closely related to the structure of the divisor D.

For example, consider the mirror to $(\mathbb{C}P^n, D)$, where D corresponds to the toric anti-canonical divisor. In this case, $R = \mathbb{C}[[q]]$ and after base changing to $\mathbb{C}((q))$, the mirror pair is given by [Aur]

$$\left((\mathbb{C}^*)^n_q := (\mathbb{C}^*)^n \times Spec(\mathbb{C}((q))), z_1 + z_2 + \cdots + z_n + \frac{q}{z_1 z_2 \cdots z_n}\right)$$

If we were instead to consider the case of $(\mathbb{C}P^n, E)$ where E is a smooth Calabi-Yau, the mirror would then be a pair (X, f), where the map

$$f : X \to \mathbb{A}^1_q := \mathbb{C} \times Spec(\mathbb{C}((q)))$$

is a *proper* fibration [AurKatOrl], [CarPumSie].

For more general pairs, the mirror to Y^0 could itself be a Landau-Ginzburg model, and the setup needs to be suitably modified. We will see interesting examples of this later. However, we will not attempt to formulate a precise version of the mirror symmetry conjecture because this would involve addressing a large number of subtle problems such as how to construct mirror pairs, how to construct cohomological field theories for general Landau-Ginzburg models, etc. Instead, this note remains entirely on "one side" of the mirror and aims to draw attention to some interesting implications that this mirror symmetry proposal should have for the symplectic geometry of such pairs (Y, D).

The pieces of the more developed theory in the "absolute" or closed case which will serve as the blueprint for our picture are summarized in section 2. In sections 3 and 4, we review symplectic cohomology as well as Seidel and Sheridan's precise formulation of the principle that compactifications give rise to deformations of the Fukaya category.

Sections 5 and 6 represent the main novelty of this paper. In them, we formulate a fairly precise prediction concerning the symplectic cohomology of Y^0, $SH^*(Y^0)$, for such pairs. This is of interest because this invariant is notoriously difficult to compute directly and there are relatively few direct symplectic computations in the literature (although see [EtguLek], [Pas], [Ueb1] for some interesting cases). Throughout the text, we compare our prediction against various predictions of mirror symmetry, some of which are partially proven.

In spite of the narrower scope of our investigation, a systematic treatment of these questions nevertheless requires one to address a large number of foundational geometric and analytic issues. Work is in progress by Sheel Ganatra and the author (see [GanPom] for some partial progress) to develop the necessary tools to study these problems systematically. There is also work in progress by Luis Diogo from the perspective of symplectic field theory to address the case when D is a smooth ample divisor [Diogo] as well as work of Borman and Sheridan [BorShe] which discusses the relationship between symplectic cohomology of $Y^{\complement}$ and quantum cohomology of the compactification Y.

However, it would be inappropriate to report on these developments until definitive results have been obtained. Moreover, the most interesting parts (section 6) of the anticipated theory remain somewhat mysterious. This text, while aiming to be logically independent of the above works in progress, explores these phenomena through examples. As a consequence, even the definitions in the "semi-positive" case are somewhat tentative. The author apologizes to the reader seeking a systematic development and hopes that the material presented is stimulating nevertheless.

§2. Compact symplectic manifolds

Let us briefly recall some features of the symplectic topology of closed symplectic manifolds. Let (Y, ω) be a closed symplectic manifold. By [FukOhOnoOht] (see also their paper in this volume), we can associate a Fukaya category, $\mathcal{F}(Y)$, to Y. The category $\mathcal{F}(Y)$ is an A_∞-category over the Novikov field, $\Lambda_{\mathbb{R}}$, with objects Lagrangian submanifolds decorated with extra structure.

We will be mostly concerned with the "closed-string state" or Hamiltonian Floer cohomology, $HF^*(Y)$. Let $H_t : \mathbb{S}^1 \times Y \to \mathbb{R}$ be a time-dependent Hamiltonian. We denote by $H_S(Y) \subset H_2(Y)$ the Hurewicz subgroup of $H_2(Y)$. The symplectic form defines a homomorphism $\phi : H_S(Y) \to \mathbb{R}$. Let $\tilde{\mathcal{L}}_0 Y$ denote the space of pairs $[x, w]$ where $x : \mathbb{S}^1 \to Y$ and w is a bounding disc up to the equivalence relation $[x, w_1] \sim [x, w_2]$ if $\omega(w_1 \# \bar{w_2}) = 0$. This defines a covering space with deck group $H_S(Y)/ker(\phi)$.

We have a functional $\mathcal{A}_H : \tilde{\mathcal{L}}_0 Y \to \mathbb{R}$

$$\mathcal{A}_H([x, w]) := -\int w^*\omega - \int H(x(t))$$

Then the Floer cohomology of this Hamiltonian is, roughly speaking, the Novikov (co)homology of this one-form. We define the Hamiltonian vector-field of H to be the unique vector-field such that $\omega(Y_H, -) = dH(-)$. Critical points of the functional $\mathcal{A}_H$ are time-one orbits of the Hamiltonian vector-field Y_H with respect to the symplectic form together with a bounding disc w. After a small perturbation, we can assume that the time-one orbits are all non-degenerate. We let $CF^*(Y, H)$ be the chain complex of formal sums

$$\sum_{[x,w]} a_{[x,w]}[x, w], \; a_{[x,w]} \in \mathbb{Q}$$

satisfying the support condition that for any $\lambda \in \mathbb{R}$, there are only finitely many nonzero $a_{[x,w]}$ with $\mathcal{A}_H([x, w]) \geq \lambda$. Any choice of cappings for our orbits will make this chain complex a finitely generated free module over a suitable Novikov ring $\Lambda_{(M,\omega)}$ (see for instance [HofSal]). For relationships with Fukaya categories it is convenient to work over the Novikov field, $\mathbb{K} = \Lambda_{\mathbb{R}}$ and we extend scalars to this field under the natural mapping $\Lambda_{(M,\omega)} \to \Lambda_{\mathbb{R}}$. Next, we pick a compatible time dependent complex structure J_t. For any two generators, $[x_0, w_0]$, $[x_1, w_1]$, the Floer differential counts isolated solutions to Floer's equation $\Sigma := \mathbb{R} \times \mathbb{S}^1 \to Y$,

$$\begin{cases} u \colon \Sigma \to Y, \\ \lim_{s \to -\infty} u(s, -) = x_0 \\ \lim_{s \to \infty} u(s, -) = x_1 \\ (du - Y_H \otimes dt)^{0,1} = 0. \end{cases} \tag{1}$$

which lift to gradient trajectories between $[x_0, w_0]$ and $[x_1, w_1]$ in $\tilde{\mathcal{L}}_0 Y$. In the case when $H_S(Y) = 0$, which we will call the "topological setting", a rigorous construction was given by Floer. Here, there is a unique equivalence class of caps for any orbit x and it is not necessary to work with Novikov coefficients. Let $\mathcal{M}(x_0, x_1)$ denote the moduli space of Floer trajectories connecting x_0 and x_1 as above. We have the following "transversality" result, which is essential for a rigorous construction of Floer cohomology.

Theorem 1. *For a C^∞ generic complex structure this is a moduli space of dimension $i(x_0, x_1)$, where $i(x_0, x_1)$ is a certain relative Conley-Zehnder type index* [FloHofSal]. *There is an $\mathbb{R}$ action on $\mathcal{M}(x_0, x_1)$ given by domain translation. We have that for a C^∞ generic complex structure the quotient space $\mathcal{M}(x_0, x_1)/\mathbb{R}$ is a manifold of dimension $i(x_0, x_1) - 1$.*

This allows us to define the Floer differential via the formula

$$\partial x_0 = \sum_{x_1} |\mathcal{M}(x_0, x_1)/\mathbb{R}| x_1 \tag{2}$$

whenever $i(x_0, x_1) = 1$. The symbol $||$ is used to note that the elements are counted with appropriate signs. The above theorem, together with a standard Gromov compactness result allows one to prove that the Floer differential is well defined, e.g. satisfies $\partial^2 = 0$. Namely, we have that:

Theorem 2. *For a generic J_t, suppose $i(x_0, x_1) = 2$. Then the moduli space $\mathcal{M}(x_0, x_1)/\mathbb{R}$ can be compactified to a manifold with boundary $\overline{\mathcal{M}(x_0, x_1)/\mathbb{R}}$ whose boundary is precisely*

$$\partial\overline{\mathcal{M}(x_0, x_1)/\mathbb{R}} = \coprod_y \mathcal{M}(x_0, y)/\mathbb{R} \times \mathcal{M}(y, x_1)/\mathbb{R}$$

where $i(x_0, y) = 1$.

A similar argument shows that the Floer cohomology is independent of H and J. In the general case, the relative index $i(x_0, x_1)$ depends on the relative homotopy class of the Floer curve. However, this poses no essential difficulty to generalizing Theorem 1. It is much more problematic to generalize Theorem 2. In the general case (where we don't impose $H_S = 0$), at the boundary of the moduli space, we may have certain limiting curves with components which are J-holomorphic spheres for almost complex structures that are *domain independent*, which makes it difficult to achieve transversality, particularly in the presence of multiply covered J-holomorphic spheres. One must argue that these configurations do not in fact contribute and the equation $\partial^2 = 0$ still holds. This is relatively straightforward in the the next least difficult case—the semi-positive case.

Definition 2.1. *A symplectic manifold (Y, ω) is semi-positive if for ever every $A \in \pi_2(Y)$ such that $3 - n \leq c_1(A) < 0$, we have that $\omega(A) \leq 0$.*

This case includes the case of Calabi-Yau projective varieties. In [HofSal], Hofer and Salamon show that the possible sphere bubbling occurs in high codimension and hence does not arise at all in the 0 and 1 dimensional moduli spaces under consideration. The completely general case was treated by [FukOno] (see also the recent [Par]), who developed a difficult theory of virtual techniques to handle this type of bubbling.

Solutions to Floer's equation are very difficult to compute explicitly, so Floer cohomology is impossible to compute directly in even the simplest examples. Nevertheless, using degeneration techniques, Fukaya and Ono [FukOno] have demonstrated an isomorphism:

$$H^*(Y, \Lambda_{\mathbb{R}}) \cong HF^*(Y, H)$$

This extends a foundational result of Floer's when $H_S(Y) = 0$. The definition of Floer cohomology is intimately connected with the dynamics of the Hamiltonian vector-field Y_H. For example, *Arnold's conjecture* on the number of time-one periodic orbits of a Hamiltonian vector-field is an immediate corollary of this isomorphism. In the semi-positive case, there is a particularly elegant way to realize this isomorphism. There is a chain map [McDSal], [PiuSalSch]:

$$\mathcal{PSS} : QH^*(Y) \to CF^*(Y, H)$$

Let $S = \mathbb{C}$ equipped with a marked point $z_0 = 0$ and with fixed cylindrical coordinates around the "cylindrical end" $U = |z| \geq 1$. Given this data, we consider a surface dependent Hamiltonian $H_{\mathbb{C}}$, which *interpolates* between the "holomorphic picture" and Floer's equation—namely it is uniformly 0 near z_0 and agrees with the usual Hamiltonian vector-field on the strip like end U. The $\mathcal{PSS}$ map counts solutions to this equation which pass through a cycle in Y at $z_0 = 0$ and which are asymptotic to Hamiltonian orbits x along the end.

Remark 2.1. *To this naive author, the PSS-isomorphism for completely general compact symplectic manifolds seems to be within range of the advanced transversality techniques of* [FukOno], [FukOhOnoOht], *or* [Par]. *The author only knows of written accounts in the technically simpler semi-positive case. As mentioned before, the isomorphism (and Arnold's conjecture) have been established by Fukaya and Ono by other means.*

There is also a map defined using solutions to a modified version of Floer's equation:

$$\mathcal{CO} : CF^*(Y, H) \to HH^*(\mathcal{F}(Y))$$

Given a Hamiltonian orbit x for this time-dependent Hamiltonian, the image of x under the above map is obtained by counting (with suitable signs) perturbed pseudo-holomorphic discs with an interior puncture converging to x, and an arbitrary number of boundary marked points along Lagrangians in Y. Composing these two maps, we obtain a map:

$$\mathcal{BB} := \mathcal{CO} \circ \mathcal{PSS} : QH^*(Y) \to HH^*(\mathcal{F}(Y))$$

This map coincides with the "bulk-boundary" map introduced by [FukOhOnoOht]. The map $\mathcal{BB}$ has a very natural geometric interpretation. Namely this map can be defined at the chain level by counting (perturbed) holomorphic discs with boundary Lagrangian labels and one interior marked point passing through a cycle in Y.

Floer cohomology also comes equipped with very rich algebraic structure, notably a product structure, which is obtained by studying a variant of equation (1) on the *pair of pants* $\Sigma := \mathbb{C}P^1 \setminus \{0, 1, \infty\}$

$$\begin{cases} u\colon \Sigma \to Y, \\ (du - Y_H \otimes \beta)^{0,1} = 0. \end{cases} \tag{3}$$

where β is a suitable closed one-form on the surface. Another elegant feature of the $\mathcal{PSS}$ map is that it clearly intertwines the *quantum product and the Floer product*, a fact which is quite difficult to prove directly.

§3. Geometry of Liouville domains and symplectic cohomology

A Liouville domain $(\bar{Y}, \theta)$ is a compact manifold with boundary equipped with a one form θ.

- $\omega = d\theta$ is a symplectic form.
- the vector field Z determined by $i_Z\omega = \theta$ points strictly outwards along the boundary.

The principal example which will be relevant to us is constructed as follows. Let Y be a projective variety, $D := D_1 \cup \cdots \cup D_k$ a normal crossings divisor and s a section of an ample line bundle $\mathcal{L} := \mathcal{O}(\Sigma_i \kappa_i D_i)$ which vanishes along D (in particular all $\kappa_i > 0$). Consider the one-form $\theta = d^c \log |s|$. An exhaustive function is given by $f_Y = -\log |s|$. For a large enough level set C, $f_Y^{-1}(-\infty, C]$ is a Liouville domain[1] with one form θ.

[1]More precisely, these examples are all examples of Weinstein domains, a very well-behaved class of Liouville domains.

Returning to the general situation, let $(\bar{Y}, \theta)$ be a Liouville domain. We attach a cylindrical end to form a space $Y^0 := \bar{Y} \cup_{\partial \bar{Y}} \partial \bar{Y} \times [1, \infty)$ and denote the canonical coordinate on the cylindrical end by r. We then equip this space with a one-form by extending the form to a form $\hat{\theta} = r(\theta|_{\partial \bar{Y}})$ on the cylindrical end. The pair $(Y^0, \hat{\theta})$ is known as the *Liouville completion* of the Liouville domain.

Definition 3.1. *We say that a family of Hamiltonians $H_\lambda(t, x) : \mathbb{S}^1 \times Y^0 \to \mathbb{R}$ is* admissible *of slope λ if $H_\lambda(t, x) = \lambda r$ near infinity.*

The boundary $\partial \bar{Y}$ is a contact manifold with contact form $\theta|_{\partial \bar{Y}}$. We let λ range over positive values which are not in the set of periods of Reeb orbits on $\partial \bar{Y}$. Next, we choose a family of admissible Hamiltonians H_λ. We may perturb the functions H_λ so that all 1-periodic orbits are non-degenerate, while keeping the functions admissible. We emphasize for the reader who is new to this subject that, perhaps contrary to what one might expect, *this theory is in many respects technically easier than the corresponding compact case.* The first simplifying feature is that for any Hamiltonian H, we may define the action directly on the loop space

$$\mathcal{A}_H : \mathcal{L}Y^0 \to \mathbb{R}$$

without passing to coverings or to the component of contractible loops. The action functional on a loop $x(t) : \mathbb{S}^1 \to Y$ is given by

$$\mathcal{A}_H(x(t)) := -\int x^*(\hat{\theta}) - \int H(x(t))$$

We may therefore form the Floer cohomology groups $HF^*(Y^0, H_\lambda)$ as follows. Let CF^*, be a chain complex

$$CF^* := \bigoplus_{\mathfrak{o}} \mathbb{K} < \mathfrak{o} >$$

where $\mathbb{K}$ is an arbitrary field and we sum over *all* non-degenerate orbits (not just those which are contractible!). We describe the differential on this complex. We must begin by picking an admissible almost complex structure J_t, which is a compatible time dependent almost complex structure satisfying

$$\theta \circ J_t = dr$$

outside of a compact set in Y^0.

We again consider solutions to Floer's equation $\Sigma := \mathbb{R} \times \mathbb{S}^1 \to Y^0$. The moduli space of solutions to (1) is denoted by $\mathcal{M}(x_0, x_1)$ and we again have that for a C^∞ generic admissible complex structure the quotient space $\mathcal{M}(x_0, x_1)/\mathbb{R}$ is of dimension $i(x_0, x_1) - 1$. When compared with the compact case, we see the crucial simplification: *because ω is exact, there are no non-constant J-holomorphic spheres.* So the sphere bubbles which caused so much trouble in the compact case do not exist.

The non-compactness of our space Y^0 does requires a bit of care and is dealt with by observing that the complex structure and Hamiltonians have been carefully chosen so that a maximum principle holds for solutions to Floer's equation [Sei2].[2] The maximum principle ensures that such solutions remain in a compact set K determined by the boundary data. A Gromov compactness theorem due to [FloHof] then ensures that if $i(x_0, x_1) = 1$, the moduli space $\mathcal{M}(x_0, x_1)/\mathbb{R}$ is compact of dimension 0. We may then define the Floer differential via the formula (2).

Suppose $\lambda_1 \leq \lambda_2$, then there are natural continuation maps [Sei2]:

$$\mathfrak{c}_{\lambda_1,\lambda_2} : HF^*(Y^0, H_{\lambda_1}) \to HF^*(Y^0, H_{\lambda_2})$$

Define symplectic cohomology to be the co-limit of this directed system

$$SH^*(Y^0, \hat{\theta}) := \varinjlim_{\lambda} HF^*(Y^0, H_\lambda)$$

It is not hard to prove that this definition is independent of the various choices made along the way [Sei2].

Remark 3.1. *The fact that the above continuation maps exist only when $\lambda_1 \leq \lambda_2$ is another manifestation of the maximum principle. This may seem like a technical issue, but is actually an essential feature of the theory.*

As in the case of closed manifolds, symplectic cohomology is equipped with a wide range of algebraic operations. For example, there is a *pair of pants product* on $SH^*(Y^0)$ given by counting solutions to (3). Let $\mathcal{W}(Y^0)$ [Abo], denote the wrapped Fukaya category, which is the natural open string generalization of the above Floer theory using Hamiltonians of "large growth". We will not go into these developments here except to note that there is a full subcategory of $\mathcal{W}(Y^0)$,

$$j : Fuk(Y^0) \to \mathcal{W}(Y^0)$$

whose objects consist of compact exact Lagrangians equipped with some extra topological data. By analogy with the compact case, there is also

[2]The conventions described here differ in several respects (sign of Hamiltonian vector field, signs in action functional, etc.) from those in [Sei2]. However, the Floer moduli spaces and hence resulting (co)homologies can be canonically identified.

a ring homomorphism

$$\mathcal{CO} : SH^*(Y^0) \to HH^*(\mathcal{W}(Y^0))$$

which is conjectured to be an isomorphism for all affine varieties. Denote by $\mathfrak{co}$ the map

$$\mathfrak{co} := j^* \circ \mathcal{CO} : SH^*(Y^0) \to HH^*(Fuk(Y^0)) \tag{4}$$

This map is not an isomorphism in general, but is for many interesting examples. This seems to be closely related to understanding when "Koszul duality" holds between the wrapped and compact Fukaya categories [EtguLek].

Let us now return to our mirror symmetry heuristics. In this section, we have used conventions that roughly correspond to setting our local ring $R = \mathbb{C}$. Then we have that any mirror to Y^0 will be in general some Landau-Ginzburg model (X, f), where X is projective over $Spec(\mathcal{A})$, and $\mathcal{A}$ is a finitely generated $\mathbb{C}$ algebra. As a consequence of mirror symmetry, we expect an equality of "closed-string" states

$$SH^*(Y^0) \cong \mathbb{H}^*\Big(\bigoplus_i \Lambda^i TX, [f, -]\Big) \tag{5}$$

On the right hand side of this equation, $\mathbb{H}^*$ denotes hypercohomology, $\Lambda^i TX$ denotes the exterior power of the tangent sheaf and $[f, -]$ is Schouten bracket with the function f. It follows from Theorem 3.2.1 of [Gro] that the right hand side of (5) is a finitely generated module over $\mathcal{A}$. This leads to the conjecture that:

Conjecture 3.1 (Version 0.1)**.** *The $SH^*(Y^0)$ is a finitely generated module over a finitely generated ring $\mathcal{A}$ over $\mathbb{C}$.*

Over the next sections, we refine the above conjecture by discussing an analogue of the map $\mathcal{BB}$, the invariants $QH^*(Y, D)$ and a description of the algebra $\mathcal{A}$.

§4. Seidel's picture

In [Sei], the author considers the case of where Y is a Calabi-Yau projective variety, D is a smooth divisor, and Y is equipped with the Fubini-Study symplectic form ω_L induced from the polarizing line bundle $\mathcal{L}$. Under these assumptions, Seidel constructs the *relative Fukaya category* $Fuk(Y, D)$ which is an A_∞ deformation of $Fuk(Y^0)$ over the ring $\mathbb{C}[[t]]$.

The objects of this category $Fuk(Y, D)$ are essentially the same as those in $Fuk(Y^0)$, however morphisms in this category now count holomorphic polygons which map to Y (as opposed to just Y^0) and which may hence intersect D. Polygons which intersect the divisor D with multiplicity d are weighted by a power t^d.

We work with almost complex structures that preserve D, $\mathcal{J}(Y, D)$, and hence the intersection number of any holomorphic map with the divisor must be positive, thus explaining the assertion that $Fuk(Y, D)$ is a deformation over the power series ring, $\mathbb{C}[[t]]$. There is also a generic fiber $Fuk_{gen}(Y, D)$ associated to this A_∞ deformation which lies over $\mathbb{C}[[t, t^{-1}]]$.

The interest in this category is the following conjecture, due to Seidel:

Conjecture 4.1. [Sei] *Suppose in addition that Y is an anti-canonical divisor in a projective variety Z, which is equipped with a Lefschetz pencil $Z \to \mathbb{C}P^1$ such that the zero fiber is Y and the base locus is D, then*

$$Fuk_{gen}(Y, D) \otimes_{\mathbb{C}[[t,t^{-1}]]} \Lambda_{\mathbb{C}} \cong Fuk(Y)$$

Here $\Lambda_{\mathbb{C}}$ denotes the Novikov field over $\mathbb{C}$. Seidel verified this conjecture when Y in a K3 surface in $\mathbb{C}P^3$ (the work of Sheridan should imply the conjecture for all Calabi-Yau hypersurfaces).

We will be most interested in the first order piece of the deformation of $Fuk(Y^0)$, which is a class α in $HH^2(Fuk(Y^0))$. Its description is remarkably similar to the bulk-boundary deformations that we have seen earlier. Namely, consider just as before a disc Σ with Lagrangian labels and an interior marked point z_0 which we can place at zero. The deformation is given by counting maps of $u : \Sigma \to Y$ such that

$$u^{-1}(D) = z_0$$

with multiplicity one. As a result, we obtain a ring homomorphism

$$BB : \mathbb{C}[\alpha] \to HH^*(Fuk(Y^0))$$

This construction can be generalized in various directions. For example, we can relax the condition that Y be Calabi-Yau at the cost that the deformation $Fuk(Y, D)$ will now live over a ring $\mathbb{C}[[t_n]]$, where t_n has non-trivial degree n. More interestingly, we can let D be a normal crossings divisor. A technically convenient setting is the following:

Definition 4.1. *We say that* (Y, D) *is a* Kähler pair *if*

- *There are at least* $n+1$ *smooth components* D_i, *where* n *is the complex dimension of* Y.
- *Each* D_i *is the zero locus of a section* s_i, *such that* $s_i \in \Gamma(\mathcal{E}^{d_i})$, *where* $\mathcal{E}$ *is a single line bundle and* $d_i > 0$.

In this setting, [She] constructs Hochschild classes $\alpha_i \in HH^*(Fuk(Y^0))$ by considering maps u as above such that:

$$u^{-1}(D_j) = \delta_{ij} z_0$$

He also constructs $Fuk(Y, D)$ under the additional assumption that Y is semi-positive.

§5. Logarithmic quantum cohomology: topological case

In this section, we consider a context which is closely connected to Floer's original restriction that $H_S(Y) = 0$. Assume that we have components $D_1, \ldots, D_k$. Let I be an indexing subset of $[1, \ldots, k]$ and $D_I := \cap_{i \in I} D_i$ (we allow $I = \emptyset$ and we set $D_\emptyset = M$). For any nonnegative integer vector $\mathbf{v} = (\mathbf{v}_1, \ldots, \mathbf{v}_k) \in (\mathbb{Z}^{\geq 0})^k$, let $|\mathbf{v}| \subset [1, \ldots, k]$ denote the support of $\mathbf{v}$, i.e. we have $\mathbf{v}_i > 0$ if and only if $i \in |\mathbf{v}|$.

In the Kahler pair setting, the key to proving that the α_i above define Hochschild classes is the fact that for any $J \in \mathcal{J}(M, D)$, any nonconstant J-holomorphic sphere must simultaneously intersect each one of the divisors with positive multiplicities. As the intersection of all of these divisors is empty because our divisor is normal crossings, it is easy to see that no sphere bubbling can occur in these moduli spaces (see the proof of Lemma 5.2. below). The fact that α_i define Hochschild cocycles is an elementary analysis of the boundary of the moduli space (as in Floer's original argument). This leads naturally to the following definition:

Definition 5.1. *A pair* (Y, D) *is* topological *if for generic* $J \in \mathcal{J}(M, D)$,

- *every* J*-holomorphic sphere* $u : \mathbb{C}P^1 \to M$ *must intersect each* D_i *non-negatively.*
- *For any nonconstant* u, *setting* $u \cdot D = \mathbf{v}_u \in (\mathbb{Z}^{\geq 0})^k$, *we have that* $D_{|\mathbf{v}_u|} = \emptyset$.

It follows immediately from the definition that any pair whose underlying variety satisfies $H_S(Y) = 0$ is topological. However there are other interesting cases.

Example 5.1. • *Kähler pairs* (Y, D) *are topological. This is because for any nonconstant sphere* u*, we have* $|\mathbf{v}_u| = [1, \dots, k]$*.*

- *Suppose that* Y *is a hypersurface of sufficiently high degree in* $\mathbb{C}P^n$*,* $n \geq 2$ *and* D *is a hyperplane section of sufficiently high degree. Then for any map* $u : \mathbb{C}P^1 \to Y$ *or* $\mathbb{C}P^1 \to D$*, we have that the expected dimension of the moduli space in any homology class* A *such that* $\omega[A] \geq 0$ *is negative. Thus, we have that for a generic* J*, the moduli space of somewhere injective spheres is empty. In such a setting, there can be no multiply covered curves either, because any curve covers a somewhere injective curve.*

We next recall some algebraic topology relevant to normal crossings compactifications. For each D_I, we have

Definition 5.2. *Define the open stratum associated to* D_I *to be*

$$\mathring{D}_I := D_I \setminus (\cup_{j \notin I}(D_j \cap D_I))$$

Over each D_I, we have a natural torus bundle S_I, which is the T^I bundle associated to the normal bundle to the stratum.

Definition 5.3. *To each open stratum* $\mathring{D}_I$*, we denote the restriction of the* $T^{|I|}$ *bundle* S_I*, to* $\mathring{D}_I$ *by* $\mathring{S}_I$*.*

If $K > I$ in the standard partial ordering, we have canonical (on the level of cohomology) restriction maps

$$r_{I,K} : H^*(\mathring{S}_I, \mathbb{C}) \to H^*(\mathring{S}_K, \mathbb{C})$$

These can either be constructed in an ad-hoc fashion by choosing suitably compatible tubular neighborhoods, or more elegantly by appealing to the *Kato-Nakayama construction* from logarithmic geometry [ACGHOSS], which is a manifold with corners whose depth $|I|$ strata are precisely the relevant torus bundles.

To a topological pair (Y, D) associate a $\mathbb{Z}/2\mathbb{Z}$ graded vector space: $QH^*(Y, D)$ as follows. We begin by fixing variables $t_1, \dots, t_k$, and for a non-negative integer vector $\mathbf{v} = (\mathbf{v}_1, \dots, \mathbf{v}_k)$, denote

$$t^{\mathbf{v}} := t_1^{\mathbf{v}_1} \cdots t_k^{\mathbf{v}_k}.$$

To any I, we let $\mathbf{v}_I$ denote the vector supported on I such that $\mathbf{v}_i = 1$, if $i \in I$. Elements of $QH^*(Y, D)$ will be linear combinations of elements of the form $\alpha t^{\mathbf{v}}$, where α is a cohomology class in $H^*(\mathring{S}_I)$ for some subset $I \subset [1,\ldots,k]$, and $\mathbf{v} = (\mathbf{v}_1,\ldots,\mathbf{v}_k)$ is a vector in Z_I, the set of non-negative integer vectors which are supported on I. Note that if $I = \emptyset$, then $\mathbf{v} = \vec{\mathbf{0}}$ and $\alpha \in H^*(Y^0)$. In other words, we define:

Definition 5.4.

$$QH^*(Y,D) := \bigoplus_{I\subset[1,\ldots,k]} \bigoplus_{\mathbf{v}\in Z_I} H^*(\mathring{S}_I)t^{\mathbf{v}} \tag{6}$$

where $S_\emptyset = Y$, *and* $S_I = \emptyset$ *if the intersection* $\cap_{i\in I} D_i$ *is empty.*

For each non-empty stratum D_I, we get canonical elements of $QH^*(Y,D)$. When $dim(D_I) > 0$, α_I is given by taking $id_I \in H^0(\mathring{S}_I)$ and letting

$$\alpha_I = id_I \cdot t^{\mathbf{v}_I}$$

When $dim(D_I) = 0$, we decompose D_I into a collection of points $D_{I,m}$. We then repeat as before, taking $id_{I,m} \in H^0(\mathring{S}_{I,m})$

$$\alpha_{I,m} = id_{I,m} \cdot t^{\mathbf{v}_I}$$

It was noticed in [She] and [Pom] that Seidel's deformations can be generalized as follows. We can consider Σ a disc with Lagrangian labels and a single interior marked point z_0 as before, and instead count maps such that:

$$u^{-1}(D_j) = d \cdot \delta_{ij} z_0$$

This gives rise to an a priori distinct Hochschild cohomology class $\alpha_{i,d}$. As it turns out, this is not actually a generalization in and of itself because we have the following lemma, which was proven in [She] and in a slightly different setting in [Pom]:

Lemma 5.1. *Let* (Y,D) *be Kähler pair. We have that* $\alpha_{i,d} \cong \alpha_i^d \subset HH^*(Fuk(Y^0))$.

Proof. The proof is by induction and almost identical to the proof by [FOOO] that the bulk-boundary map of section 2 is a ring homomorphism. Namely, one considers an auxiliary moduli space $\mathcal{R}(2,L)$ of discs with Lagrangian labels L and two interior marked points at $z_1 = t$ and $z_0 = -t$. We require that

$$u^{-1}(D_i) = (d-1)z_0 + z_1$$

and as above for $i \neq j$, we have that $u^{-1}(D_j)$ is empty. As $t \to 1$ we have the moduli space corresponding to the product of α_i^{d-1} and α_i in $HH^*(Fuk(Y^0))$. As $t \to 0$, we obtain the moduli space defining

$\alpha_{i,d}$, again due to the absence of bubbling of holomorphic spheres. Thus $\mathcal{R}(2, L)$ gives a cobordism between the moduli spaces defining the two deformations and thus gives their equality. Q.E.D.

We may generalize this idea even further as follows:

Lemma 5.2. *When* (Y, D) *is a topological pair, there is a homomorphism of vector spaces*:

$$\mathcal{BB} : QH^*(Y, D) \to HH^*(Fuk(Y^0)) \tag{7}$$

Proof. Consider as above a surface which is a punctured disc Σ with Lagrangian labels and an interior marked point at the origin z_0. Given a class of the form $\alpha t^{\mathbf{v}}$, we consider maps from Σ such that $u^{-1}(D_i) = \mathbf{v}_i z_0$. Let $\mathcal{R}(\mathbf{v}, L)$ denote this moduli space.

In this setting, we have a "refined evaluation map" at z_0, which measures the $\mathbf{v}$-jet in the direction normal to the stratum D_I. Namely, locally around a point $d_0 \in D_I$, we can find holomorphic coordinates z_i, in the directions normal to D_I such that the projection of the map to the normal direction takes the form:

$$z \to \left(a_i z^{\mathbf{v}_i} + O(z_i^{\mathbf{v}_i+1})\right)$$

Intrinsically, these coefficients a_i can be thought of as living in [Ion]

$$ev_{z_0}^* ND_i \otimes \Omega_{z_0}(\Sigma)^{\mathbf{v}_i}$$

Where Ω_{z_0} is the cotangent bundle at that point. The tangent bundle to Σ at z_0 is canonically trivialized by a tangent vector $r_0 \subset T_{z_0}\Sigma$, which points in the positive real direction. We therefore obtain for every curve an evaluation map

$$Ev_{z_0} : \mathcal{R}(\mathbf{v}, L) \to \mathring{S}_I$$

We can achieve transversality for this higher evaluation map by an argument similar to that of Proposition 6.6 in [CieMoh] (see also section 4 of [GanPom]). We therefore impose an additional constraint on our maps that

$$Ev_{z_0}(u) \in PD([\alpha])$$

where $PD([\alpha])$ is a sufficiently generic cycle Poincaré dual to α. The fact that counting such maps as before defines a Hochschild cohomology class follows by ruling out sphere bubbling exactly as in the proof of Sheridan's theorem above. Namely, any sphere bubble must intersect positively a divisor D_j, for some $j \notin |\mathbf{v}|$. As the holomorphic disc must also intersect D_j non-negatively, such bubbling is ruled out. Q.E.D.

Remark 5.1. *Using the more refined compactness arguments from* [Ion], *one can prove Lemma* 5.2 *for pairs where for generic* $J_I \in \mathcal{J}(D_I, \cup_{j\notin I}(D_j \cap D_I))$, *there are no non-constant holomorphic curves* $u : \mathbb{C}P^1 \to D_I$ *which intersect* $\cup_{j\notin I}(D_j \cap D_I)$ *in less than two distinct points.*

In [GanPom], for any topological pair, we have constructed a map

$$\mathcal{PSS}_{log} : QH^*(Y, D) \to SH^*(Y^0) \tag{8}$$

One can show that (7) factorizes as

$$\mathcal{BB} = \mathfrak{co} \circ \mathcal{PSS}_{log} : QH^*(Y, D) \to HH^*(Fuk(Y^0)) \tag{9}$$

While a detailed proof of this fact does not yet appear in the literature, the proof is again very similar to the closed case reviewed in Section 2. In general we only have a $\mathbb{Z}/2\mathbb{Z}$-grading on $QH^*(Y, D)$. In cases where ω_Y is supported on D, a choice of volume form with poles of order a_i along D_i gives rise to an isomorphism

$$\omega_Y \cong \mathcal{O}\Big(\sum_i -a_i D_i\Big)$$

We may put a $\mathbb{Z}$-grading on $QH^*(Y, D)$ via the formula:

$$|\alpha t^{\mathbf{v}}| = -|\alpha| + 2\sum_i (a_i - 1)\mathbf{v}_i \tag{10}$$

Notice that there can be several gradings on $QH^*(Y, D)$, corresponding to the fact that there can be different "R-charges" on the mirror side. The same choice gives rise to a grading on $SH^*(Y^0)$ and $HH^*(Fuk(Y^0))$. With all of these choices in place, the maps (7) and (8) are graded.

We will now sketch how to (conjecturally) equip $QH^*(Y, D)$ with a ring structure so that the maps (7) and (8) are ring maps. Let $\Sigma_{0,3}$ denote a sphere with the following data:

- Three marked points at $\psi_1 = -1, \psi_2 = 1, \psi_3 = \infty$.
- At each marked point we trivialize the tangent space with a tangent vector that points in the positive real direction.

For each i, we have $\mathbb{P}^1$ bundles over the divisors D_i given by $PD_i := P(ND_i \oplus \mathcal{O}_{D_i})$. Consider three indexing subsets I_1, I_2, I_3 such that $I_3 \leq I_1 \cup I_2$. Let $\mathbb{F}_{I_3}$ denote the $(\mathbb{P}^1)^{|I_3|}$ bundle over D_{I_3} which is given by the fiber product of the $\mathbb{P}^1$ bundles PD_i for each $i \in I_3$:

$$\mathbb{F}_{I_3} := \prod_{D_{I_3}}^{i \in I_3} PD_i|_{D_{I_3}} \tag{11}$$

As usual, we also allow $I_3 = \emptyset$ in which case we understand $\mathbb{F}_{I_3} = Y$. Each of the PD_i for each $i \in I_3$ has standard sections at 0 and ∞ which we denote by $\hat{D}_{i,0}$ and $\hat{D}_{i,\infty}$. We obtain divisors $D_{i,0}$ and $D_{i,\infty}$ in $\mathbb{F}_{I_3}$ which are fiber products of the $\hat{D}_{i,0}$ (respectively $\hat{D}_{i,\infty}$) with the remaining $\mathbb{P}^1$ bundles.[3] In other words,

$$D_{i,0} := \hat{D}_{i,0} \times_{D_{I_3}} \prod_{D_{I_3}}^{j \neq i \in I_3} PD_j|_{D_{I_3}} \tag{12}$$

and similarly for $D_{i,\infty}$. Set $D_\infty = \cup_i D_{i,\infty}$ and $D_0 := \cup_i D_{i,0}$. Let π_{I_3} denote the standard projection $\pi_{I_3} : \mathbb{F}_{I_3} \to D_{I_3}$. For every $j \notin I_3$, we also have a divisor $\mathcal{D}_j := \pi_{I_3}^{-1}(D_j \cap D_{I_3})$. Set $\mathcal{D} = \cup_j \mathcal{D}_j$. For any J, let $\mathcal{S}_J$ denote the normal torus bundle to the stratum $(\cap_{j \in J \setminus I_3} \mathcal{D}_j) \cap_{j \in J \cap I_3} D_{j,0}$ viewed as a subvariety of $\mathbb{F}_{I_3}$ and let $\mathring{\mathcal{S}}_J$ denote the restriction of $\mathcal{S}_J$ to the open piece of the stratum. We may identify the cohomology of the torus bundles $\mathring{\mathcal{S}}_J$ with the cohomology of the corresponding circle bundles $\mathring{S}_{J \cup I_3}$ above in the obvious way and we will do this implicitly in the formulae which follow. Finally, we also let $\mathcal{S}_{I_3,\infty}$ denote the torus bundle to the stratum $D_{I_3,\infty} := \cap_i D_{i,\infty}$ and $\mathring{\mathcal{S}}_{I_3,\infty}$ denote its open part.

Given three vectors $\mathbf{v}_1$, $\mathbf{v}_2$, $\mathbf{v}_3$ supported on I_i, our goal is to use moduli spaces of pseudoholomorphic spheres with three marked points to define maps

$$\Gamma_{\mathbf{v}_i} : H^*(\mathring{S}_{I_1}) \otimes H^*(\mathring{S}_{I_2}) \to H^*(\mathring{\mathcal{S}}_{I_3,\infty}) \tag{13}$$

We begin with the case where I_3 is not equal to either I_1 or I_2. We consider the moduli space of maps $\mathcal{M}_{\mathbf{v}_i}(\Sigma_{0,3}, \mathbb{F}_{I_3})$, $u : \Sigma_{0,3} \to \mathbb{F}_{I_3}$ such that u is not contained in $D_0 \cup D_\infty \cup \mathcal{D}$ and furthermore

$$u \cdot (D_0 \cup \mathcal{D}) = \mathbf{v}_1 \psi_1 + \mathbf{v}_2 \psi_2 \tag{14}$$

$$u \cdot D_\infty = \mathbf{v}_3 \psi_3 \tag{15}$$

By choosing a generic surface dependent almost complex structure $J_{s,t}$, this is a smooth manifold. This manifold is equipped with enhanced

[3]In the case $I_3 = \emptyset$, we have $D_{i,\infty} = D_{i,0} = D_i$.

evaluation maps:

$$Ev : \mathcal{M}_{\mathbf{v}_i}(\Sigma_{0,3}, \mathbb{F}_{I_3}) \to \mathring{\mathcal{S}}_{I_1} \times \mathring{\mathcal{S}}_{I_2} \times \mathring{\mathcal{S}}_{I_3,\infty}$$

Lemma 5.3. *Let $Ev_3 : \mathcal{M}_{\mathbf{v}_i}(\Sigma_{0,3}, \mathbb{F}_{I_3}) \to \mathring{\mathcal{S}}_{I_3,\infty}$ denote the composition of Ev with projection onto the third component. The map Ev_3 is proper.*

Proof. To prove this lemma, consider the projection:

$$\pi_{I_3} : \mathbb{F}_{I_3} \to D_{I_3}$$

Fix a point $z \in \mathring{\mathcal{S}}_{I_3,\infty}$ and consider a configuration u_∞ which is a limit of elements in the above moduli space passing through the point z. Let us assume that the domain of the map is some nodal domain, $\Sigma \cup \overline{\Sigma}_i$ where $\psi_3 \in \Sigma$. Supposed the induced map $\pi_{I_3} \circ u_\infty(\Sigma)$ is a non-constant map. The pair $(D_{I_3}, \cup_{j \notin I_3} D_j \cap D_{I_3})$ is itself a topological pair and therefore any non-constant map $\Sigma \to D_{I_3}$ must have at least two positive intersections with the divisors D_j. It follows from e.g. Equation 5.22 of [Ion] that these must correspond to ψ_1 and ψ_2 and thus the projection of the remaining sphere bubbles must be constant. As these must come with at least three marked points by stability, these also do not exist. Similarly, it is easy to see that in the case where the map $\pi_{I_3} \circ u_\infty$ is constant, the limiting domain must be a connected sphere (with three marked points) and the map u_∞ a fiber sphere. Q.E.D.

By a simple extension of the usual theory of pseudocycles [McDSal], we therefore obtain a map:

$$Ev_* : H^*(\mathring{\mathcal{S}}_{I_1}) \otimes H^*(\mathring{\mathcal{S}}_{I_2}) \to H^*(\mathring{\mathcal{S}}_{I_3,\infty})$$

To obtain a map of the form (13), given classes $\alpha_1 \in H^*(\mathring{S}_{I_1})$ and $\alpha_2 \in H^*(\mathring{S}_{I_2})$, we define a new cohomology class by considering

$$\Gamma_{\mathbf{v}_i}(\alpha_1 \otimes \alpha_2) := Ev_* \left(r_{|\mathbf{v}_1|,|\mathbf{v}_1| \cup |\mathbf{v}_3|}(\alpha_1) \otimes r_{|\mathbf{v}_2|,|\mathbf{v}_2| \cup |\mathbf{v}_3|}(\alpha_2) \right) \tag{16}$$

Finally, we discuss how to define $\Gamma_{\mathbf{v}_i}$ when I_3 is equal to either I_1 or I_2 (or both). Without a loss of generality, assume $I_1 = I_3$. In this case, we consider configurations of "augmented fiber spheres". The maps $\Sigma_{0,3} \to \mathbb{F}_{I_3}$ are themselves very simple. They satisfy $u \cdot \mathcal{D}$ is empty (and hence lie in the fibers of π_{I_3}) as well as

$$u \cdot D_0 = \mathbf{v}_1 \psi_1 \tag{17}$$

$$u \cdot D_\infty = \mathbf{v}'_2 \psi_2 + \mathbf{v}_3 \psi_3 \tag{18}$$

However, we require that there exists an augmentation, namely a map $u_a : S^2 \to Y$ with a marked point at $z_0 = 0$ and at $z_1 = \infty$ such that

$$u_a \cdot D = \mathbf{v}_2' z_0 + \mathbf{v}_2 z_1$$

together with a certain matching condition between the enhanced evaluations at z_0 and ψ_2, Equation 5.25 of [Ion].[4] The map $\Gamma_{\mathbf{v}_i}$ is again defined by pushing forward along the enhanced evaluation map at ψ_3 as above. To prepare for the next definition, we say that $\mathbf{v}_b < \mathbf{v}_c$ if every $\mathbf{v}_{b,i} \leq \mathbf{v}_{c,i}$ and there exists at least one i such that $\mathbf{v}_{b,i} < \mathbf{v}_{c,i}$.

Definition 5.5. *We define*:

$$\alpha_1 t^{\mathbf{v}_1} \cdot \alpha_2 t^{\mathbf{v}_2} = \left(r_{I_1, I_1 \cup I_2}(\alpha_1) \cup r_{I_2, I_1 \cup I_2}(\alpha_2)\right) t^{\mathbf{v}_1 + \mathbf{v}_2} + \sum_{\mathbf{v}_3 < \mathbf{v}_1 + \mathbf{v}_2} \Gamma_{\mathbf{v}_i}(\alpha_1 \otimes \alpha_2) t^{\mathbf{v}_3} \tag{19}$$

For the purposes of this note, the most important term in (19) is the leading term, for reasons we explain below.

Remark 5.2. *As is typical in symplectic topology, there are many possible choices of coefficient rings R that we may work over. A generalization that seems especially important for mirror symmetry is the ring $R = \mathbb{C}[H_2(Y)]$. As an R-module, we let $QH^*_R(Y, D) := QH^*(Y, D) \otimes_{\mathbb{C}} R$. The above multiplicaton rule remains the same, except that we weigh the curves we count by their homology class in Y.*

Here is the precise form of our Conjecture 3.1 for topological pairs:

Conjecture 5.1. *The map $\mathcal{PSS}_{log}$ from* (8) *is an isomorphism of rings.*

Ongoing work with Ganatra aims to establish that (8) is an additive isomorphism. We remark that we have not yet proven that the above multiplication defines an associative ring structure on $QH^*(Y, D)$. This would follow *a posteriori* from the existence of the above ring isomorphism, though it also likely follows from arguments similar to those in [McDSal] for quantum cohomology in the closed case because all of our moduli spaces are cut out transversely. In any case, taking this for granted, the above multiplication rule defines a finitely generated $\mathbb{C}$-algebra structure on $QH^*(Y, D)$. To see this, consider the filtration $F^{\leq p}(QH^*(Y, D))$ which is generated by homogeneous elements $\alpha t^{\mathbf{v}}$ such

[4]We also need to choose certain gluing data — a torsor over $\frac{\mathbb{Z}}{\gcd(\mathbf{v}'_{2,i})\mathbb{Z}}$ which records the number of ways to glue at the marked points ψ_2 and z_0.

that $\sum_i \kappa_i \mathbf{v}_i \leq p$. The associated graded algebra, $QH^{top}(Y,D)$, can be computed purely topologically via the leading term of Equation (19).

It is obvious that the algebra $QH^{top}(Y,D)$ is finitely generated. More precisely, the generators are given by taking a basis $\eta_{I,j}$ for the cohomology of each stratum $H^*(\mathring{S}_I)$ and forming the elements $\eta_{I,j}t^{\mathbf{v}_I} \subset QH^*(Y,D)$. For the formulation in the next section, it is important to notice that if we let $\mathcal{A}$ denote the subalgebra generated by the canonical elements α_I and $\alpha_{I,m}$ (when $dim(D_I)=0$), then in fact we can see that $QH^{top}(Y,D)$ is a finitely generated algebra over the ring $\mathcal{A}$.

The prototypical "topological example", is the pair of pants $\mathcal{P}$, which is the complement of $n+2$ generic hyperplanes in $\mathbb{CP}^n$. It has been achieved much attention recently in view of its role in Sheridan's spectacular proof of homological mirror symmetry for Calabi-Yau hypersurfaces [She]. The mirror to $\mathcal{P}$ is the Landau-Ginzburg model

$$(Spec(\mathbb{C}[z_1, z_2, \cdots, z_{n+2}]), W = z_1 z_2 \cdots z_{n+2}).$$

The space $\mathcal{P}$ is homotopy equivalent to an n-torus with a point removed. Let β_i be a basis for its first cohomology. The Jacobian ring is $\mathcal{J}ac(W) \cong \mathbb{C}[z_1, \cdots, z_{n+2}]/(z_1 z_2 \cdots \widehat{z_i} \cdots z_{n+2})$. Next denote by $\mathbf{v}_i$ the vector $(0, \cdots, 1, \cdots, 0)$, where the 1 is in the i-th position and let α_i be the associated element in $\mathcal{A}$.

As the strata are all connected, the elements α_i generate $\mathcal{A}$ and that the map sending $\alpha_i \to z_i$ induces an isomorphism:

$$\mathcal{A} \cong \mathcal{J}ac(W)$$

In this example, we can make a more precise comparison. There are no non-trivial quantum corrections and we have an isomorphism of rings $QH^*(\mathbb{C}P^n, D) \cong QH^{top}(\mathbb{C}P^n, D)$. On the mirror side we have elements

$$z_j \partial_{z_j} - z_{j+1}\partial_{z_{j+1}} \in H^*\left(T^{poly}_{\mathbb{C}[z_1,\cdots,z_{n+2}]}, [W,-]\right)$$

of the module $H^*(T^{poly}_{\mathbb{C}[z_1,\cdots,z_{n+2}]}, [W,-])$ which generate as an algebra over $\mathcal{J}ac(W)$, where $j \in [1, \cdots, n+1]$. With these observations in place, one can easily check that the map

$$QH^*(\mathbb{C}P^n, D) \to H^*\left(T^{poly}_{\mathbb{C}[z_1,\cdots,z_{n+2}]}, [W,-]\right)$$

defined by sending $\alpha_i \to z_i$ and $\beta_i \to z_i\partial_{z_i} - z_{i+1}\partial_{z_{i+1}}$ is an isomorphism of rings.

§6. Towards the "semi-positive" case of $QH^*(Y, D)$

We next study a case which is reminiscent of the above semi-positive situation in the sense that while there are "quantum" corrections to the differential they are relatively controlled. Let Y be a Fano manifold and D an ample normal crossings divisor such that ω_Y is supported on D with all $\kappa_i = 1$.

Definition 6.1. *We say that such a pair (Y, D) is semi-positive if either*

- *Y is an algebraic surface.*
- *Each D_i is an ample divisor.*

Let us begin with the case of a smooth divisor D. A similar formula along with a sketch of a proof using symplectic field theory appears in [BouEkhEli] and [Diogo]. In this setting, we set:

Definition 6.2.

$$QC^*(Y, D) := \bigoplus H^*(Y^0) \bigoplus_{d \geq 1} (H^*(S^1) \otimes H^*(D))t^d \tag{20}$$

This complex comes equipped with a differential d, which in this special case takes on a relatively simple form. On classes of the form $\alpha \in H^*(Y^0)$, $d(\alpha) = 0$ and we have that $d(\mathrm{id} \otimes \alpha t^d) = 0$ as well. Let e be a generator of cohomology $H^1(S^1)$. We have that

$$d(e \otimes \alpha t^d) = d_A(e \otimes \alpha t^d) + d_S(e \otimes \alpha t^d)$$

where

$$d_A : e \otimes H^*(D)t^d \to \oplus_{d' \leq d} id \otimes H^*(D)t^{d'}$$
$$d_S : e \otimes H^*(D)t^d \to H^*(Y^0)$$

We now explain how to construct these differentials. Let $\mathcal{M}_{2,d}([B])$ denote the relative moduli space of spheres $u : S^2 \to Y$ with a marked point at $z_0 = 0$ and at $z_1 = \infty$, such that $u^{-1}(D) = dz_0$ and which lie in the homology class $[B]$. Let $\mathcal{M}_{1,d}([B])$ denote the analogous moduli space with just one marked point at z_0. The moduli spaces $\mathcal{M}_{2,d}$, respectively $\mathcal{M}_{1,d}$ can be compactified in real codimension 2 either using the Deligne-Mumford compactification (as in absolute Gromov-Witten theory) or following the techniques of [IonPar] (as in relative Gromov-Witten theory). The moduli space $\mathcal{M}_{2,d}$ gives rise to a cohomology level operation:

$$\Gamma_d : H^*(D) \to H^*(Y)$$

given by taking

$$ev_{\infty,*}(ev_0^*(\alpha))$$

We set

$$d_S(e \otimes \alpha t^d) = i^* \circ \Gamma_d(\alpha)$$

where $i : Y^0 \to Y$ is the inclusion. The following trivial observation is sometimes useful in computations

Observation 6.1. If the map $H^*(Y) \to H^*(Y^0)$ is trivial, then the map d_S is automatically zero.

The differential d_A counts "augmented spheres" in the divisor D. We first consider the moduli space of spheres in D with two distinguished marked points and ℓ additional marked points in a given homology class $[B]$, $\mathcal{M}_{2,\ell}(D,[B])$. Let

$$\mathcal{M}_{2,\ell,A}([B_i],s) := \mathcal{M}_{2,\ell}(D,[B]) \times_{ev_1} \mathcal{M}_{1,d_1}([B_1]) \times_{ev_2} \cdots \times_{ev_\ell} \mathcal{M}_{1,d_\ell}([B_\ell])$$

such that

$$B \cdot D + \sum_i B_i \cdot D = s$$

For each ℓ, this moduli space $\mathcal{M}_{2,\ell,A}([B_i],s)$ is equipped with evaluations at 0 and ∞ and thus also gives rise to a cohomology level operation,[5] this time from

$$T_{\ell,s} : H^*(D) \to H^*(D)$$

We have that

$$d_A(e \otimes \alpha t^d) = (c_1(\mathcal{O}(D)) \cup \alpha)t^d + \sum_{d'<d} \sum_{\ell \geq 0} id \otimes T_{\ell,d-d'}(\alpha) t^{d'}$$

In the normal crossings case, the underlying complex generalizes very naturally

$$QC^*(Y,D) := \bigoplus_{I \subset \{1,\ldots,k\}} \bigoplus_{\mathbf{v} \in Z_I} (H^*(\mathring{D}_I) \otimes H^*(T^I))t^{\mathbf{v}} \tag{21}$$

The differential d in this general case involves a large number of terms from Gromov-Witten invariants relative to normal crossings (or logarithmic Gromov-Witten invariants) whose precise definition is left

[5]This operation incorporates certain combinatorial factors involving the multiplicities d_i and see Lemma 5.2. of [Diogo].

to future work. It should also be possible to construct the invariants algebraically using the techniques from logarithmic geometry, see [GroSie] for recent progress in this direction. However, crucial to our picture is the statement that *the cochains α_I and $\alpha_{I,m}$ defined above define cycles in $QC^*(Y, D)$ and hence cohomology classes in $QH^*(Y, D)$*. In fact, in the case when every component of the divisor is ample, something stronger is true:

$$d(H^*(\mathring{D}_I) \otimes H^s(T^I)t^{\mathbf{v}}) = 0$$

unless $|I| = k$ and $s > 0$. As before, the cases where $|I| < k$ follow because every holomorphic curve must intersect each component positively.

In the surface case, the formula simplifies considerably provided that the arithmetic genus, $p_a(D) > 0$ because none of the curves will have non-trivial components in the divisors (the reasoning here is similar to Remark 5.1). There are terms, which directly generalize those from the case of smooth divisors:

$$d_A : e \otimes H^*(\mathring{D}_i)t_i^d \to \oplus_{d' \leq d} id \otimes H^*(\mathring{D}_i)t_i^{d'}$$
$$d_S : e \otimes H^*(\mathring{D})t^d \to H^*(Y^0)$$

There is also a term

$$d_L : H^1(T^2)t^{\mathbf{v}} \to H^*(Y^0)$$

which is defined in a parallel way to d_S using the moduli space of spheres with two marked points which intersect the divisor at $z_0 = 0$ with multiplicity $\mathbf{v}$. Then we set $d = d_A + d_S + d_L$.

In this setting, defining a product structure on $QH^*(Y, D)$ would also require defining new relative Gromov-Witten invariants. Nevertheless, there is a version of Conjecture 3.1 which we can already state in this case. To each connected component of every non-empty stratum, we have seen that we can assign a generator α_I if $dim(D_I) > 0$ or $\alpha_{I,m}$ if $dim(D_I) = 0$. We let $\mathcal{A}_{naive}$ be the free commutative ring generated by α_I and $\alpha_{I,m}$ with the following relations. Assume that either I and J are disjoint or that $D_I \cap D_J$ is empty. We then impose the relations that:

(22)
$$\begin{cases} \alpha_I \cdot \alpha_J = \sum_m \alpha_{I+J,m} + \sum_{K,B} GW_{D_K}(pt, D_I \cap D_K, D_J \cap D_K, [B])\alpha_K \\ \end{cases}$$

Our formula again needs a few clarifications. The first is that if D_I or D_J is a collection of points, we have supressed the components and our formula should be thought of as applying to each component individually. We also make some remarks on the interpretation of α_{I+J}. Namely, if $D_I \cap D_J$ is empty, this is again to be regarded as zero. We allow K to be the empty set in which case we interpret D_K as Y and α_K as 1. Finally, B is in the set of non-zero classes in $H_2(D_K)$ such that we have an equality of vectors:

$$\{ B \cdot D + \mathbf{v}_K = \mathbf{v}_I + \mathbf{v}_J \tag{23}$$

The algebra $\mathcal{A}_{naive}$ also has a canonical filtration: $F^{\leq p}\mathcal{A}_{naive}$ is spanned by those generators $\prod_\ell \alpha_{I_\ell}$ such that $\sum_\ell |I_\ell| \leq p$.

In the case that D is a smooth ample divisor, we have that $\mathcal{A}_{naive} \cong \mathbb{C}[t]$. When all of the divisors D_i are individually ample, we record the following observation which will facilitate our computations. When $I+J$ above is a proper subset of $[1, 2, \ldots, k]$, any holomorphic curve must intersect every component of the divisor. We therefore conclude that the above formula specializes to:

$$\alpha_I \cdot \alpha_J = \sum_m \alpha_{I+J,m} \tag{24}$$

Remark 6.1. *Parallel to* Remark 5.3, *we note that there are versions of these objects defined over the ring* $R = \mathbb{C}[H_2(Y)]$.

Conjecture 6.1. *$QH^*(Y, D)$ can be given the structure of a finitely generated, filtered module structure over $\mathcal{A}_{naive}$ such that:*

- *The filtration $F^{\leq p}QH^*(Y, D)$ agrees with the filtration defined earlier in Section 5.*
- *The associated graded module structure on $QH^{top}(Y, D)$ is given by the action*

$$\alpha_I \cdot (\alpha t^{\mathbf{v}}) = \left(r_{|\mathbf{v}|, I \cup |\mathbf{v}|}(\alpha)\right) t^{\mathbf{v}_I + \mathbf{v}}$$

There is a ring map $\mathcal{A}_{naive} \to SH^*(Y^0)$ such that $QH^*(Y, D)$ and $SH^*(Y^0)$ are isomorphic as $\mathcal{A}_{naive}$ modules.[6]

We call this algebra $\mathcal{A}_{naive}$ since in principle the product should incorporate more relations coming from higher multiplicity relative Gromov-Witten invariants. On the other hand, in examples that we have checked

[6]We emphasize again for clarity that we have not put a ring structure on $QH^*(Y, D)$ in this note. This is likely possible, but technically demanding.

coming from mirror symmetry, this seems to be not necessary. One can see [Ueb2] for some partial results in the case of a single smooth divisor.

We will again compare our answer to several important examples in mirror symmetry. We will see that, in each example, the algebra $\mathcal{A}_{naive}$ plays the role of the ring of functions (or more generally the ring of functions on the critical locus) on the mirror.

6.1. Log Calabi-Yau examples

To begin, we consider a case which is closely related to the historical origins of the mirror conjecture. Let $Y = \mathbb{C}P^2$ and D be a smooth elliptic curve. Let us calculate the cohomology $QH^*(Y, D)$ above. An easy calclulation using the long exact sequence of pairs shows that $H^2(Y^0) \cong \mathbb{C}^2$. It follows from the formula (10) that the gradings are then concentrated in degrees $0, -1, -2, -3$ (recall that we have used homological grading conventions). We have that in each degree the pieces look like:

$$\oplus_{d\geq 1} H^0(T^2)t^d \oplus \mathbb{C} \to \oplus_{d\geq 1} H^1(T^2)t^d \oplus_{d\geq 1} H^0(T^2)[-1]t^d \to$$

$$\oplus_{d\geq 1} H^2(T^2)t^d \oplus_{d\geq 1} H^1(T^2)[-1]t^d \oplus \mathbb{C}^2[-2] \to \oplus H^2(T^2)[-1]t^d$$

From the form of our matrix, we can see that the differential from degree 0 to -1 is zero, as is the differential from degree -2 to -3. Notice that the classes in $H^2(Y^0)$ do not extend to classes in $H^2(Y)$. We therefore have by Observation 6.1 above, that the piece of the differential from

$$\oplus_{d\geq 1} H^0(T^2)[-1]t^d \to \mathbb{C}^2[-2]$$

is zero. The differential from

$$\oplus_{d\geq 1} H^0(T^2)[-1]t^d \to \oplus_{d\geq 1} H^2(T^2)t^d$$

is upper triangular with respect to the t-adic filtration and agrees with cupping with $c_1(\mathcal{O}(D))$ along the diagonal. It is therefore an isomorphism.

We therefore have that

$$QH^0(Y, D) \cong QH^{-3}(Y, D) \cong \mathcal{A}_{naive}$$
$$QH^{-1}(Y, D) \cong \mathbb{C}[t] \cdot e_1 \oplus \mathbb{C}[t] \cdot e_2$$

where e_1 and e_2 are generators of H^1 of the torus. Call the generators of $H^2(Y^0)$, x, y, then we have that

$$QH^{-2}(Y, D) \cong \mathbb{C}[t] \cdot x \oplus \mathbb{C}[t] \cdot y$$

Let $f : X \to \mathbb{A}^1$ denote the mirror of [AurKatOrl]. The map f defines an elliptic fibration with exactly three Lefschetz critical points. By properness, we have that

$$\mathbb{H}^0(\mathcal{O}_X) \cong \mathbb{H}^1(\Lambda^2 T_X) \cong \mathbb{C}[f]$$

An elementary sheaf theory computation shows that we have

$$\mathbb{H}^1(\mathcal{O}_X) \oplus \mathbb{H}^0(T_X) \cong \mathbb{H}^1(T_X) \oplus \mathbb{H}^0(\Lambda^2 T_X)$$

are both free modules of rank two over $\mathbb{C}[f]$. Thus, our prediction agrees with that coming from Equation (5). We finish this example by remarking that it can be generalized in two different ways. Suppose that (Y, D) is such that D is a smooth Calabi-Yau hypersurface in a Fano manifold Y. The mirror pair (X, f) is expected to be a proper fibration over $\mathbb{A}^1$. The functions $\Gamma(\mathcal{O}_X)$ would then agree with $\mathbb{C}[f]$, which is indeed isomorphic to our algebra $\mathcal{A}_{naive}$ in this case.

Let us now turn to an example where not all of the components D_i are ample. Let Y be a degree 5 Fano surface with an anti-canonical divisor D which consists of a ring of five -1 lines, $D_1, \cdots, D_5$. It is well-known that there are unique -1 curves E_i such that

$$D_j \cdot E_i = \delta_{ij}$$

Let us begin by explaining the calculation of $\mathcal{A}_{naive}$. The D_i curves themselves give a non-trivial Gromov-Witten invariant $GW_{D_i}(pt, pt, pt, [D_i])$. There is also a three point Gromov-Witten invariant in homology class $E_i + D_i$ which is

$$GW_Y(D_{i-1}, D_{i+1}, pt, [E_i + D_i]) = 1$$

where i ranges over $[1, 2, \ldots, 5]$ in cyclic order. Working with $R = \mathbb{C}[H_2(Y)]$ coefficients, we obtain relations:

$$\theta_{i-1}\theta_{i+1} = q^{[D_i]}\theta_i + q^{[D_i+E_i]}$$

This agrees with the mirror family obtained by [GroHacKee] using tropical considerations. It is also worth noting that in this case, the differential d cannot be computed by purely topological considerations because the curves E_i contribute non-trivially to the differential d_A.

Remark 6.2. *The examples that we have considered have been surfaces. The algebra $\mathcal{A}_{naive}$ has been calculated in a collection of higher dimensional examples such as when $Y = \mathbb{C}P^n$, D_1 is a quadric and*

$D_2, \ldots, D_n$ are hyperplanes or $Y := \mathcal{Q}^n$, an n-dimensional quadric and $D_1, D_2, \ldots, D_n$ are generic hyperplane in [Pom] *and shown to agree with mirror symmetry predictions of* [Aur].

Interestingly, in these examples, the differential $d = 0$ in the complex $QH^*(Y, D)$. It is important to mention that James Pascaleff [Pas] *has recently shown that when $Y = \mathbb{C}P^2$, D_1 is a quadric and D_2 is a hyperplane, SH^0 is indeed isomorphic to $\mathcal{A}_{naive}$ (which in the present situation is concentrated in degree zero) as rings.*

6.2. A non-commutative example

Let Y be an n-dimensional quadric $\mathcal{Q}^n$ with $n \geq 2$. Let D be a hyperplane section. It is well-known that Y^0 is exactly symplectomorphic to a cotangent bundle T^*S^n of a sphere S^n. We have by the main result of Abouzaid [Abo] that:

$$W(T^*S^n) \cong mod(C_*(\Omega_p S^n))$$

where p is a point on S^n, $\Omega_p S^n$ is the based loop space, and $C_*(\Omega_p S^n)$ denotes the differential graded algebra of chains equipped with Pontryagin product. It follows from this work that

$$SH^*(Y^0) \cong HH^*(C_*(\Omega_p S^n)) \cong H_{*-n}(LS^n)$$

Here LS^n denotes the free loop space of S^n. It also follows from Abouzaid's work that $Fuk(Y^0) \cong mod(C^*(S^n))$, where $C^*(S^n)$ denotes the differential graded algebra of cochains on S^n. The general theory of Koszul duality then shows that:

$$\phi : H_{*-n}(LS^n) \cong HH^*(C_*(\Omega_p S^n)) \cong HH^*(C^*(S^n))$$

In [Diogo], it is calculated that there is a canonical isomorphism ψ : $QH^*(Y, D) \cong H_{*-n}(LS^n)$. In [Pom], we provide some further evidence for the picture described in this note using the map

$$\mathcal{BB} : \mathbb{C}[\alpha] \to HH^*(Fuk(Y^0))$$

It is directly shown that the deformations of the Fukaya category $Fuk(Y^0)$ given by α^d coincide with the deformation given by $\phi(\alpha^d)$ in $HH^*(C^*(S^n))$.

Remark 6.3. *Similar considerations apply to other rank one homogeneous spaces such as $\mathbb{C}P^n$ or $\mathbb{H}P^n$. Additionally, these examples can also be interpreted in more traditional commutative terms see* [Pom].

References

[Abo] M. Abouzaid, A cotangent fiber generates the Fukaya Category, Adv. Math, 2011, pp. 894–939.

[ACGHOSS] D. Abramovich, Q. Chen, D. Gillam, Y. Huang, M. Olsson, M. Satriano, S. Sun, Logarithmic Geometry and Moduli, Handbook of Moduli, Farkas and Morrison eds, 2010.

[Aur] D. Auroux, Mirror symmetry and T-duality in the complement of an anticanonical divisor, J. Gökova Geom. Topol. 1, 2007, pp. 51–91.

[AurKatOrl] D. Auroux, L. Katzarkov, D. Orlov, Mirror symmetry for Del Pezzo surfaces: Vanishing cycles and coherent sheaves. Inventiones Math. 166 (2006), 537–582.

[BorShe] S. Borman, N. Sheridan, in preparation, 2017.

[BouEkhEli] F. Bourgeois, T. Ekholm, Y. Eliashberg, Effect of Legendrian surgery. Geometry and Topology 16, 2012, pp. 301–389.

[CarPumSie] M. Carl, M. Pumperla, B. Siebert, A tropical viewpoint on Landau Ginzburg models, preprint, https://www.math.uni-hamburg.de/home/siebert/preprints/LGtrop.pdf

[CieMoh] K. Cieliebak, K. Mohnke, Symplectic Hypersurfaces and Transversality in Gromov-Witten Theory, `arxiv:math/0702887`, 2007.

[Diogo] L. Diogo, Stanford University Thesis, 2012.

[EtguLek] T. Etgu, Y. Lekili, Koszul duality patterns in Floer theory, Geometry and Topology 21, 2017, pp. 3313–3389.

[FloHof] A. Floer, H. Hofer, Symplectic Homology I: Open sets in $\mathbb{C}^n$. Mathematische Zeitschrift, Volume 215, Issue 1, 1994, pp. 37–88.

[FloHofSal] A.Floer, H. Hofer, D. Salamon, Transversality in elliptic morse theory for the symplectic action. Duke Math. J., Volume 80, 1995, pp. 251–292.

[FukOno] K. Fukaya, K. Ono, Arnold Conjecture and Gromov-Witten Invariant, Topology , vol. 38, 1999, pp. 933–1048.

[FukOhOnoOht] K. Fukaya, Y.-G. Oh, H. Ohta and K. Ono, Lagrangian intersection Floer theory:anomaly and obstruction I, AMS/IP Studies in Advanced Mathematics, vol. 46, 2009.

[GanPom] S. Ganatra, D. Pomerleano, A Log PSS morphism with applications to Lagrangian embeddings, `arxiv:math/1611.06849`.

[GroHacKee] M. Gross, P. Hacking, S. Keel, Mirror symmetry for log Calabi-Yau surfaces I, 2011, `arxiv:math/1106.4977`.

[Gro] A. Grothendieck, Éleménts de géométrie algébrique III.1, Pubs. maths. de l'IHES, vol. 11, 1961.

[GroSie] M. Gross, B. Siebert, Intrinsic Mirror Symmetry and punctured Gromov-Witten invariants, `https://arxiv.org/abs/1609.00624`.

[HofSal] H. Hofer, D. Salamon, Floer Homology and Novikov Rings, The Floer Memorial Volume, 1995.

[Ion] E. Ionel, GW invariants relative normal crossings divisors, 2011, `arxiv:math/1103.3977`.

[IonPar] E. Ionel, T. Parker, Relative Gromov-Witten Invariants, Annals of Math., 157, 2003, pp. 1–52.

[McDSal] D. McDuff, D. Salamon, J-holomorphic curves and symplectic topology, Colloquium Publications, vol. 52, 2012.

[Par] J. Pardon, An algebraic approach to virtual fundamental cycles on moduli spaces of pseudo-holomorphic curves. Geom. Topol. 20 (2016), no. 2, 779–1034.

[Pas] J. Pascaleff, On the symplectic cohomology of log Calabi-Yau surfaces, 2013, `arxiv:math/1304.5298`.

[PiuSalSch] S. Piunikhin, D. Salamon, M. Schwarz, Symplectic Floer-Donaldson theory and quantum cohomology, Contact and symplectic geometry, (Cambridge, 1994), 171–200, vol. 8, 1996.

[Pom] D. Pomerleano, Curved String Topology and Fukaya categories II, University of California, Berkeley thesis, 2012, available at `arxiv:math/1208.4881`.

[Sei] P. Seidel, Fukaya categories and Deformations, Proceedings of the Beijing ICM, 2002, available at `arxiv:math/0206155`.

[Sei2] P. Seidel, A biased view of symplectic cohomology, Current Developments in Mathematics Volume 2006, 2008, pp. 211–253.

[She] N. Sheridan, Homological Mirror Symmetry for Calabi-Yau hypersurfaces in projective space, Invent. Math, 2015, 1–186.

[Ueb1] P. Uebele, Symplectic homology of some Brieskorn manifolds, Mathematische Zeitschrift, Volume 283(1), 243–274, 2016.

[Ueb2] P. Uebele, Periodic Reeb flows and products in symplectic homology, `arxiv.org/abs/1512.06208`, 2015.

University of Cambridge Wilberforce Rd, Cambridge CB3 0WA, UK
E-mail address: `d.pomerleano@imperial.ac.uk`

Advanced Studies in Pure Mathematics 83, 2019
Primitive Forms and Related Subjects — Kavli IPMU 2014
pp. 357–369

The Landau–Ginzburg/Calabi–Yau correspondence and the mirror quintic

Nathan Priddis

Abstract.

We briefly describe a deep relationship known as the Landau–Ginzburg/Calabi–Yau correspondence for the famous mirror quintic via global mirror symmetry.

§1. Introduction

The Landau–Ginzburg/Calabi–Yau correspondence was predicted by physicists nearly 20 years ago [19, 20]. The first instance of the LG/CY correspondence was proved by Chiodo–Ruan [5] for the quintic three–fold, who then formulated the correspondece more generally in mathematical terms in [7]. In fact, they predict an equivalence between the FJRW theory (see Section 4) for certain Landau–Ginzburg models and the Gromov–Witten theory (see Section 3) of the corresponding Calabi–Yau orbifolds.

Beyond the quintic, only a few other examples of the LG/CY correspondence have been proven at various levels: for simple elliptic singularities by Krawitz–Shen [13], for certain complete intersections in weighted projective space by E. Clader [8], and for Calabi–Yau hypersurfaces in weighted projective space satisfying a certain Gorenstein condition by Chiodo–Iritani–Ruan [4].

This paper will briefly describe joint work of the author with M. Shoemaker, wherein a small version of the LG/CY correspondence was proven for the mirror quintic (the full details of this result can be found in [16]). Since the delivery of this talk, the results given here have been generalized to all quotients by certain finite groups of Fermat

Received December 9, 2014.
Revised November 9, 2015.
2010 *Mathematics Subject Classification.* 14N35, 14A20, 14E16, 53D45.
Key words and phrases. LG/CY, GW theory, FJRW theory, mirror symmetry.

type hypersurfaces in weighted projective space. In all except for simple elliptic singularities, the correspondence is only given in genus zero.

§2. Notation

In order to describe this correspondence, let us establish some notation. FJRW theory takes as input (for more details, see [9, 5]) a tuple (W, G), where $W \in \mathbb{C}[x_1, \ldots, x_N]$ is a non–degenerate quasihomogeneous polynomial with an isolated singularity at the origin and with uniquely determined weights $q_i = \frac{c_i}{d}$ (with $\gcd(c_1, \ldots, c_N) = 1$)), and G is a group of $\mathrm{Aut}(W)$, the group of diagonal symmetries of W. The group G must contain the *exponential grading operator*

$$\mathfrak{j} = (e^{2\pi i q_1}, \ldots, e^{2\pi i q_N}).$$

This data determines the LG–model.

In the setting of LG/CY, we require that $\sum_{i=1}^{N} c_i = d$, and in this article, by considering $\mathrm{Aut}(W)$ as a subgroup of of diagonal matrices in $\mathrm{GL}_N(\mathbb{C})$, we also require that G injects into $\mathrm{SL}_N(\mathbb{C})$.

Under these conditions we can also consider a Calabi–Yau (orbifold) $\mathcal{X} := [X_W/\bar{G}]$, where $X_W = \{W = 0\} \subset \mathbb{P}(c_1, \ldots, c_N)$ and $\bar{G} = G/\langle \mathfrak{j} \rangle$. The LG/CY conjecture can be written schematically as

$$\mathrm{FJRW}(W, G) \cong \mathrm{GW}(\mathcal{X}).$$

We will briefly describe each part of the correspondence (see e.g. [6] for a good survey on the topic), and then describe the results for the mirror quintic.

§3. GW theory

Since our targets are orbifolds, we consider orbifold Gromov–Witten theory. For more details, see the original treatments [1] and [2].

Gromov–Witten theory concerns itself with the moduli space $\overline{M}_{g,n}(\mathcal{X}, \delta)$ of stable maps to $\mathcal{X}$ of degree $\delta \in H^2(\mathcal{X}, \mathbb{Z})$.

The state space of the theory is given by the Chen–Ruan cohomology $H^*_{CR}(\mathcal{X}, \mathbb{C})$ (see [3]). Given classes $\beta_1, \ldots, \beta_n \in H^*_{CR}(\mathcal{X}, \mathbb{C})$, we can consider the GW–invariants

$$\langle \psi^{a_1}\beta_1, \ldots, \psi^{a_n}\beta_n \rangle^{GW}_{g,n} := \sum_{\delta \in H^2(\mathcal{X}, \mathbb{Z})} Q^{\delta} \int_{[\overline{M}_{g,n}(\mathcal{X},\delta)]^{vir}} \prod_i \psi_i^{a_i} \cup ev_i^*(\beta_i)$$

by integrating against the virtual fundamental class $[\overline{M}_{g,n}(\mathcal{X}, \delta)]^{vir} \in H_*(\overline{M}_{g,n}(\mathcal{X}, \delta))$. Here ψ–classes are chern classes of the cotangent line bundle pulled back from the course underlying curve, ev_i^* are the evaluation maps $ev_i^* : H^*_{CR}(\mathcal{X}, \mathbb{C}) \to H^*(\overline{M}_{g,n}(\mathcal{X}, \delta))$, and Q is the Novikov variable.

§4. FJRW theory

FJRW theory was first introduced as a way of understanding Witten's equation

$$\bar{\partial} u_i + \overline{\frac{\partial W}{\partial u_i}} = 0$$

where u_i should be understood as sections of some appropriate (orbifold) line bundles $\mathcal{L}_i$ over an (orbifold) curve $\mathcal{C}$. In order to make the two terms of the sum compatible (so we can add them), one must impose some conditions on $\mathcal{L}_i$. This gives rise to the moduli space of *W-curves*, which we now describe. As we will of necessity be quite terse, the interested reader should consult [9] for the original construction and [6] for an alternate treatment.

Write $W = \sum_{i=1}^{s} \lambda_i \prod_{j=1}^{N} x_j^{b_{ij}}$. The moduli space of *W–orbicurves* is given by the data

$$\overline{\mathcal{W}}_{g,n,G} := \{(\mathscr{C}, p_1, \ldots, p_n, \mathcal{L}_1, \ldots, \mathcal{L}_N, \varphi_1, \ldots, \varphi_s)\} \,/\, \sim$$

(up to equivalence) satisfying the following conditions

(1) $\mathscr{C}$ is a *stable* orbifold curve with orbifold structure only at the markings and the nodes. Furthermore, we require the nodes to be *balanced* (see e.g. [9, 5]).

(2) $\mathcal{L}_j$ is an orbifold line bundle, and φ_i an isomorphism satisfying

$$\varphi_i : \mathcal{L}_1^{b_{i1}} \otimes \cdots \otimes \mathcal{L}_1^{b_{iN}} \xrightarrow{\sim} \omega_{log}.$$

Here ω_{log} is the log–canonical bundle (see e.g. [9]).

(3) The representation $G_{p_i} \to U(1)^N$ induced by the action of the local group G_{p_i} at the i–th marked point on $\bigoplus_{j=1}^{N} \mathcal{L}_j$ is faithful, and factors through G.

The moduli space decomposes into open and closed substacks as

$$\overline{\mathcal{W}}_{g,n,G} = \coprod_{\gamma_i \in G} \overline{\mathcal{W}}_{g,n,G}(\gamma_1, \ldots, \gamma_n)$$

based on the multiplicities of the action of the local group G_{p_i} on $\bigoplus_{j=1}^{N} \mathcal{L}_j$ at each marking.

The state space is

$$\begin{aligned}\mathscr{H}_{W,G} :&= \bigoplus_{\gamma \in G} H^{mid}(\mathbb{C}^{N_\gamma}, W_\gamma^{+\infty}; \mathbb{C})^G \\ &= \bigoplus_{\gamma \in G} \mathscr{H}_\gamma.\end{aligned}$$

In other words, each summand is the G–invariant part of of the cohomology of the fixed locus of γ relative to the Milnor fiber. A direct summand is called a *narrow sector* if the fixed locus of γ is trivial.

Here there are also invariants defined as integrals over the virtual fundamental class

$$\begin{aligned}&[\overline{\mathcal{W}}_{g,n,G}(\gamma_1, \ldots, \gamma_n)]^{vir} \\ &\quad \in H_*(\overline{\mathcal{W}}_{g,n,G}(\gamma_1, \ldots, \gamma_n)) \otimes \prod_{i=1}^{n} H_{mid}(\mathbb{C}^{N_{\gamma_i}}, W_{\gamma_i}^{+\infty}; \mathbb{C})^G\end{aligned}$$

given by

$$\langle \psi^{a_1}\alpha_1, \ldots, \psi^{a_n}\alpha_n \rangle_{g,n}^{(W,G)} := \int_{[\overline{\mathcal{W}}_{g,n,G}(\gamma_1,\ldots,\gamma_n)]^{vir}} \prod_{i=1}^{n} \psi_i^{a_i} \otimes \alpha_i$$

where $\alpha_i \in \mathscr{H}_{\gamma_i}$. As before the ψ–classes are pulled back from the course underlying curve.

§5. Givental's formalism

In both GW theory and FJRW theory, the invariants can be studied using the following, albeit sophisticated, geometry. For a more detailed exposition of what follows we refer the reader to Givental's original paper on the subject [10]. Let $H^\square$ denote either $H^*_{CR}(\mathcal{X})$ or $\mathscr{H}_{W,G}$. Let $\mathbf{t} = \sum_{i \in I} t^i \beta_i$ represent a point of $H^\square$ written in terms of a given basis $\{\beta_i\}_{i \in I}$. In this notation, we require $0 \in I$, and β_0 denotes the identity element in the corresponding algebra. For notational convenience denote the formal series $\sum_{k \geq 0} \mathbf{t}_k \psi^k$ as $\mathbf{t}(\psi)$. We define the genus g generating function as

$$\mathcal{F}_g = \sum_{n \geq 0} \frac{\langle \mathbf{t}(\psi), \ldots, \mathbf{t}(\psi) \rangle_{g,n}}{n!}.$$

Consider the infinite dimensional (symplectic) vector space defined by $\mathbb{H}^{\Box} := H^{\Box}((z^{-1}))$. Each point can be given in (Darbeaux) coordinates on this space as

$$f = \sum_{k\geq 0}\sum_{i\in I} q_k^i \beta_i z^k + \sum_{k\geq 0}\sum_{i\in I} p_{k,i}\beta^i(-z)^{-k-1}.$$

Because the invariants in both cases satisfy the *String Equation*, *Dilaton Equation* and *Topological Recursion Relations* (for GW theory see [18] and for FJRW theory see [9]), we can use this extra structure to rephrase the genus zero data in terms of Givental's *overruled Lagrangian cone*

$$\mathscr{L} := \left\{\mathbf{p} = d_{\mathbf{q}}\mathcal{F}_0^{\Box}\right\}$$

after the change of variables $\mathbf{q}(z) = \mathbf{t}(z) - z$ (also known as the *Dilaton shift*). Givental's cone contains all of the information of the genus zero invariants of either theory.

The Lagrangian subspace $\mathscr{L}^{\Box}$ is, in fact, a cone and it satisfies the condition that for all $f \in \mathbb{H}^{\Box}$,

$$\mathscr{L}^{\Box} \cap L_f = zL_f \tag{5.0.1}$$

where L_f is the tangent space to $\mathscr{L}^{\Box}$ at f. Hence, if we can find a "slice" of the cone transverse to the ruling, we can generate the entire cone by points in the slice together with linear combinations of derivatives of the corresponding functions. The image of the J–function

$$J^{\Box}(\mathbf{t}, z) = \beta_0 z + \mathbf{t} + \sum_{n\geq 0}\sum_{i\in I}\frac{1}{n!}\left\langle \frac{\beta_i}{z-\psi}, \mathbf{t}, \ldots, \mathbf{t}\right\rangle_{0,n+1}^{\Box}\beta^i.$$

lies in $\mathscr{L}^{\Box}$ defines such a class.

Equation (5.0.1) says that the ruling at $J^{\Box}(\mathbf{t}, -z)$ is spanned by the derivatives of $J^{\Box}(\mathbf{t}, z)$. By the string equation, $z\frac{\partial}{\partial t^0}J^{\Box}(\mathbf{t}, z) = J^{\Box}(\mathbf{t}, z)$, and since $J^{\Box}(\mathbf{t}, z)$ is transverse to the ruling, we in fact have

$$\mathscr{L}^{\Box} = \left\{z\sum c_i(z)\frac{\partial}{\partial t^i}J^{\Box}(\mathbf{t}, -z) | c_i(z) \in \mathbb{C}[z], \mathbf{t} \in H^{\Box}\right\}. \tag{5.0.2}$$

We refer the interested reader to [14] and [11] for more details.

§6. The mirror quintic

With this background in place, we can now describe the LG/CY correspondence for the mirror quintic. The complete details can be found in [16]. In this case, $W = x_1^5 + \cdots + x_5^5$, and $G = \mathrm{Aut}(W) \cap \mathrm{SL}_N(\mathbb{C})$, i.e. diagonal symmetries of determinant 1. The mirror quintic (see e.g. [12]) is defined by the Deligne–Mumford stack $\mathcal{X} = [X_W/\bar{G}]$.

For each theory we make two important restrictions. First we restrict $H^*_{CR}(\mathcal{X}, \mathbb{C})$ to the ambiant classes, and we restrict $\mathscr{H}_{W,G}$ to the narrow sectors. The reason for this is that when primitive class insertions or broad sector insertions appear in the respective invariants, the invariants vanish.

The second restriction will be to consider the "small" Lagrangian cone, i.e. we restrict to the variable s corresponding to the hyperplane class H in the *untwisted* sector of $H^*_{CR}(\mathcal{X}, \mathbb{C})$ in GW theory, and to the variable t corresponding to canonical generator ϕ_{j^2} of $\mathscr{H}_{\mathrm{j}^2}$ in FJRW theory. The reason for this restriction is that the full GW theory could not be computed in general when this result appeared.

Recall from equation (5.0.2) that in order to generate the entire (small) cone, we must consider all of the derivatives of the J–function. In GW theory, we have a basis (recall the notation at the beginning of Section 5) consisting primarily of a generator from each sector of $H^*_{CR}(\mathcal{X}, \mathbb{C})$, and powers of the (untwisted) hyperplane class. Thus for those $\gamma \in G$ indexing a sector of $H^*_{CR}(\mathcal{X}, \mathbb{C})$, we have a coordinate t^γ and

$$J^{GW}_\gamma(s, z) = z\frac{\partial}{\partial t^\gamma} J^{GW}(\mathbf{t}, z)|_{\mathbf{t}=sH}.$$

In FJRW theory, we have a basis consisting of a generator from each narrow sector $\mathscr{H}_h$. Thus for those $h \in G$ indexing the narrow sectors, we have a coordinate t^h and

$$J^{FJRW}_h(t, z) = z\frac{\partial}{\partial t^h} J^{FJRW}(\mathbf{t}, z)|_{\mathbf{t}=t\phi_{\mathrm{j}^2}}.$$

This defines the respective A–models.

We can make one more simplifying restriction. On the CY side, for $\gamma \in G$, the so-called "age"is

$$\mathrm{age}\,\gamma = \tfrac{1}{2\pi i}\log(\det \gamma).$$

For ambient classes in the twisted sectors of $H_{CR}(\mathcal{X}, \mathbb{C})$, this determines the degree of the classes. It is enough to consider $J^{GW}_\gamma(s, z)$ for $\mathrm{age}\,\gamma \leq 2$, for these determine the others.

On the LG–side, the degree of the narrow classes is shifted by the exponential grading operator, i.e. for $h \in G$ defining a narrow sector,

$$\deg \phi_h = \mathrm{age}\, h - \sum_{j=1}^{5} q_j.$$

Again, it is enough to consider $J_h^{FJRW}(t,z)$ for $\deg h \leq 2$ for these determine the others.

Because the B–model has a global moduli space, whereas the A–model does not, the LG/CY correspondence is often carried out on the B–model. On the B–side we consider the so–called I–functions, defined as the period integrals for a family of quintics $Y_v \subset \mathbb{P}^4$ given by $W_v = W - vx_1x_2\dots x_5$. The reason for considering these functions is that they determine the J–functions on the A–side. Furthermore, the I–functions are solutions to the Picard–Fuchs equations determined by this family, and thus they are hypergeometric functions, which can be written explicitly.

These I–functions are determined by integrating certain forms $\omega_v \in H^3(Y_v)$ for this deformation family over a basis of cycles (see e.g. [15] for more detail). In certain cases, these forms are related to the primitive forms of Saito–Givental theory, though in this case the Saito–Givental theory has not been generalized to the current situation (see [13] for the use of Saito–Givental theory in the B–model, also see [17] for a survey article describing the relationship).

Before we can give these functions explicitly, it will be necessary to fix some notation. In what follows, we will always desribe each element of G as tuples of the fractional powers of $e^{2\pi i}$ defining the element. For example, $\mathfrak{j} = (\frac{1}{5}, \frac{1}{5}, \frac{1}{5}, \frac{1}{5}, \frac{1}{5})$. The notation $\mathbb{1}_\gamma$ will denote the fundamental class in the γ–twisted sector of $H_{CR}(\mathcal{X}, \mathbb{C})$, and, as mentioned previously, H will denote the hyperplane class in $[\mathbb{P}^4/G]$ restricted to the hypersurface or to the fixed locus of γ as the situation requires. For any rational number $r \in \mathbb{Q}$, we will also write $\langle r \rangle \in [0,1)$ for the fractional part of r.

If we consider the family W_v in a neighborhood of $v = \infty$ (the large complex structure limit) we get a set of I–functions $I_\gamma(s,z)$ indexed by the twisted sectors of $H^*_{CR}(\mathcal{X}, \mathbb{C})$. Here $s = -5\log v$. It turns out both the J–function and the I–functions are functions of $q = e^s$. For example (see [15, 16]), we have

(i) If $\gamma = e = (0,0,0,0,0)$,

$$I_e^{\mathcal{W}}(q,z) = zq^{H/z} \left(1 + \sum_{\langle d \rangle = 0} q^d \frac{\prod\limits_{1 \leq m \leq 5d} (5H + mz)}{\prod\limits_{\substack{0 < b \leq d \\ \langle b \rangle = 0}} (H + bz)^5} \right).$$

(ii) If $\gamma = (0,0,0,r_1,r_2)$,

$$I_\gamma^{\mathcal{W}}(q,z) = \\ zq^{H/z}\mathbb{1}_\gamma\left(1+\sum_{\langle d\rangle=0} q^d \frac{\prod_{1\le m\le 5d}(5H+mz)}{\prod_{\substack{0<b\le d\\ \langle b\rangle=0}}(H+bz)^3 \prod_{\substack{0<b\le d\\ \langle b\rangle=r_2}}(H+bz) \prod_{\substack{0<b\le d\\ \langle b\rangle=r_1}}(H+bz)}\right).$$

(iii) If $\gamma = (0,0,r_1,r_1,r_2)$, let $\gamma_1 = (\langle -r_1\rangle, \langle -r_1\rangle, 0, 0, \langle r_2 - r_1\rangle)$. Then

$$I_\gamma^{\mathcal{W}}(q,z) = \\ zq^{H/z}\mathbb{1}_\gamma\left(1+\sum_{\langle d\rangle=0} q^d \frac{\prod_{1\le m\le 5d}(5H+mz)}{\prod_{\substack{0<b\le d\\ \langle b\rangle=0}}(H+bz)^2 \prod_{\substack{0<b\le d\\ \langle b\rangle=\langle 3r_2\rangle}}(H+bz)^2 \prod_{\substack{0<b\le d\\ \langle b\rangle=\langle 2r_1\rangle}}(H+bz)}\right)$$

$$+ zq^{H/z}\mathbb{1}_{\gamma_1}\left(\sum_{\langle d\rangle=r_1} q^d \frac{\prod_{1\le m\le 5d}(5H+mz)}{\prod_{\substack{0<b\le d\\ \langle b\rangle=r_1}}(H+bz)^2 \prod_{\substack{0<b\le d\\ \langle b\rangle=0}}(H+bz)^2 \prod_{\substack{0<b\le d\\ \langle b\rangle=r_2}}(H+bz)}\right)$$

The following small mirror theorem was proved for the GW theory of the mirror quintic by Lee–Shoemaker [15].

Theorem 6.1 (CY Small Mirror Theorem [15]). *For each $\gamma \in G$ with* age $\gamma \le 2$ *indexing a sector of $H^*_{CR}(\mathcal{X},\mathbb{C})$, there exists $G_\gamma(t)$ such that*

$$J_\gamma(\xi(q),z) = \frac{I_\gamma(q,z)}{G_\gamma(q)} \qquad \textit{where} \ \ \xi(q) = \frac{G_e(q)}{F(q)}.$$

Here $G_e(q)$ and $F(q)$ are explicitly determined functions that do not depend on γ. They are essentially the coefficients of z and sH, resp. in the expansion of I_e.

If on the other hand, we consider a neighborhood of $v=0$, we get a set of I–functions $I_h(t,z)$ indexed by the narrow sectors of $\mathscr{H}_{W,G}$. Here $t = v$. For example, (see [16])

(i) For $h = \mathrm{j}$,

$$tI_{\mathrm{j}}^{(W,G)}(t,z) = \sum_{k=1,2,3,4} \phi_{\mathrm{j}^k} z^{2-k} \sum_{l\ge 0} t^{k+5l} \frac{\Gamma((k+5l)/5)^5}{\Gamma(k/5)\Gamma(k+5l)}.$$

(ii) For $h = (\frac{1}{5}, \frac{1}{5}, \frac{1}{5}, \frac{3}{5}, \frac{4}{5})$ and $h_1 = (\frac{4}{5}, \frac{4}{5}, \frac{4}{5}, \frac{1}{5}, \frac{2}{5})$,

$$I_h^{(W,G)}(t,z) = z\phi_h \sum_{l\geq 0} t^{5l} \frac{\Gamma((1+5l)/5)^3\Gamma((3+5l)/5)\Gamma((4+5l)/5)}{\Gamma(1/5)^3\Gamma(3/5)\Gamma(4/5)\Gamma(1+5l)}$$
$$+ \phi_{h_1} \frac{2}{25} \sum_{l\geq 0} t^{3+5l} \frac{\Gamma((4+5l)/5)^3\Gamma((6+5l)/5)\Gamma((7+5l)/5)}{\Gamma(4/5)^3\Gamma(6/5)\Gamma(7/5)\Gamma(4+5l)}.$$

(iii) For $h = (\frac{2}{5}, \frac{2}{5}, \frac{2}{5}, \frac{3}{5}, \frac{1}{5})$ and $h_1 = (\frac{3}{5}, \frac{3}{5}, \frac{3}{5}, \frac{4}{5}, \frac{2}{5})$,

$$I_h^{(W,G)}(t,z) = z\phi_h \sum_{l\geq 0} t^{5l} \frac{\Gamma((2+5l)/5)^3\Gamma((3+5l)/5)\Gamma((1+5l)/5)}{\Gamma(2/5)^3\Gamma(3/5)\Gamma(1/5)\Gamma(1+5l)}$$
$$+ \phi_{h_1} \sum_{l\geq 0} t^{1+5l} \frac{\Gamma((3+5l)/5)^3\Gamma((4+5l)/5)\Gamma((2+5l)/5)}{\Gamma(3/5)^3\Gamma(4/5)\Gamma(2/5)\Gamma(2+5l)}.$$

(iv) For $h = (\frac{1}{5}, \frac{1}{5}, \frac{2}{5}, \frac{2}{5}, \frac{4}{5})$ and $h_1 = (\frac{3}{5}, \frac{3}{5}, \frac{4}{5}, \frac{4}{5}, \frac{1}{5})$,

$$I_h^{(W,G)}(t,z) = z\phi_h \sum_{l\geq 0} t^{5l} \frac{\Gamma((1+5l)/5)^2\Gamma((2+5l)/5)^2\Gamma((4+5l)/5)}{\Gamma(1/5)^2\Gamma(2/5)^2\Gamma(4/5)\Gamma(1+5l)}$$
$$+ \frac{\phi_{h_1}}{5} \sum_{l\geq 0} t^{2+5l} \frac{\Gamma((3+5l)/5)^2\Gamma((4+5l)/5)^2\Gamma((6+5l)/5)}{\Gamma(3/5)^2\Gamma(4/5)^2\Gamma(6/5)\Gamma(3+5l)}.$$

(v) For $h = (\frac{1}{5}, \frac{1}{5}, \frac{3}{5}, \frac{3}{5}, \frac{2}{5})$ and $h_1 = (\frac{2}{5}, \frac{2}{5}, \frac{4}{5}, \frac{4}{5}, \frac{3}{5})$,

$$I_h^{(W,G)}(t,z) = z\phi_h \sum_{l\geq 0} t^{5l} \frac{\Gamma((1+5l)/5)^2\Gamma((3+5l)/5)^2\Gamma((2+5l)/5)}{\Gamma(1/5)^2\Gamma(3/5)^2\Gamma(2/5)\Gamma(1+5l)}$$
$$+ \phi_{h_1} \sum_{l\geq 0} t^{1+5l} \frac{\Gamma((2+5l)/5)^2\Gamma((4+5l)/5)^2\Gamma((3+5l)/5)}{\Gamma(2/5)^2\Gamma(4/5)^2\Gamma(3/5)\Gamma(2+5l)}.$$

The following mirror theorem as proved by Priddis–Shoemaker.

Theorem 6.2 (LG Small Mirror Theorem [16]). *For each $h \in G$ with $\deg \phi_h \leq 2$ indexing a narrow sector of $\mathscr{H}_{W,G}$, there exists $G_h(t)$ such that*

$$J_h^{FJRW}(\tau(t), z) = \frac{I_h(t,z)}{G_h(t)} \qquad \textit{where } \tau(t) = \frac{G_{\mathrm{j}^2}(t)}{H(t)}.$$

As before $G_{\mathrm{j}^2}(t)$ and $H(t)$ are explicitly defined functions that do not depend on h. As before, these are essentially the coefficients of $\phi_{\mathrm{j}} z$ and $t\phi_{\mathrm{j}^2}$, resp. in the expansion of the I–functions.

In particular, the mirror theorems say that the I–functions generate the respective (small) Lagrangian cones. Thus, if we can relate the two sets of I functions, we obtain an equivalence between FJRW and GW theory. The two sets of I–functions are related via analytic continuation $\tilde{I}^{\mathcal{X}}_{\gamma}(t,z)$ of $I_{\gamma}(q,z)$ from a neighbor hood of $v=\infty$ to a neighborhood of $v=0$ and a symplectic transformation $\mathbb{U}$ taking one set of I–functions to the other, as in the following theorem.

Theorem 6.3 (LG/CY correspondence)**.** *There exists a linear symplectic transformation* $\mathbb{U}:\mathbb{H}^{W,G}\to\mathbb{H}^{\mathcal{X}}$ *satisfying: for each* $h\in G$ *of* $\deg\phi_h\le 2$ *indexing a narrow sector of* $\mathscr{H}_{W,G}$ *there is a uniquely determined* $\gamma\in G$ *of* age ≤ 2 *indexing a sector of* $H^*_{CR}(\mathcal{X},\mathbb{C})$ *such that*

$$\mathbb{U}\big(I^{W,G}_h(t,z)\big)=\tilde{I}^{\mathcal{X}}_{\gamma}(t,z).$$

The right hand side denotes the I–function after analytic continuation and the change of variables $q^{-5}=t$.

The symplectic transformation $\mathbb{U}$ takes the following form. We will define it on a basis of the narrow sectors of $\mathscr{H}_{W,G}$. In the untwisted sector of $H_{CR}(\mathcal{X},\mathbb{C})$, we have:

$$\mathbb{U}:\phi_{\mathfrak{j}^k}\mapsto z^{k-1}\frac{\Gamma(1+H/z)^5}{\Gamma(1+5H/z)}\frac{(-\xi)^k 2\pi i}{e^{-2\pi i H/z}-\xi^k}\frac{1}{\Gamma(1-k/5)^5}.$$

Let $h=(\frac{1}{5},\frac{1}{5},\frac{1}{5},\frac{3}{5},\frac{4}{5})$, $h_1=(\frac{4}{5},\frac{4}{5},\frac{4}{5},\frac{1}{5},\frac{2}{5})$, and $\gamma=(0,0,0,\frac{2}{5},\frac{3}{5})$. Then the transformation in these sectors is defined by

$$\mathbb{U}_h:\phi_h\mapsto \mathbb{1}_{\gamma}\frac{\Gamma(1+H/z)^3\,\Gamma(2/5+H/z)\,\Gamma(3/5+H/z)}{\Gamma(1+5H/z)}\cdot \frac{(-\xi)2\pi i}{e^{-2\pi i H/z}-\xi}\frac{1}{\Gamma(1-1/5)^3\,\Gamma(1-3/5)\,\Gamma(1-4/5)},$$

$$\mathbb{U}_{h_1}:\phi_{h_1}\mapsto z\mathbb{1}_{\gamma}\frac{\Gamma(1+H/z)^3\,\Gamma(2/5+H/z)\,\Gamma(3/5+H/z)}{\Gamma(1+5H/z)}\cdot \frac{(-\xi)^4 2\pi i}{e^{-2\pi i H/z}-\xi^4}\frac{1}{\Gamma(1-4/5)^3\,\Gamma(1-1/5)\,\Gamma(1-2/5)}$$

Let $h=(\frac{2}{5},\frac{2}{5},\frac{2}{5},\frac{3}{5},\frac{1}{5})$, $h_1=(\frac{3}{5},\frac{3}{5},\frac{3}{5},\frac{4}{5},\frac{2}{5})$, and $\gamma=(0,0,0,\frac{1}{5},\frac{4}{5})$. Then the transformation in these sectors is defined by

$$\mathbb{U}:\phi_h\mapsto \mathbb{1}_{\gamma}\frac{\Gamma(1+H/z)^3\,\Gamma(1/5+H/z)\,\Gamma(4/5+H/z)}{\Gamma(1+5H/z)}\cdot \frac{-(\xi)^2 2\pi i}{e^{-2\pi i H/z}-\xi^2}\frac{1}{\Gamma(1-2/5)^3\,\Gamma(1-3/5)\,\Gamma(1-1/5)},$$

$$\mathbb{U}_{h_1} : \phi_{h_1} \mapsto z\mathbb{1}_\gamma \frac{\Gamma(1+H/z)^3\,\Gamma(1/5+H/z)\,\Gamma(4/5+H/z)}{\Gamma(1+5H/z)} \cdot \frac{(\xi)^3 2\pi i}{e^{-2\pi i H/z}-\xi^3} \frac{1}{\Gamma(1-3/5)^3\,\Gamma(1-4/5)\,\Gamma(1-2/5)}$$

Let $h = (\frac{1}{5}, \frac{1}{5}, \frac{2}{5}, \frac{2}{5}, \frac{4}{5})$ and $h_1 = (\frac{3}{5}, \frac{3}{5}, \frac{4}{5}, \frac{4}{5}, \frac{1}{5})$, $\gamma = (0, 0, \frac{1}{5}, \frac{1}{5}, \frac{3}{5})$, $\gamma_1 = (\frac{4}{5}, \frac{4}{5}, 0, 0, \frac{2}{5})$. The transformation in these sectors is defined by

$$\begin{aligned}\mathbb{U}_h : \phi_h \mapsto\ & \mathbb{1}_\gamma \Gamma(1-1/5)^2\,\Gamma(1-3/5) \cdot \\ & \frac{(-\xi)2\pi i}{1-\xi} \frac{1}{\Gamma(1-1/5)^2\,\Gamma(1-2/5)^2\,\Gamma(1-4/5)} \\ & + \mathbb{1}_{\gamma_1} \frac{\Gamma(1/5)^2\,\Gamma(3/5)}{z} \cdot \\ & \frac{(-1)\xi^2 2\pi i}{1-\xi^2} \frac{1}{\Gamma(1-1/5)^2\,\Gamma(1-2/5)^2\,\Gamma(1-4/5)},\end{aligned}$$

$$\begin{aligned}\mathbb{U}_{h_1} : \phi_{h_1} \mapsto\ & z\mathbb{1}_\gamma \Gamma(1-1/5)^2\,\Gamma(1-3/5) \cdot \\ & \frac{(\xi)^3 2\pi i}{1-\xi^3} \frac{1}{\Gamma(1-3/5)^2\,\Gamma(1-4/5)^2\,\Gamma(1-1/5)} \\ & + \mathbb{1}_{\gamma_1} \Gamma(1/5)^2\,\Gamma(3/5) \cdot \\ & \frac{\xi^4 2\pi i}{1-\xi^4} \frac{1}{\Gamma(1-3/5)^2\,\Gamma(1-4/5)^2\,\Gamma(1-1/5)},\end{aligned}$$

Let $h = (\frac{1}{5}, \frac{1}{5}, \frac{3}{5}, \frac{3}{5}, \frac{2}{5})$ and $h_1 = (\frac{2}{5}, \frac{2}{5}, \frac{4}{5}, \frac{4}{5}, \frac{3}{5})$, $\gamma = (0, 0, \frac{2}{5}, \frac{2}{5}, \frac{1}{5})$, $\gamma_1 = (\frac{3}{5}, \frac{3}{5}, 0, 0, \frac{4}{5})$. The transformation in these sectors is defined by

$$\begin{aligned}\mathbb{U}_h : \phi_h \mapsto\ & \mathbb{1}_\gamma \Gamma(1-2/5)^2\,\Gamma(1-1/5) \cdot \\ & \frac{(-\xi)2\pi i}{1-\xi} \frac{1}{\Gamma(1-1/5)^2\,\Gamma(1-3/5)^2\,\Gamma(1-2/5)} \\ & + \mathbb{1}_{\gamma_1} \frac{\Gamma(2/5)^2\,\Gamma(1/5)}{z} \cdot \\ & \frac{(-1)\xi^3 2\pi i}{1-\xi^3} \frac{1}{\Gamma(1-1/5)^2\,\Gamma(1-3/5)^2\,\Gamma(1-2/5)},\end{aligned}$$

$$
\begin{aligned}
\mathbb{U}_{h_1}: \phi_{h_1} \mapsto & z\mathbb{1}_{\gamma}\Gamma(1-2/5)^2\,\Gamma(1-1/5)\cdot \\
& \frac{(-\xi)^2 2\pi i}{1-\xi^2}\frac{1}{\Gamma(1-2/5)^2\,\Gamma(1-4/5)^2\,\Gamma(1-3/5)} \\
& +\mathbb{1}_{\gamma_1}\Gamma(2/5)^2\,\Gamma(1/5)\cdot \\
& \frac{(-1)^2\xi^4 2\pi i}{1-\xi^4}\frac{1}{\Gamma(1-2/5)^2\,\Gamma(1-4/5)^2\,\Gamma(1-3/5)},
\end{aligned}
$$

We can restate the previous theorems schematically as in the following figure.

$$
\begin{array}{ccc}
J_h^{FJRW}(t,z) & \overset{\text{LG/CY}}{\longleftarrow\text{-}\text{-}\text{-}\text{-}} & J_\gamma^{GW}(q,z) \\
{\scriptstyle\text{mirror map}}\updownarrow & & \updownarrow{\scriptstyle\text{mirror map}} \\
I_h(t,z) & \underset{\mathbb{U}}{\overset{\text{analytic continuation}}{\longleftarrow\!\!\!-\!\!\!-\!\!\!-}} & I_\gamma(q,z)
\end{array}
$$

Finally, if we write $\widetilde{\mathcal{L}}^{\mathcal{X}}$ for the analytic continuation of the Lagrangian cone, we can express the (small) LG/CY correspondence as in the following corollary.

Corollary 6.4. *The symplectic transformation* $\mathbb{U}$ *identifies the* (*small*) *GW theory of* $\mathcal{X}$ *with the* (*small*) *FJRW theory of* (W, G), *i.e.* $\mathbb{U}(\mathcal{L}^{W,G}) = \widetilde{\mathcal{L}}^{\mathcal{X}}$.

It is conjectured that the transformation $\mathbb{U}$ relates the full genus zero theory, as well as the higher genus theories.

References

[1] Dan Abramovich, Tom Graber and Angelo Vistoli. Gromov-Witten theory of Deligne-Mumford stacks. *Amer. J. Math.*, 130(5):1337–1398, 2008.

[2] Weimin Chen and Yongbin Ruan. Orbifold Gromov-Witten theory. In *Orbifolds in mathematics and physics* (*Madison, WI, 2001*), volume 310 of *Contemp. Math.*, pages 25–85. Amer. Math. Soc., Providence, RI, 2002.

[3] Weimin Chen and Yongbin Ruan. A new cohomology theory of orbifold. *Comm. Math. Phys.*, 248(1):1–31, 2004.

[4] Alessandro Chiodo, Hiroshi Iritani and Yongbin Ruan. Landau-Ginzburg/Calabi-Yau correspondence, global mirror symmetry and Orlov equivalence. *Publ. Math. Inst. Hautes Études Sci.*, 119:127–216, 2014.

[5] Alessandro Chiodo and Yongbin Ruan. Landau-Ginzburg/Calabi-Yau correspondence for quintic three-folds via symplectic transformations. *Invent. Math.*, 182(1):117–165, 2010.

[6] Alessandro Chiodo and Yongbin Ruan. A global mirror symmetry framework for the Landau-Ginzburg/Calabi-Yau correspondence. *Ann. Inst. Fourier (Grenoble)*, 61(7):2803–2864, 2011.

[7] Alessandro Chiodo and Yongbin Ruan. LG/CY correspondence: the state space isomorphism. *Adv. Math.*, 227(6):2157–2188, 2011.

[8] Emily Clader. Landau-Ginzburg/Calabi-Yau correspondence for the complete intersections $X_{3,3}$ and $X_{2,2,2,2}$. *Adv. Math.*, 307:1–52, 2017.

[9] Huijun Fan, Tyler Jarvis and Yongbin Ruan. The Witten equation, mirror symmetry, and quantum singularity theory. *Ann. of Math.* (2), 178(1):1–106, 2013.

[10] Alexander B. Givental. Equivariant Gromov-Witten invariants. *Internat. Math. Res. Notices*, (13):613–663, 1996.

[11] Alexander B. Givental. Symplectic geometry of Frobenius structures. In *Frobenius manifolds*, Aspects Math., E36, pages 91–112. Friedr. Vieweg, Wiesbaden, 2004.

[12] B. R. Greene and M. R. Plesser. Duality in Calabi-Yau moduli space. *Nuclear Phys. B*, 338(1):15–37, 1990.

[13] Marc Krawitz and Yefeng Shen. Landau-ginzburg/calabi-yau correspondence of all genera for elliptic orbifold $\mathbb{P}^1$. arXiv:1106.6270, 2011.

[14] Y.-P. Lee. Notes on axiomatic Gromov-Witten theory and applications. In *Algebraic geometry—Seattle* 2005. *Part* 1, volume 80 of *Proc. Sympos. Pure Math.*, pages 309–323. Amer. Math. Soc., Providence, RI, 2009.

[15] Y.-P. Lee and Mark Shoemaker. A mirror thoerem for the mirror quintic. *Geom. Topol.*, 18(3):1437–1483, 2014.

[16] Nathan Priddis and Mark Shoemaker. A Landau–Ginzburg/Calabi–Yau correspondence for the mirror quintic. *Ann. Inst. Fourier (Grenoble)*, 66(3):1045–1091, 2016.

[17] Kyoji Saito. From primitive form to mirror symmetry. *RIMS Kokyuroku* 1918:59-70, 2014.

[18] Hsian-Hua Tseng. Orbifold quantum Riemann-Roch, Lefschetz and Serre. *Geom. Topol.*, 14(1):1–81, 2010.

[19] Cumrun Vafa and Nicholas Warner. Catastrophes and the classification of conformal theories. *Phys. Lett. B*, 218(1):51–58, 1989.

[20] Edward Witten. Phases of $N = 2$ theories in two dimensions. In *Mirror symmetry*, II, volume 1 of *AMS/IP Stud. Adv. Math.*, pages 143–211. Amer. Math. Soc., Providence, RI, 1997.

Institute for Algebraic Geometry
Leibniz Universität Hannover
Welfengarten 1
30060 Hannover, Germany
E-mail address: priddis@mathematics.byu.edu

Advanced Studies in Pure Mathematics 83, 2019
Primitive Forms and Related Subjects — Kavli IPMU 2014
pp. 371–387

Quantum invariants via tautological relations

Yefeng Shen

Abstract.

This is a report on the calculation of some quantum invariants, including Gromov-Witten invariants and FJRW invariants, via tautological relations on the moduli space of stable pointed curves.

§1. Cohomological Field Theories

1.1. Cohomological Field Theories

The concept of Cohomological Field Theories (CohFTs) is introduced by Kontsevich and Manin in [13] to axiomatize the properties of Gromov-Witten classes. It describes various quantum invariants, including Gromov-Witten invariants and FJRW invariants.

The definition starts with a pair (H, η), where $H \cong \mathbb{C}^r$ is a vector space of rank r and η is a symmetric nondegenerate pairing on H. We use $\mathfrak{B}$ to denote a basis of the vector space H. Let $\overline{\mathcal{M}}_{g,k}$ be the moduli space of stable k-pointed curves of genus g, where the stability is equivalent to the condition $2g - 2 + k > 0$.

Definition 1.1. A *Cohomological Field Theory* (CohFT for short) $\Lambda = \{\Lambda_{g,k}\}$ over (H, η), is a set of multi-linear maps,

$$\Lambda_{g,k} : H^{\otimes k} \longrightarrow H^*(\overline{\mathcal{M}}_{g,k}, \mathbb{C}),$$

such that the following set of axioms (CohFT axioms) are satisfied:

- (S_k-covariance) For any $\sigma \in S_k$, and $\phi_1, \ldots, \phi_k \in H$; then

$$\Lambda_{g,k}(\phi_{\sigma(1)}, \ldots, \phi_{\sigma(k)}) = \Lambda_{g,k}(\phi_1, \ldots, \phi_k).$$

Received January 15, 2015.
Revised May 12, 2015.
2010 *Mathematics Subject Classification.* 14N35, 14B05.
Key words and phrases. Quantum invariants, tautological relations.

- (Gluing tree) Let

(1) $$\rho_{tree}:\overline{\mathcal{M}}_{g_1,k_1+1}\times\overline{\mathcal{M}}_{g_2,k_2+1}\to\overline{\mathcal{M}}_{g,k}$$

be the morphism induced from gluing the last marked point of the first curve and the first marked point of the second curve, where $g=g_1+g_2, k=k_1+k_2$; then

$$\begin{aligned}&\rho^*_{tree}\big(\Lambda_{g,k}(\phi_1,\dots,\phi_k)\big)\\&\quad=\sum_{\alpha,\beta\in\mathfrak{B}}\Lambda_{g_1,k_1+1}(\phi_1,\dots,\phi_{k_1},\alpha)\eta^{\alpha,\beta}\Lambda_{g_2,k_2+1}(\beta,\phi_{k_1+1},\dots,\phi_k).\end{aligned}$$

Here $\left(\eta^{\alpha,\beta}\right)_{r\times r}$ is the inverse matrix of $\big(\eta(\alpha,\beta)\big)_{r\times r}$.

- (Gluing loop) Let

(2) $$\rho_{loop}:\overline{\mathcal{M}}_{g-1,k+2}\to\overline{\mathcal{M}}_{g,k}$$

be the morphism induced from gluing the last two marked points; then

$$\rho^*_{loop}\big(\Lambda_{g,k}(\phi_1,\dots,\phi_k)\big)=\sum_{\alpha,\beta\in\mathfrak{B}}\Lambda_{g-1,k+2}(\phi_1,\dots,\phi_k,\alpha,\beta)\eta^{\alpha,\beta}.$$

Note that $\Lambda_{0,3}$ will induce a multiplication $\bullet$ on (H,η), defined by

(3) $$\eta(\phi_i\bullet\phi_j,\phi_\ell)=\int_{\overline{\mathcal{M}}_{0,3}}\Lambda_{0,3}(\phi_i,\phi_j,\phi_\ell);$$

The multiplication $\bullet$ is Frobenius, i.e.,

$$\eta(\phi_i\bullet\phi_j,\phi_\ell)=\eta(\phi_i,\phi_j\bullet\phi_\ell).$$

In this paper, we will always consider a CohFT Λ with a *flat identity*. The flat identity is denoted by $\mathbf{1}$, and satisfies two more axioms:

- (Identity) The element $\mathbf{1}$ is a unit of the Frobenius algebra $(H,\eta,\bullet)$, and

$$\int_{\overline{\mathcal{M}}_{0,3}}\Lambda_{0,3}(\mathbf{1},\phi_1,\phi_2)=\eta(\phi_1,\phi_2).$$

- (Flatness) Let

(4) $$\pi:\overline{\mathcal{M}}_{g,k+1}\to\overline{\mathcal{M}}_{g,k}$$

be the morphism forgeting the last marked point; then

$$\Lambda_{g,k+1}(\phi_1,\dots,\phi_k,\mathbf{1})=\pi^*\Lambda_{g,k}(\phi_1,\dots,\phi_k).$$

We refer to $(H, \eta, \bullet, \mathbf{1})$ as the Frobenius algebra underlying the CohFT $\Lambda = \{\Lambda_{g,k}\}$. The CohFT is called *semisimple* if the underlying Frobenius algebra is semisimple. Based on Givental's higher genus formula [11], Teleman classified all homogenous CohFTs over semisimple Frobenius algebras in [22].

1.2. Quantum invariants

Let ψ_j be the j-th psi class in $H^*(\overline{\mathcal{M}}_{g,k})$. The quantum invariants (or correlators) with respect to Λ are defined by the following intergation:

$$(5) \qquad \langle \phi_1\psi_1^{\ell_1}, \ldots, \phi_k\psi_k^{\ell_k}\rangle_{g,k}^{\Lambda} = \int_{\overline{\mathcal{M}}_{g,k}} \Lambda_{g,k}(\phi_1, \ldots, \phi_k) \prod_{i=1}^{k} \psi_i^{\ell_i}.$$

The invariants in (5) are called primary if all $\ell_i = 0$. There is a formal Frobenius manifold structure on H, in the sense of Dubrovin [5]. Its *prepotential* is given by

$$\mathcal{F}_0^{\Lambda} = \sum \frac{1}{k!}\langle \mathbf{t}_0, \ldots, \mathbf{t}_0\rangle_{0,k}^{\Lambda}, \quad \mathbf{t}_0 = \sum_{\alpha\in\mathfrak{B}} t_{0,\alpha}\,\alpha.$$

1.3. Examples of CohFTs

We give several examples of CohFTs in this section.

1.3.1. *A trivial CohFT* We assign $\mathbb{C}^r$ the standard bi-linear pairing $\eta(\phi_i, \phi_j) = \delta_i^j$. Let $\lambda = (\lambda_1, \ldots, \lambda_N)$ be a sequence of *non-zero* complex numbers. Let

$$(6) \qquad I_{g,k}^{r,\lambda}(\phi_{i_1}, \ldots, \phi_{i_k}) := \begin{cases} \lambda_i^{g-1+\frac{k}{2}}\mathbf{e}_{g,k}, & \text{if } i = i_1 = i_2 = \cdots = i_k, \\ 0, & \text{otherwise.} \end{cases}$$

Here $\mathbf{e}_{g,k} \in H^0(\overline{M}_{g,k})$ is a unit in $H^*(\overline{M}_{g,k})$. This induces a trivial CohFT $I^{r,\lambda} := \{I_{g,k}^{r,\lambda}\}$ on $(\mathbb{C}^r, \eta)$. The Frobenius multiplication is semisimple since we have

$$\phi_i \bullet \phi_j = \delta_i^j \sqrt{\lambda_i}\,\phi_i.$$

1.3.2. *FJRW theory* FJRW theory produces quantum invariants of a singularity W, where $W : \mathbb{C}^n \to \mathbb{C}$ is a quasi-homogenous polynomial with the isolated critial point only at the orgin $0 \in \mathbb{C}^n$. Then for any nonzero $\lambda \in \mathbb{C}$, we have unique $q_i \in (0, \frac{1}{2}] \cap \mathbb{Q}$, such that

$$W(\lambda^{q_1}x_1, \ldots, \lambda^{q_n}x_n) = \lambda W(x_1, \ldots, x_n).$$

Here $q_i \in \mathbb{Q}$ is the weight of variable x_i. Furthermore, assume there is no monomials of the form xy in W. The group of diagonal symmetries of W is denoted by G_W,

(7)

$$G_W := \left\{(\lambda_1, \dots, \lambda_n) \in (\mathbb{C}^\times)^n \,\middle|\, W(\lambda_1 x_1, \dots, \lambda_n x_n) = W(x_1, \dots, x_n)\right\}.$$

Let G be a subgroup of G_W. In physics, such a pair (W, G) is referred to as a Landau-Ginzburg model. Based on a proposal of Witten [24], Fan, Jarvis, and Ruan constructed a CohFT $\Lambda^{W,G}$ for the Landau-Ginzburg A-model (W, G) [8, 9], when G contains the exponential grading element $J := \big(\exp(2\pi\sqrt{-1}q_1), \dots, \exp(2\pi\sqrt{-1}q_n)\big)$. Now the theory is called FJRW theory. The vector space $H := H^{W,G}$ of the CohFT $\Lambda^{W,G}$ is defined to be the direct sum of all G-invariant relative cohomology:

$$H^{W,G} := \bigoplus_{\gamma \in G} H_\gamma, \quad H_\gamma := H^{N_\gamma}\big(\mathrm{Fix}(\gamma); W_\gamma^\infty; \mathbb{C}\big)^G. \tag{8}$$

Here $\mathrm{Fix}(\gamma)$ is the fixed locus of γ, and $\mathbb{C}^{N_\gamma} \cong \mathrm{Fix}(\gamma) \subset \mathbb{C}^n$. W_γ is the restriction of W to $\mathrm{Fix}(\gamma)$. $W_\gamma^\infty := (\mathrm{Re}W_\gamma)^{-1}(M, \infty)$ with $M \gg 0$, where $\mathrm{Re}W_\gamma$ is the real part of W_γ. The space $H^{W,G}$ is dual to the space of Lefschetz thimbles, and it has a nondegenerate pairing η which is induced from the intersection pairing of Lefschetz thimbles.

The construction of the maps

$$\Lambda_{g,k}^{W,G} : (H^{W,G})^{\otimes k} \to H^*(\overline{\mathcal{M}}_{g,k}, \mathbb{C})$$

is highly nontrivial. Roughly speaking, they are defined via virtual counting of solutions of perturbed Witten equations, where vector space plays the role of boundary conditions for the perturbed Witten equations. $\Lambda^{W,G}$ satisfies all the axioms in Definition 1.1, and has a flat identity $\mathbf{1}$, which is a generator in H_J.

The $(H^{W,G}, \bullet)$ is called an *FJRW ring*. We call the elements in H_γ *narrow* if $\mathrm{Fix}(\gamma) = \mathbf{0} \in \mathbb{C}^n$, or *broad* otherwise. Each element $\gamma \in G$ has a unique form

$$\gamma = \Big(\exp(2\pi\sqrt{-1}\Theta_1^\gamma), \dots, \exp(2\pi\sqrt{-1}\Theta_n^\gamma)\Big) \in (\mathbb{C}^\times)^n, \quad \Theta_i^\gamma \in [0, 1) \cap \mathbb{Q}.$$

$H^{W,G}$ is a graded vector space, with the degree of a nonzero $\phi \in H_\gamma$ defined by

$$\deg \phi = N_\gamma + \sum_{i=1}^n (\Theta_i^\gamma - q_i). \tag{9}$$

According to [8], if the FJRW invariant $\langle \phi_1\psi_1^{\ell_1}, \ldots, \phi_k\psi_k^{\ell_k} \rangle_{g,k}^{W,G}$ defined via (5) for $\Lambda = \Lambda^{W,G}$ is nonzero, then the following equation holds:

$$\sum_{i=1}^{k} (\deg \phi_i + \ell_i) = (\hat{c}_W - 3)(1-g) + k. \tag{10}$$

This is called Dimension Axiom, where $\hat{c}_W$ is the *central charge* of W, defined by

$$\hat{c}_W := \sum_{i=1}^{n} (1 - 2q_i).$$

1.3.3. *Gromov-Witten theory* Let X be a projective variety, H^X be its cohomology $H^*(X)$, and η be the Poincaré pairing. Let $\overline{\mathcal{M}}_{g,k\,\beta}^X$ be the moduli space of stable maps from k-pointed stable curves of genus g and degree β to X, where $\beta \in H^2(X,\mathbb{Z})$ in the *Mori cone of* X is the image of the fundamental class of the curve under the stable map. Gromov-Witten theory studies the intersection theory on $\overline{\mathcal{M}}_{g,k,\beta}^X$. Following [13], let

$$\Lambda_{g,n,\beta}^X : (H^X)^{\otimes k} \to H^*(\overline{\mathcal{M}}_{g,k}, \mathbb{C})$$

be the Gromov-Witten classes of the target X. Then the following series

$$\Lambda_{g,k}^X(q) := \sum_{\beta} \Lambda_{g,n,\beta}^X\, q^{\beta}$$

satisfies all the axioms in Definition 1.1. In such a case we say that we have a family of CohFTs which depends formally on q. We define Gromov-Witten invariant $\langle \phi_1\psi_1^{\ell_1}, \ldots, \phi_k\psi_k^{\ell_k} \rangle_{g,k,\beta}^X$ by replacing Λ with $\Lambda_{g,k,\beta}^X$ in Definition 1.1, and a Gromov-Witten correlation function by replacing Λ with $\Lambda_{g,k}^X(q)$,

$$\left\langle\left\langle \phi_1\psi_1^{\ell_1}, \ldots, \phi_k\psi_k^{\ell_k} \right\rangle\right\rangle_{g,k} := \int_{\overline{\mathcal{M}}_{g,k}} \Lambda_{g,k}^X(q)(\phi_1, \ldots, \phi_k) \prod_{j=1}^{k} \psi_j^{\ell_j}. \tag{11}$$

Similar as the Dimension Axiom (10) in FJRW theory, the degree of the virtual fundamental cycle of $\overline{\mathcal{M}}_{g,k,\beta}^X$ gives the following Dimension Axiom in Gromov-Witten theory. For a nonzero Gromov-Witten invariant $\langle \phi_1\psi_1^{\ell_1}, \ldots, \phi_k\psi_k^{\ell_k} \rangle_{g,k,\beta}^X$, the following equation holds,

$$\sum_{i=1}^{k} (\deg_{\mathbb{C}}(\phi_i) + \ell_i) = c_1(TX)\cdot\beta + (3 - \dim_{\mathbb{C}} X)(g-1) + k. \tag{12}$$

Gromov-Witten theory also works when the target X is a symplectic orbifold [4] or a Deligne Mumford stack [1]. There the vector space H^X is replaced by Chen-Ruan cohomology $H^*_{\mathrm{CR}}(X)$. The Dimension formula above still holds.

§2. Tautological relations

Let $A^*(\overline{\mathcal{M}}_{g,k})$ be the Chow ring of $\overline{\mathcal{M}}_{g,k}$. The system of the *tautological rings* $\left(R^*(\overline{\mathcal{M}}_{g,k}) \subset A^*(\overline{\mathcal{M}}_{g,k})\right)_{g,k}$ is the smallest system of $\mathbb{Q}$-algebras that contains the psi-classes $\psi_1, \ldots, \psi_k$, and is closed under the pushforwards by natural morphisms (1), (2), and (4). Tautological relations are relations over $R^*(\overline{\mathcal{M}}_{g,k})$. In this section, we will list several tautological relations, including the most elementary one in $R^*(\overline{\mathcal{M}}_{0,4})$, a relation induced from Getzler's relation, and g-reduction.

We will express some tautological relations via *dual graphs* for boundary cycles in $\overline{\mathcal{M}}_{g,k}$. In a dual graph, a vertex (a filled circle) represents a component, with its genus labelled by a number below, an edge connecting two vertices represents a node, and a tail (or half-edge) from a vertex represents a marked point on the component. We label the tail by the corresponding marked point p_i.

Then we integrate a cohomology class $\Lambda_{g,k}(\ldots) \in H^*(\overline{\mathcal{M}}_{g,k})$ among the dual graphs. The Axioms of CohFTs allow us to obtain relations among those quantum invariants defined on each stratum. We will apply those relations to simplify some calculations of quantum invariants in the next section.

Example 2.1 (Relations in $R^*(\overline{\mathcal{M}}_{0,4})$)**.** The three boundary cycles in $H^*(\overline{\mathcal{M}}_{0,4})$ represented by the three dual graphs below are cohomologically equal:

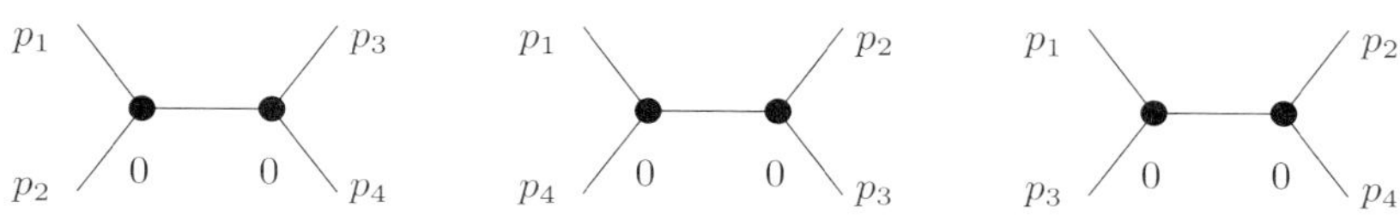

The WDVV equations are obtained by pulling back the Poincaré dual of the relations above to $\overline{\mathcal{M}}_{0,k}$ ($k \geq 4$) by $\pi_{4,k} : \overline{\mathcal{M}}_{0,k} \to \overline{\mathcal{M}}_{0,4}$, which forgets the last $k-4$ marked points (followed by stablization) and then integrating a genus zero class $\Lambda_{0,k}(\ldots)$ created by inserting some elements of H at these marked points. In terms of the prepotential $\mathcal{F}_0$, the WDVV equations are:

(13)

$$\sum_{\alpha,\beta} \frac{\partial^3 \mathcal{F}_0}{\partial t_{\phi_1}\partial t_{\phi_2}\partial t_\alpha}\eta^{\alpha,\beta}\frac{\partial^3 \mathcal{F}_0}{\partial t_\beta \partial t_{\phi_3}\partial t_{\phi_4}} = \sum_{\alpha,\beta} \frac{\partial^3 \mathcal{F}_0}{\partial t_{\phi_1}\partial t_{\phi_4}\partial t_\alpha}\eta^{\alpha,\beta}\frac{\partial^3 \mathcal{F}_0}{\partial t_\beta \partial t_{\phi_2}\partial t_{\phi_3}},$$

$t_\alpha := t_{0,\alpha}$.

Example 2.2 (Getzler's relation)**.** Getzler's relation is a linear relation among $H_4(\overline{\mathcal{M}}_{1,4}, \mathbb{Q})$. The following dual graph represents a codimension-two stratum in $\overline{\mathcal{M}}_{1,4}$:

$\Delta_{1,234} :=$

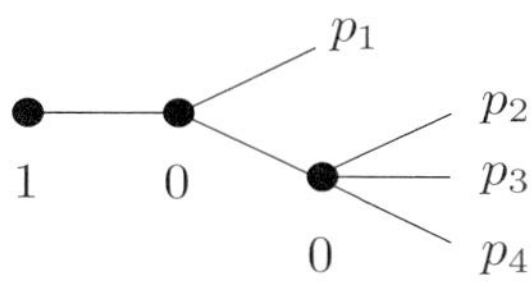

We have the following S_4-invariant of the codimension-two stratum in $\overline{\mathcal{M}}_{1,4}$,

$$\Delta_{1,234} + \Delta_{2,134} + \Delta_{3,124} + \Delta_{4,123}\,,$$

whose corresponding cycle in $H_4(\overline{\mathcal{M}}_{1,4}, \mathbb{Q})$ will be denoted by $\delta_{3,4}$. We list the corresponding unordered dual graphs for other strata below,

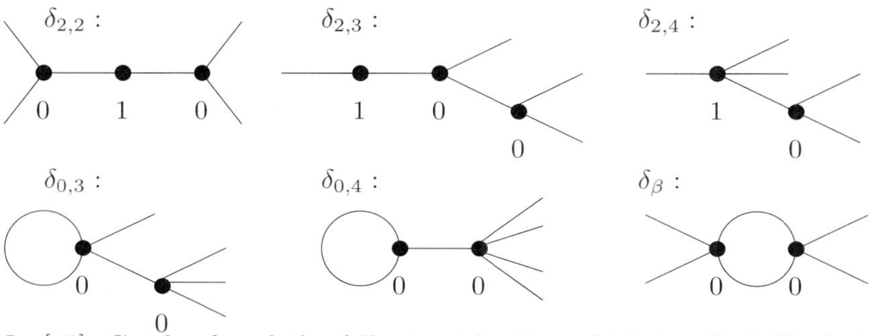

In [10], Getzler found the following identity, which is called *Getzler's relation*,

(14) $12\delta_{2,2} - 4\delta_{2,3} - 2\delta_{2,4} + 6\delta_{3,4} + \delta_{0,3} + \delta_{0,4} - 2\delta_\beta = 0 \in H_4(\overline{\mathcal{M}}_{1,4}, \mathbb{Q})$.

Example 2.3 (g-reduction)**.** At higher genus, we have the g-reduction property:

Lemma 2.4 ([12, 6])**.** *If $M(\psi, \kappa)$ is a polynomial of ψ-classes and κ-classes with* $\deg M \geq g$ *for* $g \geq 1$ *or* $\deg M \geq 1$ *for* $g = 0$*, then*

$M(\psi, \kappa)$ can be presented as a linear combination of dual graphs on the boundary of $\overline{\mathcal{M}}_{g,k}$.

§3. Applications

Historically, tautological relations are used systematically to determine the Gromov-Witten invariants, see [13, 10, 3], etc. In this section, we give some more examples of using tautological relations to simplify the calculations of Gromov-Witten invariants and FJRW invariants.

3.1. FJRW theory of A_{n-1}-singularity

Let $W = x^n$ and G be the group generated by the exponential grading element

$$J = \exp\left(\frac{2\pi\sqrt{-1}}{n}\right).$$

For the pair (W, G), its FJRW vector space is just

$$H^{A_{n-1}} := H^{W,G} = \bigoplus_{i=1}^{n-1} \mathbb{C}\gamma_i,$$

where γ_i is a generator of $H_{J^i} \cong \mathbb{C}$. All the elements are narrow and $\deg \gamma_i = \frac{i-1}{n}$ according to (9). The FJRW ring $(H^{A_{n-1}}, \bullet)$ has a unit $\gamma_1 = \mathbf{1}$. The WDVV equations and some other axioms in FJRW theory imply the following reconstruction theorem:

Theorem 3.1. [7, 8] *The prepotential $\mathcal{F}_0^{\Lambda}$ of A_{n-1}-singularity can be reconstructed from the pairing, the ring structure, and the invariant $\langle \gamma_2, \gamma_2, \gamma_{n-1}, \gamma_{n-1} \rangle_{0,4}$.*

All the elements γ_i are narrow. The primary correlators $\langle \ldots \rangle_{0,3}$ and the correlator $\langle \gamma_2, \gamma_2, \gamma_{n-1}, \gamma_{n-1} \rangle_{0,4}$ can be calculated by using the Topological Euler class Axioms for the narrow elements in FJRW theory [8, Theorem 4.1.8, (5)]. This determines the ring structure defined in (3), and as a consequence,

$$(H^{A_{n-1}}, \bullet) = \mathbb{C}[\gamma_2]/(\gamma_2^{n-1} = 0).$$

On the other hand, according to Flateness Axiom in FJRW theory (string equation), for $k \geq 4$, a primary correlator of the following form vanishes,

$$\langle \ldots, \mathbf{1} \rangle_{0,k} = 0.$$

So for $k \geq 4$, we may assume there is no insertion $\mathbf{1}$ in the correlator $\langle \ldots \rangle_{0,k}$. The WDVV equation (13) implies ([8, Lemma 6.2.6])

(15)
$$\begin{aligned}\langle \phi_1 \bullet \phi_2, \phi_3, \phi_4, \ldots \rangle_{0,k} = {} & \langle \phi_1, \phi_2 \bullet \phi_3, \phi_4, \ldots \rangle_{0,k} + \langle \phi_2, \phi_1 \bullet \phi_4, \phi_3, \ldots \rangle_{0,k} \\ & - \langle \phi_1, \phi_2, \phi_3 \bullet \phi_4, \ldots \rangle_{0,k} + \mathfrak{S}_{<k}.\end{aligned}$$

where $k \geq 3$, $\mathfrak{S}_{<k}$ is a linear combination of products of correlators with number of marked points no greater than $k-1$. Moreover, both $\mathfrak{S}_{<3} = \mathfrak{S}_{<4} = 0$.

If $\phi_1 \bullet \phi_2 \neq \mathbf{1}, \gamma_2$, then the equation above implies we can rewrite the correlator on the left hand side with correlators of less marked points and three other correlators. Besides that, the degree of the first insertion (that is, ϕ_1 or ϕ_2) of the three correlators on the right hand side of the above equation is strictly smaller than the degree of $\phi_1 \bullet \phi_2$. Thus if there are three insertions, which are neither $\mathbf{1}$ nor γ_2 in the primary correlator $\langle \ldots \rangle_{0,k}$, we can use the equation above to reduce the degree of one of the insertions, with the last $k-3$ insertions remains the same. By repeating the process, we give an algorithm in obtaining all the genus zero primary correlators from $\langle \ldots \rangle_{0,3}$ and correlators of the following type,

$$\langle \gamma_2, \ldots, \gamma_2, \phi_{k-1}, \phi_k \rangle_{0,k}.$$

For this correlator, the upper bound of its total degree is

$$\frac{k-2}{n} + \frac{2(n-2)}{n}.$$

Combine this with Dimension Axiom (10), we get

$$\left(1 - \frac{2}{n}\right) - 3 + k \leq \frac{k-2}{n} + \frac{2(n-2)}{n}.$$

Since $n \geq 2$, we must have $k \leq 4$. The equality holds if and only if the correlator is $\langle \gamma_2, \gamma_2, \gamma_{n-1}, \gamma_{n-1} \rangle_{0,4}$. Q.E.D.

3.2. Genus one invariants for elliptic orbifold curve $\mathbb{P}^1_{2,2,2,2}$

Let $X := \mathbb{P}^1_{2,2,2,2}$ be the elliptic orbifold curve with its underlying curve $\mathbb{P}^1$ and four orbifold points, each has an isotropy group $\mathbb{Z}_2$. The orbifold is elliptic since it can be obtained as a global $\mathbb{Z}_2$ quotient of an elliptic curve. The first Chern class vanishes, and $\mathbb{P}^1_{2,2,2,2}$ is a Calabi-Yau orbifold. The Chen-Ruan cohomology space of X is

$$H^*_{\mathrm{CR}}(X, \mathbb{C}) := \bigoplus_{i=1}^{4} \mathbb{C}\Delta_i \bigoplus \mathbb{C}\mathbf{1} \bigoplus \mathbb{C}P.$$

Here $\mathbf{1} \in H^0_{\mathrm{CR}}(X,\mathbb{C})$ is an unit, $P \in H^2_{\mathrm{CR}}(X,\mathbb{C})$ is a divisor class, and $\Delta_i \in H^1_{\mathrm{CR}}(X,\mathbb{C})$ is a *twisted element*, which is supported on the i-th orbifold point. The nonvanishing part of the pairing consist of

$$\eta(\mathbf{1},P)=1,\quad \eta(\Delta_i,\Delta_i)=\frac{1}{2}. \tag{16}$$

The nonvanishing part of the Chen-Ruan cup product $\bullet$ is given by

$$\mathbf{1}\bullet P = 2\,\Delta_i \bullet \Delta_i = P,\quad i=1,2,3,4. \tag{17}$$

Let the parameter t be the coordinate for the vector space spanned by the divisor class P, parameter t_i be the coordinate for the vector space spanned by the unit $\mathbf{1}$ when $i=0$ and that by the twist element Δ_i when $i=1,2,3,4$. The genus zero Gromov-Witten theory of $\mathbb{P}^1_{2,2,2,2}$ is studied in [20] via WDVV equations and direct calculations. Let $q=e^t$, then the prepotential is

(18)

$$\begin{aligned}\mathcal{F}_0^X(t_i,t) &= \text{cubic terms} + t_1t_2t_3t_4\,\langle\langle\Delta_1,\Delta_2,\Delta_3,\Delta_4\rangle\rangle_{0,4}\\ &\quad+\sum_{i=1}^{4}\frac{t_i^4}{4!}\,\langle\langle\Delta_i,\Delta_i,\Delta_i,\Delta_i\rangle\rangle_{0,4}+\sum_{i\neq j}\frac{t_i^2+t_j^2}{2!2!}\,\langle\langle\Delta_i,\Delta_i,\Delta_j,\Delta_j\rangle\rangle_{0,4}.\end{aligned}$$

According to Dimension Axiom (12), the nonzero genus-1 primary invariants are collected in the q-series $\langle\langle P\rangle\rangle_{1,1}$. In [21], the authors used the Getzler's relation [10] to express $\langle\langle P\rangle\rangle_{1,1}$ in terms of genus-0 correlation functions and their derivatives. In addition, they prove

Theorem 3.2 ([21]). *The genus-*1 *correlation function* $\langle\langle P\rangle\rangle_{1,1}$ *of* $X=\mathbb{P}^1_{2,2,2,2}$ *is reconstructed from genus-*0 *invariants in Equation* (18). *Furthermore, after the coordinate change* $Q=q^2$, *it is a quasi-modular form*

$$\langle\langle P\rangle\rangle_{1,1} = -\frac{1}{12}E_2(Q).$$

Here $E_2(Q)$ *is the usual second Eisenstein series.*

Let us explain the reconstruction part of Theorem 3.2 here. The quasi-modular form part follows from the equivalence between WDVV equations of X and Ramanujan identities of the modular subgroup $\Gamma_0(4)$ [21].

The method is as follows. We take the cohomology class $\Lambda^X_{1,4}(\Delta_1,\Delta_2,\Delta_3,\Delta_4)$ and integrate it over Getzler's relation (14). Over each stratum, we simplify the integration using axioms of CohFT. For example, when we integrate the class over the statum $\Delta_{1,234}$, Gluing Tree Axiom in Definition 1.1 implies

$$\begin{aligned}\int_{\Delta_{1,234}} \Lambda^X_{1,4}(\Delta_1,\Delta_2,\Delta_3,\Delta_4) &= \langle\langle P\rangle\rangle_{1,1}\,\eta^{P,\mathbf{1}}\,\langle\langle \mathbf{1},\Delta_1,\Delta_1\rangle\rangle_{0,3}\,\eta^{\Delta_1,\Delta_1}\,\langle\langle \Delta_1,\Delta_2,\Delta_3,\Delta_4\rangle\rangle_{0,4}\\ &= \langle\langle P\rangle\rangle_{1,1}\,\langle\langle \Delta_1,\Delta_2,\Delta_3,\Delta_4\rangle\rangle_{0,4}\,.\end{aligned}$$

Here the last equality uses (16) and (17). The intergration over the other statum in $\delta_{3,4}$ gives the same result and we summarize them to get

$$\int_{\delta_{3,4}} \Lambda^X_{1,4}(\Delta_1,\Delta_2,\Delta_3,\Delta_4) = 4\,\langle\langle P\rangle\rangle_{1,1}\,\langle\langle \Delta_1,\Delta_2,\Delta_3,\Delta_4\rangle\rangle_{0,4}\,.$$

Similarly, we use (16), (17), and all other axioms in Definition 1.1 to obtain

$$\begin{aligned}\int_{\delta_{2,2}} \Lambda^X_{1,4}(\Delta_1,\Delta_2,\Delta_3,\Delta_4) &= \int_{\delta_{2,3}} \Lambda^X_{1,4}(\Delta_1,\Delta_2,\Delta_3,\Delta_4)\\ &= \int_{\delta_{2,4}} \Lambda^X_{1,4}(\Delta_1,\Delta_2,\Delta_3,\Delta_4) = 0\,,\\ \int_{\delta_{0,3}} \Lambda^X_{1,4}(\Delta_1,\Delta_2,\Delta_3,\Delta_4) &= 16\,\langle\langle \Delta_1,\Delta_2,\Delta_3,\Delta_4\rangle\rangle_{0,4}\,\langle\langle \Delta_1,\Delta_1,\Delta_1,\Delta_1\rangle\rangle_{0,4}\\ &\quad + 48\,\langle\langle \Delta_1,\Delta_2,\Delta_3,\Delta_4\rangle\rangle_{0,4}\,\langle\langle \Delta_1,\Delta_1,\Delta_2,\Delta_2\rangle\rangle_{0,4}\,,\\ \int_{\delta_{0,4}} \Lambda^X_{1,4}(\Delta_1,\Delta_2,\Delta_3,\Delta_4) &= 6\,\langle\langle \Delta_1,\Delta_2,\Delta_3,\Delta_4,P\rangle\rangle_{0,5}\,,\\ \int_{\delta_{\beta}} \Lambda^X_{1,4}(\Delta_1,\Delta_2,\Delta_3,\Delta_4) &= 48\,\langle\langle \Delta_1,\Delta_2,\Delta_3,\Delta_4\rangle\rangle_{0,4}\,\langle\langle \Delta_1,\Delta_1,\Delta_2,\Delta_2\rangle\rangle_{0,4}\,.\end{aligned}$$

Let $\theta_q := q\frac{d}{dq}$, then Divisor Axiom in Gromov-Witten theory implies

$$\langle\langle \Delta_1,\Delta_2,\Delta_3,\Delta_4,P\rangle\rangle_{0,5} = \theta_q\,\langle\langle \Delta_1,\Delta_2,\Delta_3,\Delta_4\rangle\rangle_{0,4}\,.$$

Thus we simplify the integration over Getzler's relation and obtain

$$\begin{aligned}\langle\langle P\rangle\rangle^X_{1,1} = -\frac{2}{3}\,\langle\langle \Delta_1,\Delta_1,\Delta_1,\Delta_1\rangle\rangle_{0,4} &+ 2\,\langle\langle \Delta_1,\Delta_1,\Delta_2,\Delta_2\rangle\rangle_{0,4}\\ &- \frac{\theta_q\,\langle\langle \Delta_1,\Delta_2,\Delta_3,\Delta_4\rangle\rangle_{0,4}}{4\,\langle\langle \Delta_1,\Delta_2,\Delta_3,\Delta_4\rangle\rangle_{0,4}}.\end{aligned}$$

3.3. Chain singularity with a weight $\frac{1}{2}$-variable

Mirror symmetry between FJRW theory of invertible singularities and Saito-Givental theory of the mirror singularities is well studied in

[8, 15, 17]. For a singularity W, let E_W be its *exponent matrix*, i.e., the (i, j)-th entry of E_W is the exponent of j-th variable in the i-th monomial of W. The singularity W is called *invertible* if E_W is invertible. The *mirror singularity* W^T is constructed by letting its exponent matrix be the transpose of E_W [2].

A typical way to match the two theories at genus zero is using the ring structure and WDVV equations to recontruct the prepotentials from a few initial correlators, and compute them explicitly on both sides. In general, Topological Euler class Axioms in FJRW theory are enough for the computation. However, for some special cases as we will discuss in this section, even the FJRW ring structure is not easy to compute. Let us first recall that for invertible singularities, Krawitz proved the following mirror statement at the Frobeniu algebra level:

Theorem 3.3 ([14]). *The FJRW ring* $(H^{W,G_W}, \bullet)$ *is isomorphic to the Jacobi algebra of* W^T *if* W *contains no variable of weight* $\frac{1}{2}$.

This theorem is very useful since due to a theorem of Kreuzer and Skarke [16], all the invertible polynomials are classified as a direct sum of three atomic types, called Fermat type, chain type, and loop type respectively. Krawitz's theorem does apply to almost all the cases except for the chain singularity of the form

$$W = x_1^2 + x_1 x_2^{a_2} + \cdots + x_{n-1} x_n^{a_n}, \quad n \geq 3.$$

The FJRW ring for such W contains a generator which is broad. Thus the computation of $\langle \ldots \rangle_{0,3}^{W,G_W}$ is much more complicated. However, for some of such chain singularities, Li, Li, Saito, and the author found a way applying WDVV equations and Getzler's relation together, to simplify the calculation of $\langle \ldots \rangle_{0,3}^{W,G_W}$, and obtain the following result:

Theorem 3.4 ([17]). *For* Z_{13}*-singularity* $W := x^2 + xy^3 + yz^3$, *and* W_{13}*-singularity* $W := x^2 + xy^2 + yz^4$, *FJRW ring* $(H^{W,G_W}, \bullet)$ *is isomorphic to the Jacobi algebra of* W^T.

Since the method is similar for both cases. We only discuss here for the case $W = x^2 + xy^3 + yz^3$. The group $G_W \cong \mathbb{Z}_{18}$ is generated by

$$J = \left(\exp\left(\frac{2\pi\sqrt{-1}}{2}\right), \exp\left(\frac{2\pi\sqrt{-1}}{6}\right), \exp\left(\frac{10\pi\sqrt{-1}}{18}\right)\right).$$

The vector space is spanned by nine narrow elements $\gamma_i \in H_{J^i}$ for odd $i \in \mathbb{Z}$ such that $1 \leq i \leq 17$, and two broad elements

$$u := \gamma_{12} \in H_{J^{12}}, \quad w := \gamma_6 \in H_{J^6}.$$

The elements are chosen in a way that the nonvanishing pairing are

$$\eta(\gamma_i, \gamma_{18-i}) = \eta(u, w) = 1. \tag{19}$$

Use the degree formula (9), we get

$$\deg \gamma_1 = 0, \quad \deg \gamma_{17} = \frac{10}{9}, \quad \text{and} \quad \deg \gamma_i = \frac{19-i}{18}, \quad \text{for } 1 < i < 17. \tag{20}$$

The multiplication $\bullet$ in H^{W,G_W} can be worked out similar as in [14], except we do not know whether $u \bullet u = 0$, simply because we do not know whether the correlator

$$\langle \gamma_{13}, u, u \rangle_{0,3}^W$$

vanishes or not. Actually, it can be checked that if the above correlator is nonzero, then we can still find a ring isomorphism between H^{W,G_W} and the Jacobi algebra of $W^T = x^2y + y^3z + z^3$, by sending a scale of u to x, γ_{15} to y, and γ_{13} to z. Thus here we only prove $\langle \gamma_{13}, u, u \rangle_{0,3}^W$ is nonzero. The idea in [17] is using WDVV equations and Getzler's relation to relate this correlator to other correlators, which can be calculated via Topological Euler class Axioms in FJRW theory. We denote

$$C_1 := \langle \gamma_{13}, \gamma_{13}, w, w \rangle_{0,4}^W, \quad C_2 := \langle \gamma_7, \gamma_{13}, u, w \rangle_{0,4}^W, \quad C_3 := \langle \gamma_7, \gamma_7, u, u \rangle_{0,4}^W.$$

All these correlators contain broad insertions. However, Topological Euler class Axioms imply $\gamma_7 = \gamma_{13} \bullet \gamma_{13}$. And the WDVV equations imply

$$\langle \gamma_7, \gamma_7, u, u \rangle_{0,4}^W = \langle \gamma_7, \gamma_{13}, u, w \rangle_{0,4}^W \langle \gamma_{13}, u, u \rangle_{0,3}^W, \tag{21}$$

$$\langle \gamma_7, \gamma_{13}, u, w \rangle_{0,4}^W + \langle \gamma_{13}, \gamma_{13}, \gamma_{13}, \gamma_{17} \rangle_{0,4}^W = \langle \gamma_{13}, \gamma_{13}, w, w \rangle_{0,4}^W \langle \gamma_{13}, u, u \rangle_{0,3}^W. \tag{22}$$

If $\langle \gamma_{13}, u, u \rangle_0^W = 0$, then (21) implies $C_3 = 0$, and (22) implies

$$C_2 = -\langle \gamma_{13}, \gamma_{13}, \gamma_{13}, \gamma_{17} \rangle_{0,4}^W = -\frac{5}{18},$$

All the insertions in $\langle \gamma_{13}, \gamma_{13}, \gamma_{13}, \gamma_{17} \rangle_{0,4}^W$ are narrow and the last equality follows from Topological Euler class Axioms in FJRW theory.

On the other hand, we integrate $\Lambda_{1,4}^W(\gamma_{13}, \gamma_{13}, \gamma_7, \gamma_7)$ over Getzler's relation (14). We notice that the integration over the stratum which contains a genus-1 vertex will vanish, i.e., if $\delta = \delta_{2,2}, \delta_{2,3}, \delta_{2,4}$, or $\delta_{3,4}$, then

$$\int_\delta \Lambda_{1,4}^W(\gamma_{13}, \gamma_{13}, \gamma_7, \gamma_7) = 0.$$

The reason is as follows. According to Dimension Axiom (10), a nonzero primary genus-1 invariant of the form $\langle\phi_1,\ldots,\phi_k\rangle^W_{1,k}$ must satisfy

$$\sum_{i=1}^{k}\deg\phi_i = k.$$

Combine the degree formula in (20), among the integrations over $\delta_{2,2}$, $\delta_{2,3}$, $\delta_{2,4}$, and $\delta_{3,4}$, such equality only holds for the genus-1 stratum in $\delta_{2,2}$, where two insertions attached to the genus-1 vertex should be γ_{17} and γ_3. Thus the pairing (19) implies γ_1 and γ_{15} should appear as insertions on the genus-0 component. However, the one with the insertion γ_1 must vanish since the other insertions are chosen from the set $\{\gamma_{13},\gamma_{13},\gamma_7,\gamma_7\}$, but Topological Euler class Axioms will imply

$$\langle\gamma_1,\gamma_{13},\gamma_{13}\rangle^W_{0,3} = \langle\gamma_1,\gamma_7,\gamma_{13}\rangle^W_{0,3} = \langle\gamma_1,\gamma_7,\gamma_7\rangle^W_{0,3} = 0.$$

Thus the integration over the rest strata gives a relation among genus-0 invariants. We summarize all the integration together and get

$$-8C_2^2 - \frac{2C_2}{3} - 2C_1C_3 + \frac{8}{81} = 0$$

This contradicts with the WDVV calculation. So we must have $\langle\gamma_{13},u,u\rangle^W_{0,3}\neq 0$. Q.E.D.

3.4. Orbifold curves with $\chi\geq 0$

Let X be an orbifold curve with its Euler characteristic $\chi(X)$ nonnegative, then $H^2(X,\mathbb{Z})$ is generated by the divisor class P. We also assume X is neither an elliptic curve, nor the projective line. Such orbifold curves are classified into two types. X is called a *Fano orbifold curve* if $\chi>0$ or an *elliptic orbifold curve* if $\chi=0$. In this section, we discuss the higher genus Gromov-Witten theory of X. In [15, 18], the authors used Dimension Axiom in Gromov-Witten theory and tautological relations to prove:

Theorem 3.5 ([15, 18]). *Let X be an orbifold curve with $\chi(X)\geq 0$. Then any correlation function $\langle\langle\phi_1\psi_1^{\ell_1},\ldots,\phi_k\psi_k^{\ell_k}\rangle\rangle_{g,k}$ is uniquely determined by the prepotential of X.*

Teleman's classifiscation of homogeneous semisimple CohFTs also gives the same result. The proof in [15, 18] does not require the semisimple assumption.

Proof. Let $\beta = d\cdot P$ for some $d\in\mathbb{Z}_{\geq 0}$. We simply denote

$$\left\langle\left\langle \phi_1\psi_1^{\ell_1},\ldots,\phi_k\psi_k^{\ell_k}\right\rangle\right\rangle_{g,k} = \sum_{d\geq 0}\langle \phi_1\psi_1^{\ell_1},\ldots,\phi_k\psi_k^{\ell_k}\rangle_{g,k,d}q^d.$$

According to (12), if the correlator $\langle \phi_1\psi_1^{\ell_1},\ldots,\phi_k\psi_k^{\ell_k}\rangle_{g,k,d}$ is nonzero, then

$$(23)\quad \sum_{i=1}^k \deg_{\mathbb{C}}\phi_i + \sum_{i=1}^k \ell_i = (3-\dim_{\mathbb{C}} X)(g-1)+\chi\cdot d+k \geq 2g-2+k.$$

The last inequality follows from $\chi\geq 0$ and $\dim_{\mathbb{C}} X = 1$. Now if

$$\sum_{i=1}^k \ell_i \geq g \quad \textit{for} \quad g\geq 1, \quad \textit{or} \quad \sum_{i=1}^k \ell_i\geq 1 \quad \textit{for} \quad g=0,$$

then we can apply Lemma 2.4 of g-reduction to rewrite the correlation function as a linear combination of integrals over the corresponding homology cycles of some dual graphs, each of the dual graph lives on the boundary of $\overline{\mathcal{M}}_{g,k}$. Gluing Tree Axiom and Gluing Loop Axiom in CohFT allow us to reconstruct the correlation function using integrals over each component of the boundary strata. We can keep doing this process until on each component, the g-reduction property does not hold. On such a component, we again assume the correlator is $\langle \phi_1\psi_1^{\ell_1},\ldots,\phi_k\psi_k^{\ell_k}\rangle_{g,k,d}$, then

$$\sum_{i=1}^k \ell_i\leq g-1, \quad \textit{if} \quad g\geq 1, \quad \textit{or} \quad \sum_{i=1}^k \ell_i = 0 \quad \textit{if} \quad g=0.$$

On the other hand, since $\deg_{\mathbb{C}}\phi_i\leq 1$, the formula (23) implies that $\langle \phi_1\psi_1^{\ell_1},\ldots,\phi_k\psi_k^{\ell_k}\rangle_{g,k,d}$ must vanish unless $g=0$ and all $\ell_i=0$, or $g=1$, the correlator is $\langle P\rangle_{1,1,0}$ for $\chi>0$, or $\langle P\rangle_{1,1,d\geq 0}$ for $\chi=0$. In both cases, similar as in Section 3.2, we can integrate an appropriate Gromov-Witten class over Getzler's relation to reconstruct the genus-1 invariant from genus-0 invariants. We only explain for the $\chi>0$ case here. According to Rossi [19], if $\chi>0$, there is some $\phi_i\in H$, such that

$$\langle \phi_i,\phi_i,\phi_{i^*},\phi_{i^*}\rangle_{0,4,0}\neq 0.$$

Here ϕ_{i^*} is dual to ϕ_i, i.e., $\eta(\phi_i,\phi_{i^*}) = 1$. The integration of $\Lambda_{1,4}(\phi_i,\phi_i,\phi_{i^*},\phi_{i^*})$ over the Getzler's relation (14) vanishes on those strata with a genus-1 component except

$$\int_{\delta_{3,4}}\Lambda_{1,4}(\phi_i,\phi_i,\phi_{i^*},\phi_{i^*}) = 4\langle \phi_i,\phi_i,\phi_{i^*},\phi_{i^*}\rangle_{0,4,0}\langle P\rangle_{1,1,0}.$$

Thus the equality (14) implies $\langle P\rangle_{1,1,0}$ is reconstructed from genus-0 correlators. Q.E.D.

Remark 3.6. We remark that $\langle P\rangle_{1,1,0}$ for $\chi > 0$ can be calculated directly using virtual localization [23]. This technique here follows from the higher genus reconstruction of FJRW theory for singularities when the central charge $\hat{c}_W \leq 1$, [7, 8, 15].

§ Acknowledgements

The author would like to thank Marc Krawitz, Changzheng Li, Si Li, Todor Milanov, Yongbin Ruan, Kyoji Saito, Hsian-Hua Tseng, and Jie Zhou for collaborations. The work of the author was partially supported by NSF grant DMS-1159156.

References

[1] Abramovich, Dan; Graber, Tom; Vistoli, Angelo. *Gromov-Witten theory of Deligne-Mumford stacks.* Amer. J. Math. 130 (2008), no. 5, 1337–1398.

[2] Berglund, Per; Hübsch, Tristan. *A generalized construction of mirror manifolds.* Essays on mirror manifolds, 388–407, Int. Press, Hong Kong, 1992.

[3] Bryan, Jim; Graber, Tom; Pandharipande, Rahul. *The orbifold quantum cohomology of $\mathbb{C}^2/\mathbb{Z}_3$ and Hurwitz Hodge integrals.* Journal of Algebraic Geometry, 17 (2008), 1–28.

[4] Chen, Weimin; Ruan, Yongbin. *Orbifold Gromov-Witten theory.* Orbifolds in mathematics and physics (Madison, WI, 2001), 25–85, Contemp. Math., 310, Amer. Math. Soc., Providence, RI, 2002.

[5] Dubrovin, Boris. *Geometry of* 2d *Topological Field Theories.* Integrable Systems and Quantum Groups. Lecture Notes in Math. 1620: Springer, Berlin(1996): 120–348.

[6] Faber, C.; Pandharipande, R. *Relative maps and tautological classes.* J. Eur. Math. Soc. (JEMS) 7 (2005), no. 1, 13–49.

[7] Faber, Carel; Shadrin, Sergey; Zvonkine, Dimitri. *Tautological relations and the r-spin Witten conjecture.* Ann. Sci. Éc. Norm. Supér. (4) 43 (2010), no. 4, 621–658.

[8] Fan, Huijun; Jarvis, Tyler; Ruan, Yongbin. *The Witten equation, mirror symmetry, and quantum singularity theory.* Ann. of Math. (2) 178 (2013), no. 1, 1–106.

[9] Fan, Huijun; Jarvis, Tyler; Ruan, Yongbin. *The Witten Equation and Its Virtual Fundamental Cycle.* Book in preparation.
Preprint arXiv:0712.4025v3.

[10] Getzler, Ezra. *Intersection theory on $\overline{\mathcal{M}}_{1,4}$ and elliptic Gromov-Witten invariants.* J. Amer. Math. Soc. **10** (1997), no. 4, 973–998.

[11] Givental, Alexander B.. *Semisimple Frobenius structures at higher genus.* Internat. Math. Res. Notices 2001, no. 23, 1265–1286.

[12] Ionel, Eleny-Nicoleta. *Topological recursive relations in* $H^{2g}(\mathcal{M}_{g,n})$. Invent. Math. 148 (2002), no. 3, 627–658.

[13] Kontsevich, M.; Manin, Yu. *Gromov-Witten classes, quantum cohomology, and enumerative geometry.* Comm. Math. Phys. 164 (1994), no. 3, 525–562.

[14] Krawitz, Marc. *FJRW rings and Landau-Ginzburg Mirror Symmetry.* arXiv:0906.0796v1.

[15] Krawitz, Marc; Shen, Yefeng. *Landau-Ginzburg/Calabi-Yau correspondence of all genera for elliptic orbifold* $\mathbb{P}^1$. arXiv:1106.6270 [math.AG]

[16] Kreuzer, Maximilian; Skarke, Harald. *On the classification of quasihomogeneous functions.* Comm. Math. Phys. 150 (1992), no. 1, 137–147.

[17] Li, Changzheng; Li, Si; Saito, Kyoji; Shen, Yefeng. *Mirror symmetry for exceptional unimodular singularities.* J. Eur. Math. Soc. (JEMS) 19 (2017), no. 4, 1189–1229.

[18] Milanov, Todor; Shen, Yefeng; Tseng, Hsian-Hua. *Gromov–Witten theory of Fano orbifold curves, Gamma integral structures and ADE-Toda Hierarchies.* Geometry & Topology 20 (2016), no. 4, 2135–2218.

[19] Rossi, Paolo. *Gromov-Witten theory of orbicurves, the space of tri-polynomials and symplectic field theory of Seifert fibrations.* Math. Ann. 348 (2010), no. 2, 265–287.

[20] Satake, Ikuo; Takahashi, Atsushi. *Gromov-Witten invariants for mirror orbifolds of simple elliptic singularities.* Ann. Inst. Fourier (Grenoble) 61 (2011), no. 7, 2885–2907.

[21] Shen, Yefeng; Zhou, Jie. *Ramanujan Identities and Quasi-Modularity in Gromov-Witten Theory.* Commun. Num. Theor. Phys. 11 (2017), no. 2, 405–452.

[22] Teleman, Constantin. *The structure of 2D semi-simple field theories.* Invent. Math. 188 (2012), no. 3, 525–588.

[23] Tseng, Hsian-Hua. *On degree-0 elliptic orbifold Gromov-Witten invariants.* Int. Math. Res. Not. IMRN 2011, no. 11, 2444–2468.

[24] Witten, Edward. *Algebraic geometry associated with matrix models of two-dimensional gravity.* Topological methods in modern mathematics (Stony Brook, NY, 1991), Publish or Perish, Houston, TX, 1993, 235–269.

Department of Mathematics, University of Oregon, Eugene, OR 97403, USA.
E-mail address: `yfshen@uoregon.edu`

Advanced Studies in Pure Mathematics 83, 2019
Primitive Forms and Related Subjects — Kavli IPMU 2014
pp. 389–415

From a class of Calabi–Yau dg algebras to Frobenius manifolds via primitive forms

Atsushi Takahashi

Abstract.

It is one of the most important problems in mirror symmetry to obtain functorially Frobenius manifolds from smooth compact Calabi–Yau A_∞-categories. This paper gives an approach to this problem based on the theory of primitive forms and, in particular, reduces it to a formality conjecture of certain homotopy algebra. Under this formality conjecture, a formal primitive form for a non-negatively graded connected smooth compact Calabi–Yau dg algebra can be constructed, which enable us to have a formal Frobenius manifold.

§1. Introduction

Mirror symmetry gives an identification between two objects coming from different mathematical origins. It has been studied intensively by many mathematician for more than twenty years since it yields important, interesting and unexpected geometric information.

Almost ten years before the discovery of mirror symmetry, K. Saito studied a deformation theory of an isolated hypersurface singularity in order to generalize the theory of elliptic integrals [31]. There he developed a certain generalization of the complex Hodge theory of Calabi–Yau type and found a differential geometric structure, which he called a flat structure, on the base space of the deformation (cf. [33] for a summary). This is known as Saito's theory of primitive forms. His flat structure was also found later by Dubrovin in his study of two dimensional topological field theories in their relation with integrable systems [9], which is axiomatized under the name of a Frobenius manifold (cf. [15, 28, 30]).

Received February 4, 2015.
Revised November 19, 2016.
2010 *Mathematics Subject Classification.* 14J33, 53D37, 53D45.
Key words and phrases. Mirror Symmetry, homological mirror symmetry, primitive forms.

The classical mirror symmetry conjecture states the existence of an isomorpshism between the Frobenius manifold from the Gromov–Witten theory of a Calabi–Yau manifold and the one from the deformation theory of another Calabi–Yau manifold. In order to explain this mysterious correspondence, Kontsevich conjectured in [24] that the derived category of coherent sheaves on a Calabi–Yau manifold should be equivalent to the derived Fukaya category for another Calabi–Yau manifold, which is known as the homological mirror symmetry conjecture. In particular, he expects that the classical mirror symmetry isomorphism can naturally be induced by the homological mirror symmetry equivalence via the moduli space of A_∞-deformations.

As is already explained in [33], although Saito's construction of the Frobenius manifolds based on filtered de Rham cohomology groups, Gauß–Manin connctions, higher residue pairings and primitive forms is formulated only for isolated hypersurface singularities, several notion and the idea of this construction is of general nature. It has inspired and motivated the non-commutative Hodge theory for Calabi–Yau A_∞ categories (cf. [21, 26]), which is the basic ingredient in constructing primitive forms in this categorical language.

The present paper tries to make the contents of Appendix in [33], where we give a list how objects in the original Saito's theory of primitive forms may be generalized and rewritten, as precise as possible at present for non-negatively graded connected smooth compact Calabi–Yau dg algebras.

More precisely, in Section 2 we recall some notations and terminologies of differential graded algebras, and then we introduce the dg analogue of the weighted homogeneous polynomial with an isolated singularity at the origin. It is Proposition 2.7, the "Cartan calculus", that is the key in our story, which enable us to translate the original Saito's theory for isolated weighted homogeneous hypersurface singularities verbatim into the one for non-negatively graded connected smooth compact Calabi–Yau dg algebras. In Section 3, we show that a very good section can be constructed under a certain formality assumption which is motivated by the formality considered in [3, 7, 25], the identification of the (formal) moduli space of A_∞-deformation of the derived category of coherent sheaves with the formal neighborhood of zero in the total cohomology group of polyvectors. After studying the versal deformation of a Saito structure in Section 4, we show how a primitive form is deduced from a very good section based on the famous method developed by M. Saito [34] and Barannikov [1, 2].

Let us make some comments on recent developments in related works. The higher residue pairings on the periodic cyclic homology is constructed by Fukaya–Oh–Ohta–Ono [10] for A_∞-algebras associated to Lagrangian Floer theory on compact toric manifolds. They also show (see Chapter 3 in [10]) mirror isomorphisms of Frobenius manifolds. Shklyarov have been extensively studied the non-commutative Hodge theory for smooth compact dg categories. In particular, he constructs the Gauß–Manin connection and the higher residue pairings on the periodic cyclic homology (see [37, 38]). He also shows that they give rise to the corresponding objects in the original Saito's theory. Sheridan generalizes Shklyarov's construction to A_∞-categories in [36]. Iwanari studies a period mapping for deformations of a dg algebra as a generalization of Griffiths' period mapping for those of algebraic varieties to non-commutative setting in [16], where he shows that a curved A_∞-deformation of a smooth compact Calabi–Yau dg category is unobstructed by a different approach from the one in [21].

Acknowledgement

The author would like to thank organizers of the workshop "Primitive forms and related subjects". He is supported by JSPS KAKENHI Grant Numbers 23654009, 26610008 and 16H06337.

§2. Notations and Terminologies

In this paper, we denote by k an algebraically closed field of characteristic zero with the unit 1_k.

2.1. Differential graded algebras

On dg algebras and derived categories of modules over them, we refer the reader to [22], of which we follow the terminologies and notations.

A *differential graded k-algebra* (or simply, *dg k-algebra*) is a $\mathbb{Z}$-graded k-algebra $A = \bigoplus_{p\in\mathbb{Z}} A^p$ equipped with a differential, a k-linear map $d_A : A \longrightarrow A$ of degree one with $d_A^2 = 0$ satisfying the Leibniz rule:

$$d_A(a_1 a_2) = (d_A a_1)a_2 + (-1)^p a_1(d_A a_2), \quad a_1 \in A^p, a_2 \in A. \tag{1}$$

For a homogeneous element $a \in A^p$, let $\overline{a}$ be the degree of a, namely, $\overline{a} := p$. Denote by $H^\bullet(A, d_A) := \bigoplus_{p\in\mathbb{Z}} H^p(A, d_A)$ the cohomology of A which is a graded k-module. Throughout this paper, A denotes a dg k-algebra.

We recall some terminologies for our later use. For a dg k-algebra A, $A^e := A^{op} \otimes_k A$ denotes the enveloping dg k-algebra of A where A^{op} is the opposite dg k-algebra of A. The dg A-module is called perfect if it belongs to the smallest full triangulated subcategory of the derived category of dg A-modules which contains A and is closed under direct summands and isomorphisms.

Definition 2.1. *Let A be a dg k-algebra.*

(1) *A dg k-algebra A is called non-negatively graded if $A^p = 0$ for all negative integers p.*

(2) *A dg k-algebra A is called compact if A is a perfect dg k-module, namely, if its cohomology $H^\bullet(A, d_A)$ is finite dimensional (cf. Kontsevich–Soibelman, Definition 8.2.1 in [26]).*

(3) *A dg k-algebra A is called smooth if A is a perfect dg A^e-module (cf. Kontsevich–Soibelman, Definition 8.1.2 in [26]).*

For dg A-modules M, N, we shall denote by $\mathbb{R}\mathrm{Hom}_A(M, N)$ the $\mathbb{R}\mathrm{Hom}$-complex. Let $A^! := \mathbb{R}\mathrm{Hom}_{(A^e)^{op}}(A, A^e)$ be the inverse dualizing complex. For a smooth dg k-algebra, since A is a perfect dg A^e-module by definition, it is easy to see the following

Proposition 2.1. *Let A be a smooth dg k-algebra. The natural morphism in the derived category of dg A^e-modules*

$$A \longrightarrow \mathbb{R}\mathrm{Hom}_{A^e}(\mathbb{R}\mathrm{Hom}_{(A^e)^{op}}(A, A^e), A^e) = \mathbb{R}\mathrm{Hom}_{A^e}(A^!, A^e)$$

is an isomorphism. In particular, we have a natural isomorphism in the derived category of dg k-modules

$$A \otimes^{\mathbb{L}}_{A^e} A \cong A \otimes^{\mathbb{L}}_{A^e} \mathbb{R}\mathrm{Hom}_{(A^e)^{op}}(A^!, A^e) \cong \mathbb{R}\mathrm{Hom}_{A^e}(A^!, A). \tag{2}$$

2.2. Calabi–Yau dg algebras

Definition 2.2 (Ginzburg, Definition 3.2.3 in [14])**.** *Fix an integer $w \in \mathbb{Z}$. A smooth dg k-algebra A is called Calabi–Yau of dimension w if there exists an isomorphism*

$$A^! \cong T^{-w} A \tag{3}$$

in the derived category of dg A^e-modules.

Let A be a smooth Calabi–Yau dg k-algebra of dimension w. It is important to note that a choice of an element in $H^0(\mathbb{R}\mathrm{Hom}(A^!, T^{-w}A))$ giving the isomorphism (3) yields an isomorphism

$$\mathbb{R}\mathrm{Hom}_{A^e}(A, A) \cong \mathbb{R}\mathrm{Hom}_{A^e}(T^w A^!, A) \cong T^{-w}(A \otimes^{\mathbb{L}}_{A^e} A), \tag{4}$$

in the derived category of dg k-modules due to the isomorphism (2).

For a smooth compact dg k-algebra A, it is well-known and is easy to see that the Serre functor on the perfect derived category of dg A-modules is given by the dg A^e-module $A^* := \mathbb{R}\mathrm{Hom}_k(A, k)$ (the dg analogue of the Nakayama functor for a finite dimensional algebra with finite global dimension). It is known that the dg A^e-module $A^!$ gives its quasi-inverse (cf. Lemma 4.1 in [23]), which implies the following

Proposition 2.2. *Let A be a smooth compact Calabi–Yau dg k-algebra of dimension w. There exists an isomorphism in the perfect derived category of dg A^e-modules depending on a choice of an element in $H^{-w}(\mathbb{R}\mathrm{Hom}(A^!, A))$:*

$$T^w A \cong A^*. \tag{5}$$

2.3. Hochschild cohomology

For graded k-modules M and N, $\mathcal{G}r_k(M,N) = \bigoplus_{p\in\mathbb{Z}} \mathcal{G}r_k(M,N)^p$ denotes a graded k-module where $\mathcal{G}r_k(M,N)^p$ is a k-module consisting of graded k-linear maps from M to N of degree p. For a dg k-algebra A, set

$$C^\bullet(A) := \prod_{n\geq 0} \mathcal{G}r_k((TA)^{\otimes n}, A), \tag{6}$$

For $a \in A$, denote it by Ta when we consider it as an element of TA. Define a k-linear map $d : C^\bullet(A) \longrightarrow C^{\bullet+1}(A)$ by $(df)(1_k) := d_A(f(1_k))$ for $f \in \mathcal{G}r_k(k, A)$ and, for $f \in \mathcal{G}r_k((TA)^{\otimes n}, A)^p$, $n \geq 1$,

$$(df)(Ta_1 \otimes \cdots \otimes Ta_n) := d_A\left(f(Ta_1 \otimes \cdots \otimes Ta_n)\right)$$
$$-\sum_{i=1}^{n}(-1)^{p-1+s_{i-1}} f(Ta_1 \otimes \cdots \otimes Ta_{i-1} \otimes T(d_A a_i) \otimes Ta_{i+1} \otimes \cdots \otimes Ta_n),$$

where $s_i := \sum_{m=1}^{i}(\overline{a_m} - 1)$. Define another k-linear map $\delta : C^\bullet(A) \longrightarrow C^{\bullet+1}(A)$ by $(\delta f)(Ta_1) := (-1)^{(p-1)(\overline{a_1}-1)+\overline{a_1}} a_1 f(1_k) + (-1)^p f(1_k) a_1$ for $f \in \mathcal{G}r_k(k, A)^p$ and, for $f \in \mathcal{G}r_k((TA)^{\otimes n}, A)^p$, $n \geq 1$,

$$(\delta f)(Ta_1 \otimes \cdots \otimes Ta_{n+1}) := (-1)^{p+s_n} f(Ta_1 \otimes \cdots \otimes Ta_n) a_{n+1}$$
$$+\sum_{i=1}^{n}(-1)^{p-1+s_i} f(Ta_1 \otimes \cdots \otimes Ta_{i-1} \otimes T(a_i a_{i+1}) \otimes Ta_{i+2} \otimes \cdots \otimes Ta_{n+1})$$
$$+ (-1)^{(p-1)(\overline{a_1}-1)+\overline{a_1}} a_1 f(Ta_2 \otimes \cdots \otimes Ta_{n+1}).$$

The following proposition is well-known, which is a small variation of Chapter 9.6 in [4] (see also Section 7.1 in [26]).

Proposition 2.3. *The k-linear maps d, δ define differentials on $C^\bullet(A)$ and satisfy $d\delta + \delta d = 0$. Moreover, the dg k-module $(C^\bullet(A), \partial := d + \delta)$ is isomorphic to the dg k-module $\mathbb{R}\mathrm{Hom}_{A^e}(A, A)$ in the derived category of dg k-modules.*

Definition 2.3. *The dg k-module $(C^\bullet(A), \partial)$ is called the Hochschild cochain complex of A, whose cohomology $H^\bullet(C^\bullet(A), \partial)$ is denoted by $HH^\bullet(A)$ and is called the Hochschild cohomology of A.*

Denote by $\mathcal{T}^\bullet_{poly}(A)$ the graded k-module $H^\bullet(C^\bullet(A), \delta)$.

For $f = (f_n)_{n\geq 0} \in C^p(A)$ and $g = (g_n)_{n\geq 0} \in C^q(A)$, one can define the product $f \circ g = ((f \circ g)_n)_{n\geq 0} \in C^{p+q}(A)$ by

$$(f \circ g)_n(Ta_1 \otimes \cdots \otimes Ta_n) := \sum_{i=0}^{n} (-1)^{q\cdot s_i} f_i(Ta_1 \otimes \cdots \otimes Ta_i) g_{n-i}(Ta_{i+1} \otimes \cdots \otimes Ta_n).$$

It is also known that for $f \in C^p(A)$ and $g \in C^q(A)$ one can define the *Gerstenhaber bracket* $[f, g]_G \in C^{p+q-1}(A)$ by

$$[f, g]_G := f \circ_{-1} g - (-1)^{(p-1)(q-1)} g \circ_{-1} f, \tag{7}$$

where $f \circ_{-1} g = ((f \circ_{-1} g)_n)_{n\geq 0} \in C^{p+q-1}(A)$ is given by

$$\begin{aligned}(f \circ_{-1} g)_n(Ta_1 \otimes \cdots \otimes Ta_n) &:= \sum_{i=1}^{n} \sum_{j=1}^{n} (-1)^{(q-1)s_{i-1}} f_{n-j+1}(Ta_1 \otimes \cdots \otimes Ta_{i-1} \\ &\qquad \otimes T(g_j(Ta_i \otimes \cdots \otimes Ta_{i+j-1})) \otimes Ta_{i+j} \otimes \cdots \otimes Ta_n).\end{aligned}$$

We collect some basic properties of $\mathcal{T}^\bullet_{poly}(A)$ and $HH^\bullet(A)$, which are well-known facts or follow from a straight forward calculation.

Proposition 2.4. *The product $\circ$ on $C^\bullet(A)$ induces structures of graded commutative k-algebras on $HH^\bullet(A)$ and $\mathcal{T}^\bullet_{poly}(A)$ whose unit elements are given by the cohomology classes of $1_A \in C^0(A)$. The Gerstenharber bracket $[-,-]_G$ induces structures of graded Lie algebras on $HH^{\bullet+1}(A)$ and $\mathcal{T}^{\bullet+1}_{poly}(A)$.*

Moreover, it turns out that the tuple $(\mathcal{T}^\bullet_{poly}(A), d, \circ, [-,-]_G)$ is a differential Gerstenhaber algebra (cf. [11]). Namely, we also have

$$[X, Y \circ Z]_G = [X, Y] \circ Z + (-1)^{(\overline{X}+1)\cdot\overline{Y}} Y \circ [X, Z]_G, \quad X, Y, Z \in \mathcal{T}^\bullet_{poly}(A).$$

Let $\mathfrak{m}_A = (\mathfrak{m}^p_A)_{p\geq 0}$ be the element in $C^2(A)$ where $\mathfrak{m}^p_A := 0$ if $p \neq 1, 2$ and $\mathfrak{m}^1_A(Ta_1) := d_A a_1$, $\mathfrak{m}^2_A(Ta_1 \otimes Ta_2) := (-1)^{\overline{a_1}} a_1 a_2$. It is well-known that $[\mathfrak{m}_A, \mathfrak{m}_A]_G = 0$ since, if we write $\mathfrak{m}_A = \mathfrak{m}_A^1 + \mathfrak{m}_A^2$,

$$[\mathfrak{m}_A, \mathfrak{m}_A]_G = 0 \Leftrightarrow \begin{cases} \left[\mathfrak{m}_A^1, \mathfrak{m}_A^1\right]_G = 0 \\ \left[\mathfrak{m}_A^1, \mathfrak{m}_A^2\right]_G = 0 \\ \left[\mathfrak{m}_A^2, \mathfrak{m}_A^2\right]_G = 0 \end{cases} \Leftrightarrow \begin{cases} d_A^2 = 0 \\ \text{the Leibnitz rule (1).} \\ \text{the associativity} \end{cases}$$

It is also important to note that $\partial f = [\mathfrak{m}_A, f]_G$, $df = \left[\mathfrak{m}_A^1, f\right]_G$ and $\delta f = \left[\mathfrak{m}_A^2, f\right]_G$ for $f \in C^\bullet(A)$.

Let $\mathfrak{f}_A$ be the cohomology class of $\mathfrak{m}_A$ in $\mathcal{T}_{poly}^2(A)$, which is actually the same as the one of $\mathfrak{m}_A^1$ in $\mathcal{T}_{poly}^2(A)$. We will see later that the element $\mathfrak{f}_A$ is the dg analogue of a weighted homogeneous polynomial with an isolated singularity, which is the initial data for the original Saito's theory of primitive forms.

Indeed, by a direct calculation, we obtain the following dg analogue of the "Euler's identity" for $\mathfrak{f}_A$.

Proposition 2.5. *For $X \in \mathcal{T}_{poly}^\bullet(A)$, we have $dX = [\mathfrak{f}_A, X]_G$. In particular, we have*

$$\mathfrak{f}_A = [\mathfrak{deg}_A, \mathfrak{f}_A]_G\,, \tag{8}$$

where $\mathfrak{deg}_A$ is the cohomology class of the element $\deg_A = (\deg_A^p)_{p\geq 0} \in C^1(A)$ in $\mathcal{T}_{poly}^1(A)$ defined by $\deg_A^p := 0$ if $p \neq 1$ and $\deg_A^1(Ta_1) := \overline{a_1}\cdot a_1$.

The "Euler's identity" (8) means that $\mathfrak{f}_A$ is of degree one with respect to another grading structure on $C^\bullet(A)$ whose origin is the $\mathbb{Z}$-grading on A. This completely agrees with the fact that a weighted homogeneous polynomial in the original Saito's theory is considered as of degree two from the homological view point, e.g. when we discuss matrix factorizations, and of degree one for the compatibility with exponents.

2.4. Hochishild homology

For a dg k-algebra A, set

$$C_\bullet(A) := \coprod_{n\geq 0} A \otimes_k (TA)^{\otimes n}. \tag{9}$$

We shall denote an element of $A \otimes_k (TA)^{\otimes n}$ by a_0 if $n = 0$ and $a_0 \otimes Ta_1 \otimes \cdots \otimes Ta_n$ if $n \neq 0$. Define a k-linear map $d : C_\bullet(A) \longrightarrow C_{\bullet-1}(A)$ by $d(a_0) := d_A a_0$ and, for $n \geq 1$,

$$\begin{aligned} d(a_0 \otimes Ta_1 \otimes \cdots \otimes Ta_n) &:= d_A a_0 \otimes Ta_1 \otimes \cdots \otimes Ta_n \\ &+ \sum_{i=1}^{n} (-1)^{s_i'} a_0 \otimes Ta_1 \otimes \cdots \otimes Ta_{i-1} \otimes T(d_A a_i) \otimes Ta_{i+1} \otimes \cdots \otimes Ta_n, \end{aligned}$$

where $s'_i := \sum_{m=0}^{i}(\overline{a_m} - 1)$. Define another k-linear map $\delta : C_\bullet(A) \longrightarrow C_{\bullet-1}(A)$ by $\delta(a_0) := 0$ and, for $n \geq 1$,

$$\begin{aligned}\delta(a_0 \otimes Ta_1 \otimes \cdots \otimes Ta_n) &:= (-1)^{\overline{a_0}} a_0 a_1 \otimes Ta_2 \otimes \cdots \otimes Ta_n \\ - \sum_{i=1}^{n-1} (-1)^{s'_i} a_0 \otimes Ta_1 \otimes \cdots &\otimes Ta_{i-1} \otimes T(a_i a_{i+1}) \otimes Ta_{i+2} \otimes \cdots \otimes Ta_n \\ &+ (-1)^{\overline{a_n} + (\overline{a_n} - 1)s'_{n-1}} a_n a_0 \otimes Ta_1 \otimes \cdots \otimes Ta_{n-1}.\end{aligned}$$

Define a k-linear map $B : C_\bullet(A) \longrightarrow C_{\bullet+1}(A)$ called the *Connes differential* by

$$\begin{aligned}&B(a_0 \otimes Ta_1 \cdots \otimes Ta_n) \\ &:= \mathrm{id}_A \otimes Ta_0 \otimes \cdots \otimes Ta_n - (-1)^{\overline{a_0}-1} a_0 \otimes T(\mathrm{id}_A) \otimes Ta_1 \otimes \cdots \otimes Ta_n \\ &+ \sum_{i=1}^{n} (-1)^{\overline{a_n} - 1 + s''_i} \mathrm{id}_A \otimes Ta_i \otimes \cdots \otimes Ta_n \otimes Ta_0 \otimes \cdots \otimes Ta_{i-1} \\ &- \sum_{i=1}^{n} (-1)^{\overline{a_i} - 1 + s''_i} a_i \otimes T(\mathrm{id}_A) \otimes Ta_{i+1} \cdots \otimes Ta_n \otimes Ta_0 \otimes \cdots \otimes Ta_{i-1},\end{aligned}$$

where $s''_i := \left(\sum_{m=0}^{i-1}(\overline{a_m} - 1)\right)\left(\sum_{m=i}^{n-1}(\overline{a_m} - 1)\right)$.

The following proposition is also well-known, which is a small variation of Chapter 9.6 in [4] (see also Section 7.2 in [26]).

Proposition 2.6. *The k-linear maps d, δ, B define differentials on $C_\bullet(A)$ and satisfy $d\delta + \delta d = 0$, $dB + Bd = 0$, $\delta B + B\delta = 0$. Moreover, the dg k-module $(C_\bullet(A), \partial := d + \delta)$ is isomorphic to the dg k-module $A \otimes^{\mathbb{L}}_{A^e} A$ in the derived category of dg k-modules.*

Definition 2.4. *The dg k-module $(C_\bullet(A), \partial)$ is called the Hochschild chain complex of A, whose homology $H_\bullet(C_\bullet(A), \partial)$ is denoted by $HH_\bullet(A)$ and is called the Hochschild homology of A.*

Denote by $\Omega_\bullet(A)$ the graded k-module $H_\bullet(C_\bullet(A), \delta)$.

The pair $(\mathcal{T}^\bullet_{poly}(A), \Omega_\bullet(A))$ admits a structure of a *calculus algebra* (cf. Dolgushev–Tamarkin–Tsygan, Definition 3 in [7]), namely, we have the algebraic structures on $(\mathcal{T}^\bullet_{poly}(A), \Omega_\bullet(A))$ in the next proposition which directly follows, just by forgetting the differential d, from the result by Daletski–Gelfand–Tsygan [6] (see also Dolgushev–Tamarkin–Tsygan, Section 3 in [8]). To state the result, recall two k-linear maps $\iota_f, \mathcal{L}_f : C_\bullet(A) \longrightarrow C_\bullet(A)$ called the *contraction* and the *Lie derivative*.

For $f \in \mathcal{G}r_k((TA)^{\otimes p}, A)$, they are defined in the following way:

$$\begin{aligned}
&\iota_f(a_0 \otimes Ta_1 \cdots \otimes Ta_n) \\
:= &\pm a_0 f(Ta_1 \otimes \cdots \otimes Ta_p) \otimes Ta_{p+1} \cdots \otimes Ta_n, \\
&\mathcal{L}_f(a_0 \otimes Ta_1 \cdots \otimes Ta_n) \\
:= &f(Ta_0 \otimes \cdots \otimes Ta_{p-1}) \otimes \cdots \otimes Ta_n \\
&+ \sum_{i=1}^{n-p+1} \pm a_0 \otimes Ta_1 \cdots \otimes Tf(Ta_i \otimes \cdots \otimes Ta_{i+p-1}) \otimes \cdots \otimes Ta_n \\
&+ \sum_{i=n-p+1}^{n-1} \pm f(Ta_{i+1} \otimes \cdots \otimes Ta_n \otimes Ta_0 \otimes \cdots) \otimes \cdots \otimes Ta_i,
\end{aligned}$$

where we omit signs since we do not need their explicit expressions later.

Proposition 2.7. *The k-linear maps ι and $\mathcal{L}$ induce morphisms of graded k-modules*

$$i : \mathcal{T}^{\bullet}_{poly}(A) \longrightarrow \mathcal{G}r_k(\Omega_{\bullet}(A), \Omega_{\bullet}(A)), \quad X \mapsto i_X,$$

$$L : \mathcal{T}^{\bullet}_{poly}(A) \longrightarrow T^{-1}\mathcal{G}r_k(\Omega_{\bullet}(A), \Omega_{\bullet}(A)), \quad X \mapsto L_X,$$

satisfying, for all $X, Y \in \mathcal{T}^{\bullet}_{poly}(A)$,

$$i_X i_Y = i_{X \circ Y},\ [L_X, L_Y] = L_{[X,Y]_G},$$

$$L_X i_Y + (-1)^{\overline{X}} i_Y L_X = L_{X \circ Y},\ [i_X, L_Y] = i_{[X,Y]_G},$$

$$[B, i_X] = -L_X,\ [B, L_X] = 0,\ L_{\mathfrak{f}_A} = -d.$$

Note that $\mathcal{G}r_k(\Omega_{\bullet}(A), \Omega_{\bullet}(A))$ has both a structure of a dg k-algebra and a dg Lie algebra. Since we have

$$[d, i_X] = i_{dX},\ [d, L_X] = L_{-dX}, \quad X \in \mathcal{T}^{\bullet}_{poly}(A),$$

the k-linear maps i and L define a morphism of dg k-algebras and a morphism of dg Lie algebras.

Since A^e-modules can be considered as dg endo-functors on the derived dg category of dg A-modules (see Töen, Corollary 7.6 in [40] for the precise statement), there are "horizontal" and "vertical" associative product structures on $\mathbb{R}\mathrm{Hom}_{A^e}(A, A)$ and $\mathbb{R}\mathrm{Hom}_{A^e}(A, A)$-module structures on $\mathbb{R}\mathrm{Hom}_{A^e}(A^!, A)$ (cf. Kashiwara–Schapira, Remark 1.3.4 in [20], for endo-functors on ordinary categories). It turns out that they induce the same structures on $\mathcal{T}^{\bullet}_{poly}(A)$ and $\Omega_{\bullet}(A)$, which are the product $\circ$ and the map i given above (cf. Kaledin, Lemma 8.1 and the equation (8.2) in [18]).

The equalities in Proposition 2.7, the "Cartan calculus", play an essential role in our story, which enable us to translate the original Saito's theory for isolated hypersurface singularities verbatim into the one for non-negatively graded connected smooth compact Calabi–Yau dg algebras in this paper.

§3. Construction of a very good section

We shall consider a particular class of dg k-algebras.

Definition 3.1. *A smooth compact dg k-algebra A is called connected if $HH^0(A) = k \cdot [1_A]$, $HH^r(A) = 0$ if $r < 0$ and 1_A defines a non-zero class in $HH_0(A)$.*

We have in mind such a dg k-algebra as the dg endomorphism algebra $A_X := \mathbb{R}\mathrm{Hom}_X(G, G)$ of a split generator G of the triangulated category $\mathrm{per}(X)$ of perfect complexes on a connected smooth projective variety X over k (which is equivalent to the bounded derived category $\mathcal{D}^b(X)$ of coherent sheaves). It is known that

$$HH^r(A_X) \cong \bigoplus_{r=p+q} H^q(X, \wedge^p \mathcal{T}_X), \tag{10}$$

$$HH_r(A_X) \cong \bigoplus_{r=p-q} H^q(X, \Omega_X^p), \tag{11}$$

where $\mathcal{T}_X$ is the tangent sheaf on X and Ω_X^p is the sheaf of regular p-forms on X (cf. Section 8 in [40], Section 9.2 in [26] and also [12, 24]). Here, G can be chosen as $G = \bigoplus_{i=0}^{\dim X} \mathcal{L}^i$ with a very ample invertible sheaf $\mathcal{L}$ on X (cf. Theorem 4 in [29]) and $\mathbb{R}\mathrm{Hom}_X(G, G)$ can be defined by the Čech cochain complex. Under this choice, the dg k-algebra A_X will be non-negatively graded.

3.1. Formality for certain homotopy calculus algebras

Let A be a connected smooth compact Calabi–Yau dg k-algebra. On the algebraic structure on the pair $(C^\bullet(A), C_\bullet(A))$, the following proposition directly follows from Corollary 1 in [7]:

Proposition 3.1. *The pair $(C^\bullet(A), C_\bullet(A))$ has a structure of homotopy calculus algebra.*

We omit the definition of homotopy calculus algebras due to limitations of space since we shall not use it in later discussions. Motivated by the description of Section 8.4 by Kontsevich in [25] and based on Theorem 5 in [7], we expect the following conjecture:

Conjecture 1. *As homotopy calculus algebras,* $(C^\bullet(A), C_\bullet(A))$ *is quasi-isomorphic to the calculus algebra* $(\mathcal{T}^\bullet_{poly}(A), \Omega_\bullet(A))$.

Note in particular that Conjecture 1 implies that the dg Lie algebra $(C^{\bullet+1}(A), [\mathfrak{m}_A, -]_G, [-,-]_G)$ is quasi-isomorphic to the dg Lie algebra $(\mathcal{T}^{\bullet+1}_{poly}(A), [\mathfrak{f}_A, -]_G, [-,-]_G)$ as an L_∞-algebra.

Remark 3.1. *We expect that the quasi-isomorphism might relate categorical (co)chain complexes of a Calabi–Yau manifold to the geometric ones used by Barannikov and Kontsevich in* [1, 2, 3]*. In other example, such a statement as Corollary* 4.6 *by Shklyarov in* [37] *which relates categorical and geometric mixed complexes for isolated hypersurface singularities is of our expectation.*

Remark 3.2. *It is an important problem to recover Hodge numbers of a smooth projective variety from its derived category of coherent sheaves. For this purpose, it will be necessary to have a* $\mathbb{Z} \times \mathbb{Z}$*-grading structure on the Hochschild homology group. This conjecture also contains an approach to this problem.*

Assumption 1. *Conjecture* 1 *holds for a connected smooth compact Calabi–Yau dg* k*-algebra* A.

We also expect that Conjecture 1 implies that dg k-algebras as above which are dg Morita equivalent yield isomorphic Saito structures (filtered de Rham cohomology, Gauß–Manin connection and the higher residue pairings which will be introduced in Section 4).

3.2. Hochschild cohomology of Calabi–Yau dg algebras

From now on, A always denotes a non-negatively graded connected smooth compact Calabi–Yau dg k-algebra. Under Assumption 1, there are isomorphisms of graded k-modules

$$HH^\bullet(A) \cong H^\bullet(\mathcal{T}^\bullet_{poly}(A), d), \quad HH_\bullet(A) \cong H_\bullet(\Omega_\bullet(A), d).$$

In particular, an element of $HH_w(A)$ giving the isomorphism (3) corresponds to a non-zero element $v_1 \in H_w(\Omega_\bullet(A), d)$. Since the dg k-algebra A is connected, we have $H_w(\Omega_\bullet(A), d) \cong HH_w(A) \cong HH^0(A) = k \cdot [1_A]$, which implies that $H_w(\Omega_\bullet(A), d) = k \cdot v_1$.

Since the differentials on $\mathcal{T}^\bullet_{poly}(A)$ and $\Omega_\bullet(A)$ are inner, namely, we have $d = [\mathfrak{f}_A, -]_G$ on $\mathcal{T}^\bullet_{poly}(A)$ and $d = -L_{\mathfrak{f}_A}$ on $\Omega_\bullet(A)$, we expect that the following holds under Assumption 1.

Conjecture 2. *Let v_1 be a non-zero element in $H_w(\Omega_\bullet(A), d)$. Under Assumption* 1, *the contraction map*

$$(C^\bullet(A), \partial) \longrightarrow (C_{w-\bullet}(A), \partial), \quad X \mapsto \iota_X v_1,$$

induces a morphism of dg k-modules

$$(\mathcal{T}^\bullet_{poly}(A), d) \longrightarrow (\Omega_{w-\bullet}(A), d), \quad X \mapsto i_X v_1,$$

which is an isomorphism.

Assumption 2. *Conjecture* 2 *holds for a non-negatively graded connected smooth compact Calabi–Yau dg k-algebra A.*

In particular, under Assumption 2 we have the isomorphism

$$H^p(\mathcal{T}^\bullet_{poly}(A), d) \cong H_{w-p}(\Omega_\bullet(A), d),\ p \in \mathbb{Z}, \quad X \mapsto i_X v_1, \tag{12}$$

of graded k-modules.

Definition 3.2. *Set*

$$Jac(\mathfrak{f}_A) := H^\bullet(\mathcal{T}^\bullet_{poly}(A), d). \tag{13}$$

We call the graded k-module $Jac(\mathfrak{f}_A)$ the Jacobian ring *of A.*

It is known that if A is a smooth compact Calabi–Yau dg k-algebra of dimension w then A^e is a smooth compact Calabi–Yau dg k-algebra of dimension $2w$. Since $Jac(\mathfrak{f}_A) \cong HH^\bullet(A) \cong H^\bullet(\mathbb{R}\mathrm{Hom}_{A^e}(A, A))$, we have the following

Proposition 3.2. *Fix a non-zero element $v_1 \in HH_w(A)$. Let $v_1^{\otimes 2}$ be the element in $HH_{2w}(A^e)$ corresponding to $v_1 \otimes v_1$ under the Künneth formula $HH_{2w}(A^e) \cong HH_w(A^{op}) \otimes_k HH_w(A) = HH_w(A) \otimes_k HH_w(A)$. Then the induced isomorphism by $v_1^{\otimes 2}$*

$$T^{2w} A^e \longrightarrow (A^e)^*$$

in the perfect derived category of the dg k-algebra $(A^e)^e$ equips $Jac(\mathfrak{f}_A)$ with a structure of a finite dimensional graded commutative Frobenius k-algebra. Namely, we have a non-degenerate graded symmetric bilinear form $\eta_{\mathfrak{f}_A, v_1^{\otimes 2}} : Jac(\mathfrak{f}_A) \otimes_k Jac(\mathfrak{f}_A) \to T^{-2w} k$ such that

$$\eta_{\mathfrak{f}_A, v_1^{\otimes 2}}(X \circ Y, Z) = \eta_{\mathfrak{f}_A, v_1^{\otimes 2}}(X, Y \circ Z), \quad X, Y, Z \in Jac(\mathfrak{f}_A). \tag{14}$$

3.3. Filtered de Rham cohomology and the degeneration of Hodge to de Rham

Definition 3.3. *Let u be a formal variable of degree two. Define a graded $k((u))$-module $\mathcal{H}_{\mathfrak{f}_A}$, called the filtered de Rham cohomology, by*

$$\mathcal{H}_{\mathfrak{f}_A} := H_\bullet(T^{-w}\Omega_\bullet(A)((u)), d+uB) \tag{15}$$

and for any integer $p \in \mathbb{Z}$ the graded $k[[u]]$-submodules $\mathcal{H}_{\mathfrak{f}_A}^{(-p)}$ of $\mathcal{H}_{\mathfrak{f}_A}$

$$\mathcal{H}_{\mathfrak{f}_A}^{(-p)} := H_\bullet(T^{-w}\Omega_\bullet(A)[[u]]u^p, d+uB). \tag{16}$$

Define a graded k-module $\Omega_{\mathfrak{f}_A}$ by

$$\Omega_{\mathfrak{f}_A} := H_\bullet(T^{-w}\Omega_\bullet(A), d). \tag{17}$$

Proposition 3.3. *For all $p \in \mathbb{Z}$, there exists an exact sequence of graded k-modules*

$$0 \longrightarrow \mathcal{H}_{\mathfrak{f}_A}^{(-p-1)} \longrightarrow \mathcal{H}_{\mathfrak{f}_A}^{(-p)} \overset{r^{(-p)}}{\longrightarrow} \Omega_{\mathfrak{f}_A} \longrightarrow 0. \tag{18}$$

Proof. For non-negatively graded smooth compact dg k-algebras, it follows that $\mathrm{Im}(B) \cap HH_\bullet(A) = 0$ from the degeneration of Hodge to de Rham conjecture proven by Kaledin, Theorem 5.5 in [17]. Under Assumption 1 we have $HH_\bullet(A) \cong T^w\Omega_{\mathfrak{f}_A}$, which implies $\mathrm{Im}(B)\cap\Omega_{\mathfrak{f}_A} = 0$ and hence $r^{(-p)}$ is surjective. Q.E.D.

Remark 3.3. *The assumption that A is non-negatively graded is no more necessary in this proposition since Kaledin proves in [19] the degeneration of Hodge to de Rham conjecture without this assumption.*

The $k[[u]]$-submodules $\left\{\mathcal{H}_{\mathfrak{f}_A}^{(-p)}\right\}_{p\in\mathbb{Z}}$ of $\mathcal{H}_{\mathfrak{f}_A}$ define an increasing filtration

$$\cdots \subset \mathcal{H}_{\mathfrak{f}_A}^{(-p-1)} \subset \mathcal{H}_{\mathfrak{f}_A}^{(-p)} \subset \cdots \subset \mathcal{H}_{\mathfrak{f}_A},$$

such that the multiplication of u induces an isomorphism of k-modules

$$u : \mathcal{H}_{\mathfrak{f}_A}^{(-p)} \cong \mathcal{H}_{\mathfrak{f}_A}^{(-p-1)}.$$

3.4. Gauß–Manin connection on $\mathcal{H}_{\mathfrak{f}_A}$

Set $\mathcal{T}_{\widehat{\mathbb{A}}^1_u} := k[[u]]\frac{d}{du}$, where $\frac{d}{du}$ is the derivation on $k[[u]]$ satisfying $\frac{d}{du}(u) = 1$. Define a morphism of graded k-modules $\nabla : \mathcal{T}_{\widehat{\mathbb{A}}^1_u} \otimes_k \Omega_\bullet(A)((u)) \to \Omega_\bullet(A)((u))$ by

$$\nabla_{\frac{d}{du}} := \frac{d}{du} - \frac{1}{u^2} i_{\mathfrak{f}_A}. \tag{19}$$

Remark 3.4. *Following the idea in* [21, 26] *based on the work of Getzler* [13], *Shklyarov studied the Gauß–Manin connection on the periodic cyclic homology and shows that it gives rise to the one up to a weight shift in the original Saito's theory. See* [37, 38]. *The Gauß–Manin connection given above is different from Shklyarov's although both are analogues of the one for isolated hypersurface singularities. The difference is explained as follows.*

In order to simplify the discussion, take a smooth compact Calabi–Yau variety X of dimension w and the space of smooth (p,q)-forms $\mathcal{A}^{p,q}(X)$ for example. On $\mathcal{A}^{p,q}(X)$, Shklyarov's Gauß–Manin connection will yield $(q-p)/2$ as a degree (cf. Introduction in [37]*), while ours is expected to yield q. When $p+q=w$, both coincides up to the weight factor $-w/2$. In this way, one may understand why Shklyarov's Gauß–Manin connection with a weight shift coincides with the one in Saito's theory.*

Shklyarov's generalization is natural and does not need our Conjecture 1 *at all, nevertheless, we are interested in the above generalization of the Gauß–Manin connection since, in the mirror dual side, it is necessary to use p or q (and impossible to use $(p+q)/2$) as the degree of an element in $H^q(Y,\Omega_Y^p)$ (where Y is a mirror dual of X) in formulating the Virasoro conjecture for the gravitational quantum cohomology (cf. Chapter* 10. 1. 4 *in* [5] *and Chapter VI.* 2 *in* [28]).

The following proposition is very simple, however, it will be the key in our construction of a primitive form. It is the dg analogue of the corresponding fact used in the original Saito's theory for weighted homogeneous polynomials when we lift a homogeneous basis of the Jacobian ring multiplied with the standard holomorphic volume form to a very good section.

Proposition 3.4. *The morphism of graded k-modules ∇ is a connection which satisfies*

$$\left[\nabla_{u\frac{d}{du}}, d+uB\right] = d+uB.$$

Therefore, ∇ induces a connection on $\mathcal{H}_{\mathfrak{f}_A}$. Moreover, we have

$$\nabla_{u\frac{d}{du}}\left(\mathcal{H}^{(0)}_{\mathfrak{f}_A}\right) \subset \mathcal{H}^{(0)}_{\mathfrak{f}_A}.$$

Proof. Use equalities in Proposition 2.7. We have

$$\left[\nabla_{u\frac{d}{du}}, d+uB\right] = uB - \frac{1}{u}\left[i_{\mathfrak{f}_A}, d+uB\right] = uB - L_{\mathfrak{f}_A} = d+uB$$

Together with the equation (8), we also have

$$u\frac{d}{du} - \frac{1}{u}i_{\mathfrak{f}_A} = u\frac{d}{du} - \frac{1}{u}\left[i_{\mathfrak{deg}_A}, L_{\mathfrak{f}_A}\right] = u\frac{d}{du} + L_{\mathfrak{deg}_A} + \frac{1}{u}\left[d + uB, i_{\mathfrak{deg}_A}\right]$$

Q.E.D.

Definition 3.4. *The connection ∇ on $\mathcal{H}_{\mathfrak{f}_A}$ is called the Gauß–Manin connection.*

3.5. Exponents

For $a_0 \otimes Ta_1 \otimes \cdots \otimes Ta_n \in C_\bullet(A)$ such that $a_i \in A^{\overline{a_i}}$, the endomorphism $\mathcal{L}_{\deg_A}$ on $C_\bullet(A)$ can be calculated as

$$\mathcal{L}_{\deg_A}(a_0 \otimes Ta_1 \otimes \cdots \otimes Ta_n) := \left(\sum_{i=0}^{n} \overline{a_i}\right) \cdot (a_0 \otimes Ta_1 \otimes \cdots \otimes Ta_n)\,.$$

This obviously commutes with the operator δ on $C_\bullet(A)$ and hence defines an endomorphism on $\Omega_\bullet(A)$, which is exactly $L_{\mathfrak{deg}_A}$.

Proposition 3.5. *The endomorphism of graded k-modules $L_{\mathfrak{deg}_A}$ on $\Omega_\bullet(A)$ induces a k-linear endomorphism on $\Omega_{\mathfrak{f}_A}$.*

Proof. The equalities in Proposition 2.7 and the equation (8) yield

$$\left[L_{\mathfrak{deg}_A}, d\right] = -\left[L_{\mathfrak{deg}_A}, L_{\mathfrak{f}_A}\right] = -L_{[\mathfrak{deg}_A, \mathfrak{f}_A]_G} = -L_{\mathfrak{f}_A} = d,$$

which implies that $L_{\mathfrak{deg}_A}(\Omega_{\mathfrak{f}_A}) \subset \Omega_{\mathfrak{f}_A}$. Q.E.D.

Denote by N_A the endomorphism of graded k-modules on $\Omega_{\mathfrak{f}_A}$ induced by $L_{\mathfrak{deg}_A}$. Define graded k-submodules $\Omega_{\mathfrak{f}_A}^{p,q}$ of $\Omega_{\mathfrak{f}_A}$ by

$$\Omega_{\mathfrak{f}_A}^{p,q} := \{\omega \in \Omega_{\mathfrak{f}_A} \mid \overline{\omega} = w - p + q,\ N_A\omega = q\omega\}, \quad p, q \in \mathbb{Z}.$$

and set

$$F_q := \bigoplus_{p\in\mathbb{Z}, r\leq q} \Omega_{\mathfrak{f}_A}^{p,r}, \quad q \in \mathbb{Z}.$$

Definition 3.5. *The graded k-submodules $\{F_q\}_{q\in\mathbb{Z}}$ form an increasing filtration of $\Omega_{\mathfrak{f}_A}$*

$$(20) \qquad 0 \subset \cdots \subset F_q \subset F_{q+1} \subset \cdots \subset \Omega_{\mathfrak{f}_A},$$

which is called the Hodge filtration of $\Omega_{\mathfrak{f}_A}$.

Since A is non-negatively graded, we have $F_q = 0$ for $q < 0$.

Definition 3.6. *The Hodge numbers are*

$$h^{p,q}(A) := \dim_k \Omega^{p,q}_{\mathfrak{f}_A}, \quad p, q \in \mathbb{Z}.$$

The integer q with $h^{p,q}(A) \neq 0$ is called an exponent. The set of exponents is the multi-set

$$\left\{ q * h^{p,q}(A) \mid p, q \in \mathbb{Z},\ \Omega^{p,q}_{\mathfrak{f}_A} \neq 0 \right\},$$

where by $u * v$ we denote v copies of the integer u.

3.6. Pairing on the Hochschild homology

Since A is Calabi–Yau of dimension w, an element $v_1 \in HH_w(A)$ yields an isomorphism of graded k-modules

$$Jac(\mathfrak{f}_A) \cong \Omega_{\mathfrak{f}_A}, \quad X \mapsto i_X v_1. \tag{21}$$

Move the k-bilinear form on $Jac(\mathfrak{f}_A)$ to $\Omega_{\mathfrak{f}_A}$ by this isomorphism.

Definition 3.7. *Define a k-bilinear form $J_{\mathfrak{f}_A} : \Omega_{\mathfrak{f}_A} \otimes_k \Omega_{\mathfrak{f}_A} \to T^{-2w}k$ by*

$$J_{\mathfrak{f}_A}(i_X v_1, i_Y v_1) := (-1)^{w\cdot \overline{Y}} \eta_{\mathfrak{f}_A, v_1^{\otimes 2}}(X, Y), \quad X, Y \in Jac(\mathfrak{f}_A). \tag{22}$$

Note that $J_{\mathfrak{f}_A}$ does not depend on the choice of v_1. Moreover, it induces a perfect pairing

$$J_{\mathfrak{f}_A} : \Omega^{p,q}_{\mathfrak{f}_A} \otimes_k \Omega^{w-p,w-q}_{\mathfrak{f}_A} \longrightarrow k.$$

Some elementary homological algebras for $Jac(\mathfrak{f}_A)$ and $\Omega_{\mathfrak{f}_A}$ yields the following

Proposition 3.6. *The Hodge numbers satisfy*

(1) $h^{p,q}(A) = 0$ *if* $p < 0$ *or* $q < 0$.
(2) $h^{w,0}(A) = 1$.
(3) $h^{w-p,q}(A) = h^{p,w-q}(A)$.

Proof. The property (1) holds since A is non-negatively graded. The property (2) follows from the fact that $[1_A] \in HH^0(A)$ is mapped to v_1 under the isomorphism $HH^0(A) \cong HH_w(A)$ and that $v_1 \in \Omega^{w,0}_{\mathfrak{f}_A}$. The perfectness of $J_{\mathfrak{f}_A}$ implies the property (3). Q.E.D.

This proposition means that 0 is the minimal exponent and the duality of exponents holds, namely, if q is an exponent then $w - q$ is also an exponent.

3.7. Existence of a very good section

Proposition 3.7. *Let $r^{(0)} : \mathcal{H}^{(0)}_{\mathfrak{f}_A} \longrightarrow \Omega_{\mathfrak{f}_A}$ be the k-linear map of degree zero in the exact sequence of graded k-modules in Proposition* 3.3:

$$0 \longrightarrow \mathcal{H}^{(-1)}_{\mathfrak{f}_A} \longrightarrow \mathcal{H}^{(0)}_{\mathfrak{f}_A} \stackrel{r^{(0)}}{\longrightarrow} \Omega_{\mathfrak{f}_A} \longrightarrow 0.$$

Then there exists a section $s^{(0)} : \Omega_{\mathfrak{f}_A} \longrightarrow \mathcal{H}^{(0)}_{\mathfrak{f}_A}$ of $r^{(0)}$ such that

$$\nabla_{u\frac{d}{du}} \left(s^{(0)} \left(\Omega_{\mathfrak{f}_A} \right) \right) \subset s^{(0)} \left(\Omega_{\mathfrak{f}_A} \right).$$

Proof. Set $l_A := \dim_k Jac(\mathfrak{f}_A)$. Then Proposition 3.5 implies the existence of a k-basis $\{v_1, \ldots, v_{l_A}\}$ of $\Omega_{\mathfrak{f}_A}$ such that $N_A v_i = q_i \cdot v_i$ for some $q_i \in \mathbb{Z}$ for $i = 1, \ldots, l_A$, where v_1 is the element in $\Omega^{w,0}_{\mathfrak{f}_A}$ as in the previous section.

There exists a section $s^{(0)} : \Omega_{\mathfrak{f}_A} \longrightarrow \mathcal{H}^{(0)}_{\mathfrak{f}_A}$ so that $s^{(0)}\left(\Omega_{\mathfrak{f}_A}\right)$ is inside of $H_\bullet(T^{-w}\Omega_\bullet(A) \otimes_k k[u], d + uB)$ since k is a field and the dg k-algebra A is non-negatively graded. Moreover, the equalities $\left[\nabla_{u\frac{d}{du}}, d + uB\right] = d + uB$, $\left[L_{\mathfrak{deg}_A}, d\right] = d$ and $\left[L_{\mathfrak{deg}_A}, B\right] = 0$, which in particular means that d (resp. B, u) is of degree 1 (resp. $0, 1$) with respect to the grading given by $L_{\mathfrak{deg}_A}$, enable us to choose $\omega_{il} \in \Omega_\bullet(A)$ satisfying $L_{\mathfrak{deg}_A}\omega_{i,l} = (q_i - l) \cdot \omega_{i,l}$ so that $s^{(0)}(v_i) = \sum_{l=0}^{m} [\omega_{i,l}] u^l$ for some $m \in \mathbb{Z}_{\geq 0}$. Therefore, $\nabla_{u\frac{d}{du}} s^{(0)}(v_i) = q_i \cdot s^{(0)}(v_i)$. Q.E.D.

A section $s^{(0)}$ of $r^{(0)}$ in Proposition 3.7 is called a *very good section* after M. Saito (see Introduction and Proposition 3.1 in [35]).

3.8. Higher residue pairings

Once a very good section $s^{(0)}$ for $r^{(0)}$ is given, we have an isomorphism

$$\Omega_{\mathfrak{f}_A}((u)) \cong \mathcal{H}_{\mathfrak{f}_A}, \quad \sum_{p=-\infty}^{\infty} v_p u^p \mapsto \sum_{p=-\infty}^{\infty} s^{(0)}(v_p) u^p,$$

of $k((u))$-modules.

Definition 3.8. *Fix a very good section $s^{(0)}$ for $r^{(0)}$ and define a k-bilinear form $K_{\mathfrak{f}_A} : \mathcal{H}_{\mathfrak{f}_A} \otimes_k \mathcal{H}_{\mathfrak{f}_A} \to k((u))$ by*

$$K_{\mathfrak{f}_A}\left(\omega(u), \omega'(u)\right) := J_{\mathfrak{f}_A}\left(\omega(u), \omega'(-u)\right) u^w, \tag{23}$$

where $J_{\mathfrak{f}_A}$ on the right hand sided denotes the pairing on $\Omega_{\mathfrak{f}_A}((u))$ which is the $k((u))$-linear extension of $J_{\mathfrak{f}_A}$ on $\Omega_{\mathfrak{f}_A}$. The pairing $K_{\mathfrak{f}_A}$ is called the higher residue pairings after K. Saito, Section 4 in [32].

It follows from some elementary calculation that for all $h(u) \in k((u))$

$$h(u)K_{\mathfrak{f}_A}(\omega_1, \omega_2) = K_{\mathfrak{f}_A}(h(u)\omega_1, \omega_2) = K_{\mathfrak{f}_A}(\omega_1, h(-u)\omega_2),$$

and for all $\omega_1, \omega_2 \in \mathcal{H}_{\mathfrak{f}_A}$

$$u\frac{d}{du}K_{\mathfrak{f}_A}(\omega_1, \omega_2) = K_{\mathfrak{f}_A}(\nabla_{u\frac{d}{du}}\omega_1, \omega_2) + K_{\mathfrak{f}_A}(\omega_1, \nabla_{u\frac{d}{du}}\omega_2).$$

Since these properties with the extension of scalars and the flatness of the pairing with respect to vector fields on the deformation space characterize uniquely the higher residue pairings for the universal unfolding of an isolated hypersurface singularity (see p.45 – p.46 in [34]), we just denote it by $K_{\mathfrak{f}_A}$ without mentioning the dependence on the chosen very good section $s^{(0)}$. To be honest, we have to define independently the higher residue pairings from the good section and then show that it is equal to the above one. This is also one of the problems to be solved.

Remark 3.5. *K. Saito and the author asked the categorical analogue of the higher residue pairings in Appendix in* [33]. *The higher residue pairings on the periodic cyclic homology is constructed by Fukaya–Oh–Ohta–Ono* [10] *for A_∞-algebras associated to Lagrangian Floer theory on compact toric manifolds and by Shklyarov* [38] *for smooth proper dg categories. Shklyarov's construction is generalized to A_∞-categories by Sheridan* [36].

Proposition 3.7 imply the following.

Proposition 3.8. *Let $s^{(0)}$ be a very good section for $r^{(0)}$. The graded k-submodule $S := s^{(0)}(\Omega_{\mathfrak{f}_A}) \otimes_k k[u^{-1}]u^{-1}$ of $\mathcal{H}_{\mathfrak{f}_A}$ satisfies*

$$\mathcal{H}_{\mathfrak{f}_A} = \mathcal{H}^{(0)}_{\mathfrak{f}_A} \oplus S,\ u^{-1}S \subset S,\ \nabla_{u\frac{d}{du}} S \subset S,\ K_{\mathfrak{f}_A}(S, S) \subset k[u^{-1}]u^{w-2}.$$

§4. Deformation

In this section, A always denotes a non-negatively graded connected smooth compact Calabi–Yau dg k-algebra of dimension w satisfying Assumption 1 and Assumption 2.

4.1. Versal deformation

Under Assumption 2, we can define a morphism $\Delta : \mathcal{T}^\bullet_{poly}(A) \longrightarrow \mathcal{T}^{\bullet-1}_{poly}(A)$ of graded k-modules by $i_{\Delta(X)}v_1 := Bi_X v_1$, $X \in \mathcal{T}^\bullet_{poly}(A)$. Note that Δ does not depend on the particular choice of v_1. Obviously, it satisfies $\Delta^2 = 0$ and hence it defines a differential on $\mathcal{T}^\bullet_{poly}(A)$.

The differential Δ does not satisfy the Leibniz rule with respect to the product $\circ$, however, it is a part of a rich structure on $\mathcal{T}^{\bullet}_{poly}(A)$ as given below.

Proposition 4.1. *The tuple* $(\mathcal{T}^{\bullet}_{poly}(A), d, \circ, [-,-]_G, \Delta)$ *is a dGBV algebra. Namely, we have*

$$[X,Y]_G = (-1)^{\overline{X}}\Delta(X\circ Y) - (-1)^{\overline{X}}\Delta(X)\circ Y - X\circ\Delta(Y),\ X,Y\in\mathcal{T}^{\bullet}_{poly}(A).$$

Proof. The equality

$$\begin{aligned} & i_{[X,Y]_G}v_1 = [i_X, L_Y]\,v_1 \\ &= -i_X\,[B, i_Y]\,v_1 + (-1)^{\overline{X}\cdot(\overline{Y}+1)}\,[B,i_Y]\,i_Xv_1 \\ &= -i_XBi_Yv_1 + (-1)^{\overline{X}\cdot(\overline{Y}+1)}Bi_Yi_Xv_1 - (-1)^{\overline{Y}}\cdot(-1)^{\overline{X}\cdot(\overline{Y}+1)}i_YBi_Xv_1 \\ &= -i_{X\circ\Delta(Y)}v_1 + (-1)^{\overline{X}\cdot(\overline{Y}+1)}i_{\Delta(Y\circ X)}v_1 + (-1)^{(\overline{X}+1)\cdot(\overline{Y}+1)}i_{Y\circ\Delta(X)}v_1 \\ &= -i_{X\circ\Delta(Y)}v_1 + (-1)^{\overline{X}}i_{\Delta(X\circ Y)}v_1 - (-1)^{\overline{X}}i_{\Delta(X)\circ Y}v_1 \end{aligned}$$

yields the statement. Q.E.D.

Denote by $\mathcal{O}_M$ the completed symmetric algebra $k[[T^2Jac(\mathfrak{f}_A)]]$ of $T^2Jac(\mathfrak{f}_A)$ and by $\mathfrak{m}$ the maximal ideal in $k[[T^2Jac(\mathfrak{f}_A)]]$. Note that $\mathcal{O}_M$ is isomorphic to the completed symmetric algebra $k[[HH^{\bullet+2}(A)]]$ of $HH^{\bullet+2}(A)$ under Assumption 1.

Let $t_1,\dots,t_{l_A}$ be the dual coordinates for the basis $\{v_1,\dots,v_{l_A}\}$ as in the proof of Proposition 3.7. Denote by $\mathcal{T}_M$ a graded $\mathcal{O}_M$-free module of derivations on $\mathcal{O}_M$, which satisfies $\mathcal{T}_M \cong \bigoplus_{i=1}^{l_A}\mathcal{O}_M\frac{\partial}{\partial t_i}$.

Proposition 4.2. *The dGBV algebra* $(\mathcal{T}^{\bullet}_{poly}(A), d, \circ, [-,-]_G, \Delta)$ *is smooth formal. Namely, there exists a solution* $\gamma(t)$ *to the Maurer–Cartan equation in formal power series with values in* $\mathcal{T}^{\bullet}_{poly}(A)$,

$$d\gamma(t) + \frac{1}{2}[\gamma(t),\gamma(t)]_G = 0,\quad \gamma(t)\in\mathcal{T}^{\bullet}_{poly}(A)\widehat{\otimes}_k\mathfrak{m}, \tag{24}$$

satisfying the following properties:

(1) *For* $i=1,\dots,l_A$, *the element*

$$\left[\frac{\partial\gamma(t)}{\partial t_i}\right]\in\mathcal{T}^{\bullet}_{poly}(A)\widehat{\otimes}_k\mathfrak{m}\,/\,\mathcal{T}^{\bullet}_{poly}(A)\widehat{\otimes}_k\mathfrak{m}^2$$

considered as an element in $\mathcal{T}^{\bullet}_{poly}(A)$ *form a* k*-basis of* $Jac(\mathfrak{f}_A)$.

(2) *The solution $\gamma(t)$ is homogeneous in the sense that*

$$\gamma(t) = \sum_{i=1}^{l_A}(1-q_i)t_i\frac{\partial\gamma(t)}{\partial t_i} + [\mathfrak{d}eg_A, \gamma(t)]_G\,. \tag{25}$$

Proof. We can apply the Terilla's result, Theorem 2 in [39], since our dGBV algebra $(\mathcal{T}^\bullet_{poly}(A), d, \circ, [-,-]_G, \Delta)$ satisfies his "degeneration of the spectral sequence" condition due to the Hodge to de Rham degeneration of our filtered de Rham cohomology $\mathcal{H}_{\mathfrak{f}_A}$ (Proposition 3.3). Q.E.D.

Let $\gamma(t)$ be as in Proposition 4.2. Define a formal power series $\mathfrak{F}_A$ with values in $\mathcal{T}^\bullet_{poly}(A)$ as

$$\mathfrak{F}_A := \mathfrak{f}_A \otimes 1 + \gamma(t). \tag{26}$$

Then it follows that $[\mathfrak{F}_A, \mathfrak{F}_A]_G = 0$ from the Maurer–Cartan equation (24). Define an $\mathcal{O}_M$-endomorphism d_γ on $\mathcal{T}^\bullet_{poly}(A)\widehat{\otimes}_k\mathcal{O}_M$ by

$$d_\gamma X := [\mathfrak{F}_A, X]_G\,, \quad X \in \mathcal{T}^\bullet_{poly}(A)\widehat{\otimes}_k\mathcal{O}_M. \tag{27}$$

It follows that $d_\gamma^2 = 0$ since $[\mathfrak{F}_A, \mathfrak{F}_A]_G = 0$.

Proposition 4.3. *For all $X, Y \in \mathcal{T}^\bullet_{poly}(A)\widehat{\otimes}_k\mathcal{O}_M$, we have*

$$d_\gamma [X,Y]_G = [d_\gamma(X), Y]_G + (-1)^{\overline{X}}[X, d_\gamma(Y)]_G\,,$$

$$d_\gamma(X\circ Y) = d_\gamma(X)\circ Y + (-1)^{\overline{X}} X\circ d_\gamma(Y).$$

Namely, the triple $(d_\gamma, \circ, [-,-]_G)$ equip $\mathcal{T}^\bullet_{poly}(A)\widehat{\otimes}_k\mathcal{O}_M$ with a structure of differential Gerstenharber algebra.

Proof. The statement follows since the bracket $[-,-]_G$ satisfies the Jacobi and the Poisson identities. Q.E.D.

Definition 4.1. *The graded $\mathcal{O}_M$-module*

$$Jac(\mathfrak{F}_A) := H^\bullet(\mathcal{T}^\bullet_{poly}(A)\widehat{\otimes}_k\mathcal{O}_M, d_\gamma), \tag{28}$$

is called the Jacobian ring of $\mathfrak{F}_A$.

Note that the property $[\mathfrak{F}_A, \mathfrak{F}_A]_G = 0$ implies

$$d_\gamma(\mathfrak{F}_A) = 0, \quad d_\gamma\left(\frac{\partial\mathfrak{F}_A}{\partial t_i}\right) = \left[\mathfrak{F}_A, \frac{\partial\mathfrak{F}_A}{\partial t_i}\right]_G = 0,\ i = 1,\ldots,l_A.$$

Proposition 4.4. *The morphism $\rho : \mathcal{T}_M \longrightarrow Jac(\mathfrak{F}_A)$ of graded $\mathcal{O}_M$-modules defined by*

$$\rho : \mathcal{T}_M \longrightarrow Jac(\mathfrak{F}_A), \quad \frac{\partial}{\partial t_i} \mapsto \left[\frac{\partial \mathfrak{F}_A}{\partial t_i}\right], \; i = 1, \ldots, l_A, \tag{29}$$

is an isomorphism.

The isomorphism (29) enable us to introduce two particular elements of $\mathcal{T}_M$ which play important roles later.

Definition 4.2. *The element $e \in \mathcal{T}_M$ such that $\rho(e) = [1_A]$ is called the primitive vector field. The element $E \in \mathcal{T}_M$ such that $\rho(E) = [\mathfrak{F}_A]$ is called the Euler vector field.*

The equation (25) in Proposition 4.2 implies the following.

Proposition 4.5. *We have*

$$E = \sum_{i=1}^{l_A} (1 - q_i) t_i \frac{\partial}{\partial t_i} \tag{30}$$

and the "Euler's identity":

$$\mathfrak{F}_A = E\mathfrak{F}_A + [\mathfrak{deg}_A, \mathfrak{F}_A]_G \,. \tag{31}$$

Consider the $\mathcal{O}_M$-linear extensions of the morphisms i and L of graded k-modules defined in Proposition 2.7 and define a morphism d_γ of graded $\mathcal{O}_M$-modules on $\Omega_\bullet(A)\widehat{\otimes}_k\mathcal{O}_M$ as

$$d_\gamma := -L_{\mathfrak{F}_A}, \tag{32}$$

which is a deformation by γ of the boundary operator d on $\Omega_\bullet(A)$. The equalities in Proposition 2.7 lead the following.

Proposition 4.6. *We have*

$$d_\gamma^2 = 0, \quad [B, i_{\mathfrak{F}_A}] = -L_{\mathfrak{F}_A} = d_\gamma, \quad [B, d_\gamma] = 0,$$

$$[d_\gamma, i_X] = i_{d_\gamma X}, \; X \in \mathcal{T}^\bullet_{poly}(A)\widehat{\otimes}_k\mathcal{O}_M.$$

In particular, d_γ defines a boundary operator on $\Omega_\bullet(A)\widehat{\otimes}_k\mathcal{O}_M$.

4.2. Deformed filtered de Rham cohomology $\mathcal{H}_{\mathfrak{F}_A}$

Definition 4.3. *Let u be a formal variable of degree two. Define a graded $k((u))\widehat{\otimes}\mathcal{O}_M$-module, called the deformed filtered de Rham cohomology, by*

$$\mathcal{H}_{\mathfrak{F}_A} := H_\bullet(T^{-w}\Omega_\bullet(A)((u))\widehat{\otimes}_k\mathcal{O}_M, d_\gamma + uB) \tag{33}$$

and for any integer $p \in \mathbb{Z}$ the graded $\mathcal{O}_M[[u]]$-submodules of $\mathcal{H}_{\mathfrak{F}_A}$

$$\mathcal{H}_{\mathfrak{F}_A}^{(-p)} := H_\bullet(T^{-w}\Omega_\bullet(A)[[u]]u^p\widehat{\otimes}_k\mathcal{O}_M, d_\gamma + uB). \tag{34}$$

Define a graded $\mathcal{O}_M$-module $\Omega_{\mathfrak{F}_A}$ by

$$\Omega_{\mathfrak{F}_A} := H_\bullet(T^{-w}\Omega_\bullet(A)\widehat{\otimes}_k\mathcal{O}_M, d_\gamma). \tag{35}$$

Proposition 3.3 and Nakayama's Lemma imply the following.

Proposition 4.7. *For all $p \in \mathbb{Z}$, there exists an exact sequence of graded $\mathcal{O}_M$-modules*

$$0 \longrightarrow \mathcal{H}_{\mathfrak{F}_A}^{(-p-1)} \longrightarrow \mathcal{H}_{\mathfrak{F}_A}^{(-p)} \longrightarrow \Omega_{\mathfrak{F}_A} \longrightarrow 0.$$

4.3. Gauß–Manin connection on $\mathcal{H}_{\mathfrak{F}_A}$

Set $\mathcal{T}_{\widehat{\mathbb{A}}_u^1\times M} := \mathcal{O}_M[[u]]\frac{d}{du} \oplus \mathcal{O}_M[[u]] \otimes_{\mathcal{O}_M} \mathcal{T}_M$ and define a morphism of graded k-modules $\nabla^\gamma : \mathcal{T}_{\widehat{\mathbb{A}}_u^1\times M} \otimes_k \left(\Omega_\bullet(A)\widehat{\otimes}_k\mathcal{O}_M((u))\right) \to \Omega_\bullet(A)\widehat{\otimes}_k\mathcal{O}_M((u))$ by

$$\nabla^\gamma_{\frac{d}{du}} := \frac{d}{du} - \frac{1}{u^2}i_{\mathfrak{F}_A}, \quad \nabla^\gamma_{\frac{\partial}{\partial t_i}} := \frac{\partial}{\partial t_i} + \frac{1}{u}i_{\frac{\partial \mathfrak{F}_A}{\partial t_i}}, \ i = 1, \ldots, l_A.$$

Proposition 4.8. *The morphism of graded k-modules ∇^γ is a flat connection which satisfies*

$$\left[\nabla^\gamma_{u\frac{d}{du}}, d_\gamma + uB\right] = d_\gamma + uB, \quad \left[\nabla^\gamma_{\frac{\partial}{\partial t_i}}, d_\gamma + uB\right] = 0, \ i = 1, \ldots, l_A.$$

Therefore, ∇^γ induces a connection on $\mathcal{H}_{\mathfrak{F}_A}$. Moreover, we have

$$\nabla^\gamma_{\frac{\partial}{\partial t_i}}\left(\mathcal{H}_{\mathfrak{F}_A}^{(-1)}\right) \subset \mathcal{H}_{\mathfrak{F}_A}^{(0)}, \ i = 1, \ldots, l_A, \quad \nabla^\gamma_{u\frac{d}{du}+E}\left(\mathcal{H}_{\mathfrak{F}_A}^{(0)}\right) \subset \mathcal{H}_{\mathfrak{F}_A}^{(0)}.$$

Proof. It is clear that ∇^γ is a connection, which is flat since

$$\left[\nabla^\gamma_{\frac{d}{du}}, \nabla^\gamma_{\frac{\partial}{\partial t_i}}\right] = \frac{1}{u^2}i_{\frac{\partial \mathfrak{F}_A}{\partial t_i}} - \frac{1}{u^2}i_{\frac{\partial \mathfrak{F}_A}{\partial t_i}} - \frac{1}{u^3}\left[i_{\mathfrak{F}_A}, i_{\frac{\partial \mathfrak{F}_A}{\partial t_i}}\right] = 0,$$

$$\left[\nabla^{\gamma}_{\frac{\partial}{\partial t_i}}, \nabla^{\gamma}_{\frac{\partial}{\partial t_j}}\right] = \frac{1}{u} i_{\frac{\partial^2 \mathfrak{F}_A}{\partial t_i \partial t_j}} - (-1)^{(-\overline{t_i})\cdot(-\overline{t_j})} \frac{1}{u} i_{\frac{\partial^2 \mathfrak{F}_A}{\partial t_j \partial t_i}} + \frac{1}{u^2}\left[i_{\frac{\partial \mathfrak{F}_A}{\partial t_i}}, i_{\frac{\partial \mathfrak{F}_A}{\partial t_j}}\right] = 0.$$

The equalities in Proposition 4.6 yield

$$\left[\nabla^{\gamma}_{u\frac{d}{du}}, d_\gamma + uB\right] = uB - \frac{1}{u}\left[i_{\mathfrak{F}_A}, d_\gamma + uB\right] = uB - L_{\mathfrak{F}_A} = d_\gamma + uB,$$

$$\left[\nabla^{\gamma}_{\frac{\partial}{\partial t_i}}, d_\gamma + uB\right] = -\frac{\partial}{\partial t_i} L_{\mathfrak{F}_A} - \frac{1}{u}\left[i_{\frac{\partial \mathfrak{F}_A}{\partial t_i}}, L_{\mathfrak{F}_A}\right] + \left[i_{\frac{\partial \mathfrak{F}_A}{\partial t_i}}, B\right]$$

$$= \frac{\partial}{\partial t_i}\left[B, i_{\mathfrak{F}_A}\right] + \left[i_{\frac{\partial \mathfrak{F}_A}{\partial t_i}}, B\right] = (-1)^{-\overline{t_i}}\left[B, i_{\frac{\partial \mathfrak{F}_A}{\partial t_i}}\right] + (-1)^{1-\overline{t_i}}\left[B, i_{\frac{\partial \mathfrak{F}_A}{\partial t_i}}\right] = 0.$$

It is obvious from the definition that $\nabla^{\gamma}_{\frac{\partial}{\partial t_i}}\left(\mathcal{H}^{(-1)}_{\mathfrak{F}_A}\right) \subset \mathcal{H}^{(0)}_{\mathfrak{F}_A}$, $i = 1, \ldots, l_A$. By Proposition 4.5 and the equalities in Proposition 4.6,

$$\begin{aligned} & u\frac{d}{du} - \frac{1}{u} i_{\mathfrak{F}_A} + E + \frac{1}{u} i_{E\mathfrak{F}_A} \\ = {} & u\frac{d}{du} + E - \frac{1}{u}\left[i_{\mathfrak{d}eg_A}, L_{\mathfrak{F}_A}\right] \\ = {} & u\frac{d}{du} + E + L_{\mathfrak{d}eg_A} + \frac{1}{u}\left[d_\gamma + uB, i_{\mathfrak{d}eg_A}\right], \end{aligned}$$

which gives the last statement. Q.E.D.

Definition 4.4. *The connection ∇^γ on $\mathcal{H}_{\mathfrak{F}_A}$ is called the Gauß–Manin connection.*

§5. Primitive forms and Frobenius structures

Let $s^{(0)} : \Omega_{\mathfrak{f}_A} \longrightarrow \mathcal{H}^{(0)}_{\mathfrak{f}_A}$ be a very good section and let $\{v_1, \ldots, v_{l_A}\}$ be elements of $\Omega_{\mathfrak{f}_A}$ as in the proof of Proposition 3.7. Set $\zeta_i := s^{(0)}(v_i)$ for $i = 1, \ldots, l_A$.

5.1. Fundamental solution to the Gauß–Manin connection

By the equalities in Proposition 2.7 and Proposition 4.6 we obtain the following.

Proposition 5.1. *For all $\omega \in T^{-w}\Omega_\bullet(A)((u))\widehat{\otimes}_k \mathcal{O}_M$, we have*

$$(d_\gamma + uB)\left(e^{-\frac{i_{\gamma(t)}}{u}}\omega\right) = e^{-\frac{i_{\gamma(t)}}{u}}(d + uB)\,\omega$$

Moreover, we have

$$\nabla^{\gamma}_{\frac{\partial}{\partial t_j}}\left[e^{-\frac{i_{\gamma(t)}}{u}}\zeta_i\right] = 0, \quad i, j = 1, \ldots, l_A,$$

$$\nabla^{\gamma}_{u\frac{d}{du}}\left[e^{-\frac{i_{\gamma(t)}}{u}}\zeta_i\right]=q_i\cdot\left[e^{-\frac{i_{\gamma(t)}}{u}}\zeta_i\right],\quad i=1,\dots,l_A,$$

where $\left[e^{-\frac{i_{\gamma(t)}}{u}}\zeta_i\right]$ *denotes the equivalence class of* $e^{-\frac{i_{\gamma(t)}}{u}}\zeta_i$ *in* $\mathcal{H}_{\mathfrak{F}_A}$.

Therefore, we may identify $\mathcal{H}_{\mathfrak{f}_A}\widehat{\otimes}_k\mathcal{O}_M$ with $\mathcal{H}_{\mathfrak{F}_A}$ as follows (cf. Proposition 1.3 in [35]).

Proposition 5.2. *There is an isomorphism*

$$\mathcal{J}_{\mathfrak{F}_A}:\mathcal{H}_{\mathfrak{F}_A}\xrightarrow{\cong}\mathcal{H}_{\mathfrak{f}_A}\widehat{\otimes}_k\mathcal{O}_M,\quad \omega\mapsto e^{\frac{i_{\gamma(t)}}{u}}\omega, \tag{36}$$

which is compatibles with $k((u))\widehat{\otimes}_k\mathcal{O}_M$*-module structures, Gauß–Manin connections and multiplications of* u *on both sides.*

This idea together with the equation (37) below is the essential part of the construction of primitive forms, which is implicit in M. Saito's paper [34], later re-discovered by Barannikov [1, 2] and recently used by C. Li–S. Li–K. Saito [27] and M. Saito [35].

5.2. Primitive forms

Pull the $\mathcal{O}_M$-linear extension of the higher residue parings $K_{\mathfrak{f}_A}$ back to $\mathcal{H}_{\mathfrak{F}_A}$ by the isomorphism $\mathcal{J}_{\mathfrak{F}_A}$ and denote it by $K_{\mathfrak{F}_A}$. We can now introduce the notion of primitive forms.

Definition 5.1 (cf. K. Saito, Definition 3.1 in [31]). *Let* $r\in k$. *An element* $\zeta\in\mathcal{H}^{(0)}_{\mathfrak{F}_A}$ *is called a formal primitive form with the minimal exponent* r *for the tuple* $(\mathcal{H}^{(0)}_{\mathfrak{F}_A},\nabla^{\gamma},K_{\mathfrak{F}_A})$ *if it satisfies following five conditions;*

(1) $u\nabla^{\gamma}_{e}\zeta=\zeta$ *and* ζ *induces an* $\mathcal{O}_M$*-isomorphism:*

$$\mathcal{T}_M[[u]]\cong\mathcal{H}^{(0)}_{\mathfrak{F}_A},\quad \sum_{p=0}^{\infty}\delta_p u^p\mapsto\sum_{p=0}^{\infty}(u\nabla^{\gamma}_{\delta_p}\zeta)u^p. \tag{P1}$$

(2) *We have*

$$K_{\mathfrak{F}_A}(u\nabla^{\gamma}_{\delta}\zeta,u\nabla^{\gamma}_{\delta'}\zeta)\in k\cdot u^w,\quad \delta,\delta'\in\mathcal{T}_M. \tag{P2}$$

(3) *We have*

$$\nabla^{\gamma}_{u\frac{d}{du}+E}\zeta=r\zeta. \tag{P3}$$

(4) *There exists a connection* $\not\nabla$ *on* $\mathcal{T}_M$ *such that*

$$u\nabla^{\gamma}_X\nabla^{\gamma}_Y\zeta=\nabla^{\gamma}_{X\circ Y}\zeta+u\nabla^{\gamma}_{\not\nabla_X Y}\zeta,\quad X,Y\in\mathcal{T}_M. \tag{P4}$$

(5) *There exists an $\mathcal{O}_M$-endomorphism $N : \mathcal{T}_M \longrightarrow \mathcal{T}_M$ such that*

$$u\nabla^{\gamma}_{\frac{d}{du}}(u\nabla^{\gamma}_{X}\zeta) = -\nabla^{\gamma}_{E\circ X}\zeta + u\nabla^{\gamma}_{NX}\zeta, \quad X \in \mathcal{T}_M. \tag{P5}$$

We obtain a formal primitive form by applying a famous construction of primitive forms developed by M. Saito and Barannikov.

Theorem 1. *There exists a formal primitive form ζ with the minimal exponent zero for the tuple $(\mathcal{H}^{(0)}_{\mathfrak{F}_A}, \nabla^{\gamma}, K_{\mathfrak{F}_A})$.*

Proof. If a graded k-submodule S of $\mathcal{H}_{\mathfrak{f}_A}$ satisfies

$$\mathcal{H}_{\mathfrak{f}_A} = \mathcal{H}^{(0)}_{\mathfrak{f}_A} \oplus S,\ u^{-1}S \subset S,\ \nabla_{u\frac{d}{du}}S \subset S,\ K_{\mathfrak{f}_A}(S,S) \subset k[u^{-1}]u^{w-2},$$

then the unique element $\zeta \in \mathcal{H}^{(0)}_{\mathfrak{F}_A}$ such that

$$\mathcal{J}_{\mathfrak{F}_A}(\zeta) = \mathcal{J}_{\mathfrak{F}_A}(\mathcal{H}^{(0)}_{\mathfrak{F}_A}) \cap \left(\zeta_1 + S\widehat{\otimes}_k\mathcal{O}_M\right), \tag{37}$$

becomes a formal primitive form, whose minimal exponent is zero due to the second property of Proposition 3.6. Proposition 3.8 shows that such an S exists. Q.E.D.

5.3. Frobenius structures

By a standard procedure, the existence of a formal primitive form implies the existence of a formal Frobenius structure (cf. Theorem 7.5 in [33], which can be generalized to graded cases verbatim).

Theorem 2. *Let ζ be a formal primitive form with the minimal exponent zero for the tuple $(\mathcal{H}^{(0)}_{\mathfrak{F}_A}, \nabla, K_{\mathfrak{F}_A})$. Define an $\mathcal{O}_M$-bilinear form $\eta_{\zeta^{\otimes 2}} : \mathcal{T}_M \otimes_{\mathcal{O}_M} \mathcal{T}_M \longrightarrow \mathcal{O}_M$ by*

$$\eta_{\zeta^{\otimes 2}}(X,Y) := (-1)^{w\cdot\overline{Y}} K_{\mathfrak{F}_A}(u\nabla_X\zeta, u\nabla_Y\zeta)u^{-w}. \tag{38}$$

Then the tuple $(\circ, \eta_{\zeta^{\otimes 2}}, e, E)$ gives a formal Frobenius structure on $\mathcal{T}_M$ such that $Lie_E(\circ) = \circ$ and $Lie_E(\eta_{\zeta^{\otimes 2}}) = (2-w)\eta_{\zeta^{\otimes 2}}$.

References

[1] S. Barannikov, *Quantum Periods*, I: *Semi-Infinite Variations of Hodge Structures*, Internat. Math. Res. Notices **23** (2001), 1243–1264.

[2] S. Barannikov, *Non-Commutative Periods and Mirror symmetry in Higher dimension*, Commun. Math. Phys. **228** (2002), 281–325.

[3] S. Barannikov and M. Kontsevich, *Frobenius Manifolds and Formality of Lie Algebras of Polyvector Fields*, Internat. Math. Res. Notices **4** (1998), 201–215.

[4] H. Cartan, S. Eilenberg, *Homological Algebra*, Princeton University Press, Princeton, N. J., (1956).

[5] D. Cox, S. Katz, *Mirror Symmetry and Algebraic Geometry*, Mathematical Surveys and Monographs, vol. **68**, American Mathematical Society, Providence, RI, 1999.

[6] Yu. Daletski, I. Gelfand and B. Tsygan, *On a variant of noncommutative geometry*, Soviet Math. Dokl., **40**, 2 (1990), 422–426.

[7] V. A. Dolgushev, D. E. Tamarkin and B. L. Tsygan, *Formality of the homotopy calculus algebra of Hochschild (co)chains*, arXiv:0807.5117.

[8] V. A. Dolgushev, D. E. Tamarkin and B. L. Tsygan, *Formality theorems for Hochschild complexes and their applications*, Lett. Math. Phys. **90** (2009), no. 1–3, 103–136.

[9] B. Dubrovin, *Geometry of 2D Topological Field Theories*, in Integrable system and Quantum Groups, Montecatini, Terme 1993. Lecture Note in Math. **1620**, Springer–Verlag 1996, 120–348.

[10] K. Fukaya, Y.-G. Oh, H. Ohta, K. Ono, *Floer theory and mirror symmetry on compact toric manifolds*, Astérisque **376** (2016).

[11] M. Gerstenhaber, *The cohomology structure of an associative ring*, Annals of Math., **78** (1963), 267–288.

[12] M. Gerstenhaber, S. Schack, *Algebraic cohomology and deformation theory*, NATO ASI Ser. C (1988), 11–264.

[13] E. Getzler, *Cartan homotopy formulas and the Gauss-Manin connection in cyclic homology*, Quantum deformations of algebras and their representations (Ramat-Gan, 1991/1992; Rehovot, 1991/1992), 65–78, Israel Math. Conf. Proc., 7, Bar-Ilan Univ., Ramat Gan, 1993.

[14] V. Ginzburg, *Calabi–Yau algebras*, arXiv:math/0612139.

[15] C. Hertling, *Frobenius manifolds and moduli spaces for singularities*, Cambridge University Press, 2002.

[16] I. Iwanari, *Period mappings for noncommutative algebras*, arXiv:1604.08283.

[17] D. Kaledin, *Non-commutative Hodge-to-de Rham degeneration via the method of Deligne-Illusie*, Pure Appl. Math. Q. **4** (2008), no. 3, Special Issue: In honor of Fedor Bogomolov. Part 2, 785–875.

[18] D. Kaledin, *Homological methods in Non-commutative Geometry*, Lecture notes for the course "Homological methods in Non-commutative Geometry", University of Tokyo, October 2007–March 2008. http://imperium.lenin.ru/ kaledin/tokyo/

[19] D. Kaledin, *Spectral sequences for cyclic homology*, arXiv:1601.00637.

[20] M. Kashiwara and P. Schapira, *Categories and sheaves*, Fundamental Principles of Mathematical Sciences, **332**. Springer–Verlag, Berlin, 2006. x+497 pp.

[21] L. Katzarkov, M. Kontsevich and T. Pantev, *Hodge theoretic aspects of mirror symmetry*, Proceedings of Symposia in Pure Mathematics **78** (2008), 87–174.

[22] B. Keller, *On differential graded categories*, International Congress of Mathematicians. Vol. II, Eur. Math. Soc., Zürich (2006), 151–190.

[23] B. Keller, *Calabi–Yau triangulated categories*, Trends in Representation Theory of Algebras (Zürich), A. Skowroński, ed., Eur. Math. Soc. (2008), 467–489.

[24] M. Kontsevich, *Homological algebra of mirror symmetry*, In Proceedings of the ICM (Zürich, 1994), Birkhäuser, Basel, (1995), 120–139.

[25] B. Kontsevich, *Deformation quantization of Poisson manifolds*, Letters in Mathematical Physics **66** (2003), 157–216.

[26] M. Kontsevich and Y. Soibelmann, *Notes on $\mathcal{A}_\infty$-algebras, $\mathcal{A}_\infty$-categories and non-commutative geometry*. I, Homological mirror symmetry, Lecture Notes in Phys., Springer, Berlin, **757** (2009), 153–219.

[27] C. Li, S. Li and K. Saito, *Primitive forms via polyvector fields*, arXiv:1311.1659.

[28] Y. Manin, *Frobenius manifolds, quantum cohomology, and moduli spaces*, American Math.Society Colloquium Publ. v.**47**, 1999.

[29] D. Orlov, *Remarks on generators and dimensions of triangulated categories*, Moscow Math. J. **9** (2009), no. 1, 153–159.

[30] C. Sabbah, *Déformations isomonodromiques et variétés de Frobenius*, 2002, EDP Sciences.

[31] K. Saito, *Period mapping associated to a primitive form*, Publ. RIMS, Kyoto Univ. **19** (1983), 1231–1264.

[32] K. Saito, *The higher residue pairings $K_F^{(k)}$ for a family of hypersurface singular points*, Advanced Studies in Pure Mathematics **40** (1983), part 2, 441–463.

[33] K. Saito and A. Takahashi *From Primitive Forms to Frobenius Manifolds*, Proceedings of Symposia in Pure Mathematics **78** (2008), 31–48.

[34] M. Saito, *On the structure of Brieskorn lattice*, Annales de l'institut Fourier, **39** (1989), 27–72.

[35] M. Saito, *On the structure of Brieskorn lattice*, II, arXiv:1312.6629.

[36] N. Sheridan, *Formulae in noncommutative Hodge theory*, arXiv:1510.03795.

[37] D. Shklyarov, *Non-commutative Hodge structures: Towards matching categorical and geometric examples*. Trans. Amer. Math. Soc. **366** (2014), 2923–2974.

[38] D. Shklyarov, *Matrix factorizations and higher residue pairings*, arXiv:1306.4878.

[39] J. Terllia, *Smoothness theorem for differential BV algebras*, arXiv: 0707.1290.

[40] B. Töen, *The homotopy theory of dg-categories and derived Morita theory*, Invent. Math. **167** (2007), no. 3, 615–667.

Department of Mathematics, Graduate School of Science, Osaka University, Toyonaka Osaka, 560-0043, Japan
E-mail address: `takahashi@math.sci.osaka-u.ac.jp`

Advanced Studies in Pure Mathematics

A SERIES OF UP-TO-DATE GUIDES OF LASTING INTEREST
TO ADVANCED MATHEMATICS

Volume 1 Algebraic Varieties and Analytic Varieties.
Edited by S. Iitaka. February, 1983

Volume 2 Galois Groups and their Representations.
Edited by Y. Ihara. December, 1983

Volume 3 Geometry of Geodesics and Related Topics.
Edited by K. Shiohama. June, 1984

Volume 4 Group Representations and Systems of Differential Equations.
Edited by K. Okamoto. March, 1985

Volume 5 Foliations.
Edited by I. Tamura. February, 1986

Volume 6 Algebraic Groups and Related Topics.
Edited by R. Hotta. March, 1985

Volume 7 Automorphic Forms and Number Theory.
Edited by I. Satake. February, 1986

Volume 8 Complex Analytic Singularities.
Edited by T. Suwa and P. Wagreich. February, 1987

Volume 9 Homotopy Theory and Related Topics.
Edited by H. Toda. February, 1987

Volume 10 Algebraic Geometry, Sendai, 1985.
Edited by T. Oda. July, 1987

Volume 11 Commutative Algebra and Combinatorics.
Edited by M. Nagata and H. Matsumura.
October, 1987

Volume 12 Galois Representations and Arithmetic Algebraic Geometry.
Edited by Y. Ihara. November, 1987

Volume 13 Investigations in Number Theory.
Edited by T. Kubota. March, 1988

Volume 14 Representations of Lie Groups, Kyoto, Hiroshima, 1986.
Edited by K. Okamoto and T. Oshima. February, 1989

Volume 15 Automorphic Forms and Geometry of Arithmetic Varieties.
Edited by K. Hashimoto and Y. Namikawa. July, 1989

Volume 16 Conformal Field Theory and Solvable Lattice Models.
Edited by M. Jimbo, T. Miwa and A. Tsuchiya.
June, 1988

Volume 17 Algebraic Number Theory—in honor of K. Iwasawa.
Edited by J. Coates, R. Greenberg, B. Mazur and I. Satake. August, 1989

Volume 18 I—Recent Topics in Differential and Analytic Geometry.
Edited by T. Ochiai. November, 1990
II—Kähler Metric and Moduli Spaces.
Edited by Edited by T. Ochiai. November, 1990

Volume 19 Integrable Systems in Quantum Field Theory and Statistical Mechanics.
Edited by M. Jimbo, T. Miwa and A. Tsuchiya.
November, 1989

Volume 20 Aspects of Low Dimensional Manifolds.
Edited by Y. Matsumoto and S. Morita.
December, 1992

Volume 21 Zeta Functions in Geometry.
Edited by N. Kurokawa and T. Sunada.
December, 1992

Volume 22 Progress in Differential Geometry.
Edited by K. Shiohama. September, 1993

Volume 23 Spectral and Scattering Theory and Applications.
Edited by K. Yajima. February, 1994

Volume 24 Progress in Algebraic Combinatorics.
Edited by E. Bannai and A. Munemasa. May, 1996

Volume 25 CR-Geometry and Overdetermined Systems.
Edited by T. Akahori, G. Komatsu, K. Miyajima, M. Namba and K. Yamaguchi. July, 1997

Volume 26 Analysis on Homogeneous Spaces and Representation Theory of Lie Groups, Okayama-Kyoto.
Edited by M. Kashiwara, T. Kobayashi, T. Matsuki, K. Nishiyama and T. Oshima. April, 2000

Volume 27 Arrangements – Tokyo 1998.
Edited by M. Falk and H. Terao. August, 2000

Volume 28 Combinatorial Methods in Representation Theory.
Edited by M. Kashiwara, K. Koike, S. Okada, I. Terada, and H.-F. Yamada. December, 2000

Volume 29 Singularities – Sapporo 1998.
Edited by J.-P. Brasselet and T. Suwa.
December, 2000

Volume 30 Class Field Theory – Its Centenary and Prospect.
Edited by K. Miyake. April, 2001

Volume 31 Taniguchi Conference on Mathematics Nara '98.
Edited by M. Maruyama and T. Sunada. May, 2001

Volume 32 Groups and Combinatorics — in memory of Michio Suzuki.
Edited by E. Bannai, H. Suzuki, H. Yamaki and T. Yoshida. October, 2001

Volume 33 Computational Commutative Algebra and Combinatorics.
Edited by T. Hibi. February, 2002

Volume 34 Minimal Surfaces, Geometric Analysis and Symplectic Geometry.
Edited by K. Fukaya, S. Nishikawa and J. Spruck.
June, 2002

Volume 35 Higher Dimensional Birational Geometry.
Edited by S. Mori and Y. Miyaoka. July, 2002

Volume 36 Algebraic Geometry 2000, Azumino.
Edited by S. Usui, M. Green, L. Illusie, K. Kato, E. Looijenga, S. Mukai and S. Saito. November, 2002

Volume 37 Lie Groups, Geometric Structures and Differential Equations — One Hundred Years after Sophus Lie —.
Edited by T. Morimoto, H. Sato and K. Yamaguchi.
December, 2002

Volume 38 Operator Algebras and Applications.
Edited by H. Kosaki. January, 2004

Volume 39 Stochastic Analysis on Large Scale Interacting Systems.
Edited by T. Funaki and H. Osada. March, 2004

Volume 40 Representation Theory of Algebraic Groups and Quantum Groups.
Edited by T. Shoji, M. Kashiwara, N. Kawanaka, G. Lusztig and K. Shinoda. March, 2004

Volume 41 Stochastic Analysis and Related Topics in Kyoto
In honour of Kiyosi Itô.
Edited by H. Kunita, S. Watanabe and Y. Takahashi.
March, 2004

Volume 42 Complex Analysis in Several Variables
– Memorial Conference of Kiyoshi Oka's Centennial Birthday, Kyoto/Nara 2001.
Edited by K. Miyajima, M. Furushima, H. Kazama, A. Kodama, J. Noguchi, T. Ohsawa, H. Tsuji and T. Ueda. June, 2004

Volume 43 Singularity Theory and Its Applications.
Edited by S. Izumiya, G. Ishikawa, H. Tokunaga, I. Shimada and T. Sano. January, 2007

Volume 44 Potential Theory in Matsue.
Edited by H. Aikawa, T. Kumagai, Y. Mizuta and N. Suzuki. August, 2006

Volume 45 Moduli Spaces and Arithmetic Geometry (Kyoto, 2004).
Edited by S. Mukai, Y. Miyaoka, S. Mori, A. Moriwaki and I. Nakamura. January, 2007

Volume 46 Singularities in Geometry and Topology 2004.
Edited by J.-P. Brassele and T. Suwa. February, 2007

Volume 47 Asymptotic Analysis and Singularities.
1—Hyperbolic and dispersive PDEs and fluid mechanics.
2—Elliptic and parabolic PDEs and related problems.
Edited by H. Kozono, T. Ogawa, K. Tanaka, Y. Tsutsumi and E. Yanagida. October, 2007

Volume 48 Finsler Geometry, Sapporo 2005
— In Memory of Makoto Matsumoto.
Edited by S. V. Sabau and H. Shimada.
November, 2007

Volume 49 Probability and Number Theory — Kanazawa 2005.
Edited by S. Akiyama, K. Matsumoto, L. Murata and H. Sugita. December, 2007

Volume 50 Algebraic Geometry in East Asia — Hanoi 2005.
Edited by K. Konno and V. Nguyen-Khac.
February, 2008

Volume 51 Surveys on Geometry and Integrable Systems.
Edited by M. Guest, R. Miyaoka and Y. Ohnita.
November, 2008

Volume 52 Groups of Diffeomorphisms
in honor of Shigeyuki Morita
on the occasion of his 60th birthday.
Edited by R. Penner, D. Kotschick, T. Tsuboi,
N. Kawazumi, T. Kitano and Y. Mitsumatsu.
November, 2008

Volume 53 Advances in Discrete Dynamical Systems.
Edited by S. Elaydi, K. Nishimura, M. Shishikura and
N. Tose. June, 2009

Volume 54 Algebraic Analysis and Around
in honor of Professor Masaki Kashiwara's 60th birthday.
Edited by T. Miwa, A. Matsuo, T. Nakashima and
Y. Saito. February, 2009

Volume 55 Noncommutativity and Singularities
Proceedings of French–Japanese symposia held at IHÉS in 2006.
Edited by J.-P. Bourguignon, M. Kotani, Y. Maeda and
N. Tose. July, 2009

Volume 56 Singularities — Niigata–Toyama 2007.
Edited by J.-P. Brasselet, S. Ishii, T. Suwa and
M. Vaquie. November, 2009

Volume 57 Probabilistic Approach to Geometry.
Edited by M. Kotani, M. Hino and T. Kumagai.
March, 2010

Volume 58 Algebraic and Arithmetic Structures of Moduli Spaces
(Sapporo 2007).
Edited by I. Nakamura and L. Weng. June, 2010

Volume 59 New Developments in Algebraic Geometry, Integrable
Systems and Mirror Symmetry (RIMS, Kyoto, 2008).
Edited by M.-H. Saito, S. Hosono and K. Yoshioka.
August, 2010

Volume 60 Algebraic Geometry in East Asia — Seoul 2008.
Edited by J. Keum, S. Kondō, K. Konno and K. Oguiso.
September, 2010

Volume 61 Exploring New Structures and Natural Constructions
in Mathematical Physics.
Edited by K. Hasegawa, T. Hayashi, S. Hosono and
Y. Yamada. October, 2011

Volume 62 Arrangements of Hyperplanes—Sapporo 2009.
Edited by H. Terao and S. Yuzvinsky. March, 2012

Volume 63 Galois–Teichmüller Theory and Arithmetic Geometry. Edited by H. Nakamura, F. Pop, L. Schneps and A. Tamagawa. October, 2012

Volume 64 Nonlinear Dynamics in Partial Differential Equations. Edited by S.-I. Ei, S. Kawashima, M. Kimura and T. Mizumachi. April, 2015

Volume 65 Algebraic Geometry in East Asia — Taipei 2011. Edited by J. A. Chen, M. Chen, Y. Kawamata and J. Keum. April, 2015

Volume 66 Singularities in Geometry and Topology 2011. Edited by V. Blanlœil and O. Saeki. May, 2015

Volume 67 Variational Methods for Evolving Objects. Edited by L. Ambrosio, Y. Giga, P. Rybka and Y. Tonegawa. June, 2015

Volume 68 School on Real and Complex Singularities in São Carlos, 2012. Edited by O. Saeki, R. N. Araújo dos Santos, T. Nishimura and V. H. Jorge Pérez. July, 2016

Volume 69 Development of Moduli Theory—Kyoto 2013. Edited by O. Fujino, S. Kondō, A. Moriwaki, M.-H. Saito and K. Yoshioka. July, 2016

Volume 70 Minimal Models and Extremal Rays (Kyoto, 2011). Edited by J. Kollár, O. Fujino, S. Mukai and N. Nakayama. September, 2016

Volume 71 Schubert Calculus — Osaka 2012. Edited by H. Naruse, T. Ikeda, M. Masuda and T. Tanisaki. December, 2016

Volume 72 Geometry, Dynamics, and Foliations 2013 In honor of Steven Hurder and Takashi Tsuboi on the occasion of their 60th birthdays. Edited by T. Asuke, S. Matsumoto and Y. Mitsumatsu. August, 2017

Volume 73 Hyperbolic Geometry and Geometric Group Theory. Edited by K. Fujiwara, S. Kojima and K. Ohshika. September, 2017

Volume 74 Higher Dimensional Algebraic Geometry in honour of Professor Yujiro Kawamata's sixtieth birthday. Edited by K. Oguiso, C. Birkar, S. Ishii and S. Takayama. November, 2017

Volume 75 Algebraic Varieties and Automorphism Groups.
Edited by K. Masuda, T. Kishimoto, H. Kojima, M. Miyanishi and M. Zaidenberg. December, 2017

Volume 76 Representation Theory, Special Functions and Painlevé Equations — RIMS 2015.
Edited by H. Konno, H. Sakai, J. Shiraishi, T. Suzuki and Y. Yamada. June, 2018

Volume 77 The 50th Anniversary of Gröbner Bases.
Edited by T. Hibi. August, 2018

Volume 78 Singularities in Generic Geometry.
Edited by S. Izumiya, G. Ishikawa, M. Yamamoto, K. Saji, T. Yamamoto and M. Takahashi. September, 2018

Volume 79 Mathematics of Takebe Katahiro and History of Mathematics in East Asia.
Edited by T. Ogawa and M. Morimoto.
December, 2018

Volume 80 Operator Algebras and Mathematical Physics.
Edited by M. Izumi, Y. Kawahigashi, M. Kotani, H. Matui and N. Ozawa. May, 2019

Volume 81 Asymptotic Analysis for Nonlinear Dispersive and Wave Equations.
Edited by K. Kato, T. Ogawa and T. Ozawa.
October, 2019

Volume 82 Differential Geometry and Tanaka Theory — Differential Systems and Hypersurface Theory—.
Edited by T. Shoda and K. Shibuya. November, 2019

Volume 83 Primitive Forms and Related Subjects — Kavli IPMU 2014.
Edited by K. Hori, C. Li, S. Li and K. Saito.
December, 2019

TO BE CONTINUED

Good days ahead. Friends going zigzag: Avoid hesitation in making decisions.

ZODIAC:

Of any sign of the zodiac: Will have big fortune in the future. Own sign of the zodiac: Will win a lottery. Relative's sign in the zodiac: Will have arguments with relatives. Family's sign of the zodiac: Riches. Children's sign of the zodiac: Will be greatly loved by the children.

ZOO:

Of a zoo: Select friends carefully. Going to the zoo alone: Will be molested. Going to the zoo with family and children: Good hopes for the future. Husband and wife going to the zoo: Hopes will not be realized. Going to the zoo with sweetheart: Danger and misfortune. Friends going to the zoo: Do not confide your secrets.